Practical Asterisk 1.4 and 1.6

Practical Asterisk 1.4 and 1.6

Stefan Wintermeyer
Stephen Bosch

✦Addison-Wesley

Upper Saddle River, NJ • Boston • Indianapolis • San Francisco
New York • Toronto • Montreal • London • Munich • Paris • Madrid
Capetown • Sydney • Tokyo • Singapore • Mexico City

Many of the designations used by manufacturers and sellers to distinguish their products are claimed as trademarks. Where those designations appear in this book, and the publisher was aware of a trademark claim, the designations have been printed with initial capital letters or in all capitals.

The authors and publisher have taken care in the preparation of this book, but make no expressed or implied warranty of any kind and assume no responsibility for errors or omissions. No liability is assumed for incidental or consequential damages in connection with or arising out of the use of the information or programs contained herein.

The publisher offers excellent discounts on this book when ordered in quantity for bulk purchases or special sales, which may include electronic versions and/or custom covers and content particular to your business, training goals, marketing focus, and branding interests. For more information, please contact:

> U.S. Corporate and Government Sales
> (800) 382-3419
> corpsales@pearsontechgroup.com

For sales outside the United States, please contact:

> International Sales
> international@pearson.com

Visit us on the Web: informit.com/aw

Library of Congress Cataloging-in-Publication Data

Wintermeyer, Stefan.
    Practical Asterisk 1.4 and 1.6 : from beginner to expert / Stefan Wintermeyer, Stephen Bosch.
        p.   cm.
    Includes index.
    ISBN 978-0-321-52566-6 (pbk. : alk. paper)   1. Internet telephony.
2. Asterisk (Computer file)   3. Telephone—Private branch exchanges.
I. Bosch, Stephen.   II. Title.
    TK5105.8865.W56 2009
    021.005 dc22   2009039861

> Pearson Education, Inc.
> Rights and Contracts Department
> 501 Boylston Street, Suite 900
> Boston, MA 02116
> Fax: (617) 671-3447

ISBN-13: 978-0-321-52566-6
ISBN-10:   0-321-52566-3

Text printed in the United States on recycled paper at Edwards Brothers in Ann Arbor, Michigan.

Second printing, January 2010

Editor-in-Chief
Mark L. Taub

Acquisitions Editor
Debra Williams Cauley

Development Editor
Michael Thurston

Technical Editors
L.D. Paniak
John Kennedy

Managing Editor
John Fuller

Full-Service Production Manager
Julie B. Nahil

Project Editor
Kim Arney

Copy Editor
Keith Cline

Indexer
Jack Lewis

Proofreader
Linda Begley

Cover Designer
Chuti Prasertsith

Compositor
Kim Arney

For grandaunt Helga.
—SW

For my father, who showed me
how much fun tinkering could be.
—SB

Contents

Foreword

You are reading the English version of my second Asterisk book. I wrote the book in German, and Stephen Bosch translated it into English and adapted the content for a North American reader. Nonetheless, this book was the result of collaboration. The first version was published under the GNU Free Documentation License (GFDL), and more than 100 people contributed by sending error reports or by writing new sections. A special thanks goes to my colleague, Philipp Kempgen, who wrote many sections for this new edition. You can find the German edition at www.das-asterisk-buch.de and the English version at www.the-asterisk-book.com.

I encourage you to take a closer look at the current online edition, especially when it comes to hardware and driver-related information.

The past few years have been active ones in the VoIP world. My work with Asterisk has brought me to many new places, such as Huntsville, Alabama (home of Digium, the developer of Asterisk); Toronto, Canada (home of Sangoma, a manufacturer of telephony interface hardware); and Bucharest, Romania (home of the Yate project, an open source telephony engine). I learned much about Asterisk, but also a lot about Yate (http://yate.null.ro) and FreeSwitch (www.freeswitch.org), both interesting open source VoIP alternatives.

Asterisk has come a long way since I first used it. I didn't use that very first installation for very long; I didn't understand the concept, and documentation was typical for open source software (i.e., "read the source code"). The same may be said of my first experience with Linux. Today, I wouldn't dream of being without either of them.

Asterisk has changed my way of thinking about PBXes. In the future, everyone will be more likely to think of a PBX as software rather than hardware.

The transition from the old, hardware-oriented way of doing things to the new software-oriented way is not without its challenges. In my daily work as an Asterisk consultant, I often see conflict between the former data guys and the former telephony guys. I hope to bring their respective worlds closer together with this book.

I welcome your feedback, praise, and criticism. Please send it to me at stefan.wintermeyer@amooma.de

—Stefan Wintermeyer

It has been a special privilege to work with Stefan on the translation of this book into English. Although Stefan speaks English quite well, he knew that a competent English version would need a dedicated translator. To be entrusted with this task has been an honor. I was also charged with some of the adaptations and changes for the English edition, including the chapter on ISDN.

A technical translation is a special kind of beast. Details are critical, and there isn't much room for nuance, so I hope that the book you have in your hands proves to be both readable and helpful. Like Stefan, I welcome your feedback. Please send it to me at sbosch@vodacomm.ca.

—Stephen Bosch

Preface

Perhaps it was folly to strive to make a book for everyone, but only audacious goals pave the path to excellence. This book does indeed try to be useful for both the new user and the expert.

Chapter 1 provides guideposts for the beginner, journeyman, and expert. Chapter 2 offers detailed instructions for the configuration of a basic system, with pointers to the installation instructions in the appendix. Chapter 3 is an introduction to the dialplan, truly the heart of Asterisk, which leads into Chapter 4, a practical business case example that lays out what is involved in designing a system for actual deployment.

The next chapters go into more depth for the more experienced Asterisk user. Chapter 5 covers the use of dialplan applications and functions for building voice applications. Chapter 6 looks at the new Asterisk Extensions Language, which introduces familiar programming language format to the dialplan. With it, you can build concise, readable, and maintainable dialplans. Asterisk uses a variety of network and IP telephony protocols in its work, which we review in Chapter 7. Chapter 8 looks at the audio codecs and wire transports Asterisk uses to connect to devices and telephone networks, including ISDN, and takes a brief look at the associated hardware.

Then we look at the feature set of Asterisk. Asterisk provides a fully functional voicemail system. Learn how to configure and use it in Chapter 9. Chapter 10 shows you how to build interactive voice response (IVR) systems using Asterisk.

Useful applications need a database, and Asterisk provides one, the Asterisk database, covered in Chapter 11. Call queues, arguably the backbone of any call center, are covered in Chapter 12.

The very brief Chapter 13 looks at two basic but common business telephone features: call transfer and pickup. Teleconferencing can be a powerful collaboration tool, and Asterisk has built-in conferencing functions, which are addressed in Chapter 14.

There are many ways to have Asterisk interact with external programs. Chapter 15 shows you how to control Asterisk from external applications, scripts, or the system shell with shell commands, call files, or the Asterisk Manager Interface. Chapter 16 introduces the Asterisk Gateway Interface, Asterisk's principal method for interacting with external applications.

You can extend Asterisk for faxing with IAXmodem and Hylafax. Learn how to do this in Chapter 17.

Finally, Chapter 18 addresses a frequent request: setting up busy lamp field (extension monitoring) and hints for SIP telephones. We look at Aastra and Polycom examples.

The appendixes in this book are extensive and are meant to be used frequently. Appendix A contains very detailed instructions for installing Asterisk on Debian Etch. Appendices B, C, D, and E have command and function summaries for the dialplan and the AGI and AMI interfaces. Appendix F provides examples for using dialplan templates.

Although Asterisk 1.4 is now the production-ready version, 1.6 is in development and can be used. Appendix G tells you what you need to know to migrate successfully.

Appendix H covers the renaming of Zaptel to DAHDI. This has implications for operation, which are addressed here.

Which should you choose, IAX or SIP? It's a matter of debate. We provide some input in Appendix I.

You can read sequentially or follow your interests. Just remember that some of the examples build on previous examples.

Acknowledgments

The first edition of this book was published in 2006 in Germany. In the interim, we received lots of useful feedback from the Asterisk community. We want to say a big thank you to everybody who helped us.

In the first edition, we acknowledged everybody by name, but we've decided not to do that this time because the e-mail addresses were harvested by spammers and many of the contributors have expressly asked to remain anonymous.

Nevertheless, we want to thank you sincerely for all your feedback and effort! Without your help, the book could not have become as good as it has.

Acknowledgments from Stefan Wintermeyer

First and foremost, I want to thank my editor, Boris Karnikowski. Without his patience, not even the first edition would have seen the light of day. The second edition was incrementally delayed 12 months. I am also thankful that he agreed to allow the book to be published under the GNU Free Documentation License. There are few publishers and editors who have the courage to take such a step. Thanks, Boris!

I also want to thank my colleagues at AMOOMA, Peter Kozak and Sören Sprenger. Special thanks go to Philipp Kempgen. The depth and breadth of this book owes itself to your efforts.

Friederike Daenecke's editing work was miraculous. She went through the entire manuscript in a little under three weeks.

Kevin Fleming of Digium provided very prompt and invaluable technical assistance.

The staff at Sangoma were also very helpful. In particular, thanks go to Konrad Hammel and Doug Vilim.

Special thanks go to my wife. Without her support, I wouldn't be able to keep my head above water on these many projects.

Acknowledgments from Philipp Kempgen

Thanks to my family, Jeannie, Coco, Kanji, and all those who get the short end of the stick a little too often.

Acknowledgments from Stephen Bosch

Thanks to Debra Williams Cauley for your humane and generous spirit, your (apparently inexhaustible!) patience, and your persistence. You are a real professional. To you and everyone else at Pearson: I am grateful.

About the Authors

Stefan Wintermeyer is the founder and director of AMOOMA GmbH, a leading provider of Asterisk telephony solutions in Europe. He was Germany's first Digium-Certified Asterisk Professional and continues to share his knowledge in conferences, seminars, and workshops.

Stephen Bosch is an Asterisk consultant and writer who divides his time between Aschaffenburg, Germany, and Calgary, Alberta, Canada. When not making telephone calls with Asterisk, he sails.

How to Get the Most Out of This Book

First, thank you for buying this book. It's only through the interest of readers like you that a project such as this is possible.

We wrote this book because Asterisk doesn't come with an instruction manual. Many of the resources recommended by the Asterisk community are in the form of wikis and websites, which are inconsistently updated and often contain incorrect information. We want this book to serve as that missing manual for Asterisk. In other words, we sought to create the kind of book we were looking for: a readable text that can be read in chapter order or used as a handbook.

1.1 What Is Asterisk?

Asterisk is a telephone system, but different. It's not hardware. It's software that you can download via the Internet. The only thing you need to get started is an ordinary PC, the Asterisk software, this book, and interest in doing so.

Asterisk works with a variety of telephony technologies, including analog, ISDN, and VoIP. With an Asterisk system, an employee can work at home with an IP phone and make calls in the office system as though she were in the office. The possibilities are limitless, both to extend the functionality of your system and to save money. Telephony remains somewhat complex, however, and a move to Asterisk is not without its challenges. Therefore, you need to familiarize yourself with both the general technology and with Asterisk itself.

Let's explore the difference between Asterisk and a conventional telephone system by way of an analogy from childhood.

Most of us grew up with toys of two basic types. With the first type, everything came "ready to go" (for example, teddy bears or Hot Wheels cars). The second type offered some room for variation and creativity (for example, Lego blocks or Lincoln Logs). With the "ready to go" toys, the fun was right there when you opened the package, but you were limited to play as conceived by the toy company. With the blocks/logs, you didn't get a finished toy. To get one, you had to come up with a design and then build it, but you could build anything you wanted, within the limitations of the pieces available.

That, in a nutshell, is exactly how it is with Asterisk. It is delivered in pieces. Think of a design, and suddenly anything is possible. Best of all, you can improve the resulting "toy" any time with new pieces.

Both Asterisk and conventional telephone systems have their pros and cons.

The main advantages of a conventional telephone system include the following:

- Unpack and you're done! You can start making calls almost immediately.
- Limited instructions are required.

The main disadvantages of a conventional telephone system include the following:

- Expansion is usually very expensive, and not all functions are supported or even available.
- It's a black box. You have no way to find out what's happening inside, which can be vital when troubleshooting.

The main advantages of a software-based Asterisk system include the following:

- You can build new solutions to problems any time. All you need is an appropriate design.
- You can combine components to create new systems and solutions.

The main disadvantages of Asterisk include the following:

- You start with just a bunch of components. Building them into a system requires some patience.
- You need some skill. Success doesn't come quickly.

1.2 Who Should Read This Book?

Not everybody has to start at square one, nor does every interested person want to become an Asterisk professional. This book takes that into account and presents material that will prove to be a valuable reference for beginners, novices, and experts alike.

1.2.1 The Beginner

Absolute beginners should read this book in chapter order because the chapters build on each other. At a minimum, read Chapter 2, "Installation and 'Hello World,'" and 5, "Dialplan Fundamentals," and you'll be able to put together increasingly complex telephone systems.

Beginners often have questions about what equipment is best to buy and which service to order, so we've provided this brief FAQ for those who need some guidance.

Q: *How do I choose a VoIP provider? What do I need to watch out for?*

A: When starting out, choose a SIP provider that offers free accounts so that you can gather some experience. Which provider you choose for the long term can depend on many factors. Here are just a few things to consider:

- Does the provider offer only SIP or also IAX? Do you need IAX?
- How good is the support? For VoIP, good support can be critical.
- How are services priced? This is not always obvious.

Q: *Which digital interface card should I use?*

A: Get comfortable using only SIP trunks and extensions to start. Once you understand the dialplan and contexts, you can proceed to Chapter 8, "Making Connections," where you will find tips for choosing the right digital card for your application.

Q: *Which phones are standard, and what do they cost?*

A: There aren't really "standard" phones. You have several options:

- A SIP telephone. (This is the simplest option.)
- An analog telephone using an ATA (analog telephone adapter) connected to the network.

- An analog telephone using an FXS port on an analog interface card.
- A digital telephone (such as a Nortel Meridian phone) using a suitable digital telephone adapter connected to the network.
- An ISDN phone using a port on a T1 card. These phones are sold almost exclusively in Europe where ISDN termination is much more common. Before the arrival of VoIP, they were the most feature-filled devices; currently, however, they are functionally indistinguishable from VoIP phones and can cost three times as much. If you have the option, it's best to go with a SIP telephone.
- Exotic constructs. You can, for example, connect a channel bank to a T1 interface card and attach analog phones to the channel bank. Such constructions generally prove useful in retrofit applications. If you're starting from scratch, keep it simple.

As for pricing, this changes constantly. There are countless suppliers. Inquire frequently and compare. Quality VoIP hardware is becoming more and more affordable.

1.2.2 The Journeyman

It is impossible to guess what an intermediate Asterisk user knows and doesn't know. Even if you have good command of a subject, there are always aspects to it that might be new to you. If you have some practical experience, you may want to skip Chapter 2 or just skim it. Personally, though, we recommend that you read it anyway. After all, reviewing the fundamentals almost always pays off.

Leaf through all the other chapters afterward. When writing this book, we focused on practical examples. Just reading the sample configurations may give you some new ideas.

1.2.3 The Expert

You can skip Chapter 2 and just go on a discovery tour of the book. You'll find numerous practical examples you can integrate immediately into your own practice; the examples may even give you a few ideas about how to expand your system.

Beyond that, we hope this book will be a valuable reference. All the important application and function parameters are included and explained, usually with working examples. We placed particular emphasis on comparisons between the different Asterisk versions (1.2 to 1.6).

1.3 Updates and Versions of the Book

This book is being developed just as a typical open source project would be. What you are holding in your hands is a printed version of the stable branch. You can access updates and errata at www.the-asterisk-book.com.

You'll find the current SVN version there. That'll be the next edition, but for now it is on the bleeding edge. The procedures contained there may not have been tested and may not even be recommended. Therefore, a warning: Use that material only if you know what you're doing.

1.4 Reader Contributions and Feedback

We intend this book to be a living text, and so have released it under the GNU Free Documentation License (see Appendix J, "GNU Free Documentation License"). You can find the most current version of the book at www.the-asterisk-book.com.

If you have suggestions for improvements, include them in an e-mail to stefan.wintermeyer@amooma.de. If you have more comprehensive improvements to contribute, attach them in ASCII text format or directly in DocBook format to your e-mail.

Thanks in advance for your interest in this project!

This book will never be perfect or ever really "finished." The subject is too complex and changes too rapidly. Remember that you can always find the most current (and most bleeding-edge) version of the book at www.the-asterisk-book.com. Accordingly, this FAQ is intended for the web version of the book. If you find an error in this book, please check the website first to see whether the error has already been fixed.

Q: *I tried something from the book, but it didn't work as described. Should I report this?*

A: Yes, absolutely! When reporting, provide the URL and any relevant error messages.

Q: *How can I report an error?*

A: The easiest way is via e-mail to the author, Stefan Wintermeyer (stefan.wintermeyer@amooma.de).

Q: *In what format is the book written?*

A: The source for the book is written in DocBook format. You can learn more about DocBook at www.docbook.org.

Q: *Why isn't there a wiki, where anybody can make changes?*

A: First, because the book is printed, a print deadline has to be set. A constantly changing text makes it impossible for the editors to keep up and maintain consistency in the resulting printed material. Second, there are already numerous Asterisk wikis on the Internet. Their chief advantage (quick and easy changes) also leads to their chief disadvantage: plenty of painfully outdated and outright incorrect information. Using DocBook and managing changes means we have a more structured, more readable text that is as accurate as human beings can make it. If the entire text were in a wiki, it would make export for printing very difficult.

Q: *What editor do I need to write in DocBook format?*

A: That's almost a "religious" question. Some people like emacs, others like vi. The author and contributors use XMLmind XML Editor most of the time. You can obtain it from www.xmlmind.com/xmleditor.

Q: *Can I obtain the original DocBook files?*

A: Yes. If you have extensive changes to make, you are welcome to work on the original files. Write an e-mail to the author, Stefan Wintermeyer (stefan.wintermeyer@amooma.de).

Q: *You didn't cover XYZ! Unbelievable. It's so important! What were you thinking?*

A: Contact the author and briefly describe the subject. (If you're willing to write a chapter and make it available under the GNU FDL, let Stefan know. He'll send you a sample DocBook chapter.)

Q: *How is the website generated?*

A: A script pulls the DocBook sources from the SVN and uses them to generate the HTML pages.

1.5 Additional Resources

You have a problem and weren't able to find the answer in this book? Here are your options:

- **www.voip-info.org**
  This site is dedicated to VoIP in general and to Asterisk, and it offers solutions to most problems. Unfortunately, many of the articles contain errors. If you find one, fix it! The next visitor will be grateful.

- **IRC channel**

  There is an Asterisk IRC channel, #asterisk at irc.freenode.net. It's sometimes possible to get a quick answer to your question there, but be warned: Some channel participants have a low tolerance for laziness. You improve your chances of success by thinking your question through carefully. Generally, the motto "explaining your problem is half the solution" applies.

  If you are new to IRC, you can learn more about it at http://en.wikipedia.org/wiki/Internet_Relay_Chat.

- **Mailing lists**

  The Digium mailing lists are not for everybody. The list has very high volume, and a new list user can be quickly overwhelmed. Still, it has a few significant advantages: It's read by Digium staff, and they often answer questions. You can also get more detailed answers than is typical for IRC. You can find out how to subscribe to the mailing lists at www.asterisk.org/support and http://lists.digium.com.

- **E-mail me**

  If all else fails, you can always try contacting Stefan (stefan.wintermeyer@amooma.de). Just be warned that because of the volume of e-mail received, you might not receive a reply for several days.

2

Installation
and "Hello World"

Some newcomers to Asterisk say it takes at least two days of studying web pages and documentation before you can get an Asterisk server to do anything at all. If you don't like delving into the theoretical underpinnings of a complicated piece of software like Asterisk and would rather see something practical and working as soon as possible, this chapter is the place to start. In contrast to the rest of the book, this chapter should be read, and the examples followed, in sequence.

Note: The subject of this book is Asterisk, not Linux, and we assume that you have basic Linux administration skills. You should understand how to use a text editor and how to configure a network adapter. Commands are to be executed as the superuser root unless otherwise noted.

For the purposes of this introduction, it doesn't matter whether you install Asterisk 1.4 or 1.6. All the basics described here apply to both versions. For production installations, the more stable version is recommended, so these examples feature Asterisk 1.4.

2.1 Installing Asterisk on the Server

As to which Linux distribution or Asterisk version to choose, opinions differ (as is so often the case in the open source world). Some stick with distribution-specific packages (e.g., .rpm or .deb), whereas others compile Asterisk from source code. For the examples in this book, *it is best to install from source code*. The reasons are simple: The versions found in distribution packages are almost never current,[1] and packages often use nonstandard configurations.

Refer to Appendix A, "Installation Instructions for Asterisk 1.4," and install Asterisk 1.4.x on a fresh install of Debian Linux or even KNOPPIX. For practice, you might consider doing this in an x86 virtual machine (e.g., VMware). In either case, be it a physical server or a virtual machine, the system should have a sound card and an output device (e.g., speakers) so that you can make test calls to the console to determine whether your system is working.

Warning: Should you decide to try the install with a different distribution or with a different installation method and run into problems, try following the method described here first. It is often easier to work with a tested installation than it is to find the errors in an "off menu" configuration or environment.

2.2 Calling "Hello World" from the CLI

In this first example, we create a simple "Hello World" dialplan and call it from the Asterisk console, or CLI (command-line interface).

2.2.1 Configuring Asterisk

After a standard install, you should find these files in the /etc/asterisk directory:

```
debian:/usr/src# cd /etc/asterisk
debian:/etc/asterisk# ls
adsi.conf               cdr_tds.conf        indications.conf    privacy.conf
adtranvofr.conf         codecs.conf         logger.conf         queues.conf
agents.conf             dnsmgr.conf         manager.conf        res_odbc.conf
alarmreceiver.conf      dundi.conf          meetme.conf         rpt.conf
alsa.conf               enum.conf           mgcp.conf           rtp.conf
asterisk.adsi           extconfig.conf      misdn.conf          sip.conf
asterisk.conf           extensions.ael      modem.conf          sip_notify.conf
cdr.conf                extensions.conf     modules.conf        skinny.conf
cdr_custom.conf         features.conf       musiconhold.conf    telcordia-1.adsi
cdr_manager.conf        festival.conf       osp.conf            voicemail.conf
cdr_odbc.conf           iax.conf            oss.conf            vpb.conf
cdr_pgsql.conf          iaxprov.conf        phone.conf          zapata.conf
debian:/etc/asterisk#
```

---

1. You can't really blame the distributors for this. New versions of Asterisk come out almost every month. No distributor on Earth could match that pace.

That's a long list, but don't worry; we care about only one of them for our example: `extensions.conf`. To keep things simple, we'll move the sample `extensions.conf` file created by **make samples** to `/var/tmp/asterisk-etc-backup/` (so that we can retrieve it later if required):

```
debian:/etc/asterisk# mkdir -p /var/tmp/asterisk-etc-backup
debian:/etc/asterisk# mv extensions.* /var/tmp/asterisk-etc-backup/
debian:/etc/asterisk#
```

Using your favorite console text editor[2] enter the following text into `/etc/asterisk/extensions.conf`:

```
[default]
exten => 1001,1,Answer()
exten => 1001,2,Playback(hello-world)
exten => 1001,3,Hangup()
```

2.2.2 Starting Asterisk and Calling "Hello World"

You might be surprised that just four lines are enough to configure Asterisk. Perhaps you thought Asterisk was more complicated than that. In any case, let's give it a try. Start Asterisk with the command **asterisk -c** (the **-c** switch gives us the console):

```
debian:/etc/asterisk# asterisk -c
Asterisk 1.4.21, Copyright (C) 1999 - 2008 Digium, Inc. and others.
Created by Mark Spencer <markster@digium.com>
[...]
[ Booting...
[ Reading Master Configuration ]
[...]
Asterisk Ready.
*CLI>
```

With this console, you can operate a running Asterisk server and give it commands interactively and in real time. Let's try generating a call to our "Hello World" extension with **console dial 1001**:

```
*CLI> console dial 1001
*CLI>  << Console call has been answered >>
 << Hangup on console >>

*CLI>
```

---

2. If you don't have one, I recommend nano. Installing it in Debian is easy as superuser with **apt-get -y install nano**. Open the file you want to edit with **nano *<filename>***. The most important commands are displayed in the bottom portion of the screen.

The command **console dial 1001** calls extension 1001. This extension answers and plays the `hello-world.gsm` sound file from the `/var/lib/asterisk/sounds` directory.

Note: console dial behaves like a very simple telephone. You can use it to call an extension and listen to it via the sound card.

2.2.2.1 What Is an Extension?

An extension is a programming unit in a dialplan. Every extension consists of at least one line, written in the following format:

```
exten => extension_name,priority,application
```

Here, *priority* describes the sequence of the individual extension elements. Our extension 1001 has three priorities:

```
exten => 1001,1,Answer()
exten => 1001,2,Playback(hello-world)
exten => 1001,3,Hangup()
```

The applications are self-explanatory:

- **Answer()**

  Answers and opens a new Asterisk channel (see Appendix B, "Dialplan Applications").

- **Playback(hello-world)**

  Plays the file `hello-world.gsm` in the current channel (see Appendix B).

- **Hangup()**

  Hangs up and closes the channel (see Appendix B).

2.2.2.2 Increasing Verbosity

When you are debugging Asterisk, you'll often find it helpful to increase the verbosity of the console messages. When Asterisk is started with **asterisk -c**, the verbose level is set to 0 (the allowed range is 0 to 10). You can increase this to level 5 from the console with the command **core set verbose 5**, which is a good level to use for debugging:

```
*CLI> core set verbose 5
Verbosity was 0 and is now 5
*CLI>
```

Now, when you enter the command **console dial 1001**, you see details about the dialplan execution:

```
*CLI> console dial 1001
   == Console is full duplex
*CLI>    -- Executing [1001@default:1] Answer("Console/dsp", "") in new
      stack
<< Console call has been answered >>
   -- Executing [1001@default:2] Playback("Console/dsp", "hello-world") in
      new stack
   -- <Console/dsp> Playing 'hello-world' (language 'en')
*CLI>    -- Executing [1001@default:3] Hangup("Console/dsp", "") in new
      stack
   == Spawn extension (default, 1001, 3) exited non-zero on 'Console/dsp'
   << Hangup on console >>

*CLI>
```

You can also set the verbose level to 5 at start time with the switch **-vvvvv** (five v's).

Warning: Don't use **core set debug** to debug the dialplan! This is intended for developers doing Asterisk debugging. It is not very helpful for dialplan debugging, however.

2.2.3 Stopping Asterisk

Enter **stop now**, and Asterisk stops:

```
*CLI> stop now
debian:/etc/asterisk#
```

2.3 Calling "Hello World" with a SIP Phone

Having tested our extension 1001 from the console, we take the next logical step and try the call from a SIP telephone. If you don't have a hardware SIP phone, you can use a software SIP phone that you install on a client computer. Many of these are freely available on the Internet.

Warning: If you want to install the software SIP phone and Asterisk on the same machine, you must set a custom port number for the SIP phone (e.g., 5061) because 5060 is already being used by Asterisk. Beginners, however, should start with a hardware SIP phone if at all possible. When starting out, minimize the potential for problems by not building unnecessary complexity into your test installation.

2.3.1 Configuring the SIP Phone and sip.conf

Before you can use a SIP phone with Asterisk, you need to create an account for it in your Asterisk configuration. We will move the sample `sip.conf` file to our backup directory, `/var/tmp/asterisk-etc-backup/`, in the same way we previously moved `extensions.conf`, with the command **mv sip.conf /var/tmp/asterisk-etc-backup/**:

```
debian:/etc/asterisk# mv sip.conf /var/tmp/asterisk-etc-backup/
debian:/etc/asterisk#
```

Now we create a new `/etc/asterisk/sip.conf` and enter the following:

```
[general]
port=5060
bindaddr=0.0.0.0

[2000]
type=friend
secret=1234
host=dynamic
```

Your SIP telephone must now be configured with the following account information:

- **User:** 2000
- **Password:** 1234
- **SIP registrar:** IP address of your Asterisk server
- **SIP proxy:** IP address of your Asterisk server

Tip: Some phones require that every field contain information, even if it is not relevant. Unfortunately, there are no general instructions that will apply to all phones. Experiment!

2.3.2 Starting Asterisk and the Phone

Now we start Asterisk in verbose level 5 with **asterisk -vvvvvc**:

```
debian:/etc/asterisk# asterisk -vvvvvc
Asterisk 1.4.21, Copyright (C) 1999 - 2008 Digium, Inc. and others.
[...]
```

Next, we start the SIP phone and wait for the registration message in the Asterisk console:

```
*CLI> -- Registered SIP '2000' at 47.6.3.4 port 5060
expires 120 -- Unregistered SIP '2000'
```

When you call extension 1001 from the SIP phone, you will hear the hello-world.gsm file played back to you.

2.3.2.1 Calling the Phone from the Asterisk Console

Because we can call an extension with **console dial** and we have successfully attached a SIP phone to the system, it should be possible to call that SIP phone from the console. To do this, we need to add an extension to /etc/asterisk/extensions.conf:

```
[default]
exten => 1001,1,Answer()
exten => 1001,2,Playback(hello-world)
exten => 1001,3,Hangup()

exten => 2000,1,Dial(SIP/2000)
```

To apply these changes, you must either restart Asterisk or reload the dialplan. To restart Asterisk, enter **stop now** in the Asterisk console and **asterisk -vvvvvc** in the Linux shell. To reload the dialplan from within the running Asterisk, enter **dialplan reload** in the Asterisk console. Now you can dial the SIP phone with **console dial 2000**:

```
*CLI> console dial 2000
```

The dialplan application **Dial()** sets up a connection to a telephone. It uses a parameter consisting of two parts: the first, **SIP**, describes the technology used for establishing the connection (the SIP VoIP protocol in our example). The second part defines the target device using that technology (in this case, **2000**). When using **Dial()**, no **Answer()** or **Hangup()** is required. Because we do not know in advance whether the called station will even accept the call, **Dial()** has additional intelligence for opening and closing the channel.

In our example, the extension 2000 corresponds with the SIP target 2000, but this is, strictly speaking, coincidental. You could also write the extension like this:

```
exten => 55,1,Dial(SIP/2000)
```

Reload the dialplan, and now you can call the same SIP telephone with **console dial 55**.

2.3.2.2 Comments in the Configuration

Because the number sign (#) corresponds to a dual-tone multi-frequency (DTMF) tone, it cannot be used as the comment character. Instead, Asterisk configuration files use the semicolon (;) for indicating comments, as follows:

```
[default]
; Extension 1001 is used for testing
; of all phones.
;
exten => 1001,1,Answer() ; answering...
exten => 1001,2,Playback(hello-world)
exten => 1001,3,Hangup() ; hanging up...

; Extension 2000 calls SIP telephone
; 2000.
;
exten => 2000,1,Dial(SIP/2000)
```

2.4 Building a Minimal Phone System with Two SIP Phones

What does the simplest possible working Asterisk system look like? Two phones and one Asterisk server.

We've already worked with all the individual pieces of this puzzle, and now we just have to put them together. To start, we configure two SIP phones in /etc/asterisk/sip.conf:

```
[general]
port=5060
bindaddr=0.0.0.0

[2000]
type=friend
secret=1234
host=dynamic

[2001]
type=friend
secret=1234
host=dynamic
```

Next, we have to make extensions that call the SIP phones in /etc/asterisk/extensions.conf. We will leave the "Hello World" example in place for testing purposes; we can verify that each phone is working by dialing 1001 and listening for the "Hello World" message. The resulting dialplan looks like this:

```
[default]
exten => 1001,1,Answer()
```

```
exten => 1001,2,Playback(hello-world)
exten => 1001,3,Hangup()

exten => 2000,1,Dial(SIP/2000)
exten => 2001,1,Dial(SIP/2001)
```

Restart Asterisk and the phones (which you configure following the instructions in the previous section). After the phones have registered, you can call one phone from the other, or you can call the test extension from either phone.

2.4.1 Configuring Voicemail

Asterisk already includes a working voicemail module. We need only configure it for use via the /etc/asterisk/voicemail.conf file. First, we move the existing sample file to our sample backup directory:

debian:/etc/asterisk# **mv voicemail.conf /var/tmp/asterisk-etc-backup/**

Now we create a new /etc/asterisk/voicemail.conf file and enter the following:

```
[general]
format = wav

[default]
2000 => 4711,Joe Bloggs,jbloggs@example.com
2001 => 0815,Daisy Duke,daisy.duke@hazzard.com
```

Now the mailboxes are configured (yes, it really is that easy). Each entry starts with the access password, then the full name of the user, and finally the user's e-mail address. The final step is to add a few more lines to /etc/asterisk/extensions.conf to attach this voicemail functionality to our telephones. Don't forget to add **,20** to **Dial()**:

```
[default]
exten => 1001,1,Answer()
exten => 1001,2,Playback(hello-world)
exten => 1001,3,Hangup()

exten => 2000,1,Dial(SIP/2000,20)
exten => 2000,2,VoiceMail(2000,u)

exten => 2001,1,Dial(SIP/2001,20)
exten => 2001,2,VoiceMail(2001,u)

exten => 2999,1,VoiceMailMain(${CALLERID(num)},s)
```

Done! Start Asterisk with **asterisk -vvvvvc** and call one phone from the other. (In a running Asterisk, typing **reload** in the Asterisk CLI is sufficient to apply any changes to the configuration files.) After 20 seconds of ringing (the reason for the `,20` in `Dial()`), you are transferred to the voice mailbox. If the called station is busy, the call goes directly to the voice mailbox. Dial 2999 from any phone and you can access the voicemail menu for that phone.

Note: If your Asterisk server has a working Mail Transfer Agent (MTA) (e.g., sendmail or postfix), new voicemails are sent as e-mail attachments directly to the e-mail addresses specified in `voicemail.conf`.

If you want to secure your mailboxes with passwords, or would like to review the voicemail menus, see Chapter 9, "Voicemail."

2.5 Rights Administration with Contexts

So far, we have been able to call any extension from any telephone. For a small, private system, this is sufficient. Large systems and systems connecting to other telephone networks need some way to manage calling rights (that is, the rules that determine which phones and users are allowed to make calls to where). Asterisk does this via *contexts*.

Think of a context as a kind of category. Each category contains rules about what can be dialed. To access the rules in a context, the phone must be a member of that context or have been directed to that context by another rule.

2.5.1 The Originating Context

We haven't yet specified a context for the SIP phones we configured. In this case, Asterisk assumes the `[default]` context. If we want to specify specific contexts for specific telephones, we use the following syntax in `sip.conf`:

```
context = ContextName
```

We can set a context for all the phones in the [general] section, but this context can be overwritten on a per-telephone basis. Let's take a look at a couple of examples.

2.5.1.1 A Context Example

ABC Co. has the SIP phones 10 and 11. They are both in the [internal] context. The `sip.conf` file looks like this:

```
[general]
port=5060
bindaddr=0.0.0.0

[10]
type=friend
secret=1234
host=dynamic
context=internal  ; <-- context

[11]
type=friend
secret=1234
host=dynamic
context=internal  ; <-- context
```

You can achieve the same result more simply, though:

```
[general]
port=5060
bindaddr=0.0.0.0
context=internal  ; <-- context

[10]
type=friend
secret=1234
host=dynamic

[11]
type=friend
secret=1234
host=dynamic
```

2.5.1.2 Example with Multiple Phones

ABC Co. has the SIP phones 10, 11, 12, and 20. The phones 10, 11, and 12 are standard staff phones in the [abc] context, and 20 is a courtesy phone at reception in the [visitor] context:

```
[general]
port=5060
bindaddr=0.0.0.0
context=abc  ; <-- context

[10]
type=friend
secret=1234
host=dynamic

[11]
type=friend
```

```
secret=1234
host=dynamic

[12]
type=friend
secret=1234
host=dynamic

[20]
type=friend
secret=1234
host=dynamic
context=visitor  ; <-- this device-specific context
; overrides the default context in the [general] section.
```

2.5.2 Call Destination: Contexts in extensions.conf

The dialplan, extensions.conf, is broken up into sections, or contexts. Each context is specified by a name in square brackets.

The following is a sample extensions.conf containing the three contexts, [default], [building-mgr], and [apple-pie]:

```
[default]
exten => 1001,1,Answer()
exten => 1001,2,Playback(hello-world)
exten => 1001,3,Hangup()

[building-mgr]
exten => 2000,1,Dial(SIP/2000,20)
exten => 2000,2,VoiceMail(2000,u)

exten => 2001,1,Dial(SIP/2001)
exten => 2001,2,VoiceMail(2001,u)

[apple-pie]
exten => 2999,1,VoiceMailMain(${CALLERID(num)},s)
```

If the context is not specified for a SIP phone in sip.conf, Asterisk assumes the [default] context in extensions.conf.

2.6 Calls to and from the Public Switched Telephone Network

In this section, we connect our mini-system to the public switched telephone network (PSTN). Once we've done that, we can make and receive calls using our attached SIP phones.

2.6.1 Calling the PSTN

At this point, you have a working telephone system, which, although exciting, is not really useful, because it has no connection with the outside world. With just ten more minutes and a working Internet connection, you can have a working connection to the PSTN. To do this, you need an account with a SIP provider.

We will configure Asterisk so that you can make calls to the PSTN with the phones 2000 and 2001. First, the provider account must be defined in /etc/ asterisk/sip.conf:

```
[general]
port = 5060
bindaddr = 0.0.0.0
context = other

register => 17984512232:UHDZJD@my-voip-provider.com/17984512232
;                        ^          ^          ^                          ^
;                        |          |          |                          |
;                      user password  provider                         user

[2000]
type=friend
context=my-phones
secret=1234
host=dynamic

[2001]
type=friend
context=my-phones
secret=1234
host=dynamic

[ext-sip-account]
type=friend
context=from-voip-provider
username=17984512232
fromuser=17984512232
secret=UHDZJD
host=my-voip-provider.com
fromdomain=my-voip-provider.com
qualify=yes
insecure=port,invite
nat=yes
```

You must obtain the username (17984512232 in our example) and password (UHDZJD in our example) from the SIP provider; often you can do this through the provider's customer website. Asterisk needs this information to register with the provider and make calls.

Warning: Most SIP providers charge a per-minute rate for local calls, and many require prepayment. An advantage is that there is no monthly flat rate for most SIP accounts.

Next we need an additional dialplan rule to allow outbound calls:

```
[other]

[my-phones]

exten => 2000,1,Dial(SIP/2000,20)
exten => 2000,2,VoiceMail(2000,u)

exten => 2001,1,Dial(SIP/2001,20)
exten => 2001,2,VoiceMail(2001,u)

exten => 2999,1,VoiceMailMain(${CALLERID(num)},s)

exten => _X.,1,Dial(SIP/${EXTEN}@ext-sip-account)
```

After these new entries have been entered, save the file and start Asterisk as before, with **asterisk -vvvvvc** so that we get the Asterisk console. Wait a few seconds for the SIP phones to register. Now simply dial a number.

Tip: When dialing through most VoIP providers, you need to dial the complete number, including the predial digit (1 in North America) and area or city code, even if the call is a local call in your calling area. For North America, this means you dial the predial digit, followed by the full ten-digit number including area code, even in regions that do not already have ten-digit local dialing. In our example, we provide an open dialing pattern that will pass on the digits exactly as dialed to the provider. (Later on, we show you some techniques that you can use in /etc/asterisk/extensions.conf so that you don't need to dial the full number for local calls in areas where it is not normally required.)

If everything is working as it should, you will hear the remote line ringing and be able to observe the call progress in the console.

It's a bit early to explain exactly how this works, but you'll read more about that later.[3]

---

3. Not too much at once! For now, all you need to know is that the ${EXTEN} variable always contains the number dialed by the caller for the specific instance (see Chapter 5, "Dialplan Programming").

2.6.2 Taking Calls from the PSTN

The last step is a small one: We want to be able to take incoming calls via our SIP provider on extension 2000. To do this, we need to add another context to /etc/asterisk/extensions.conf:

```
[other]

[my-phones]
exten => 2000,1,Dial(SIP/2000,20)
exten => 2000,2,VoiceMail(2000,u)

exten => 2001,1,Dial(SIP/2001,20)
exten => 2001,2,VoiceMail(2001,u)

exten => 2999,1,VoiceMailMain(${CALLERID(num)},s)

exten => _X.,1,Dial(SIP/${EXTEN}@ext-sip-account)

[from-voip-provider]
exten => 17984512232,1,Dial(SIP/2000)
```

In our example, the number 17984512232 is the PSTN number (also called a DID or DN; more on that later) given to your account by your SIP provider. That the DID corresponds to the username is coincidental; it doesn't have to.

You can, of course, configure voicemail for calls coming in from the PSTN:

```
[other]

[my-phones]
exten => 2000,1,Dial(SIP/2000,20)
exten => 2000,2,VoiceMail(2000,u)

exten => 2001,1,Dial(SIP/2001,20)
exten => 2001,2,VoiceMail(2001,u)

exten => 2999,1,VoiceMailMain(${CALLERID(num)},s)

exten => _0[1-9].,1,Dial(SIP/${EXTEN}@ext-sip-account)

[from-voip-provider]
exten => 17984512232,1,Dial(SIP/2000,20)
exten => 17984512232,2,VoiceMail(2000,u)
```

If you were so inclined, you could just leave things like this and start using your new mini-PBX. But what fun would that be? This chapter was only meant to show you how quickly you can build a working Asterisk system. In the coming chapters, we fill in the gaps and show you just how much you can really do with Asterisk.

3

Dialplan Fundamentals

The dialplan is the heart of Asterisk, and everything it does begins here.

In Asterisk 1.4, two important files in `/etc/asterisk` make up the dialplan. The first is `extensions.conf`, which uses the original and Digium-recommended priority model. The second is `extensions.ael`, which uses the newer Asterisk Extensions Language; we'll look at that in more detail in a separate chapter. For now, we'll use the traditional priority model because even in 1.4 and 1.6 `extensions.ael` is converted into priority format and added to `extensions.conf` when Asterisk is started.[1]

Note: Should you use .conf or .ael? Digium says ".conf is the default." For certain applications, however, AEL can make for a more readable and maintainable dialplan.

3.1 Contexts

The Asterisk dialplan is divided into sections, and each section is called a context. Any dialplan must begin with a `[general]` context where global configuration entries reside, but the subsequent contexts can have any name. Physical devices

---

1. The exact definition of *dialplan* is a matter of some debate in the telephony world. In this book, we use the word in its Asterisk-specific sense.

are bound to the dialplan through contexts. These devices are usually telephones, but can be other types of hardware (for example, SIP or Zap devices). The configuration for every device, be it a softphone, hardphone, or outgoing trunk, must specify the default context for that device. Here's an example from a `sip.conf` file:

```
[2000]
type=friend context=internal-phones
secret=1234
host=dynamic
```

This SIP device called 2000 always initiates calls in the `internal-phones` context. This means that if a caller uses this phone to dial a number, Asterisk will look in the `internal-phones` context for an extension matching that number. If no matching extension is present, nothing happens.

Warning: A good understanding of contexts is essential for effective Asterisk programming and administration. If you're not comfortable with contexts, follow the step-by-step example for a simple PBX system in Chapter 2, "Installation and 'Hello World.'"

It pays to take the time necessary to become comfortable with the concept.

Contexts are defined by a name inside square brackets ([*name*]). Ideally, the name should be relevant and help to describe the intended use for the context. This name will also be used to refer to the context elsewhere, be it in other contexts or in other Asterisk configuration files. All lines following a context name are considered part of that context, until the next context name is encountered:

```
[general]

[internal-phones]
Rules, commands, etc.

[widgets]
Rules, commands, etc.
```

3.2 Extensions

Individual entries in `extensions.conf` are called *extensions*. Individual extensions are executed by Asterisk every time a call is initiated for that extension, but `extensions.conf` is only read into Asterisk once, at start time. This means that changes are not applied until the dialplan is reloaded or Asterisk restarted.

Note: An exception is the Asterisk RealTime Architecture (ARA). In an ARA system, the dialplan is stored in a database (e.g., MySQL) and read into Asterisk for each call, not simply when Asterisk is started. Therefore, an administrator can make dialplan changes on a running Asterisk server that take effect immediately. Nevertheless, this approach is not without significant disadvantages. You can learn more about ARA at www.voip-info.org/wiki/view/Asterisk+RealTime.

You can also refresh the dialplan during operation from the CLI (command-line interface) by entering the command **reload now** (which reloads all the configurations) or **extensions reload** (which reloads only the dialplan).

An extension consists of the following parts:

- **Extension:** A name or number
- **Priority:** A kind of program line number
- **Application:** An instruction that tells Asterisk what it should do with the call

The following is a template example of how you specify an extension line in `extensions.conf`:

```
exten => extension,priority,application
```

A real statement would look like this:

```
exten => 123,1,Answer()
```

Warning: The first priority in an extension must always be 1 (one); otherwise, Asterisk will never call the extension. Subsequent priorities must be incremented by 1 and not more. Asterisk does not recognize larger increments.

3.2.1 Fundamental Applications

To build the dialplan examples in this chapter, we need the following basic applications (all of which are described in greater detail in Appendix B, "Dialplan Applications"):

- `Answer()`

  The **Answer()** application does just that. It answers a call. When a channel rings, **Answer()** tells Asterisk to "lift the virtual receiver."

- `Hangup()`

  Hangup() is the opposite of **Answer()**. An active connection is terminated, and Asterisk "hangs up" the virtual receiver.

- **Playback(*soundfile*)**

  This tells Asterisk to play a specified sound file. By default, it plays files found in /var/lib/asterisk/sounds/, but you can also specify another source directory. No file extension is specified because the directory may contain the same sound in different formats. Asterisk will select the most appropriate format (more on that later).

- **Wait(*number*)**

  Wait() defines a pause; *number* indicates the number of seconds to pause.

- **NoOp(*string*)**

  This application does nothing. **NoOp** means "no operation." It is useful, however, when you are trying to troubleshoot the dialplan. When **NoOp(*string*)** is executed, Asterisk prints *string* on the CLI, but only if the verbosity level is set to 3. (You can do this easily by entering the command **set verbose 3** in the CLI.)

- **VoiceMail(*mailbox,u*)**

  Lets the caller leave a voice message in the mailbox specified.

- **VoiceMailMain()**

  Provides access to the voicemail system. The mailbox owner will use this to retrieve her messages.

3.2.2 Priorities

A typical extension is composed of multiple entries. Each entry has a priority so that Asterisk knows in what order it should execute the entries. If you've ever worked with early versions of BASIC, you might be familiar with line numbers; priorities work in much the same way, but with one important distinction. They are always executed in numeric order from smallest to largest, but there can be no skipping! If Asterisk executes an entry of priority n, then it will look for the next entry at $n + 1$. If it cannot find an entry at $n + 1$, it stops executing without displaying an error in the CLI.

3.2.2.1 A "Hello World" Example

The following extension will be invoked when a phone with the default context widgets dials 8888.

Asterisk picks up the line, plays the hello-world sound file (which is installed with Asterisk), and hangs up.

```
[widgets]
exten => 8888,1,Answer()
exten => 8888,2,Playback(hello-world)
exten => 8888,3,Hangup()
```

3.2.2.2 n Priority

To make it easier to work with priorities, Asterisk versions from 1.2 onward support the n priority. The n priority is like automatic line numbering; when Asterisk is running through the dialplan and encounters an entry with priority n, it simply executes it as though it were equivalent to the previous priority, plus 1. This is useful when you have extensions with many entries and you need to add or remove an entry, because it saves you from having to renumber the entire extension. The following example illustrates what a standard extension would look like:

```
exten => 1234,1,Answer()
exten => 1234,2,Wait(2)
exten => 1234,3,Playback(hello-world)
exten => 1234,4,Wait(2)
exten => 1234,5,Hangup()
```

You can define the same extension with the n priority:

```
exten => 1234,1,Answer()
exten => 1234,n,Wait(2)
exten => 1234,n,Play(hello-world)
exten => 1234,n,Wait(2)
exten => 1234,n,Hangup()
```

You can start using the n priority at any point in the extension, as long as all the subsequent entries also use it:

```
exten => 1234,1,Answer()
exten => 1234,2,Wait(2)
exten => 1234,3,Play(hello-world)
exten => 1234,n,Wait(2)
exten => 1234,n,Hangup()
```

3.3 Pattern Matching

In `extensions.conf`, patterns let us specify multiple number combinations in a single line.

Using what we know so far, we need to write a separate extension for each telephone number. As the system expands, this leads to unwieldy and error-prone dialplans. Suppose, for our example, we need numbers 100 to 109 to play the hello-world sound file. Our `extensions.conf` would look like this:

```
[general]

[widgets]
exten => 100,1,Answer()
exten => 100,2,Playback(hello-world)
exten => 100,3,Hangup()
```

```
exten => 101,1,Answer()
exten => 101,2,Playback(hello-world)
exten => 101,3,Hangup()

exten => 102,1,Answer()
exten => 102,2,Playback(hello-world)
exten => 102,3,Hangup()

exten => 103,1,Answer()
exten => 103,2,Playback(hello-world)
exten => 103,3,Hangup()

exten => 104,1,Answer()
exten => 104,2,Playback(hello-world)
exten => 104,3,Hangup()

exten => 105,1,Answer()
exten => 105,2,Playback(hello-world)
exten => 105,3,Hangup()

exten => 106,1,Answer()
exten => 106,2,Playback(hello-world)
exten => 106,3,Hangup()

exten => 107,1,Answer()
exten => 107,2,Playback(hello-world)
exten => 107,3,Hangup()

exten => 108,1,Answer()
exten => 108,2,Playback(hello-world)
exten => 108,3,Hangup()

exten => 109,1,Answer()
exten => 109,2,Playback(hello-world)
exten => 109,3,Hangup()
```

If we use a *pattern*, the same dialplan becomes instantly more compact and elegant:

```
[general]

[widgets]
exten => _10X,1,Answer()
exten => _10X,2,Playback(hello-world)
exten => _10X,3,Hangup()
```

The _10X extension describes the number range from 100 to 109.

Note: The terms *pattern* and *regular expression* are often casually interchanged. In general, what we are using in Asterisk is a *pattern,* although many programmers would use the term *regular expression* also.

3.3.1 Pattern Elements

Dialplan patterns always begin with the underscore (_) character:

```
exten => _pattern,priority,application
```

An Asterisk dialplan pattern can have the following elements:

- **[abc]**

  The digits a, b, and c. For example, to match 34, 37, and 38

  ```
  exten => _3[478],1,NoOp(Test)
  ```

- **[a-b]**

  Any digit in the range a to b. For example, to match any number between 31 and 35

  ```
  exten => _3[1-5],1,NoOp(Test)
  ```

  (e.g., `[25-8]` is also acceptable for the digits 2,5,6,7,8)

- **X**

  Any digit from 0 to 9. For example, to match any number between 300 and 399

  ```
  exten => _3XX,1,NoOp(Test)
  ```

- **Z**

  Any digit from 1 to 9. For example, to match any number between 31 and 39

  ```
  exten => _3Z,1,NoOp(Test)
  ```

- **N**

  Any digit from 2 to 9. For example, to match any number between 32 and 39

  ```
  exten => _3N,1,NoOp(Test)
  ```

- *****

  Matches the asterisk (*) key

  ```
  exten => _*7,1,NoOp(Test)
  ```

- **#**

  Matches the number sign or pound key (#)

  ```
  exten => _#7,1,NoOp(Test)
  ```

- **.**

  Any number of digits of any kind. For example, to match all numbers that begin with 011

  ```
  exten => _011.,1,NoOp(Test)
  ```

Warning: Don't use the _. pattern! If you do, the pattern will include special extensions such as i, t, and h, and that will lead to unpredictable and probably undesirable behavior. Use _X. or _X if you need broad pattern matching.

- **!**

    This special "wildcard" character will match as soon as the number dialed is unambiguous; i.e., when the number being dialed cannot match any other extension in the context. Once a match is made, the outgoing line is picked up and dialing proceeds in real time with direct feedback (this is known as *overlap dialing*).

Warning: A common error is to forget the underscore (_) character at the beginning of the pattern. This convention is necessary because SIP devices, as configured in `sip.conf`, can have alphanumeric names. (For example, in Asterisk, 333, loadingdock, and A31 are all acceptable names for a SIP device.) It also means that if you forget to use the underscore, your extension will never match *and* you will never see an error message informing you of your mistake.

3.3.2 Testing a Pattern Using dialplan show

An example dialplan looks like this:

```
[general]

[my-phones]
exten => 23,1,Answer()
exten => 23,2,Playback(hello-world)
exten => 23,3,Hangup()
```

We can call **dialplan show** from the CLI (invoked with **asterisk -r** if Asterisk is already running) to verify that our dialplan has been loaded:

```
*CLI> dialplan show
[ Context 'default' created by 'pbx_config' ]

[ Context 'my-phones' created by 'pbx_config' ]
    '23' =>           1. Answer()                              [pbx_config]
                      2. Playback(hello-world)                 [pbx_config]
                      3. Hangup()                              [pbx_config]

[ Context 'parkedcalls' created by 'res_features' ]
    '700' =>          1. Park()                                [res_features]

-= 2 extensions (4 priorities) in 3 contexts. =-
*CLI>
```

The output includes all the dialplan rules that Asterisk knows about. Notice that there is a `parkedcalls` context that we haven't seen before; this is activated by default in `features.conf` and needn't concern us further. What if we are only interested in the `my-phones` context? We can make our request more specific with **dialplan show my-phones**:

```
*CLI> dialplan show my-phones
[ Context 'my-phones' created by 'pbx_config' ]
  '23' =>           1. Answer()                                [pbx_config]
                    2. Playback(hello-world)                   [pbx_config]
                    3. Hangup()                                [pbx_config]

-= 1 extension (3 priorities) in 1 context. =-
*CLI>
```

The command **dialplan show** can also be used to show what Asterisk will do if we dial a specific number. Suppose we want to dial 25 from a phone in the my-phones context. We can see what will happen with the command **dialplan show 25@my-phones**:

```
*CLI> dialplan show 25@my-phones
There is no existence of 25@my-phones extension
*CLI>
```

Nothing happens because there is no match for 25 in the context. If we dial 23 instead, we get this output:

```
*CLI> dialplan show 23@my-phones
[ Context 'my-phones' created by 'pbx_config' ]
  '23' =>           1. Answer()                                [pbx_config]
                    2. Playback(hello-world)                   [pbx_config]
                    3. Hangup()                                [pbx_config]

-= 1 extension (3 priorities) in 1 context. =-
*CLI>
```

If we want to check 23 against all the accessible contexts, we use **dialplan show 23@**:

```
*CLI> dialplan show 23@
[ Context 'my-phones' created by 'pbx_config' ]
  '23' =>           1. Answer()                                [pbx_config]
                    2. Playback(hello-world)                   [pbx_config]
                    3. Hangup()                                [pbx_config]

-= 1 extension (3 priorities) in 1 context. =-
*CLI>
```

Let's expand our dialplan with an additional context by editing `extensions`
`.conf` like so:

```
[general]

[my-phones]
exten => 23,1,Answer()
exten => 23,2,Playback(hello-world)
exten => 23,3,Hangup()

[department-q]
exten => _2X,1,Answer()
exten => _2X,2,Playback(hello-world)
exten => _2X,3,Hangup()
```

Now we go back to the CLI and, after reloading the dialplan with the **reload**
command, run **dialplan show 23@**:

```
*CLI> dialplan show 23@
[ Context 'department-q' created by 'pbx_config' ]
    '_2X' =>              1. Answer()                          [pbx_config]
                          2. Playback(hello-world)             [pbx_config]
                          3. Hangup()                          [pbx_config]

[ Context 'my-phones' created by 'pbx_config' ]
    '23' =>               1. Answer()                          [pbx_config]
                          2. Playback(hello-world)             [pbx_config]
                          3. Hangup()                          [pbx_config]

-= 2 extensions (6 priorities) in 2 contexts. =-
*CLI>
```

All the matching extensions are displayed. Let's try it with **dialplan show 23@**:

```
*CLI> dialplan show 25@
[ Context 'department-q' created by 'pbx_config' ]
    '_2X' =>              1. Answer()                          [pbx_config]
                          2. Playback(hello-world)             [pbx_config]
                          3. Hangup()                          [pbx_config]

-= 1 extension (3 priorities) in 1 context. =-
*CLI>
```

There is only one match, in context `department-q`. In this example, if you dial
25 from a phone in the `my-phones` context, you still won't hear the "Hello World"
message. Extension 25 only works for phones in the `department-q` context.

3.3.3 Pattern-Matching Order

Although very powerful, pattern matching in your Asterisk dialplan can be tricky.
It is easy to assume that Asterisk runs through the dialplan in a completely

sequential manner. This is generally the case, but it does prioritize patterns based on the quality of the match.

The reason for this is simple: More than one pattern might match a dialed number. If two extensions match a dialed number, Asterisk will always choose the better match. Before deciding which extension matches best, it processes the entire context.

Let's look at an example with some ambiguous extensions:

```
[sales]
exten => _12X.,1,NoOp{12X}
exten => 12345,1,NoOp(12345}
exten => _1234.,1,NoOp{1234.}
```

It is not immediately clear which extension is executed when we dial 12345. To find out, we use **dialplan show 12345@sales**:

```
*CLI> dialplan show 12345@sales
[ Context 'sales' created by 'pbx_config' ]
  '12345' =>        1. NoOp(12345})                      [pbx_config]
  '_1234.' =>       1. NoOp{1234.}()                     [pbx_config]
  '_12X.' =>        1. NoOp{12X}()                       [pbx_config]

-= 3 extensions (3 priorities) in 1 context. =-
*CLI>
```

Asterisk shows all the hits, but gives extension 12345,1,NoOP{12345} first priority. The highest priority extension is always displayed at the top.

Let's try it with 12346 using the command **dialplan show 12346@sales**:

```
*CLI> dialplan show 12346@sales
[ Context 'sales' created by 'pbx_config' ]
  '_1234.' =>       1. NoOp{1234.}()                     [pbx_config]
  '_12X.' =>        1. NoOp{12X}()                       [pbx_config]

-= 2 extensions (2 priorities) in 1 context. =-
*CLI>
```

Again, the pattern with the best match to the dialed digits is listed first.

Important: The order in which the patterned extensions appear in the dialplan makes no difference. Patterned extensions are matched strictly in order of match precision.

Digium changed the expected behavior for the _. pattern in Asterisk 1.2. Although the pattern is the most general and should be therefore assigned the lowest priority, the behavior is opposite the expected behavior. In Asterisk 1.2, the extension _. always gets the highest priority!

Note: Note that the **show dialplan** command will work in Asterisk 1.4 but is deprecated. From here on out, examples for Asterisk 1.2 use **show dialplan**, and **dialplan show** is used for examples in Asterisk 1.4.

Let's try adding the extension _. to our previous dialplan example:

```
[sales]
exten => _12X.,1,NoOp{12X}
exten => 12345,1,NoOp(12345}
exten => _1234.,1,NoOp{1234.}

exten => _.,1,NoOp{Bingo}
```

When we try testing 12346 with **dialplan show 12346@sales**, we get the following output:

```
*CLI> dialplan show 12346@sales
[ Context 'sales' created by 'pbx_config' ]
    '_1234.' =>           1. NoOp{1234.}()                        [pbx_config]
    '_12X.' =>            1. NoOp{12X}()                          [pbx_config]
    '_.' =>               1. NoOp{Bingo}()                        [pbx_config]

-= 3 extensions (3 priorities) in 1 context. =-
*CLI>
```

In Asterisk 1.2, **show dialplan 12346@sales** gives a very different result:

```
*CLI> show dialplan 12346@sales
[ Context 'sales' created by 'pbx_config' ]
    '_.' =>               1. NoOp{Bingo}()                        [pbx_config]
    '_1234.' =>           1. NoOp{1234.}()                        [pbx_config]
    '_12X.' =>            1. NoOp{12X}()                          [pbx_config]

-= 3 extensions (3 priorities) in 1 context. =-
*CLI>
```

This is why it is preferable to use _X. as the wildcard pattern (if we use a wildcard pattern at all!). The following dialplan example is processed identically in Asterisk 1.2 and 1.4:

```
[sales]
exten => _12X.,1,NoOp{12X}
exten => 12345,1,NoOp(12345}
exten => _1234.,1,NoOp{1234.}

exten => _X.,1,NoOp{Bingo}
```

The priorities appear as follows in both versions:

```
*CLI> dialplan show 12346@sales
[ Context 'sales' created by 'pbx_config' ]
    '_1234.' =>         1. NoOp{1234.}()                          [pbx_config]
    '_12X.' =>          1. NoOp{12X}()                            [pbx_config]
    '_X.' =>            1. NoOp{Bingo}()                          [pbx_config]

-= 3 extensions (3 priorities) in 1 context. =-
*CLI>
```

3.4 Include Statements

Includes are a powerful tool for simplifying and organizing larger dialplans. Using an include statement, you can include other contexts in the current context. Here is a more general example:

```
include => name-of-the-other-context
```

And here is how you might see an include statement used in a larger dialplan:

```
[general]

[sales]
include => internal
include => external

[internal]
exten => 2000,1,Dial(SIP/2000)

[external]
exten => 17005551212,1,Dial(SIP/5551212)
```

3.4.1 Order of Execution When Using Include Statements

Asterisk will always look for a match in the current context before referencing any included contexts. If a matching entry is found, that entry is used. If no matching entry is found, Asterisk will look for a match in the first included context, then the next, and so on. It is also possible to have nested includes; that is, includes *within* includes.

You can verify what entry Asterisk is using to handle a call by entering **dialplan show *number@name-of-context*** in the Asterisk CLI.

Note: Users of Asterisk 1.2 use **show dialplan** rather than **dialplan show**.

Again, here is an example of an `include` statement:

```
[general]

[sales]
include => internal
include => external

[internal]
exten => 2000,1,Dial(SIP/2000)

[external]
exten => 17005551212,1,Dial(SIP/5551212)
```

Say we want to understand how Asterisk is handling a call to 2000 in the `sales` context. To do that, we enter **dialplan show 2000@sales** in the CLI:

```
*CLI> dialplan show 2000@sales
[ Included context 'internal' created by 'pbx_config' ]
    '2000' =>            1. Dial(SIP/2000)                              [pbx_config]

-= 1 extension (1 priority) in 1 context. =-
*CLI>
```

If we then expand the `sales` context like so

```
[general]

[sales]
include => internal
include => external

exten => 2000,1,Answer()
exten => 2000,2,Playback(hello-world)
exten => 2000,3,Hangup()

[internal]
exten => 2000,1,Dial(SIP/2000)

[external]
exten => 17005551212,1,Dial(SIP/5551212)
```

We will see this CLI output:

```
*CLI> dialplan show 2000@sales
[ Context 'sales' created by 'pbx_config' ]
    '2000' =>            1. Answer()                                    [pbx_config]
                         2. Playback(hello-world)                       [pbx_config]
                         3. Hangup()                                    [pbx_config]
[ Included context 'internal' created by 'pbx_config' ]
    '2000' =>            1. Dial(SIP/2000)                              [pbx_config]

-= 2 extensions (4 priorities) in 2 contexts. =-
*CLI>
```

Asterisk will play the `hello-world` sound file and not send the call to 2000, even though the `include` occurs first in the dialplan. This is because Asterisk will always look for a match in the current context before checking the included contexts.

3.4.2 Time-Conditional Include Statements

An `include` statement can be made conditional upon the time of day. This makes it easy to implement different day and night behaviors. The general syntax of a time-based `include` statement is shown here:

```
include => context|<time>|<day>|<day-of-month>|<month>
```

The day and month are specified using the first three letters of the full name. For example, weekdays are specified `mon`, `tue`, `wed`, `thu`, `fri`, `sat`, `sun`, and months are specified `jan`, `feb`, `mar`, `apr`, and so on. The time is specified in 24-hour format.

Suppose, for example, that you have a business that is open from 9:00 a.m. until 9:00 p.m. Monday to Friday and from 9:00 a.m. to 6:00 p.m. Saturday. The dialplan would look like this:

```
; Day
include => open|09:00-21:00|mon-fri|*|*
include => open|09:00-18:00|sat|*|*
include => closed

[open]
exten => 2000,1,Dial(SIP/2000)

[closed]
exten => 2000,1,VoiceMail(2000,u)
```

3.5 The ${EXTEN} Variable and the ${CALLERID(num)} Function

Though we'll examine variables (see Chapter 5, "Dialplan Programming") and functions (see Appendix C, "Dialplan Functions") in more detail later, we want to introduce two simple and intuitive elements here: the system variable `${EXTEN}` and the function `${CALLERID(num)}`.

3.5.1 ${EXTEN}

Asterisk automatically puts the dialed number in the `${EXTEN}` system variable. Instead of

```
exten => 2000,1,Dial(SIP/2000)
```

we can use

```
exten => 2000,1,Dial(SIP/${EXTEN})
```

and get the same result. This doesn't make much difference for a single line, but if you use this in conjunction with pattern matching it saves a lot of time and makes for a much more readable dialplan.

To make all the SIP phones with extension numbers 2000 to 2999 dialable, only one line in `extensions.conf` is required:

```
exten => _2XXX,1,Dial(SIP/${EXTEN})
```

3.5.2 ${CALLERID(num)}

The function `${CALLERID(num)}` always returns the calling party's number. This is especially practical when used with `VoiceMailMain()` because it can accept the mailbox number as a parameter. This way, you can make each user's voicemail box accessible with a single line:

```
exten => 99,1,VoiceMailMain(${CALLERID(num)})
```

You can find a more complete discussion about `${CALLERID(num)}` in Appendix C, "Dialplan Functions."

Case Study: A Typical Business Telephone System

4

Building on what we covered in Chapter 2, "Installation and 'Hello World,'" in this chapter we put together a more comprehensive system for a fictional business. Doing so will show you how to handle the requirements you're likely to be faced with in real installations.

Our fictional company is Widgets, Inc., a widget manufacturing business. We've been charged with the task of planning a phone system for the new office, which houses the following departments:

- Management (2 people)
- Reception (3 people)
- Domestic sales (6 people)
- International sales (4 people)
- Production (50 people)
- Shipping (10 people)
- IT (2 people)
- Building manager (1 person)

Every employee has his or her own phone.

What costs should Widgets, Inc. anticipate? That depends very much on management's objectives. The following should be considered:

- Should the Asterisk server be fault tolerant? Will a cold-standby server be sufficient, or do we need a hot standby?

- What server hardware are we buying? For most systems, a generic computer will suffice. It doesn't have to be the newest and most expensive.

- Which phones are we buying? Phones can range in price from $30 to $300 and up.

Does Asterisk actually make sense for Widgets, Inc., or is this just a make-work project for the IT department? This depends on what the traditional PBX vendors are offering and Widgets, Inc. needs. Although it's not unusual for traditional vendors to produce surprise discounts when they discover an Asterisk system is being considered, there can be hidden costs, sometimes in the form of fixed long-term contracts that force the client to buy from that vendor for several years.

One of the advantages of Asterisk (and open source software generally) is the wide choice and availability of support and consulting services. You can do it all yourself, or you can hire a consultant to do it all for you. If you find the services of your existing consultant unsatisfactory, you can switch to another. This is usually difficult with traditional vendors because they lock you in with a contract and because the proprietary nature of the product also means there is some degree of sales territory exclusivity.

Whether Asterisk makes economic sense has to be calculated on a case-by-case basis. Experience shows that it usually does. If you consider the extensibility, freedom to upgrade (or not), and scalability of Asterisk in your calculations, the mid- and long-term capital and operating costs of an Asterisk system usually prove highly affordable.

What about a packaged Asterisk system, such as FreePBX, Trixbox, PBX-in-Flash, Elastix, or Gemeinschaft? "Prerolled" Asterisk systems have their advantages (rapid installation and configuration, simple interfaces for commonly used features, a GUI) and their disadvantages (lack of transparency, configuration restrictions, smaller installed base). A well-written dialplan for a generic Asterisk installation can serve as a template for new installations, and an experienced Asterisk specialist can have new, complex systems up and running in the same time or less than it takes to set up a prerolled Asterisk. Whatever you choose, you're better off knowing how Asterisk works under the hood before you take the prerolled route. That's the purpose of this book!

4.1 The Numbering Plan

The business has 78 employees. Although it's possible to get by with two-digit extension numbers, we'll use three; this leaves us some breathing room in the event the company expands or we need more extensions (for example, for fax machines or for group numbers). See Table 4.1.

Table 4.1 Numbering Plan for Widgets, Inc.

Number Range	Description	Explanation
911	Emergency number	Emergency numbers should be absolutely intuitive. In the case of fire or serious injury, no employee should have to think twice about what number to dial.[a]
150–159	Building manager	The building manager is responsible for all building operations.
160–169	IT	The IT department is responsible for computer and network operations.
200–229	Management	The executive offices have 2xx extensions.
230–269	Reception	Because reception will be fielding calls mostly to management, they have numbers in the range 230 to 269. Numbers from 270 to 299 are kept in reserve.
300–349	Domestic sales	Sales gets the entire 3xx number block, to ensure adequate reserve in case of expansion. The first 50 numbers are allocated to domestic sales.
350–399	International sales	The second 50 numbers are allocated to international sales.
400–449	Shipping	The first half of the 4xx number block is allocated to shipping.
500–599	Production	Production is the largest department and gets the entire 5xx block.
800–899	Internal services	Special services and nonstandard extensions (e.g., voicemail or teleconferencing).

a. 9-1-1 is the universal emergency number for the North American Numbering Plan (NANP). If you are configuring systems for an international business, you may want to consider adding 9-9-9 and 1-1-2, as well, because these are standard emergency numbers for the United Kingdom and continental Europe, respectively.

4.2 Choosing the Infrastructure

The new building is getting a modern structured cabling system, so it makes sense to equip each workstation with a SIP phone. This saves us the grief and extra expense of running telephone wiring in parallel to the network wiring. Fax machines and cordless telephones can be connected to the system using analog telephone adapter (ATA) devices.

Warning: Faxing in Asterisk is nevertheless problematic. VoIP connections, depending on how they are compressed, can have less bandwidth available than a fax connection requires. This can result in fax transmissions being only partially received, or worse, failing completely. If you are only using VoIP internally, you must ensure that the codecs you choose can adequately support fax. The better option is to use an analog telephony card to provide internal analog extensions purely for fax use, or, ideally, build fax server functionality into Asterisk.

We install the Asterisk server in a climate-controlled server room with controlled access. Only authorized personnel are to have access to the server.[1]

4.2.1 Network

The network is such a significant aspect of Asterisk's functionality that it demands a more thorough discussion.

Telephony is a synchronous medium, which means that the receiver must accept the sender's transmission the instant it is sent. Human beings can tolerate a delay (also known as latency) of up to 300ms (milliseconds), although the delay can be perceptible when over 100ms and distracting when over 200ms. Most people find anything over 500ms unacceptable. This is the principal challenge that IP telephony must address: In IP telephony, we are attempting to carry a synchronous data stream over an asynchronous medium. TCP/IP, developed with funding from

---

1. In the past, it was common for PBX and key system manufacturers to build secret backdoors into their systems, mainly for the purpose of providing timely support in the event of an outage. It was inevitable that clever but unscrupulous parties would figure out a way to exploit these backdoors to make free long-distance and international calls, mostly at the expense of public institutions too large to have a clear picture of how their systems were configured. In today's world, the security of your systems, whether they are Asterisk or anything else, should always be a paramount concern.

the U.S. Defense Advanced Research Projects Agency (DARPA), was designed to maximize stability and robustness, but it achieves this at the expense of the kind of real-time performance we would expect of a telephone connection.

When an e-mail suffers an occasional pause in transmission, it is materially unaffected. A voice conversation, by contrast, is highly time dependent. If someone says "hi" to us, we expect it to be followed by something relevant and intelligible immediately (such as our name), because that is the way people speak to each other in person. That we can even carry voice traffic on an IP connection is owing to the high performance of modern network hardware. The high bandwidth of modern networks coupled with the relatively small bandwidth required by a single telephone-grade voice conversation means that voice traffic can be carried in a pseudo-real-time fashion.

Nevertheless, network conditions vary, both by site and in time. A big network with many endpoints will carry a lot of traffic; networks also tend to be busier during normal business hours (say from 8 a.m. to 6 p.m.) while employees send and receive e-mail, surf the web, download programs and documents, etc. This happens to be the time they make most of their telephone calls, and things can reach a critical point very quickly. If the available bandwidth drops below the minimum needed, we will start to see packet delay or loss. When this happens, the effect on voice conversations is drastic and immediate, resulting in dropouts, clipped words, clicks, pops, and other strange noises. Anybody who has any experience with wireless telephone service knows how distracting this can be. To deal with these eventualities, commercial-grade routers and most managed switches support some kind of traffic classification through setting a TOS (Type of Service) flag in the packet header.

By setting a TOS flag in voice packets, we can prioritize traffic passing through the router or switch. This feature is useful for managing traffic inside our network and across the network boundary, but it has no impact once the packets are on the Internet service provider's (ISP) network or on the public Internet. Only once the ISP guarantees low latency and bandwidth for real-time traffic can we have an assurance of call quality (called quality of service, or QoS). If you will be supporting a multibranch operation and plan to connect them with VoIP, you should consider this requirement carefully and discuss it with your ISP.

The Asterisk system for Widgets, Inc. has at least 78 physical stations. In the worst case, we will have 39 simultaneous conversations (which happens if all the employees call each other). Depending on the codec used, we can expect a maximum bandwidth of 6,500kbps (kilobits per second), which should be no problem for a typical 100Mbps (megabits per second) LAN on structured, twisted-pair cable.

All the same, this network will be carrying other traffic, and a couple of big file transfers can be enough to hit performance limits. How do we get the 6,500kbps value? The standard G.711 compression standard includes two companding algorithms: u-law (used in North America and Japan) and a-law (used in Europe).[2]

These provide excellent sound quality, but need at least 64kbps bandwidth. A single connection consists of an incoming and outgoing channel, each at 32kbps; 39 connections add up as follows: $2 \times 39 \times 32$kbps = 2,496kbps. In addition to the channel bandwidth required, there will be TCP/IP network overhead, so that each channel will use about 80kbps in total. Now our total is 3,120kbps; round up, and we have 3,500kbps. While this is only a worst-case scenario, when it comes to voice systems, it is better to be safe than sorry![3]

Tip: If you have doubts about whether you have sufficient bandwidth available to support all your extensions, consider starting with a lossier codec, such as Global System for Mobile (GSM). At between 13kbps and 15kbps per connection, it needs only a fifth of the bandwidth of u-law and delivers adequate sound quality for most applications. After you've verified the stability of your system, you can try switching groups of extensions over to u-law and wait to see whether you have network problems as a result. Just remember highly compressive codecs like GSM are more processor intensive, so you need to ensure you have adequate CPU resources available on your Asterisk server.

4.2.2 Server Hardware

The choice of server hardware for the Asterisk server is subject to many of the same considerations. The need for synchronous, real-time data transfer requires timely handling of the necessary processes in Asterisk. Most of the computing effort comes in the coding and decoding of digital sound streams; this effort is even greater if you have equipment that uses different codecs and Asterisk has to transcode between them.[4] The tremendous variety and quantity of hardware available means there is no hard-and-fast rule for calculating processor requirements. As a guide, however, we can start with Table 4.2.

---

2. These two algorithms are used for carrying voice traffic in ISDN channels (such as those used in T1 and T3 connections).

3. The website www.asteriskguru.com/tools/bandwidth_calculator.php has an Asterisk bandwidth calculator with bandwidth values for various codecs that can help you in the planning process.

4. Transcoding is the process by which an audio stream in one codec is translated into another codec. For example, a connection between a device supporting only GSM and one supporting only G.711 will have to be transcoded; this is normally done by Asterisk.

Table 4.2 Server Hardware Benchmarks

CPU	Simultaneous Connections	Endpoints
1.0GHz	15	40
1.5GHz	40	100
2.0GHz	80	200

By *CPU*, we mean a typical CPU from an off-the-shelf PC (e.g., Intel or AMD). We've calculated the number of simultaneous connections based on the reasonable assumption that not all the phones will be active at one time, but rather between one-third and one-half. Note that this is just a rough estimate, with a considerable margin of safety. If little transcoding, conferencing, or echo cancellation is required, the number of simultaneous connections supported would be higher. That said, the exact limits for any given installation can be reliably determined only through direct testing and experimentation.

Tip: The Linux program top provides a simple tool for monitoring processor loads on a running server, but only at the current point in time. Processor and process loads are dynamic and change dramatically from one moment to the next, so point measurements have limited usefulness. You can try monitoring the performance of your Asterisk servers using monitoring tools like Nagios (www.nagios.org). If you prefer to use native Linux tools, you can do simple monitoring with a cron job running a bash script with this content:

```
w | head -n 1 >> /var/log/serverloads
```

By running this cron job every minute, you will build up a set of data points over time which you can use to evaluate server performance.[5]

Widgets, Inc. has 78 employees, and an estimate of 40 simultaneous connections is reasonable, so a computer with a reasonably current CPU (e.g., Pentium 4, AMD Athlon) at a clock speed of 1.5GHz should suffice. In general, the newer the CPU and system, the better the performance. Processor clock speed is no longer an accurate measure of CPU performance, particularly as applied to the newer dual-core

---

5. This cron job would indeed be a very simple analysis tool. If you are comfortable with shell scripting, you can write more comprehensive scripts, which can give you more detailed information, by using output from tools such as top and uptime, processing it, and dumping the processed output into a comma-delimited text file that you can graph in a spreadsheet program.

processors, and your server will likely be able to handle more calls with a newer processor at the same clock speed. Dual-processor machines will improve performance correspondingly.

For the voice mailboxes, we need about 0.1MB (megabytes) of disk space per minute of message time.[6]

If we allot 30 minutes of message storage to each user, we can expect to use a total of about 2GB (gigabytes) of disk space. Because voice message data is valuable business data and can even be critical, the storage should be redundant and backed up regularly. A RAID 1 array (two disks or disk sets in a mirror configuration) will provide the necessary redundancy.

Warning: Note that software RAID burdens the CPU significantly. A professional installation should use hardware RAID if at all possible.

Memory requirements for Asterisk are not onerous, and 512MB RAM will suffice. With 1GB RAM, Asterisk can handle very large business units. Any additional steps taken to enhance the reliability and performance of the server, such as the installation of redundant network adapters, commercial-grade telephone interface cards, and uninterruptible power supplies, will go a long way to increasing user and decision-maker confidence in the system. If your background is in IT services, you are facing a whole new level of user expectations. Users will usually tolerate a brief e-mail outage; the phones, however, must always work. Readers with a background in circuit-switched telephony will know exactly what we are talking about. Hell hath no fury like an executive without a working desk set.

4.3 Base Configuration

We'll configure Asterisk so that all the employees can call each other and have their own mailbox. When that's done, the system can be extended with additional functions and modules.

---

6. This is for wav49 or GSM format message files. If you decide to use uncompressed WAV format files, you will use many times the space (up to ten times as much). Because disk space is cheap, however, this is not normally an issue, and more installations are using uncompressed WAV because of the higher sound quality. Remember, though, that more disk space used means more storage required for backup.

4.3.1 A sip.conf for 100 Users

Configuring /etc/asterisk/sip.conf for 100 users is no different from configuring for 2 users; there are no shortcuts here. Each SIP extension must be configured individually. What follows is an abridged version of the sip.conf for Widgets, Inc. Lines beginning with the semicolon (;) are treated as comments and ignored by Asterisk. The file is composed in the same way as in our simple example in Chapter 2. In general, we've avoided assigning numbers to individual users that end in 0 because we want to reserve those numbers for group phone numbers. The building manager is an exception, however, because we don't anticipate any need for a group phone number there.

```
[general]
port = 5060
bindaddr = 0.0.0.0
context = other

; Building Manager
;
[150] ; normal office phone
context=building-mgr
secret=1234
callerid="Building Manager" <150>
type=friend
host=dynamic

[151] ; cordless phone
context=building-mgr
secret=1234
callerid="Building Manager" <150>
type=friend
host=dynamic

; Internal IT department
;
[161] ; Tech 1
context=it
secret=1234
type=friend
host=dynamic

[162] ; Tech 2
context=it
secret=1234
type=friend
host=dynamic

[165] ; Cordless phone
context=it
secret=1234
```

```
type=friend
host=dynamic

; Executive offices
;
[201] ; Manager 1
context=management
secret=1234
type=friend
host=dynamic

[202] ; Manager 2
context=management
secret=1234
type=friend
host=dynamic

; Reception
;
[231] ; Assistant 1
context=reception
secret=1234
type=friend
host=dynamic

[232] ; Assistant 2
context=reception
secret=1234
type=friend
host=dynamic

[233] ; Assistant 3
context=reception
secret=1234
type=friend
host=dynamic

; Domestic Sales
;
[301] ; Account Agent 1
context=domestic-sales
secret=1234
type=friend
host=dynamic

[302] ; Account Agent 2
context=domestic-sales
secret=1234
type=friend
host=dynamic

[303] ; Account Agent 3
context=domestic-sales
```

```
secret=1234
type=friend
host=dynamic

[304]  ; Account Agent 4
context=domestic-sales
secret=1234
type=friend
host=dynamic

[305]  ; Account Agent 5
context=domestic-sales
secret=1234
type=friend
host=dynamic

[306]  ; Account Agent 6
context=domestic-sales
secret=1234
type=friend
host=dynamic

;  International Sales
;
[351]  ; Account Agent 1
context=international-sales
secret=1234
type=friend
host=dynamic

[352]  ; Account Agent 2
context=international-sales
secret=1234
type=friend
host=dynamic

[353]  ; Account Agent 3
context=international-sales
secret=1234
type=friend
host=dynamic

[354]  ; Account Agent 4
context=international-sales
secret=1234
type=friend
host=dynamic

; Shipping
;
[401]  ; Shipping Agent 1
context=shipping
secret=1234
```

```
type=friend
host=dynamic

; The remaining 9 entries for shipping
; are omitted here for the sake of brevity.

; Production
;

[501] ; Assembly line 1
context=production
secret=1234
type=friend
host=dynamic

; The remaining 49 entries for production
; are omitted here for the sake of brevity.
```

You can make `sip.conf` more concise with templates. For more information, refer to Appendix F, "Configuration Templates."

Our example `sip.conf` has two instances of the `callerid` variable. This variable sets what is displayed on the called party's telephone:

```
callerid = "Building Manager" <150>
```

In our example, both the building manager's phones (150 and 151) use the caller ID 150 and the caller ID name Building Manager. This ensures that returned calls always go to extension 150.

4.3.2 The Dialplan

In Chapter 2, we wrote a simple dialplan for two phones. In the case study in this chapter, we have more users. Writing a dialplan with 100 entries is not exactly difficult, but it is tedious, and maintenance could become a problem. It's much better to use pattern matching in `extensions.conf` using the patterns described earlier in this book.

4.3.2.1 extensions.conf

The numbering plan[7] for Widgets, Inc. translates into a readable, and maintainable `extensions.conf`, which looks like this:

```
[other]

[building-mgr]
```

---

7. With the exception of the emergency numbers. We'll deal with those as soon as we've established a connection to the outside world.

```
    include => internal-calls
    include => voicemail-easy

    [it]
    include => internal-calls
    include => voicemail-easy
    ;
    ; The IT department should be able to
    ; access all voicemail boxes for
    ; debugging purposes
    ;
    include => voicemail-normal

    [management]
    include => internal-calls
    include => voicemail-easy

    [reception]
    include => internal-calls
    include => voicemail-easy

    [domestic-sales]
    include => internal-calls
    include => voicemail-easy

    [international-sales]
    include => internal-calls
    include => voicemail-easy

    [shipping]
    include => internal-calls
    include => voicemail-easy

    [production]
    include => internal-calls
    include => voicemail-easy

    [internal-calls]
    exten => _[1-5]XX,1,Dial(SIP/${EXTEN},60)
    exten => _[1-5]XX,2,VoiceMail(${EXTEN},u)

    [voicemail-easy]
    ;
    ; Users don't have to enter their extension number
    ; when retrieving voicemail.
    ;
    exten => 800,1,VoicemailMain(${CALLERID(num)})

    [voicemail-normal]
    ; Prompts for a mailbox number.
    exten => 801,1,VoicemailMain()
```

The extensions in the `[internal-calls]` context mean all calls to numbers 100 to 599 are passed to the `Dial()` application exactly as dialed, connecting the calling party to the corresponding SIP phone. If the user dials 800, she is connected to the mailbox for that phone. Only the IT department can call 801, at which point the system prompts for a mailbox number. This is done so that the IT department, which is now responsible for supporting the telephone system, can access all the mailboxes for debugging purposes.

In actual practice, our dialplan would have to be somewhat more complex. The numbering plan has number ranges (e.g., 270 to 299) that don't contain any telephones. Our example ignores this, but this could lead to confusion in an actual deployment. If a user dials an unassigned number, the call goes straight to a voice mailbox that may or may not exist; even if it does, the user could leave messages that are never retrieved.

To prevent this from happening, the `[internal-calls]` context would look like this:

```
[internal-calls]
exten => _1[5-6]X,1,Dial(SIP/${EXTEN},60)
exten => _1[5-6]X,2,VoiceMail(${EXTEN},u)

exten => _2[0-6]X,1,Dial(SIP/${EXTEN},60)
exten => _2[0-6]X,2,VoiceMail(${EXTEN},u)

exten => _[358]XX,1,Dial(SIP/${EXTEN},60)
exten => _[358]XX,2,VoiceMail(${EXTEN},u)

exten => _4[0-4]X,1,Dial(SIP/${EXTEN},60)
exten => _4[0-4]X,2,VoiceMail(${EXTEN},u)
```

For Widgets, Inc., we accept this risk for the sake of simplicity and use the following configuration:

```
[internal-calls]
exten => _[1-5]XX,1,Dial(SIP/${EXTEN},60)
exten => _[1-5]XX,2,VoiceMail(${EXTEN},u)
```

4.3.2.2 voicemail.conf

The `voicemail.conf` remains as described in Chapter 9, "Voicemail," except with a few additional entries restricting message length and defining the server e-mail address:

```
[general]
format = gsm
serveremail = voicemail@widgets.com
maxmessage = 600
```

```
[local]
150 => 999999,Bob Boilerman,building@widgets.com
200 => 999999,Charles Crucial,charles.crucial@widgets.com
201 => 999999,John Smart,john.smart@widgets.com


; We omit the remaining entries for brevity
```

Note: The passwords used in the preceding example are for demonstration purposes only. You'll want to use stronger passwords in an actual production installation.

4.4 What Next?

The preceding configuration for Widgets, Inc. serves as a planning example for all systems of similar size. You can use this configuration to experiment with many of the configuration examples in this book.

This system allows telephone calls only and includes a simple voicemail system. A professional installation usually demands features such as teleconferencing and queues. If this is what you need, you have two basic options: You can reinvent the wheel and build a dialplan that includes those features from scratch, or you can use a standard framework such as FreePBX (www.freepbx.com) or Gemeinschaft (www.amooma.de/gemeinschaft) and configure it according to your requirements. (At the moment, Gemeinschaft is available only in German. Volunteers willing to help internationalize it are welcome! Contact Stefan at stefan.wintermeyer@amooma.de.)

5

Dialplan Programming

In Asterisk, functions or programs can be implemented either externally, through an Asterisk Gateway Interface (AGI) script (in much the same way that a Common Gateway Interface [CGI] script can add functionality to a web page) or internally, through functions and applications in the dialplan. This chapter focuses strictly on the internal functions; AGI is treated in depth in a separate chapter.

The dialplan is defined in the `extensions.conf` configuration file. The dialplan itself looks much like a BASIC program. The administrator can implement features and call flow using a simple scripting language.

5.1 Programming "How-To"

Whatever your experience level with Asterisk (whether you're an administrator, programmer, telephone specialist, or hobbyist), you need a basic level of programming skill and a grasp of the fundamentals to take full advantage of Asterisk. This how-to should give you the basic understanding you need to make useful dialplans, through the use of plenty of examples and with frequent reference to Appendix B, "Dialplan Applications." You will probably recognize some of the material from other chapters.

5.1.1 Program Structure

Each telephone number defined in the Asterisk dialplan (/etc/asterisk/extensions.conf) is really a small program. In Asterisk, the program is called an *extension*. An extension looks like this:

```
exten => 1001,1,Answer()
exten => 1001,n,Playback(hello-world)
exten => 1001,n,Hangup()
```

Priorities may also be numbered sequentially:

```
exten => 1001,1,Answer()
exten => 1001,2,Playback(hello-world)
exten => 1001,3,Hangup()
```

The two extensions depicted here are functionally identical. If you use n, however, it makes adding and deleting entries in the extension much easier later on.

5.1.2 Using Set()

Use the application **Set()** to create and change variables:

```
exten => 1002,1,Set(Favoriteanimal = "Tiger")
exten => 1002,n,Set(Favoritenumber = 23)
```

Use the syntax ${VARIABLENAME} to read and print variables. You can print variable values on the CLI with NoOp() (with verbosity level 3 and up):

```
exten => 1003,1,NoOp(${Favoriteanimal})
exten => 1003,n,NoOp(${Favoritenumber})
```

There are different kinds of variables:

- **Global variables**

  Valid anywhere in the dialplan and created or modified with
  Set(<variable>=<content>,g):

  ```
  exten => 1004,1,Set(READABLEANYWHERE = 23,g)
  exten => 1004,n,NoOp(${READABLEANYWHERE})
  ```

- **Channel variables**

  Valid only in the current channel (a channel could be a connection between two people having a phone conversation). Created or modified with
  Set(<variable>=<content>) (without the g):

  ```
  exten => 1005,1,Set(READABLEHEREONLY= 42)
  exten => 1005,n,NoOp(${READABLEHEREONLY})
  ```

- **System variables**

  These dynamic variables are set by Asterisk and may be called in the dialplan without needing to create them. A frequently used system variable is ${EXTEN}:

  ```
  exten => 1006,1,NoOp(Dialed number: ${EXTEN})
  ```

For a more in-depth look at variables, see the "Variables" section, later in this chapter.

Note: For all the commands discussed in this chapter, you can find syntax and options in Appendix B.

5.1.3 Labels and Goto()

Goto() lets you jump from one dialplan entry to another. If you are using n priorities, this can be problematic. The solution is to use labels to tag specific entries and then call the entry by label in **Goto()**.

You can use **Goto()** within an extension, between extensions, or between contexts.

- You might use it within an extension:
  ```
  exten => 1007,1,Answer()
  exten => 1007,n(Start),Wait(1)
  exten => 1007,n,Playback(hello-world)
  exten => 1007,n,Goto(Start)
  ```

- Between extensions:
  ```
  exten => 1008,1,Answer()
  exten => 1008,n,Goto(1009,Ping)

  exten => 1009,1(Ping),Playback(hello-world)
  exten => 1009,n,Wait(2)
  exten => 1009,n,Goto(1010,Pong)

  exten => 1010,1(Pong),Playback(tt-weasels)
  exten => 1010,n,Wait(2)
  exten => 1010,n,Goto(1009,Ping)
  ```

- Between contexts:
  ```
  [hq]
  exten => 1011,1,Answer()
  exten => 1011,n,Playback(hello-world)
  exten => 1011,n,Goto(sales,1012,1)
  ```

```
[sales]
exten => 1012,1,Playback(hello-world)
exten => 1012,n,Hangup()
```

5.1.4 While() Loops

Loops let you perform operations repeatedly, which is useful for reading out sequences. Use **While()** to run loops in the dialplan:

```
exten => 1013,1,Answer()
exten => 1013,n,Set(i=1)
exten => 1013,n,While($[${i} < 10])
exten => 1013,n,SayNumber(${i})
exten => 1013,n,Wait(1)
exten => 1013,n,Set(i=$[${i} + 1])
exten => 1013,n,EndWhile()
exten => 1013,n,Hangup()
```

5.1.5 GotoIf() Conditional

You can jump to other parts of the dialplan, if a specific condition is met, with **GotoIf()**:

```
exten => 1014,1,Answer()
exten => 1014,n,Set(Favoritestation = 0815)
exten => 1014,n,NoOp(Check to see if ${Favoritestation} is calling.)
exten => 1014,n,GotoIf($[${CALLERID(num) = ${Favoritestation}]?yes,no)

exten => 1014,n(yes),Playback(hello-world)
exten => 1014,n,Hangup()

exten => 1014,n(no),Playback(tt-monkeys)
exten => 1014,n,Hangup()
```

5.1.6 Gosub() Subroutines

With **Gosub()**, the call is directed to a subroutine; it can be returned to the initiating priority with **Return()**:

```
exten => 1015,1,Gosub(cid-set)
exten => 1015,n,Dial(SIP/${EXTEN})

exten => 1015,n(cid-set),Set(CALLERID(all)=Widgets Inc <8005551212>)
exten => 1015,n,Return()
```

5.2 Variables

A variable is a placeholder for an actual value. Exactly what that value is depends on the kind of variable. In Asterisk, variables can contain numbers, letters, and strings (sequences of letters and numbers). Variables are useful because they let us create rules for call flow that apply in changing circumstances and make it easier to accommodate future changes in the telephone application or system. In Asterisk, variables have varying scope. There are local variables (called *channel variables* in Asterisk), which can only set values for the current, active channel, and global variables, which set values for all channels. We should already be familiar with some of the variables Asterisk sets from our exposure to them as configuration parameters in the Asterisk configuration files (such as `sip.conf`, for example). We also have the freedom to define our own variables and use them in configuration files.

5.2.1 Expanding Variables in an Extension

The value of a variable can be obtained using the syntax `${VARIABLENAME}`. There are variables that are automatically set by Asterisk. For example, the called number is always stored in the Asterisk system variable `${EXTEN}`. Using patterns and variables, it is often possible to dramatically compress a long dialplan.

Without variables, a set of extensions would have to be represented like this:

```
exten => 100,1,Dial(SIP/100)
exten => 101,1,Dial(SIP/101)
exten => 102,1,Dial(SIP/102)
exten => 103,1,Dial(SIP/103)
exten => 104,1,Dial(SIP/104)
exten => 105,1,Dial(SIP/105)
exten => 106,1,Dial(SIP/106)
exten => 107,1,Dial(SIP/107)
exten => 108,1,Dial(SIP/108)
exten => 109,1,Dial(SIP/109)
```

Using a variable, however, lets us make multiple extensions succinct:

```
exten => _10X,1,Dial(SIP/${EXTEN})
```

5.2.2 General Considerations

When working with variables, you need to keep a few things in mind. First, variables are not case sensitive. String variables should be enclosed in quotes as a matter of practice. Reserved characters need to be escaped if you're going to use them in a variable, and integer variables have size limits. Let's explore these considerations in detail.

5.2.2.1 Capitalization

Variable names needn't be in all uppercase as in our examples, nor are user-defined variables case sensitive. It is a good idea to use uppercase variable names nonetheless because it makes the variables easier to identify and the dialplan code easier to read. Unfortunately, this means you cannot distinguish variable names based on case. For example, `${FOO}` is considered the same as `${foo}`.

Important: Asterisk system variables such as `${EXTEN}` must always be uppercase.

5.2.2.2 Using Double Quotes with String Variables

String variables (meaning variables that contain text and not numbers) should be defined using double quotes, though Asterisk will still accept them without double quotes. The following two entries are functionally identical:

```
exten => 1234,1,Set(FRUIT=Apple)
exten => 1234,2,Set(FRUIT="Apple")
```

If the string contains commas or spaces, you must use double quotes:

```
exten => 1234,1,Set(FRUITTYPES="Apple, Pear, etc.")
```

This is why it is a good idea to get into the habit of using them for any string variables you define.

5.2.2.3 Reserved Characters

Sometimes a variable will contain reserved characters (characters that have special functions and are interpreted differently). For example, if you want a variable to contain the underscore character (_), you must use an "escape" character to tell the dialplan interpreter that it should ignore the reserved character. The following characters must be escaped when used in a variable:

```
[ ] $ " \
```

The escape character in `extensions.conf` is the backslash (\):

```
exten => 1234,1,Set(AMOUNT="\$10.00")
```

Similarly, if you want to use the backslash character in a variable, you must escape it:

```
exten => 1234,1,Set(ROOMNUMBER="48\\10")
```

5.2.2.4 Limits on Integer Size

If a variable contains an integer, it can have no more than 18 digits. Anything larger will cause an error, which will be recorded in the log file.

Tip: If you need to work with larger integers or floating point numbers, you can use an AGI script (see Chapter 16, "Asterisk Gateway Interface").

5.2.3 Defining Global Variables in extensions.conf

Global variables are defined at the beginning of `extensions.conf`. You must place them in the special `[globals]` context, which follows `[general]`:

```
[general]

[globals]
RINGTIME=90

[from-internal]
exten => _XXX,1,Dial(SIP/${EXTEN},${RINGTIME})
exten => _XXX,n,VoiceMail(${EXTEN})
```

5.2.4 Defining Variables with Set()

`Set()` is used to define a variable inside an extension (see also the application description in Appendix B). The general syntax is `Set(<variable1>=<value1>[,<variable2>=<value2>][,<option>])`.

Setting option `g` makes the variable global. Without it, the variable is treated as a local channel variable:

```
; Set a global variable:
exten => 10,1,Set(RINGTIME=90,g)

; Set a local channel variable:
exten => 10,2,Set(FAVORITEFRUIT="Apple")

; Set two channel variables at once:
exten => 10,3,Set(VAR1=10,VAR2=23)

; Print variables to the CLI
exten => 10,4,NoOp(RINGTIME = ${RINGTIME})
exten => 10,5,NoOp(FAVORITEFRUIT = ${FAVORITEFRUIT})
exten => 10,6,NoOp(VAR1 = ${VAR1})
exten => 10,7,NoOp(VAR2 = ${VAR2})
```

5.2.5 Inheritance of Channel Variables

If new channels are spawned while a conversation is in progress, they will have their own channel variables.

5.2.5.1 Single-Level Inheritance

Sometimes you want to have a channel variable persist into the spawned channel. You can do this by prefixing the variable with an underscore (_) character. When the variable is inherited by the spawned channel, Asterisk automatically removes the prefix. This ensures that the variable is inherited only once.

The variable in this example will be passed to any dependent channels:

```
exten => 1234,1,Set(_CAKE="Marble cake")
```

5.2.5.2 Multilevel Inheritance

If you need unlimited inheritance of a channel variable, you can do this by prefixing the variable with two underscore characters (__). Variables prefixed in this way will always be inherited by spawned channels.

Warning: Asterisk makes no distinction between variable names that are preceded with an underscore and those that are not. In the following example, a variable with multilevel inheritance ("__CAKE") is rendered uninheritable by the subsequent entry:

```
exten => 1234,1,Set(__CAKE="Marble cake")
exten => 1234,n,Set(CAKE="Marble cake")
```

In this example, the variable CAKE will be inherited by any and all spawned channels:

```
exten => 1234,1,Set(__CAKE="Sponge cake")
```

When calling an inherited variable, it doesn't matter if it is called with a prefix or not. These entries will give the same output in the CLI:

```
exten => 1234,1,NoOp(${__CAKE})
exten => 1234,n,NoOp(${CAKE})
```

5.2.6 System Channel Variables

The following list describes the more important system channel variables. These variables may be read but not overwritten by entries in extensions.conf, as they are predefined by Asterisk. You can find a complete list of all the predefined variables in doc/README.variables (Asterisk 1.2) and doc/channelvariables.txt (Asterisk 1.4). Deprecated variables are not included in this list. For example, the variable ${CALLERIDNUM} (previously commonly used) is not in this list; it is preferable to use the Asterisk function ${CALLERID(num)} instead.

Tip: It is a good practice to replace dialplan code that depends on deprecated variables or functions with code that uses the recommended replacements. This will reduce the chance of an installation breaking when you upgrade Asterisk.

System variables relevant to specific Asterisk functions are covered again in their respective chapters.

Some of the "variables" described here are not really variables but in fact built-in *functions*. In practice, they often play a similar role, so they are listed here for convenience:

- `${ANSWEREDTIME}`

  The total elapsed time for the active connection (in other words, the number of seconds since the conversation started)

- `${BLINDTRANSFER}`

  The name of the channel on the other side of a blind transfer

- `${CHANNEL}`

  Name of the current channel

- `${CONTEXT}`

  Name of the current context

- `${EPOCH}`

  Current UNIX time (total number of seconds elapsed since the beginning of the UNIX "epoch," which began at midnight UTC, January 1, 1970)

- `${EXTEN}`

  Currently dialed extension

- `${ENV(VARIABLENAME)}`

  Environment variable *VARIABLENAME*

- `${HANGUPCAUSE}`

  Cause of connection hang-up

- `${INVALID_EXTEN}`

  Used in the `i` extension and contains the dialed extension

- `${PRIORITY}`

  Current priority in the current extension

- `${TRANSFER_CONTEXT}`

  Context of a transferred call

- `${UNIQUEID}`

  The unique ID for the current connection

- `${SYSTEMNAME}`

  The system name as defined by `systemname` in `/etc/asterisk/asterisk.conf`

5.2.7 Manipulating Variables

Variables are most useful when we can change their contents at execution time. This gives us the flexibility to create complex and powerful behavior in our Asterisk system.

5.2.7.1 Substrings

In general, a string consists of a sequence of individual characters. The size of a string is determined by the number of characters contained in it. For example, the string apple tree has ten characters; we must include the space. Any string can be broken into substrings. For example, apple, tree, app, and le tre are all valid substrings of apple tree. In theory, a string can be of any length; this entire book could be contained in a single string, although it would be impractical. Manipulation of strings is an important technique in programming applications. Asterisk lets you manipulate strings and substrings using the colon (:) character. Using the colon character, you can extract a specified portion of an existing string variable:

```
${VARIABLENAME[:start[:length]]}
```

5.2.7.1.1 Examples

Many telephone systems require that a prefix digit be dialed to get an outside line. (In North America, this is usually 9.) The target number, however, cannot include this prefix digit. If we dial 9-1-202-7075000, we can store the actual outside number in the ${OUTGOINGNUMBER} using the following dialplan entry.[1]

```
exten => _9X.,1,Set(OUTGOINGNUMBER=${EXTEN:1})
```

If the *length* option is omitted, the rest of the string is taken automatically.

What if we only need the last seven digits of the dialed number? I this case, we use a negative number for the start parameter. The following entry would store 7075000 from our preceding example in the variable ${LOCALNUMBER}.

Note: Under the original rules of the NANP (North American Numbering Plan) the last seven digits of a number constituted the local portion of the number. That is, if your telephone number was in the same area code as the dialed number, you needed to dial only seven digits. To accommodate population growth and density, the NANP has been extended with *overlay dialplans*. In areas with overlay

---

1. For our curious readers, this is the general information number for the Library of Congress in Washington, D.C.

plans, two telephone lines on the same street, or even in the same building, may have different area codes. You are in an overlay area if you are required to dial ten digits for local calls. In this case, a local number filter as depicted above will not be appropriate. Also, many area codes covering larger areas still have portions of the number space that are treated as long distance.

```
exten => _0X.,1,Set(LOCALNUMBER=${EXTEN:-7})
```

We can also capture just the area code:

```
exten => _0X.,1,Set(AREACODE=${EXTEN:2:3})
```

Note: Obviously, readers in other parts of the world (e.g., the United Kingdom, Australia, or elsewhere) will have different national dialplans that impact how outside numbers should be processed. Some countries have area and city codes that are variable in length; in those cases, this kind of number filtering will not be practical.

Here, then, is how we might extract useful information from a dialed number:

```
exten => _9X.,1,Set(AREACODE=${EXTEN:2:3})
exten => _9X.,n,Set(LOCALNUMBER=${EXTEN:5})
```

5.3 Special Extensions

Because all the programming logic must occur via extensions, we need some additional system-defined extensions.

5.3.1 The h Extension

The h is the standard "hang-up" extension. The h extension, if it is configured, is called when a caller hangs up the phone. Note that as soon as this happens, the content of ${EXTEN} changes to h.

Suppose we want the global variable CONNECTIONS to reflect the number of currently active conversations at any given time. This means we need the value of CONNECTIONS to increase by one every time a connection is initiated and decrease by one every time someone hangs up. The following dialplan illustrates the basic idea:

```
[global]
CONNECTIONS=0
```

```
[from-internal]
exten => _X.,1,Set(CONNECTIONS=$[${CONNECTIONS} + 1]|g)
exten => _X.,2,Dial(SIP/${EXTEN})


exten => h,1,Set(CONNECTIONS=$[${CONNECTIONS} - 1]|g)
```

5.3.2 The i Extension

To make a context gracefully handle every conceivable circumstance, we use the special extension i (which stands for invalid), which handles all dialed numbers that are not explicitly handled within the context. Again, as with the h extension, when the i extension is invoked, ${EXTEN} will no longer contain the dialed number. To get the dialed number while in the i extension, use ${INVALID_EXTEN}.

In our example business, Widgets, Inc. employees in department B can only dial extensions 100 to 199. Callers dialing any other numbers hear this message: "I'm sorry. That is not a valid extension. Please try again."

```
[department-b]
exten => _1XX,1,Dial(${EXTEN})

exten => i,1,NoOp(An invalid number ${INVALID_EXTEN} was dialed.)
exten => i,2,Answer()
exten => i,3,Playback(invalid)
exten => i,4,Hangup()
```

5.3.3 The o and a Extensions

If operator yes is not in voicemail.conf, the call will be directed to the o extension (o is for operator, kids!) if the caller presses zero (0).

Pressing * (asterisk) will direct the call to the a extension (abort).

5.3.4 The t and T Extensions

The t and T extensions are for handling timeouts in the context.

5.3.4.1 t Extension

If there is no input in an interactive voice response (IVR) menu within a certain time, the t extension is called.

```
[mainmenu]
exten => 10,1,Answer()
exten => 10,n,Background(marryme)           ; "Marry me? Press 1 for
                                              yes, 2 for no."
```

```
exten => 1,1,Playback(thank-you-cooperation) ; 1 => "Thank you."
exten => 1,n,Hangup()

exten => 2,1,Playback(hangup-try-again)      ; 2 => "Hang up and try
                                                     again."
exten => 2,n,Hangup()

exten => t,1,Hangup()                         ; no input => hang up
```

5.3.4.2 T Extension

The T extension is called after the absolute timeout has been exceeded. You can set this timeout value with `Set(TIMEOUT(absolute)=<seconds>)`.

Warning: Be careful not to have any spaces before and after the equal sign (=) character.

The timer starts whenever the timeout value is set. (In other words, it does not automatically start with the connection; it must be started explicitly with the `Set()` command.) `Set(TIMEOUT(absolute)=0)` deactivates the absolute time-out. In the following example, the timeout is set to 120 seconds. If the call lasts longer than that, the T extension is called, which plays back a message that says "Thank you for calling" and hangs up:

```
exten => 20,1,Answer()
exten => 20,2,Set(TIMEOUT(absolute)=120)
exten => 20,3,Playback(hello-world)
exten => 20,4,Wait(1)
exten => 20,5,Goto(3)

exten => T,1,Wait(1)
exten => T,2,Playback(thank-you-for-calling)
exten => T,3,Wait(1)
exten => T,4,Hangup()
```

5.3.5 The s Extension

The first entry in any extension is always the name or number dialed by the caller. When a call comes in from the public switched telephone network (PSTN), however, Asterisk doesn't know what was dialed or whom the caller is trying to reach. For any scenario in which we cannot determine the number dialed, we use the s extension. In this example, the sound of monkeys is played back to calls that come in via the PSTN:

```
exten => s,1,Answer()
exten => s,2,Wait(1)
```

```
exten => s,3,Play(tt-monkeys)
exten => s,4,Wait(1)
exten => s,5,Hangup()
```

Warning: If you are using an ATA device (analog telephone adapter) you don't need the s extension. You can configure the dialed number in the configuration interface (often a web interface) for the ATA.

5.4 Macros

A macro is a kind of subroutine. It can contain complex workflows but is called through a single entry. This reduces repetition in the dialplan and makes it cleaner and smaller. A simple example might look like this:

```
[macro-incoming]
exten => s,1,Dial(SIP/${MACRO_EXTEN},10)
exten => s,n,VoiceMail(${MACRO_EXTEN})
```

This macro would be called like so:

```
[sales]
exten => _2XXX,1,Macro(incoming)

[building-mgr]
exten => _2XXX,1,Macro(incoming)
```

The effect is not quite so impressive with a two-line macro as it would be with a much longer macro, but the advantages of such an approach should be clear.

The use of macros tends to divide the Asterisk user community into two groups. One thinks that macros make the dialplan easier to understand, the other believes that they make the dialplan confusing. We encourage you to draw your own conclusions!

When defining macros, take care to note the following points:

- When defining a macro, only one extension (the s extension) is allowed.
- The original ${EXTEN} and ${CONTEXT} variables cannot be used inside a macro. We must use ${MACRO_EXTEN} and ${MACRO_CONTEXT} instead.
- When calling a macro, additional comma- (,) or pipe-separated (|) arguments can be supplied. These arguments are called within the macro with ${ARGn} (where n is a positive integer indicating which argument in the sequence).
- A macro is defined in square brackets ([macro-macroname]) and is called with the Macro() application in an extension.

More information on macros may be found in Appendix B, "Dialplan Applications."

Tip: The application `MacroExclusive()` ensures that the specific macro can only be called once at any given time. If another channel is calling the macro, no other channel can call it until it completes (see Appendix B).

5.5 Deprecated Features

For a long time, priority jumping was a standard way of moving a call about the dialplan. Specific applications (e.g., `Dial()`) would elevate the priority by 101 under certain circumstances. This feature is now officially deprecated. Basically, this means that although it is currently still supported, eventually it *will* be removed. Those who continue to use it are making their dialplans vulnerable to failure after an upgrade.

When working with Asterisk, you will encounter dialplan configuration examples on the Internet. (It is not uncommon to stumble upon grossly outdated examples that use deprecated constructs.) In professional practice, you should use the most current code constructions and versions of Asterisk. The software is always being updated, but deprecated features are not maintained.

In summary, if you want the operation of your Asterisk systems to be as predictable as possible, keep your code current!

Asterisk Extension Language

As of version 1.4, Asterisk offers two ways of composing the dialplan: the traditional `extensions.conf` format, and the newer AEL, or Asterisk Extension Language, which uses the filename `extensions.ael`.

AEL will be of particular interest to those Asterisk users who find large `extensions.conf` files difficult to read. AEL looks more like a real scripting language. They are not mutually exclusive, either. It's possible to use both, defining contexts in one or the other file. If both files are present, `extensions.conf` will just overwrite identical portions of `extensions.ael` when the dialplan is loaded.

Asterisk comes packaged with a command-line utility for converting AEL files into the traditional `extensions.conf` format called `aelparse`.

Note: When referring to AEL, we mean AEL2. The first version was experimental.

6.1 CLI Commands for AEL

The most important Asterisk CLI command for AEL is **ael reload**. It reads `extensions.ael` into memory and is useful if you've made changes to `extensions.ael` and want to apply them without restarting Asterisk.

```
*CLI> ael reload
```

The **dialplan show** command displays the aggregate dialplan as loaded by Asterisk from `extensions.conf` and `extensions.ael`. The source of each dialplan rule is displayed in a line, in square brackets, preceding the rule. For rules originating from `extensions.conf`, the line looks like this:

```
[ Context 'office-incoming' created by 'pbx_config' ]
```

For rules originating from `extensions.ael`, the line looks like this:

```
[ Context 'queues' created by 'pbx_ael' ]
```

6.2 aelparse

Asterisk versions as of 1.4 come packaged with the command-line utility `aelparse`. You can use this utility to quickly convert `extensions.ael` files into `extensions.conf` format, should that ever be necessary. Programmers may think of it as a kind of test compiler for AEL, a somewhat higher-level "language" than that used in `extensions.conf`.

By default, `aelparse` reads `/etc/asterisk/extensions.ael` unless called with the option -d, in which case it looks for `extensions.ael` in the local directory. If it is invoked without option -n, it will spit out a lot of information that is mostly useful to the Asterisk developers. Use -n for more concise output, or -q for output that shows only warnings and errors.

Tip: `aelparse` is useful for hunting syntax errors in `extensions.ael`.

With option -w, `aelparse` dumps the output into the file `extensions.conf.aeldump` in the local directory:

```
asterisk:/etc/asterisk# aelparse -q -w
LOG: lev:4 file:ael2_parse line:543 func: main 19 contexts, 25 extensions,
62 priorities
```

6.3 Comparing extensions.conf with extensions.ael

The following examples show you how AEL compares with the standard `extensions.conf`. In doing so, we're assuming that you're already comfortable with dialplan programming.

First, you'll see right away that the formatting of extensions is different. The structure of `extensions.ael` is a lot more like a programming language. For example, the following two code blocks are functionally equivalent. We've made the equivalent portions bold to illustrate this.

Here's a typical context, as it would be written in `extensions.conf`:

```
[my-phones]
exten => 20,1,Answer()
exten => 20,n,Playback(beep)
exten => 20,n,Hangup()
```

The same result can be achieved in `extensions.ael` with this:

```
context my-phones {
   20 => {
     Answer();
     Playback(beep);
     Hangup();
   }
}
```

6.3.1 Line Termination

AEL commands must always end with the semicolon (;) character, because multiple commands may appear on a single line. For readability, however, most people put each command on a separate line.

6.3.2 Contexts, Extensions, and Priorities

The description of contexts, extensions, and priorities is what makes AEL different from its predecessor. AEL uses curly braces ({ ... }). Numbered priorities (1, n) are no longer required, and so we are spared the dubious charms of BASIC style programming in which every line must be numbered, as well as the tedious effort of writing the same extension number for every priority. In the simple examples that follow, the AEL versions often appear longer, but its advantages can really make themselves felt with large dialplans.

Here's a simple set of extensions in `extensions.conf`:

```
[internal-users]

exten => 21,1,Dial(SIP/anna)
exten => 21,n,VoiceMail(anna)

exten => 22,1,Dial(SIP/lisa)
exten => 22,n,VoiceMail(lisa)

exten => _3X,1,Dial(SIP/${EXTEN})
```

The same extensions would look like this in `extensions.ael`:

```
context internal-users {
    21 => {
        Dial(SIP/anna);
        VoiceMail(anna);
    }
    22 => {
        Dial(SIP/lisa);
        VoiceMail(lisa);
    }
    _3X => {
        Dial(SIP/${EXTEN});
    }
}
```

For extensions that need only a single command, you can skip one level of curly braces and your dialplan would still work:

```
context default {
    23 => Playback(hello-world);
}
```

Don't get used to this practice, though. It's best to include them always, even when they might be superfluous:

```
context default {
    23 => {
        Playback(hello-world);
    }
}
```

Most extensions do require multiple commands and consistency is your friend when writing code. There are only a few cases where the short form is appropriate (for example, `jump`).

Warning: The opening curly brace { of a code block must appear on the same line, not on a line of its own!

6.3.3 Comments

In `extensions.conf`, comments are preceded by a semicolon:

```
; a comment
exten => 10,1,Dial(SIP/anna) ; Dial
```

Comments in AEL are denoted with two forward slashes, //, as follows:

```
// a comment
10 => {
Dial(SIP/anna); // Dial
}
```

Warning: Programmers: Don't use the C comment style (/* ... */). Every comment line must begin with its own // comment marker.

6.3.4 Includes

We can include contexts within contexts. A typical use of includes in `extensions.conf` would look like this:

```
[sales]
exten => 2001,1,Dial(SIP/anna)
exten => 2002,1,Dial(SIP/brock)

[warehouse]
exten => 3001,1,Dial(SIP/lisa)

[day]
include => sales
include => warehouse
```

Doing the same in AEL works this way:

```
context sales {
  2001 => {
    Dial(SIP/anna);
  }
  2002 => {
    Dial(SIP/brock);
  }
}

context warehouse {
  3001 => {
    Dial(SIP/lisa);
  }
}

context day {
  includes {
    sales;
    warehouse;
  }
}
```

Warning: Be careful to write `includes` and not `include` in AEL!

6.3.5 Global Variables

In AEL, global variables are set in a special `globals` block.

`extensions.conf`	`extensions.ael`
`[globals]`	`globals {`
`CAKE=marblecake`	`  CAKE=marblecake;`
`RINGTIME=60`	`  RINGTIME=60;`
	`}`

6.3.6 Expressions and Variable Assignment

AEL handles expressions in control structures (such as `if()`, `while()`, the break condition in `for()` and the right side of variable assignments) as though they were contained in a `$[...]` expression.

This seems complicated at first glance, but this behavior is common in other languages. For variable assignments in Asterisk, it is atypical, however, and can easily lead to strange errors. (AEL isn't a full programming language yet!) For this reason, we recommend against writing variable assignments this way:

```
context test {
  123 => {
    result=10/2;
    NoOp(result is ${result});
  }
}
```

For predictable results, it's better to use `Set()`, as you might in `extensions.conf`:

```
context test {
  123 => {
    Set(result=$[ 10 / 2 ]);
    NoOp(result is ${result});
  }
}
```

For constructions which use `if()`, `while()`, and so forth, however, the new behavior is nevertheless desirable because it spares having to use the clunky `$[ ... ]`. In `extensions.conf`, we would use `ExecIf()`:

```
exten => 50,1,Set(a=test)
exten => 50,n,ExecIf($["${a}" = "101"],SayDigits,123)
```

In `extensions.ael`, the end result is cleaner:

```
50 => {
  Set(a=test);
  if ("${a}" = "test") {
    SayDigits(123);
  }
}
```

6.3.7 Labels, goto, and jump

The hardened `extensions.conf` programmer is forced to get used to moving around dialplan by jumping to priorities or predefined labels using applications such as `Goto()`, `GotoIf()`, `Gosub()`, and `GosubIf()`:

```
[example]

; go to a label in the same extension
;
exten => 10,1(begin),NoOp()
exten => 10,n,Wait(1)
exten => 10,n,SayNumber(1)
exten => 10,n,NoOp(endlessloop)
exten => 10,n,Goto(begin)

; go to a label in a different extension
; in the same context
exten => 20,1,SayNumber(20)
exten => 20,n,Goto(10,begin)

; go to a label in a different
; context
exten => 30,1,SayNumber(30)
exten => 30,n,Goto(cntxt2,40,forty)

[cntxt2]

exten => 40,1(forty),NoOp()
exten => 40,n,SayNumber(40)

exten => 50,1,Goto(40,1)
exten => 60,1,Goto(example,10,1)
```

In AEL, a label always appears inside an extension and on its own line. Don't forget the colon (:) at the end of the line:

```
context example {

  // go to a label in the same extension
  //
```

```
   10 => {
    begin:
     Wait(1);
     SayNumber(10);
     NoOp(endlessloop);
     goto begin;
   }

   // go to a label in a different extension
   // in the same context
   //
   20 => {
     SayNumber(20);
     goto 10|begin;
   }

   // go to a label in a different context
   //
   30 => {
     SayNumber(30);
     goto cntxt2|40|forty;
   }
 }

context cntxt2 {

   40 => {
    forty:
     SayNumber(40);
   }
   50 => jump 40;
   60 => jump 10@cntxt1;
 }
```

In the preceding example we see a different syntax for the "goto" construction. In extensions.conf, we had to use the Goto() application, but in AEL, it's been replaced by the command goto. (If you really want to keep using Goto(), you can, but why?)

A comparison of the syntax for Goto() and goto shows us they are functionally the same:

```
extensions.conf        Goto([[context,]extension,]label)
extensions.ael         goto [[context|]extension|]label
```

The concept of jumping to a specific priority is moot in AEL, because priorities aren't used; still, you might have a need to jump to another extension, be it in the same or another context. For this purpose, AEL includes the additional command jump.

```
extensions.conf                              Goto([context,]extension,1)
extensions.ael (replaces priority jumping)   jump extension[@context]
```

In the following sections, you'll see that AEL doesn't really need to rely on goto statements for control flow, because it has real control structures.

6.3.8 Conditionals

Conditionals are structures that let you control program execution.

AEL has both **if** and **switch** commands. This is a big advantage because it makes an AEL dialplan much more readable. The more complex your program logic, the more obvious this advantage becomes.

6.3.8.1 if

In extensions.conf, we have to label targets explicitly:

```
exten => 90,1,Dial(SIP/anna)
exten => 90,n,GotoIf($["${DIALSTATUS}" = "BUSY"]?b:n)

exten => 90,10(b),Answer()
exten => 90,11,Playback(hello-world)
exten => 90,12,Voicemail(anna,b)
exten => 90,13,Goto(end)

exten => 90,20(n),Dial(SIP/lisa)
exten => 90,21,Playback(beeperr)
exten => 90,22,Goto(end)
exten => 90,30(end),NoOp(done)
```

In extensions.ael, the **if** statement has fewer parentheses:

```
90 => {
  Dial(SIP/anna);
  if ("${DIALSTATUS}" = "BUSY") {
    Answer();
    Playback(hello-world)
    Voicemail(anna,b);
  }
  else {
    Dial(SIP/lisa);
    Playback(beeperr);
  }
  NoOp(done);
}
```

Someone used to programming in AEL will find jumping around a dialplan with GotoIf() cumbersome, to say nothing of the confusion caused by a statement like GotoIf($["${DIALSTATUS}" = "BUSY"]?b:n). Are all the brackets closed?

Warning: Again, the opening curly brace must stand on the same line as the `if` statement, not on its own line.

6.3.8.2 switch

In AEL, the **switch** command lets you easily build code that can handle more than one case. In `extensions.conf`, a construction which does the same thing as a **switch** command is a confusing mess:

```
exten => 70,1,Dial(SIP/anna)
exten => 70,n,Goto(70-${DIALSTATUS},10)
exten => 70,n(end),NoOp(done)

exten => 70-BUSY,10,NoOp(busy)
exten => 70-BUSY,11,Goto(end)

exten => 70-NOANSWER,10,NoOp(no answer)
exten => 70-NOANSWER,11,Goto(end)

exten => _70-.,10,NoOp(something else)
exten => _70-.,11,Goto(end)
```

In `extensions.ael`, by contrast, the result is clean and elegant:

```
70 => {
  Dial(SIP/anna);
  switch ("${DIALSTATUS}") {
    case "BUSY":
      NoOp(busy);
      break;
    case "NOANSWER":
      NoOp(no answer);
      break;
    default:
      NoOp(something else);
  }

  NoOp(done);
}
```

Trying to do a *nested* `if` or `switch` in `extensions.conf` would be incomprehensible and impossible to maintain. Now we can really start to see the advantages of AEL!

Warning: Don't forget the `break` statements in the `case` branches! If you omit these, Asterisk will just run them sequentially, be it to the next case branch or to `default`.

In addition to case comparison, `switch` also lets you use `pattern` statements to do explicit pattern matching. Here's what pattern matching looks like in our `extensions.conf`:

```
exten => _70,1,NoOp(dialed: ${EXTEN})
exten => _70,n,Goto(70-${EXTEN},10)

exten => 70-703,10,NoOp(703)
exten => 70-703,11,Goto(end)

exten => 70-704,10,NoOp(704)
exten => 70-704,11,Goto(end)

exten => _70-70[5-8],10,NoOp(70[5-8]);
exten => _70-70[5-8],11,Goto(end)

exten => _70-.,10,NoOp(something else)
exten => _70-.,11,Goto(end)

exten => 70,n(end),NoOp(done)
```

In `extensions.ael`, the pattern syntax is the same, but it is preceded by a `pattern` statement and treated just like a case:

```
_70. => {
  NoOp(dialed: ${EXTEN});
  switch (${EXTEN}) {
    case 703:
      NoOp(703);
      break;
    case 704:
      NoOp(704);
      break;
    pattern 70[5-8]:
      NoOp(70[5-8]);
      break;
    default:
      NoOp(something else);
  }

  NoOp(done);
}
```

6.3.8.3 ifTime

The standard application `GotoIfTime()` (see Appendix B, "Dialplan Applications") is replaced with the `ifTime` command in AEL:

```
20 => {
  ifTime (08:00-18:00|mon-fri|*|*) {
    Dial(SIP/20);
```

```
  } else {
    Playback(announcement-closed);
    Voicemail(20,s);
  }
}
```

The time syntax is the same as for GotoIfTime(). You can accomplish the same thing with the IFTIME() function (see Appendix C, "Dialplan Functions"), although the result is not as readable:

```
20 => {
  if (${IFTIME(08:00-18:00|mon-fri|*|*?1:0)}) {
    Dial(SIP/20);
  } else {
    Playback(announcement-closed);
    Voicemail(20,s);
  }
}
```

6.3.8.4 random

The random(){...} command takes a value from 1 to 99. This value is the percentage chance that the subsequent code block will be executed:

```
20 => {
  random (42) {
    NoOp(42 % Chance);
  } else {
    NoOp(58 % Chance);
  }
}
```

You can accomplish the same objective with an if command and the RAND() function with slightly more code:

```
20 => {
  if (${RAND(0,100)} < 42) {
    NoOp(42 % Chance);
  } else {
    NoOp(58 % Chance);
  }
}
```

6.3.9 Loops

Like conditionals, loops are control structures.

AEL has classic **for** and **while** loops, which behave just as they do in other programming languages.

6.3.9.1 while

A `while` command in AEL is used in much the same way as the `While()` and `EndWhile()` applications.

The `break` and `continue` statements found in some other languages can also be used in AEL loops. This corresponds to the `ExitWhile()` and `ContinueWhile()` in `extensions.conf`:

```
exten => 40,1,Set(x=5)
exten => 40,n,While($[${x} <= 9])
exten => 40,n,NoOp(x is ${x})
exten => 40,n,ExecIf($[${x} > 5],ExitWhile)
exten => 40,n,Playback(beep)
exten => 40,n,Set(x=$[${x} + 1])
exten => 40,n,EndWhile()
exten => 40,n,NoOp(done)
```

In `extensions.ael`, the `break` statement jumps to the end of the loop block, `continue` to the start. In this example, we use `break`:

```
30 => {
  x=0;
  while (${x} <= 9) {
    NoOp(x is ${x});
    if (${x} > 5) {
      break;
    }
    Playback(beep);
    y=${x} + 1;
  }
  NoOp(done);
}
```

Note: In this example, we needed more lines in AEL; if we had added another command to the `if` condition in our `while` loop, `ExecIf()` would not be enough anymore and we would be forced to use a more complex construction with `GotoIf()`. Our `extensions.conf` would be a lot longer.

6.3.9.2 for

AEL also gives us `for` loops. This has no corresponding application in the `extensions.conf` format, though every `for` loop may be written as a `while` loop:

```
exten => 40,1,Set(x=5)
exten => 40,n,While($[${x} <= 5])
exten => 40,n,NoOp(x is ${x})
```

```
exten => 40,n,Playback(beep)
exten => 40,n,Set(x=$[${x} + 1])
exten => 40,n,EndWhile()
exten => 40,n,NoOp(done)
```

Even here, `extensions.ael` is more compact:

```
40 => {
  for (x=0; ${x}<=5; x=${x}+1) {
    NoOp(x is ${x});
    Playback(beep);
  }
  NoOp(done);
}
```

6.3.10 Macros

In `extensions.conf`, macros or macro-like functions can be achieved with `Macro()` or `Gosub()`. Our example uses `Macro()`:

```
[macro-countdown]
exten => s,1,Set(c=${ARG1})
exten => s,n,While($[ ${c} > 0])
exten => s,n,SayNumber(${c})
exten => s,n,Set(c=$[ ${c} - 1 ])
exten => s,n,EndWhile()

[default]
exten => 123,1,Macro(countdown,3)
exten => 124,1,Macro(countdown,5)
```

In AEL, we don't have to choose. It has the `macro` command.

```
macro countdown( count ) {
  for (c=${count}; ${c}>0; c=${c}-1) {
    SayNumber(${c});
  }
}

context default {
  123 => {
    &countdown(3);
  }
  124 => &countdown(5);
}
```

The AEL compiler converts `macro` into a `Gosub()` subroutine internally; fortunately, it's not one we have to maintain.

6.3.11 Hints

We cover the implementation of hints in AEL in depth in Chapter 18, "Busy Lamp Field, Hints, Pickup." Here's a simple example to demonstrate the proper syntax. In extensions.conf, hints look like this:

```
[internal-users]
exten => 21,hint,SIP/anna
exten => 21,1,Dial(SIP/anna)

exten => 22,hint,SIP/lisa
exten => 22,1,Dial(SIP/lisa)
```

In AEL, the hint can encapsulate a reference to the device:

```
context internal-users {
  hint(SIP/anna) 21 => {
    Dial(SIP/anna);
  }
  hint(SIP/lisa) 22 => {
    Dial(SIP/lisa);
  }
}
```

Here is a similar example, this time using a pattern. In extensions.conf, that would look like this:

```
[internal-users]
exten => 21,hint,SIP/21
exten => 22,hint,SIP/22


exten => _2X,1,Dial(SIP/${EXTEN})
```

In extensions.ael, it would look like this:

```
context internal-users {
  hint(SIP/21) 21 => {}
  hint(SIP/22) 22 => {}

  _2X => {
    Dial(SIP/${EXTEN});
  }
}
```

Notice that the pattern is now outside the hint parentheses.

6.3.12 Filtering by Caller ID

In `extensions.conf`, the ever-popular ex-girlfriend extension looks like this:

```
exten => 10/6135303122,1,NoOp(ex-girlfriend)
exten => 10/6135303122,n,Busy()
exten => 10,1,Dial(SIP/dave)
exten => 10,n,Voicemail(dave)
```

The AEL syntax that accomplishes the same thing looks like this:

```
10/6135303122 => {
  NoOp(ex-girlfriend);
  Busy();
}
10 => {
  Dial(SIP/dave);
  Voicemail(dave);
}
```

If a call comes in from the ex-girlfriend (with the caller ID 6135303122), it is routed to `Busy()`. Everybody else gets connected to Dave's phone.

Patterns are allowed here too. With `/_613`, you can match the entire area code.

6.4 Choosing between extensions.ael and extensions.conf

Asterisk application developers are left to decide which format is the right one. The development road map once hinted that AEL would replace `extensions.conf`, but since then Digium has been careful about declaring in favor of one over the other. One thing is certain: You cannot understand Asterisk well unless you understand `extensions.conf`, not least because the bulk of the documentation available on the Web is written for `extensions.conf`.

If you don't spend a lot of time doing dialplan programming, you're probably better off sticking with `extensions.conf`, at least for the foreseeable future. For application developers who work with Asterisk a lot, however, AEL has too many time- and headache-saving features to justify not trying it out. Even if Digium were to end support for AEL (which is highly unlikely), there is always `aelparse` to enable a return to the old `.conf` world.

7

Protocols

This chapter covers the techniques by which we actually get voice data from one point to another. The session protocols for this purpose are really a kind of "grammar" that governs this transmission, and are in turn built upon the transport layer protocols that make up the TCP/IP protocol stack: namely TCP and UDP. You should have a fundamental understanding of these protocols when troubleshooting certain problems, which is why we are describing them in some detail here. This understanding is not strictly necessary to implement Asterisk, but is nevertheless valuable. As such, this chapter is intended as a reference for the event it is needed.

7.1 Network Protocols

TCP is a protocol that ensures that all the information transmitted by the sender arrives at the receiver complete and in the correct order. This reliability is at the cost of relatively higher transmission overhead, which, during times of congestion, is more likely to lead to interruptions or delays in transmission. By contrast, its sister protocol UDP invests almost nothing in ensuring that everything is received; instead, it merely sends out every packet as soon as it is ready to be delivered without regard to whether it arrives intact or even in the right sequence. TCP ensures a secure, complete transfer of data by calculating a checksum on each packet and comparing the checksums when the packet is received, as well as attaching a sequence number to each packet and asking for receipt confirmation from the destination host. The destination host can ask for a retransmit if a packet is corrupt or missing. This is how HTTP (World Wide Web) traffic is transmitted.

By contrast, streaming data (as used in video and audio transmission) doesn't actually need this level of rigor because it is time sensitive. If a video packet is missed, there is no sense in retransmitting it because by the time it arrives, the video will have played on past the point of the missing data. For this reason, UDP is used for streaming data. There is no delivery guarantee, but the transport overhead is much less than in TCP. As a general guideline, you can define the two protocols this way: TCP is for "when it absolutely has to get there intact, no matter how long it takes" and UDP is for "when it has to get there fast and we don't care if we lose a few packets."

Both protocols are of importance in VoIP applications, because connections are established using TCP, whereas the voice data itself is transmitted using UDP. Particularly where audio data packets are concerned, quantity is indeed more important than quality.

7.1.1 Transmission Control Protocol

The Transmission Control Protocol (TCP) is a protocol governing the transmission of data between computers. The operating systems of all modern computers understand and support TCP and use it to exchange data with other computers on a network. It is a reliable, connection-oriented transport protocol, and forms part of the Internet Protocol family.

TCP was developed by Robert E. Kahn and Vinton G. Cerf, beginning initially as a research project in 1973 and culminating in the publication, in 1981, of the Internet standards document RFC[1] 793, which forms the basis for all the TCP standards documents which followed, and, like many of its companion protocols, remains a "living" protocol that is always being extended and updated, through new RFC documents, to meet new needs and circumstances.

TCP sets up a virtual connection between two endpoints in a network. Data may be transmitted in either direction. TCP is almost always encapsulated in IP (Internet Protocol); for this reason, you will often hear the term TCP/IP used. It allows bidirectional data transmission, no matter which endpoint originated the connection; it automatically detects and corrects data loss, and automatically adjusts the data transmission rate to accommodate network conditions. These features, along

---

1. Request For Comments. These documents begin as simple proposals addressed to peers in the Internet engineering community and subject to revision; when a draft proposal has been accepted it is annotated and becomes part of the body of standards.

with the fact the standard is open, have made TCP the most widely used data transmission protocol, and today it is used for the World Wide Web, e-mail, peer-to-peer file sharing, and other popular network services.

7.1.1.1 TCP Connections

TCP is an end-to-end (circuit-like), full-duplex (meaning it allows the transmission of data in both directions at the same time) connection. Alternatively, connections may be assembled as two, half-duplex connections in which information may travel in both directions, albeit not at the same time; data may be returned for the purpose of connection management. Connection management itself is handled by the TCP protocol stack, software that is typically attached to, or forms part of, the core of the operating system. In the case of Linux, the TCP stack is in the kernel itself; this stack is accessed by network software such as e-mail clients, web browsers, and servers.

Each TCP connection consists of two endpoints, with each endpoint establishing a pair of identifying characteristics: namely, the unique IP address of the endpoint, and the port. This pair forms a software interface that allows bidirectional data transfer, and is sometimes called a socket. The computers participating in the connection are identified by their IP addresses, while the particular connection is identified by the port number. As a result, it is possible for a web server, for example, to have multiple connections to a remote endpoint from its port 80; it keeps track of these multiple connections because each connection uses a different port on the remote endpoint.

Ports are 16-bit numbers, and so range from 0 to 65535. The ports from 0 to 1023 are reserved and assigned by Internet Assigned Numbers Authority (IANA) for specific applications. For example, web traffic, which uses HTTP, travels over the port reserved for HTTP traffic, port 80. The use of particular ports is not binding; an administrator may operate an FTP server on a port other than 21. For publicly accessible services, however, client software will expect to find its specific service on the IANA-specified port, and special, private services are usually operated on unassigned ports above 1024.

7.1.1.2 Connection Setup and Teardown

A web server publishing a service to the outside world sets up a socket consisting of its IP address and a port, and waits for clients to initiate connections to this socket. A socket in this state is said to be listening, and indeed a display of the active sockets (typically through a command such as **netstat -an**) will show some in the LISTEN state. If a client wants to connect, it sets up a socket of its own with its IP address and a randomly selected, unused port number, and uses this socket to connect to the listening socket on the server. When the connection is established, the roles of client and server are indistinguishable and basically symmetrical. Either side may terminate the connection. (When done gracefully, this is sometimes called teardown.) Data may still be transferred by the opposite endpoint while the connection is being torn down.

The three-way handshake is the manner by which two TCP endpoints establish and maintain a connection so as to minimize data loss.

The client initiating a connection sends the server a TCP SYN (synchronize, identified with the SYN flag bit set to 1) packet with a sequence number x. Sequence numbers are used to set the correct order for packets, ensure that no packets are missing, and check for duplicates when the packet stream is reassembled at the server.

The type of packet (SYN) and the sequence number are set in the TCP header and the sequence number is selected randomly (how this is done depends on the TCP implementation). The server replies with a SYN/ACK (synchronize-acknowledge, identified with both SYN and ACK flags set to 1) packet having a (similarly randomly selected) sequence number y. At the same time, it increments the sequence number of the initiating SYN packet by one ($x + 1$) and places this number in the acknowledgment field of the packet's TCP header.

Finally, the client completes the handshake by replying to the SYN/ACK with an ACK packet of its own, incrementing the sequence number by one ($y + 1$). This process is also called forward acknowledgment. For extra security, the client also wraps the received $x + 1$ sequence number in this ACK reply. If all goes well, the connection is established.

Graceful teardown follows a similar process. Instead of packets with SYN flag bits set, the FIN (finish) flag bit is set. This indicates to the receiver that the sender will not be transmitting any further data. This is acknowledged with an ACK packet. The recipient of the initial FIN packet then responds with an ACK packet of its own, which is likewise acknowledged. Although four packets are exchanged, this is still treated as a three-way handshake because the FIN-ACK combination from the server to the client is considered a "one-way" operation. (The FIN and ACK message can also be contained in a single packet.)

The maximum segment lifetime (MSL) is the maximum time that a TCP segment may exist in the network before it is discarded (normally 2 minutes). After the transmission of the last ACK, the client enters a wait state equal to two MSL periods; during this time, all late-arriving segments are discarded. This is to ensure that no late-arriving segments are misinterpreted as the initiation of new connections. This also ensures that the connection is indeed closed gracefully; if the ACK $y + 1$ goes missing, a wait timer on the server expires and the LAST_ACK segment is retransmitted.

7.1.1.3 TCP Header Structure

A TCP segment consists of two parts: the header, containing the transmission information such as port and sequence number; and the payload, which is the real information we want to send (see Figure 7.1), which usually includes application layer protocol information (such as HTTP or SSH). The values are transmitted in network (big endian) byte order.

7.1.1.4 Data Transmission

A TCP segment is typically no larger than 1,500 bytes. It must be small enough to fit in the protocol data unit[2] of the protocol layer immediately beneath it in the stack, in this case, IP. An IP packet may be as large as 64 kilobytes, but because IP is practically always transmitted over Ethernet, most IP packets are smaller than the maximum Ethernet frame size of 1,500 bytes. Both TCP and IP specify a header of 20 bytes, respectively, so the payload is limited to 1,460 bytes. Some connections (particularly some forms of DSL service) use PPP (Point-to-Point Protocol), which requires an additional 8 bytes of header information, meaning that the TCP/IP

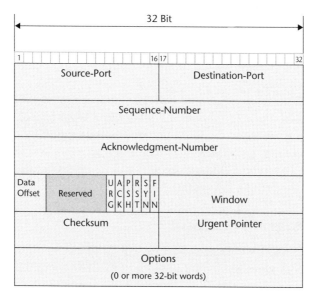

Figure 7.1 TCP header structure.

---

2. *Protocol data unit* is a term used to refer to the unit of transmission corresponding to a given network layer. In the session layer, the PDU is called a segment (TCP segment); in the network layer, it is a packet (IP packet); in the link layer, it is a frame (Ethernet frame).

packet can be a maximum of 1,492 bytes, leaving a maximum segment size (or MSS) of 1,452 bytes. This corresponds to about 3.2% overhead.

The sender and receiver negotiate the maximum segment size before data transmission begins, using the options field in the TCP header. The application wanting to send data (for example, a web server) places a data block of 10 kilobytes in the transmit buffer. To send this 10kB data block, it is divided into multiple packets, each with its own TCP header, and transmitted as TCP segments, a process called segmentation.

Segments usually leave in sequential order, though this is not strictly necessary, because the sequence numbering allows the receiver to reassemble them in the correct order. Received segments are checked for integrity by calculating a checksum. If it is correct, the packet is acknowledged. If it is incorrect, the acknowledgment is withheld and the sender retransmits the packet.

7.1.1.5 Data Integrity and Reliability

TCP allows a bidirectional, byte-oriented, reliable data stream between two endpoints. The underlying network layer protocol (IP) is packet switched, and packets can get lost, mangled, or arrive in the wrong order. TCP was designed to deal with these inherent uncertainties, and so it verifies data integrity using a checksum provided in the packet header, establishes the correct packet order using sequence numbers, and discards duplicate packets. The sender retransmits packets that are not acknowledged within a certain time. The received packets are assembled in the correct order in the receive buffer to form a data stream which is passed to the appropriate application.

Of course, the data transfer can be interrupted, corrupted, or terminated at any time. The connection will expire after a timeout has been exceeded. The successful establishment of a connection does not guarantee that a lasting and reliable transfer will follow. Network speed and stability may be sufficient for negotiating the connection while being insufficient to sustain it. Network conditions can also change such that they cause a working connection to fail.

7.1.2 User Datagram Protocol

In contrast to TCP, the User Datagram Protocol (UDP) is a minimal, connectionless session layer protocol. Development of UDP began in 1977, after the limits of TCP for the transmission of time-sensitive data became apparent. A protocol that handled only addressing was needed, without the overhead that went with guaranteed delivery.

7.1.2.1 Operation

To ensure that data sent via UDP reaches the appropriate application, source and destination ports are specified in the UDP header. This process-to-process communication is called application *multiplexing* and *de-multiplexing* (sometimes muxing and demuxing).

7.1.2.2 Characteristics

UDP establishes a connectionless, potentially unreliable transfer service. This means that neither transmission nor order of reception is guaranteed; that is, UDP sends data on a "best effort" basis. Applications that use UDP must take these characteristics into account and implement data correction where it is appropriate and necessary. Because no connection setup is required prior to data transmission (as is the case with TCP), the endpoints can begin exchanging data more quickly. This relatively low overhead makes UDP particularly useful for applications that exchange only small amounts of data, such as Domain Name System (DNS).

In addition, the absence of handshaking means that there is less variation in latency, or jitter. If a packet is lost in a TCP transaction, precious time is needed while the recipient waits for the sender to retransmit it, and the momentary latency can increase dramatically. This is important for multimedia applications that are sensitive to delay. In Voice over IP (VoIP) applications, these kinds of delays lead to sound artifacts and dropouts.

Connectionless protocols like UDP do not stop the flow of data if packets go missing. It is up to the receiving end to make sense of the gaps in the data stream when packets are lost, and most do this by momentarily reducing the resolution or quality of the resulting audio or video.

UDP, like TCP, is constrained by the protocol data unit size of the underlying network stack layers. IP packets cannot be larger than 65,535 bytes, of which the IP and UDP headers consume at least 28 bytes, leaving a maximum of 65,507 bytes for the actual payload. Such datagrams are usually fragmented in IP. In practice, UDP packets are rarely more than a few kilobytes in size. IP will discard packets in the case of transmission error or overflow, so datagrams can go missing.

7.2 Channels

In Asterisk, a channel is the connection between two endpoints (usually human beings in conversation). The following channel types exist:

- **Agent**
  An automatic call distribution (ACD) agent channel
- **CAPI**
  An ISDN channel

- **Console**

  A Linux console channel driver for sound cards that can be manipulated through OSS and ALSA

- **H.323**

  A VoIP protocol

- **IAX**

  A VoIP protocol

Note: In principle, there are two versions of IAX (1 and 2). Today, anyone referring to IAX is almost certainly speaking about IAX2 (in other words, IAX version 2). IAX version 1 has been deprecated and is no longer used.

- **Local**

  A loopback into another context

- **MGCP**

  A VoIP protocol

- **mISDN**

  An ISDN channel

- **NBS**

  Network Broadcast Sound

- **phone**

  Linux Telephony channel

- **SIP**

  A VoIP protocol

- **Skinny**

  A VoIP protocol (also called SCCP, or Skinny Client Control Protocol)

- **vISDN**

  An ISDN channel

- **VOFR**

  Voice over frame relay Adtran style

- **VPB**

  Voicetronix hardware channel driver for PSTN connections

- **Zap**

  Digium hardware channel driver for PSTN connections; also used for hardware from competing manufacturers

Most examples in this book assume SIP connections. The reason for this is simple: At the moment, there are more SIP-capable VoIP telephones than telephones for any other VoIP protocol. The other significant VoIP protocol for Asterisk is IAX. This chapter covers these two important protocols in some further detail, but may also be used as parameter reference for configuration.

7.3 Peers, Users, and Friends

When you configure IP connections in Asterisk, you will have to specify whether the connected device is a peer, user, or friend. The distinction between the classifications is not always clear in the Asterisk documentation.

Simply put, the classifications work like this. Asterisk...

- *Accepts calls from* a user
- *Makes calls to* a peer
- *Makes calls to* and *accepts calls from* a friend

In practice, friend is used the most.

7.4 IAX Versus SIP

Every Asterisk administrator is eventually faced with the question of whether to build primarily on SIP or IAX. Although the dispute over protocol has a religious tone at times, we think that from a purely technical standpoint, IAX is, in certain important respects, at least slightly superior to SIP. In many installations, IAX will mean many fewer implementation problems than one would expect with SIP. The primary reason for this is the way that SIP handles Network Address Translation (NAT). SIP connections are problematic because they require the opening of an entire range of ports in the firewall to work. The ability of IAX to multiplex conversations over a single connection (IAX trunk) is also more efficient because it reduces overhead. SIP, by contrast, establishes a separate connection for each conversation.

If IAX is available and possible (that is, the sets support it), IAX should be used; otherwise, you should use SIP.

Mark Spencer, the inventor of Asterisk, wrote an e-mail to the Asterisk users mailing list in 2004 that attempts to address the question (Figure 7.2).

Date: Mon, 5 Jul 2004 18:59:52 -0500 (CDT)
From: Mark Spencer <markster@digium.com>

Let me summarize some differences between SIP and IAX, and it might help you make a decision about what is best for you.

1. IAX is more efficient on the wire than RTP for any number of calls, any codec. The benefit is anywhere from 2.4k for a single call to approximately tripling the number of calls per megabit for G.729 when measured to the MAC level when running trunk mode.
2. IAX is information-element encoded rather than ASCII encoded. This makes implementations substantially simpler and more robust to buffer overrun attacks since absolutely no text parsing or interpretation is required. The IAXy runs its entire IP stack, IAX stack, TDM interface, echo canceler, and callerid generation in 4k of heap and stack and 64k of flash. Clearly this demonstrates the imple-mentation efficiency of its design. The size of IAX signaling packets is phenome-nally smaller than those of SIP, but that is generally not a concern except with large numbers of clients frequently registering. Generally speaking, IAX2 is more efficient in its encoding, decoding and verifying information, and it would be extremely difficult for an author of an IAX implementation to somehow be incompatible with another implementation since so little is left to interpretation.
3. IAX has a very clear layer2 and layer3 separation, meaning that both signaling and audio have defined states, are robustly transmitted in a consistent fashion, and that when one end of the call abruptly disappears, the call WILL terminate in a timely fashion, even if no more signaling and/or audio is received. SIP does not have such a mechanism, and its reliability from a signaling perspective is obvi-ously very poor and clumsy requiring additional standards beyond the core RF3261.
4. IAX's unified signaling and audio paths permit it to transparently navigate NAT's and provide a firewall administrator only a *single* port to have to open to per-mit its use. It requires an IAX client to know absolutely nothing about the net-work that it is on to operate. More clearly stated, there is *never* a situation that can be created with a firewall in which IAX can complete a call and not be able to pass audio (except of course if there was insufficient bandwidth).
5. IAX's authenticated transfer system allows you to transfer audio and call control off a server-in-the-middle in a robust fashion such that if the two endpoints can-not see one another for any reason, the call continues through the central server.
6. IAX clearly separates Caller*ID from the authentication mechanism of the user. SIP does not have a clear method to do this unless Remote-Party-ID is used.

Figure 7.2 Mark Spencer's e-mail on IAX.

7. SIP is an IETF standard. While there is some fledgling documentation courtesy Frank Miller, IAX is not a published standard at this time.
8. IAX allows an endpoint to check the validity of a phone number to know whether the number is complete, may be complete, or is complete but could be longer. There is no way to completely support this in SIP.
9. IAX always sends DTMF out of band so there is never any confusion about what method is used.
10. IAX support transmission of language and context, which are useful in an Asterisk environment. That's pretty much all that comes to mind at the moment.

Mark

PS:
I Guess there must be some advantages to SIP (or we should call the writers of it stupid).

So here a few questions to elaborate how IAX handles:

1. Bandwidth indications
2. New codecs
3. extensibility
4. Call Hold and other complex scenarios
5. Video telephone

I have gotten the impression this has all been better arranged in SIP

Figure 7.2 *Continued*

In the intervening years, IAX has itself become a well-documented and open protocol. IAX 2 was accepted as informational (nonstandards track) RFC 5456 in February 2009.

8

Making Connections

Asterisk finds its real power when you connect it to other systems, telephone devices, and networks. It accomplishes this through audio codecs and a variety of transport media, such as digital trunks (T1, E1) and traditional analog lines. In this chapter, we look more closely at codecs, ISDN, and analog telephony.

8.1 Codecs

A discussion of codecs, as they are used in telephony, will put someone on one of two sides of a large mountain. On one side of the mountain, we have the situation in North America, which has very little ISDN penetration, at least at the retail level. The average North American telephone customer has only ever used analog telephones, and so the stack of digital codecs is seen from the base of a mountain yet to be scaled: At the summit is an appealing array of digital codecs that offer exceptional sound quality, but represent the unknown.

On the other side of the mountain, we have the situation in Europe, where ISDN has been ubiquitous for so long that analog seems like the dark ages and nobody thinks twice about codecs or call quality. The average European telephone customer under the age of 35 has only ever used digital telephones, and so the stack of digital codecs is seen from a mountain she has already been to: The summit was nice, we took some pictures, now let's make phone calls.

In either case, the codecs are far from the surface of anyone's consciousness. Nevertheless, choice of codecs has implications for Asterisk implementation. Whether you are in North America, Europe, or elsewhere in the world, you will still be faced with this choice.

As a general guideline, the most common digital codec used in telephony is the ISDN codec G.711 (a-law or u-law companding; u-law[1] is used in the United States and Japan, while a-law is the standard in Europe).

The limited use of digital telephony in the United States is at least part of the reason why Voice over IP (VoIP) has caught on much more than in Europe. The sound quality offered by a clean, IP-based connection is often vastly superior to that of the aging analog infrastructure.

8.1.1 What Does a Codec Do?

Simply put, a codec takes an analog audio signal and converts it into a digital bitstream. This consists of a series of steps. First, the signal is digitized into a time-coded bitstream, and then this bitstream is compressed using a compression algorithm so that it uses less bandwidth when transmitted. This compression can be nearly lossless (e.g., G.711) or lossy (e.g., GSM). Without a codec, no audio can be transmitted. Which codec one ought to choose depends on the bandwidth available and the available processing capacity. Use of any codec is always a compromise between these two considerations.

This chapter does not cover the function of codecs in too much detail. For our purposes, an overview will suffice.

8.1.2 Performance

The biggest problem with high-compression codecs is the processing they require. Using the Asterisk console, enter the command **core show translation** to see a table of the actual transcoding times required for the various codecs:

```
debian*CLI> core show translation
        Translation times between formats (in milliseconds) for one second of
data Source Format (Rows) Destination Format (Columns)
```

---

1. Actually, this is µ-law (the Greek letter mu, used in physics as a prefix meaning "micro").

	g723	gsm	ulaw	alaw	g726aal2	adpcm	slin	lpc10	g729	speex	ilbc	g726	g722
g723	–	–	–	–	–	–	–	–	–	–	–	–	–
gsm	–	–	2	2	3	2	1	9	–	34	26	3	–
ulaw	–	6	–	1	3	2	1	9	–	34	26	3	–
alaw	–	6	1	–	3	2	1	9	–	34	26	3	–
g726aal2	–	7	3	3	–	3	2	10	–	35	27	1	–
adpcm	–	6	2	2	3	–	1	9	–	34	26	3	–
slin	–	5	1	1	2	1	–	8	–	33	25	2	–
lpc10	–	9	5	5	6	5	4	–	–	37	29	6	–
g729	–	–	–	–	–	–	–	–	–	–	–	–	–
speex	–	13	9	9	10	9	8	16	–	–	33	10	–
ilbc	–	9	5	5	6	5	4	12	–	37	–	6	–
g726	–	7	3	3	1	3	2	10	–	35	27	–	–
g722	–	–	–	–	–	–	–	–	–	–	–	–	–

Codecs not available on your system are marked with a dash (-). It is easy to see which codec demands the most processing resources: speex.

To minimize load, use the minimum possible number of codecs. If set A and set B both use the same codec in a call to each other, Asterisk need only pass through the packets. If set A and set B use different codecs, Asterisk has to transcode each packet, which involves decoding the incoming packet and then encoding it using the other codec before sending it out to the destination set, and it has to do this in *each direction*. You can see how this becomes onerous quickly.

8.1.3 Configuring Codecs

How codecs are configured depends on the technology. IAX sets are configured in iax.conf, and SIP sets are configured in sip.conf. If you want to set all SIP sets to use G.711, you would write the following in sip.conf:

```
[general]
disallow=all
allow=alaw
allow=ulaw
```

If you have a specific telephone (call it 2000) that should use only GSM, you can set this as follows :

```
[general]
disallow=all
allow=alaw
allow=ulaw

[2000]
disallow=all
allow=gsm
```

8.1.4 Commonly Used Codecs

The CLI command **core show translation**, which displays the table shown earlier, will show you which codecs are available on your system (namely, all those with numbers in their respective columns). The following list briefly examines the most commonly used codecs in Asterisk:

- **GSM**

  An Asterisk "classic" and familiar to anyone with GSM mobile phone service. The bandwidth is 13.3kbps, and the quality is acceptable, if not exactly overwhelming.

- **iLBC**

  iLBC is a sort of secret weapon for use in low-bandwidth situations. The sound quality is excellent, and the bandwidth is limited to between 13.3kbps and 15kbps (depending on frame size). Skype uses a variant of this codec. The problem with iLBC is that it is processor intensive; that is also why it is not implemented in very many hardphones.

- **G.711**

  Another classic, from the suite of ISDN codecs. This codec offers superb sound quality and requires 64kbps of bandwidth. This is standard for most calls traveling in corporate intranets and is also widely used for calls over the public Internet.

- **G.722**

  The chances are good that G.722 will eventually replace G.711; at the highest resolution, it offers a much better sound quality than G.711, but while requiring the same bandwidth (only 64kbps). Unfortunately, it not, as yet, supported by many telephones.

- **G.726**

  Offers the same sound quality as G.711 while needing only half the bandwidth (32kbps). In contrast to G.711, only the data that has changed relative to the previous voice packet is transmitted (similar to the way MPEG video works). The result is the same information transmitted with half the bandwidth. G.726 is not too burdensome, but still requires more processing than G.711.

- **G.729a**

  G.729a is a patented codec that needs only 8kbps and offers sound quality comparable to GSM. Licenses for Asterisk can be purchased from Digium (www.digium.com). It is not exceptional either in voice quality or performance, but offers a fair balance and is implemented in many VoIP telephones. As such, it is a good option when you need to conserve bandwidth.

- **G.729.1**

  G.729.1 is an extension offering bitstream interoperability with existing G.729 codecs, as well as a high level of configurability (it supports 8kpbs to 32kbps streams). Of particular interest are the wide-band extensions, which can offer very close to CD-quality sound; as yet, only a few phones support them. To use the wideband extensions, all the devices in the media path must support them.

8.1.5 Bandwidth and Trunking

Anyone wanting to pass multiple simultaneous calls over a network connection (e.g., when two Asterisk systems are connected to each other) will soon have a bandwidth problem. Depending on the constellation of codec and frame size, it is easily possible for the IP overhead to exceed the payload. For example, using SIP and GSM on a 2Mbps data connection, you can support a maximum of 35 simultaneous calls, which isn't exactly a lot. In such a case, you would be better served by an ISDN PRI, where the sound quality will be significantly better.

Trunking is great because it lets you wrap multiple voice chunks in a single connection, so that you are not incurring an overhead penalty for each call. In our example above using the same 2Mbps connection, we can support 77 simultaneous GSM calls. If we want to improve the sound quality, we can do this and still support 44 G.726 calls.

Trunking is supported only in IAX2 (see Chapter 7, "Protocols") and should only be used with at least two simultaneous calls, because the overhead for a single call would be measurably higher than without trunking.

If you want to assess your bandwidth requirements more closely, see the bandwidth calculator at www.asteriskguru.com/tools/bandwidth_calculator.php and try out various scenarios.

8.2 Integrated Services Digital Network

ISDN is a twisted-pair, digital multiplexing standard for PSTN and data applications. Although used widely around the world, it is not well known in North America. For most data applications, ISDN has been displaced by ADSL, coaxial cable, and FTTH (fiber-to-the-home) technologies, which offer much larger bandwidth at competitive rates. It remains in use for voice and video applications. We'll look at those here.

ISDN itself is a giant topic, and because it is an older technology, it is well covered in other texts. Do not expect much depth here. We discuss ISDN only in the context of its relevance to Asterisk.

8.2.1 ISDN Basics

An ISDN connection provides two basic channel types: B and D. (More specialized channel types exist but are not in wide use, and are not covered here.) A single B (bearer) channel can be equated, roughly, with a single analog telephone line: You can carry on one conversation on that channel. Where it differs is that a single physical line pair can carry multiple B channels, making better use of the existing wiring. The data is transported in digital and not analog format, which means you can't connect an analog telephone to the line and expect to hear anything. The digital transport usually delivers better sound quality when everything is properly configured.

The D (data) channel carries signaling information about the B channels on the circuit.

The differences between the B and D channels are as follows:

- The B channel is, for our purposes, a voice channel. (In a non-Asterisk installation, it could carry data.) A B channel provides 64kbps of bandwidth.

- The D channel is data only, but is dedicated for signaling purposes. Depending on the interface type, the D channel can be 16kbps or 64kbps. The ISDN device—be it a telephone, PBX, or interface card—uses the D channel for call setup and teardown.

The following sections cover the two major types of ISDN connection: the basic rate interface and primary rate interface.

8.2.1.1 ISDN Basic Rate Interface

The entry-level ISDN service is called basic rate interface, or BRI, and is used in homes (in the world outside of North America) and small businesses. It provides two 64kbps B channels and one 16kbps D channel (which is why some refer to it as 2B+D) over a single line pair (two wires). You can attach ISDN phones, BRI ports on PBX devices, ISDN-BRI terminal adapters (which provide FXS ports for analog telephones), or ISDN-BRI adapter cards to a BRI.

To install and use BRI, you need to know what signaling type the carrier expects. There are two types:

- **2B1Q**

  This (2 binary, 1 quaternary) is the signaling type used in North America. It is a linear coding type that supports a loop length of up to 5,500 meters (about 18,000 feet).

- **4B3T**

  This (4 binary, 3 ternary) is the signaling type used in the rest of the world, including Europe. It is a block coding type that supports a loop length of up to 4,200 meters (about 13,700 feet).

8.2.1.2 BRI in the United States

ISDN never really caught on at the subscriber level in North America. Although there are many opinions why, the most common explanation given is that the size of the existing network made implementing ISDN to the suburbs prohibitively expensive. In North America, average loop lengths are much longer when compared to Europe, where ISDN circuits now make up more than 80% of installations.

Still, sometimes BRI makes sense, and in most jurisdictions it is a tariffed service, which means that some carrier (usually the incumbent telephone company) must provide it if it is technically feasible. Here are some reasons why you might consider BRI:

- You are faxing and your analog lines are poor quality.

  Faxing does not work well over IP (see Chapter 17, "Fax Server"). Analog is an option, but if your analog lines are noisy, you should consider BRI.

- You need the sound quality and call-signaling features that only ISDN provides, but don't need enough channels to justify PRI.

  Features such as Explicit Call Transfer, DNIS (Dialed Number Identification Service), and COLP (COnnected Line identification Presentation) are ISDN services that can't easily be replicated on analog lines. If you need these but don't need more than six to ten channels, BRI is a real option. One thing is beyond dispute: In the circuit-switched, PSTN world, ISDN offers unparalleled sound quality.

Of course, the economics will vary depending on your location and the carrier. Be warned that most carriers that offer BRI don't actively market it, so you may face challenges in getting anyone in the sales department to acknowledge that the service exists.

8.2.1.3 ISDN Primary Rate Interface

You can think of the PRI as the bigger, flashier brother of the BRI. In North America, PRI is provided over a T1 (23 B channels, 1 64kbps D channel) with 2 or 3 pairs (4 or 6 wires); in Europe, the E1 (30 B channels, 1 64kbps D channel). Again, you can only connect ISDN devices to a PRI circuit. Examples include the PRI interface on a PBX, a channel bank (which provides FXS ports for analog telephones), or an ISDN-PRI interface card. The PRI is intended for larger installations.

Nothing says a PRI must use all 24 channels. Some carriers provide partial PRI service (usually ten channels), sometimes at considerable cost savings.

PRI always implies ISDN. PRI service is supplied via a T1, but not all T1 circuits are PRI. Non-ISDN T1 circuits (also called in-band T1) can still be found in North America. In such circuits, call setup and teardown signaling is sent in the channel (CAS, or channel associated signaling) instead of via a separate D channel. This steals some bandwidth from each channel but makes all 24 channels available.

For voice applications, PRI is on its way to replacing non-PRI T1 connections completely. Its broad support and extremely fast signaling make it the easy choice.

8.2.2 Choosing an ISDN Card

This is a difficult question, and the answer depends on your specific circumstances and requirements. Here are some things you need to consider:

- A card that's excellent today may be mediocre tomorrow, particularly when compared with newer product in the marketplace.
- Card performance can depend heavily on the driver software, which can change over time.
- Support quality is subjective.
- Performance requirements will depend on the intended use. Some products that would not be a good choice for the business market may be perfectly acceptable for the residential market.

Actually *using* an ISDN card in an Asterisk server is really quite simple, even if installation and configuration aren't always trivial. There are a few different ways to get an ISDN card running with Asterisk. We touch on a few variations below.

Important: All the installation examples in this book use a Debian Linux system. The installation procedure will differ on other Linux systems and may not work at all on BSD systems. If you are starting out, we recommend you follow the instructions completely and begin with a fresh Debian installation.

8.2.2.1 Determining Your Quality Needs

Most private Asterisk users are not hung up on sound quality and availability. Cost is likely the primary consideration here. If you aren't planning to fax over the card, and you only need two B channels, any generic HFC-based BRI card will do. The cards are cheap and widely available. You've been warned, however: Installation can be challenging with these cards, and you shouldn't expect amazing sound quality. If you can live with these limitations, this is a real option.

For business systems, the bar rises substantially. If you are replacing an older system you must at least match it in quality, unless you like doing battle with users and clients. If you value your career in this business, you'd best steer clear of generic cards.

8.2.2.1.1 Hardware Echo

Sadly, the echo problem remains. Any professional system connected to the outside world must have the highest-quality echo cancellation available, and this is generally achieved through a hardware echo canceler integrated into the card itself. The advantage over software echo cancellation is clear: The performance of hardware echo cancelers is historically better, and they don't need to steal CPU cycles to achieve it.

Because echo cancellation is not easy to do well, these modules are accordingly more expensive. Nevertheless, in business applications, hardware echo cancellation is not negotiable for any system you intend to connect to the PSTN. Cards providing PRI to internal servers can get away without it because echo in local trunks is rare, but if you're buying a card anyway, spend the extra money. You never know how you might need to use the card in the future.

Tail length is a performance parameter to consider when evaluating hardware echo cancelers. It determines the length of the reference sample when calculating echo cancellation. The rule of thumb is that the longer the tail length, the better the echo cancellation. In practice, it's difficult to hear a difference beyond 128 milliseconds.

8.2.2.1.2 Server Load and Interrupts

In the early years of Asterisk, there was no affordable echo cancellation, so Mark Spencer and others wrote software to perform the cancellation. This necessitated breaking the in- and outbound audio streams into 1ms chunks. For this reason, all the standard ISDN cards in use today set interrupts at 1ms intervals for each B channel.

This historical artifact can lead to problems in some applications. You probably won't notice this if you use a single PRI or BRI connected to modern server hardware. If you're using a compact, embedded system, however, even a simple BRI with two B channels can cause problems under load. If you have a large system with multiple PRIs and interface cards, similar problems can occur. Enough interrupts can bring the most powerful system to its knees.

If you are faced with either of these circumstances, you have these options:

- Find out how much a single system can handle and cluster them.
- Install an ISDN-SIP gateway. These are black-box systems with an ISDN interface on one side and a SIP interface on the other, which connects to your Asterisk server. Be careful, though: Some of these systems are just Asterisk boxes themselves! You want real, dedicated translation.
- Buy an ISDN card that solves the interrupt problem at the driver level. For example, Sangoma produced a special driver for its cards that sets an interrupt at 10ms intervals on a by-port, rather than by-B-channel, basis.

8.2.2.1.3 Internal Analog Fax Devices

If you're connected to the PSTN via an ISDN connection but still use analog fax devices internally, you might run into another timing problem, particularly with long faxes. This is caused by different clock timings on the audio channels. The ISDN cards use the PSTN for their clock source; the analog interface cards use their own, usually internal (and often cheap and unstable) clock source. The result is a clock differential that is managed using a buffer. After two pages (on average), the buffer empties, and you lose a bit of audio, often enough to overwhelm the fax error correction. The result is usually a black line or gap in the fax image. In the best case, nothing is visible; in the worst case, the transmission ends prematurely.

Here are some things you can do to address faxing problems:

- If you can do without a fax machine, going to software faxing on the server is a cost-effective alternative (see an in-depth explanation in Chapter 17).
- An ISDN terminal adapter attached to the same ISDN card as the incoming service will also solve the problem. If you attach the device to a different ISDN card in the same machine, you need to verify that the cards will operate in sync.
- Cards from some manufacturers can be synced to each other via an internal cable that provides the clock signal from the ISDN card to the analog card.

8.2.2.2 Manufacturers and Drivers

There seem to be as many ISDN card manufacturers as people with an opinion about them. The products described here are a subjective selection, but have been chosen because they are used widely. Further exploration of the subject could fill another book.

Important: Hardware and driver information provided here was current at the time of printing. Check this book's website (www.the-asterisk-book.com), where we will post updates.

8.2.2.2.1 Digium

As the parent company and home of Asterisk, Digium (www.digium.com) has offered digital interface cards for several years now. (The early cards were analog only.) The advantage for the Digium cards is obvious: You always have drivers updated for the most current Asterisk release. In the past, these drivers have been middling in quality and there have been problems. With the new DAHDI driver generation, Digium is promising better performance.

8.2.2.2.2 Sangoma

Sangoma (www.sangoma.com) is a popular alternative manufacturer in North America, and was, until recently, purely a hardware company. They make both analog and ISDN interface cards and have a reputation for good technical support.

Open source purists may take exception to Sangoma's use of a binary-only ISDN stack (for BRI). Sangoma's response is that the commercial stack is certified for ISDN and has demonstrably fewer issues than the open source mISDN stack.

8.2.2.2.3 Generic HFC-Based BRI Cards

These cards have been sold in Europe for many years now, usually at bargain-basement prices. Support is rarely if ever provided, and at $100 to $150, it can hardly be expected. They are almost always used with mISDN.

8.2.3 Media Gateways

Just as you might use an ATA in place of an analog interface card for regular phones, you could use a "media gateway" for ISDN connections. This gateway converts an ISDN connection into one or more SIP accounts and vice versa. These are usually "black box" solutions, which are configured via a web GUI or over SSH.

Purely analog media gateways are simply ATAs (analog telephone adapters).

8.2.3.1 Advantages

One major advantage is that a media gateway is often easier to configure than an interface card. You don't have to open up the server and install the card, either.

Configuring Asterisk is simpler because you only have to configure SIP. Finally, you don't have to run telephone lines of any kind to the server; you install the gateway device in the wire room, connect it to the network patch panel, and put the server anywhere on the network that is practical.

8.2.3.2 Disadvantages

Media gateways are generally sold on price and not on the quality of installed components, so manufacturers are tempted to use lower-quality interface cards, which can be problematic (see the "Choosing an ISDN Card" section, earlier in this chapter). Asterisk works best when using the ISDN clock; the clock signal is very stable, and a stable clock signal is vital, especially for conferencing. A media gateway does not pass this signal over Ethernet; the Asterisk server must rely on the system clock, which will vary in stability. Updating a media gateway is also more difficult, if it's possible at all.

Whether one should choose an interface card or a media gateway is a regular topic of discussion in the Asterisk community. Both sides make good arguments, and you can run a successful system with either. What is best for you depends on the circumstances. If you need stable clock (you are conferencing or you have existing ISDN cards in your system), you'll need an ISDN card. If you have physical access problems, or you need to maximize the simplicity of the system, go with a media gateway.

8.3 Analog Telephony

In spite of the increasing adoption of digital and IP transports, the classic analog line pair still plays a huge role in communications. Most residential land lines in North America are still analog, and many businesses rely on analog lines to connect their small PBXes and key systems. Analog can have a place in installations that have fax devices, which work poorly over IP.

Two types of device can be attached to a telephone line:

Note: The name describes what the device is attaching to, which can be counterintuitive.

- **FXO:** Foreign Exchange Office

  A Foreign Exchange Office (FXO) device is any end-user device that attaches to a line and *does not* provide its own line battery voltage. The standard telephone we are all familiar with qualifies as an FXO device, but fax machines and modems are also FXO devices. A phone is an FXO device because it attaches to a Foreign Exchange Office, also called a telephone exchange.

- **FXS:** Foreign Exchange Station

  A Foreign Exchange Station (FXS) is any device that provides line battery voltage. This almost always means a dial tone and ringing voltage also. You need an FXS device if you need to provide service to analog phones, fax machines, and modems. The telephone outlet on the wall is a typical FXS "device."

8.3.1 Connecting Analog Devices

FXS devices provide significant voltage; a telephone line in the "on-hook" state can have a voltage as high as 52 volts DC, and a ringing line can have a voltage between 90 volts and 110 volts AC! Keep the AC ringing voltage in mind when you are working on an exposed, active line pair with bare fingers. While the ringing signal is unlikely to do you any harm, the bite can make you quite uncomfortable! If you want to avoid this, wear gloves.

Warning: If you attach an FXS module on an interface card to a telephone outlet, it can permanently damage your card or even cause the interruption of service from the telephone company! Use caution when connecting your card to the line. Make absolutely sure you are connecting an FXO device to your incoming phone line pair. Some interface cards actually light up the port with a colored LED to indicate whether the port is FXS (usually green) or FXO (usually red). Failing that, mark the ports with a label or marker before you install the interface card in the server.

In the telephony world, polarity is described by the terms *tip* and *ring*, which correspond to the parts of the plug used by telephone operators in the early days of the telephone system.

- **Tip**

  This is the positive (+), or ground (the tip of the plug).
- **Ring**

  This is the negative (–), or battery (the ring, just behind the tip). Ring voltage against ground can vary from –48 VDC to –52 VDC.

If the polarity between the device and the line is reversed, it can cause problems in some cases. For example, if you have an older dual-tone multi-frequency (DTMF) phone (such as a Western Electric model 2500), you won't be able to dial out; reversed polarity has been known to cause noise problems also. Even if most newer devices ignore polarity, it is good practice to keep polarity consistent throughout your wiring system. This way, you can be guaranteed that your phone system will work properly with any standard device.

8.3.2 Analog Telephone Adapter

The analog telephone adapter (ATA) allows you to connect an analog telephone device to an IP network (either SIP or IAX2, and almost always Ethernet). The advantages are the simplicity and low cost when compared with FXS modules on analog interface cards (as little as $50). Also, FXS modules can suffer from noise caused by nearby components in the Asterisk server; an ATA does not suffer from this problem. A major disadvantage of an ATA is that it won't work any better with fax.

An ATA has one Ethernet port and at least one RJ-11/RJ-12 telephone port, sometimes two. An ATA can be configured via an attached telephone or through a web interface. ATAs may connect directly to a phone or be connected to the building wiring to support multiple extensions sharing a single line.

A SIP ATA is treated just like any other SIP telephone, and you'll need to create an entry for it in `sip.conf`.

Voicemail

In this chapter, we review the capabilities of Asterisk's built-in voicemail system. Note that this is different from an interactive voice response (IVR) system, which is covered in depth in Chapter 10, "Interactive Voice Response."

Voicemail is an essential component of any professional telephone system.[1]

9.1 Example Implementations

The examples that follow should give you a quick overview of typical configurations. Specifics and special features are covered in section 9.3, "Dialplan Applications."

9.1.1 An Example Home System

In the course of modernizing their home, the Robinson family has decided to install an Asterisk system, including a voicemail system with a mailbox for every family member.

---

1. As a humorous homage to the popular Nortel Meridian PBX voicemail system called Meridian Mail, the Asterisk developers have named the Asterisk voicemail system *Comedian Mail*.

Each family member needs a mailbox. The extension numbers and mailbox parameters are shown in Table 9.1.

The `voicemail.conf` looks like this:

```
[general]
format = wav
attach = yes

[default]
; Syntax for new entries looks like this:
; MailboxNumber => password,name,e-mail,pager,options
; (usually, the MailboxNumber is the same as the Extension)
200 => 1234,Dave Robinson
201 => 1234,Colleen Robinson
202 => 1234,Matthew Robinson,matt@robinsonfamily.name
203 => 1234,Lisa Robinson,lisa@robinsonfamily.name,,delete=yes
```

In `extensions.conf`, we send an unanswered call to voicemail like so:

```
[robinson-family]
; If nobody picks up within 30 seconds, the call is sent
; to voicemail
; If the extension is busy, the call is sent to voicemail
exten => _20[0-3],1,Set(TARGETNO=${EXTEN})
exten => _20[0-3],n,Dial(SIP/${EXTEN},30)
exten => _20[0-3],n,Goto(s-${DIALSTATUS},1)
; the above line routes the call to the status priority
; (NOANSWER,BUSY,CHANUNAVAIL,CONGESTION,ANSWER)
exten => s-NOANSWER,1,VoiceMail(${TARGETNO},u)
; Person at extension "is unavailable" message
exten => s-BUSY,1,VoiceMail(${TARGETNO},b)
; Person at extension "is busy" message
exten => s-ANSWER,1,Hangup()
```

Table 9.1 The Robinson Family Mailboxes

Name	Extension	Notes
Dave Robinson	200	Standard voice mailbox.
Colleen Robinson	201	Standard voice mailbox.
Matthew Robinson	202	Normal voice mailbox with e-mail notification. (In this case, voice messages are attached as an audio file to an e-mail and sent to a specified e-mail address.)
Lisa Robinson	203	Normal voice mailbox with e-mail notification with deletion. (In this case, voice messages are attached as an audio file to an e-mail, sent to a specified e-mail address, but the original message is deleted from the Comedian Mail system immediately.)

```
; To be safe, clean up the call after an answer by hanging up
exten => _s-.,1,Goto(s-NOANSWER,1)
; Handle any unhandled status the same way we handle NOANSWER

; Check your voicemail from your own extension by
; dialing "250"
exten => 250,1,VoiceMailMain(${CALLERID(num)})
```

A more elegant implementation can use a macro:

```
[robinson-family]
exten => _20[0-3],1,Macro(normal|SIP/${EXTEN}|${EXTEN})

exten => 250,1,VoiceMailMain(${CALLERID(num)})

[macro-normal];
; ${ARG1} - extension(s) being called (e.g. SIP/123&SIP/124)
; ${ARG2} - Mailbox (usually the same as ${MACRO_EXTEN})
exten => s,1,Dial(${ARG1},30)
; ring extension for a maximum of 30 seconds
exten => s,n,Goto(s-${DIALSTATUS},1)
; go to status priority (NOANSWER,BUSY,CHANUNAVAIL,
; CONGESTION,ANSWER)
exten => s-NOANSWER,1,VoiceMail(${ARG2},u)
; Person at extension "is unavailable" message
exten => s-BUSY,1,VoiceMail(${ARG2},b)
; Person at extension "is busy" message
exten => s-ANSWER,1,Hangup()
; To be safe, clean up the call after an answer by hanging up
exten => _s-.,1,Goto(s-NOANSWER,1)
; Handle any unhandled status the same way we handle NOANSWER
```

9.1.2 An Example Business System

A business like Widgets, Inc. will need a more comprehensive voicemail system. In this example, you'll be able to see how much you can do with Comedian Mail.

The default settings for each voice mailbox are as follows:

- Voicemails are saved in WAV format.
- Each mailbox is limited to a maximum of 200 messages.
- The maximum length of a voice message is 5 minutes.
- Voice messages are stored on the system and also sent to the user as an e-mail attachment.

Beyond these default settings, individual departments have additional needs and wants for their mailboxes, as you can see in Table 9.2.

Table 9.2 Widgets, Inc. Mailboxes

Mailbox	Department/Title	Notes
150	Building manager on duty	Message notifications are sent only to a pager, not e-mail.
		Callers may listen to their messages before sending and, if necessary, re-record them.
		After the recipient has listened to the message, she can return the call directly from the voicemail menu.
160–169	IT	In the IT department, every staff member has his own mailbox.
		Calls route to voice mail only if nobody in the department answers. No messages can be left if all extensions are busy.[a]
802	Sales (Domestic)	No e-mail notifications are sent.[b]
		No password is required to listen to messages.
803	Sales (International)	No e-mail notifications are sent.
		No password is required to listen to messages.
201	Division Manager	Callers may listen to their messages before sending and, if necessary, re-record them.
		After the recipient has listened to the message, she can return the call directly from the voicemail menu.
202	Assistant Manager	Callers may listen to their messages before sending and, if necessary, re-record them.
		After the recipient has listened to the message, she can return the call directly from the voicemail menu.
804	Reception	After the recipient has listened to the message, she can return the call directly from the voicemail menu.

a. This is a safety measure to prevent a complete denial of service in the event of a major IT outage; if the entire company staff call in and leave a message, the IT department would never be able to listen to them all.

b. We don't want e-mail sent to a specific person because there is more than one staff member in the sales group. We might, however, configure all the extensions in the group to illuminate the Message Waiting Indicator light if there are new messages in the mailbox.

The `voicemail.conf` for Widgets, Inc. looks like this:

Warning: Statements in all the code examples are to be entered on a single line unless otherwise indicated. Lines are wrapped in the book due to space limitations, and are designated with an arrow.

```
[general]
; Messages are stored in higher-quality WAV format.
format = wav

; E-mail notifications sent by the system
```

```
; have voicemail@widgets-inc.biz as the From: address
serveremail = voicemail@widgets-inc.biz

; We set a maximum of 200 messages per mailbox.
maxmsg = 200

; The maximum message length is 5 minutes. The message length
; is set in seconds (5 x 60 = 300)
maxmessage = 300

; We set the text for the e-mail notifications.
; The entire text must fit on one line!
emailbody = Hello, ${VM_NAME},\n\nyou have a new voicemail message from
↳${VM_CALLERID} in your mailbox ${VM_MAILBOX}. To listen to
↳your new messages, dial 800.\n\n-- Asterisk Voicemail System\n

; Text for the pager notifications.
; The entire text must fit on one line!
pagerbody = New voicemail from ${VM_CALLERID} at ${VM_DATE}.

; Attach messages to e-mail notifications?
attach = yes

[default]
; Syntax for new entries looks like this:
; MailboxNumber => password,name,e-mail,pager,options
150 => 1234,Building Manager,,pager.buildingmgr@widgets-inc.biz,,
↳review=yes|callback=internal-extensions
802 => 1234,Sales (Domestic)
803 => 1234,Sales (International)
201 => 1234,Chelsea Important,bob.important@widgets-inc.biz,,
↳review=yes|callback=internal-extensions
202 => 1234,Rick Important,rick.important@widgets-inc.biz,,
↳review=yes|callback=internal-extensions
804 => 1234,Reception,reception@widgets-inc.biz,,review=yes
```

Calls are sent to voicemail in `extensions.conf` like so:

```
[building-mgr]
include => internal-extensions
include => voicemail-buildingmgr

[it]
include => internal-extensions
include => voicemail-easyaccess
include => voicemail-generalaccess

[managers]
include => internal-extensions
include => voicemail-easyaccess

[reception]
```

```
include => internal-extensions
include => voicemail-easyaccess

[sales-domestic]
include => internal-extensions
include => voicemail-sales-domestic

[sales-international]
include => internal-extensions
include => voicemail-sales-international

[shipping]
include => internal-extensions
include => voicemail-easyaccess

[production]
include => internal-extensions
include => voicemail-easyaccess

[others]

[macro-simple];
; ${ARG1} - extension(s) being called (e.g. SIP/123&SIP/124)
; ${ARG2} - Mailbox (usually the same as ${MACRO_EXTEN})
exten => s,1,Dial(${ARG1},30)
; ring extension for a maximum of 30 seconds
exten => s,n,Goto(s-${DIALSTATUS},1)
; go to status priority (NOANSWER,BUSY,CHANUNAVAIL,CONGESTION,ANSWER)
exten => s-NOANSWER,1,VoiceMail(${ARG2},u)
; Person at extension "is unavailable" message
exten => s-BUSY,1,VoiceMail(${ARG2},b)
; Person at extension "is busy" message
exten => s-ANSWER,1,Hangup()
; To be safe, clean up the call after an answer by hanging up
exten => _s-.,1,Goto(s-NOANSWER,1)
; Handle any unhandled status the same way we handle NOANSWER

[internal-extensions]
; If the building manager on duty does not answer the phone,
; the call is routed to mailbox 150:
exten => _15X,1,Macro(simple|SIP/${EXTEN}|150)

; IT has normal mailboxes:
exten => _16X,1,Macro(simple|SIP/${EXTEN}|${EXTEN})

; Each manager has his or her own mailbox:
exten => _20[1-2],1,Macro(simple|SIP/${EXTEN}|${EXTEN})

; The reception staff have a common mailbox:
exten => _2[3-6]X,1,Macro(simple|SIP/${EXTEN}|804)

; The domestic sales group has a common mailbox:
```

```
exten => _3[0-4]X,1,Macro(simple|SIP/${EXTEN}|802)

; The international sales group has a common mailbox:
exten => _3[5-9]X,1,Macro(simple|SIP/${EXTEN}|803)

; No voice mail on the other extensions.
exten => _[4-5]XX,1,Dial(SIP/${EXTEN},30)

[voicemail-easyaccess]
exten => 800,1,VoiceMailMain(${CALLERID(num)})

[voicemail-generalaccess]
exten => 801,1,VoiceMailMain()

[voicemail-buildingmgr]
exten => 800,1,VoiceMailMain(150)

[voicemail-sales-domestic]
exten => 800,1,VoiceMailMain(802,s)

[voicemail-sales-international]
exten => 800,1,VoiceMailMain(803,s)
```

The `extensions.conf` for Widgets, Inc. is already starting to look a bit more complicated. This is because we are using different mailbox features for different users and departments. There are normal individual mailboxes for some staff and group mailboxes for sales and reception; on top of that, sales staff must be able to access messages without having to enter a password. This must be as easy and transparent as possible for individual staff members, so we set the standard voice mailbox access number to 800 for everyone.

9.3 Dialplan Applications

We can use two voicemail applications in the dialplan (`extensions.conf`):

- **VoiceMail()**

  This application sends a caller to the voicemail system, where she will be asked to leave a message.

- **VoiceMailMain()**

  This application lets recipients check their voice messages and record new voicemail prompts.

9.3.1 VoiceMail

This command prompts the caller to leave a voice message.

The **VoiceMail()** command is always called from the dialplan (`extensions.conf`), as in this example:

```
exten => 2000,2,VoiceMail(2000,u)
```

The command uses the following syntax:

```
VoiceMail(mailbox[@context][,u|b|s])
```

- *mailbox*

  This is the mailbox number. This does not have to be the same as the extension the caller dialed; nevertheless, this is a sensible practice, particularly in larger installations.

- *@context*

  Mailboxes may be implemented in a specific context. If no context is provided, the [default] context is used.

  If the caller presses 0 while listening to the prompt, the application will jump to extension o (the small letter o) in the specified context.

  If the caller presses * while listening to the prompt, the application will jump to extension a (the small letter a) in the specified context.

- [u|b|s]

  - u

    Causes the "unavailable" message to be played. The pathname for this message is `/var/lib/asterisk/sounds/vm-isunavail`.`gsm`.[2]

  - b

    Causes the "busy" to be played. The pathname for this message is `/var/lib/asterisk/sounds/vm-rec-busy.gsm`.

  - s

    Suppresses playback of the "unavailable" or "busy" notifications, plays a beep, and begins recording.

---

2. If you are using a prepackaged Asterisk, the path may be different (for example, `/usr/share/asterisk/sounds`).

If there is no mailbox configured in `voicemail.conf` for the given number but there is an $n+101$ priority, Asterisk jumps to this priority and continues executing there.

9.3.2 VoiceMailMain

This command gives users access to their messages and lets them record prompts.

The **VoiceMailMain()** command is always called from the dialplan (`extensions.conf`). For example:

```
exten => 300,1,VoiceMailMain()
VoiceMailMain([mailbox][@context][,s|p|g(#)])
```

- *mailbox*

  This is the mailbox number. If no mailbox number is provided, Asterisk prompts for it.

- *@context*

  Specifies the voicemail context (in `voicemail.conf`) for the mailbox.

- [s|p|g(#)]

  - s

    Disables the password requirement.

  - p

    The user is asked for a mailbox number. The number entered is attached as a suffix to the contents of [mailbox]; for example, if the user enters 123, [mailbox]123 is called. This lets you easily configure mailbox groups.

  - g(#)

    Adjusts the gain (in decibels) when recording voicemail prompts.

A complete description of the voice menus for **VoiceMailMain()** is difficult because they depend on the installed prompts. The main functions are described in Table 9.3.

Table 9.3 Menu Structure for VoiceMailMain()

1	Play messages			
	3	Advanced options		
		1	Reply	
		2	Call back	
		3	Envelope	
		4	Outgoing call	
	4	Play previous message		
	5	Repeat the current message		
	6	Play the next message		
	7	Delete the current message		
	8	Forward the message to another mailbox		
	9	Save the message in a folder		
	*	Help; during message playback, rewind		
	#	Exit; during message playback, skip forward		
2	Change folders			
0	Mailbox options			
	1	Record your unavailable message		
	2	Record your busy message		
	3	Record your name		
	4	Record your temporary message		
		Recording options		
		1	Accept	
		2	Review	
		3	Re-record	
*	Help			
#	Exit			

9.4 voicemail.conf

Voicemail is configured in `voicemail.conf`. The file has three sections, which are described in detail in this section:

- `[general]`
  Global configuration options go here.

- `[zonemessages]`
  For organizations spread across time zones, time-zone-specific options go here.

- Defined contexts

  If the organization has multiple divisions or departments, you can maintain them in separate voicemail contexts.

9.4.1 [general]

Global configurations for the voicemail system go here. The available options for this section are described in detail here:

- `attach = [yes|no]`

  Sets whether voice message files are attached to e-mail notifications. Default: yes. The format defined by the first entry in the format= line is used.

- `callback=[context]`

  Specifies the context through which callbacks from the voicemail system are made. If unset, callbacks cannot be made from voicemail. Default: unset.

- `charset=[charset]`

  Specifies the character set used to encode the text of the e-mail. In this example, ISO Latin-1 is set:

  `charset=ISO-8859-1`

- `delete=[yes|no]`

  Sets whether messages are deleted once the e-mail notification with attachment has been sent. This saves hard disk space on the server if users are only to receive messages via e-mail.

- `directoryintro=[filename]`

  Specifies the filename of a sound file (found in the default sound file directory `/var/lib/asterisk/sounds`) to be used rather than the default sound file for the Dial-by-Name system.

- `emailsubject=[subject]`

  Specifies the subject line to be used in e-mail notifications. For information about variables, see `emailbody`. The text need not be enclosed in quotation marks:

  `emailsubject=New message from ${VM_CALLERID}`

- `pbxskip=[yes|no]`

  Asterisk prefixes [PBX] to the subject line of every e-mail notification. This is intended to make filtering e-mail notifications into a specific folder more straightforward, but can be suppressed with yes if undesired.

- emailbody=[*email_text*]

  This specifies the body text of the e-mail notifications and is limited to 512 characters.

  You may use variables in the subject and body of e-mail notifications:

  - ${VM_NAME}

    Name of mailbox owner

  - ${VM_DUR}

    Message length

  - ${VM_MSGNUM}

    Message number

  - ${VM_MAILBOX}

    Mailbox number

  - ${VM_CALLERID}

    Name and number of caller

  - ${VM_CIDNUM}

    Number of caller

  - ${VM_CIDNAME}

    Name of caller

  - ${VM_DATE}

    Date and time of call

  - ${VM_MESSAGEFILE}

    Name of the sound file that contains the voice message

All the text needs to be on one line. If you need carriage returns, use the \n combination:

```
emailbody=Hi, ${VM_NAME}!\n\nYou have a new message from ${VM_CALLERID} in
mailbox ${VM_MAILBOX}.
```

- serveremail=[*fromaddress*]

  This specifies the e-mail address that appears in the From field of e-mail notifications. Here is an example:

```
serveremail=voicemail@widgets-inc.biz
```

- fromstring=[*fromname*]

  Specifies the envelope sender (From header) in e-mail notifications. This is useful for mail filtering and sorting.

- `mailcmd=[`*`mailer`*`]`

  Specifies the application (including absolute path) to be used for sending e-mail notifications. Here are two more common examples; note that you can append options:

  ```
  mailcmd=/usr/sbin/sendmail -t
  mailcmd=/usr/exim/bin/exim -t
  ```

- `externnotify=[`*`application`*`]`

  Specifies the application (including absolute path) called by Asterisk when a new message arrives. You can use it to call custom scripts:

  ```
  externnotify=/usr/bin/local/send-sms.sh
  ```

- `externpass=[`*`application`*`]`

  Specifies the application (including absolute path) called by Asterisk when a user changes his password. You can use it to call custom scripts:

  ```
  externpass=/usr/bin/local/password-notify.sh
  ```

- `forcegreetings=[yes|no]`

  Sets whether the user will be forced to record a new greeting when logging in to the system for the first time. The default is no.

- `forcename=[yes|no]`

  Sets whether the user will be forced to record her name when logging in to the system for the first time. The default is no.

- `format=[gsm|wav|wav49]`

  Defines the codecs used to save voicemail messages. If more than one codec is specified, the message is stored once in each of the formats specified. This can lead to a shortage of available disk space but means further transcoding will not be needed if a user using a different codec is retrieving the messages.

  If `attach=yes`, the first format specified is used for the attachment to the e-mail notification.

  ```
  format=gsm|wav
  ```

  To save messages only in WAV format, use:

  ```
  format=wav
  ```

Caution: If this setting is changed during operation, existing files in other formats must be deleted manually by the system administrator.

- `searchcontexts=[yes|no]`

  Asterisk only looks for mailboxes in the specified context. Setting `search-contexts=yes` makes Asterisk search in all contexts. The default is `no`.

- `maxmsg=[number_of_messages]`

  Defines the maximum number of messages a single mailbox may hold. When this limit is reached, no further messages can be recorded, and the caller hears the message `/var/lib/asterisk/sounds/vm-mailboxfull.gsm`. The default is `100`.[3]

- `maxmessage=[length_in_seconds]`

  Sets the maximum duration of a message. There is no limit by default.

- `minmessage=[length_in_seconds]`

  Sets the minimum duration of a message. The default is `0`.
  Example:

  ```
  minmessage=5
  ```

- `maxgreet=[length_in_seconds]`

  Sets the maximum duration of greeting messages. There is no limit by default.

- `maxsilence=[length_in_seconds]`

  Sets the number of seconds of silence that the system will allow before it assumes that the caller has finished.

- `silencethreshold=[threshold]`

  Specifies the maximum sound level that Asterisk considers silence when measuring `maxsilence=`. The lower the number, the more sensitive the detection. The default is `128`.[4]

- `maxlogins=[no_of_login_attempts]`

  Sets the maximum number of failed login attempts (e.g., the user enters an incorrect password) before Asterisk hangs up. The default is `3`.

- `skipms=[milliseconds]`

  Sets the number of milliseconds the "skip forward" and "rewind" keys will jump forward or back during message playback. The default is `3000`.

---

3. The path depends on the installation; some packages install the sounds in `/usr/share/asterisk/sounds`.

4. There is no indication in the Asterisk source code what the range or unit for this value is. The only way to set the value sensibly is through trial and error.

- `usedirectory=[yes|no]`
  Gives callers access to the Dial-by-Name system. The default is `no`.

- `saycid=[yes|no]`
  Sets whether the telephone number of the caller is announced when the message is heard. The default is `no`.

- `cidinternalcontexts=[contextA,contextB,...]`
  Defines (through a comma-separated list) which contexts are treated as internal when the originating telephone number is announced as per `saycid=`. For contexts defined as internal only the caller's extension is announced. There is no default.

- `pagerfromstring=[sendername]`
  Indicates the sender for pager notifications. See `fromstring=`.

- `pagersubject=[subject]`
  Specifies the subject for pager notifications. See `emailsubject=`.

- `pagerbody=[body_text]`
  Specifies the message body for pager notifications. See `emailbody=`.

There are other parameters in addition to those documented here; nevertheless, these are those most frequently used. A complete listing of the available parameters, with brief descriptions, may be found in the sample version of `voicemail.conf`.

9.4.2 [zonemessages]

In larger enterprises, voicemail users can be spread across time zones; the `[zonemessages]` section allows for time announcements that correspond to the user's time zone.[5]

```
zonename=timezone | format
```

- `zonename`
  This is an arbitrary identifier used to identify the time zone. Only small letters (a–z), numbers (0–9), dash (-), and underscore (_) characters are allowed.

---

5. Note that time zones defined here must still be activated for individual users.

- *timezone*

  Time zone information is stored in /usr/share/zoneinfo for most Linux
  distributions. Typically, the time zone files are sorted by continent, then
  region or city, separated by a /.

  Examples:

  ```
  Canada/Mountain
  Europe/Berlin
  Australia/Sydney
  ```

- *format*

  Defines the format for time announcements. This is done with the variables,
  which are described in Table 9.4.

  Other allowed variables include ${ANY_VARIABLE} (the contents of the
  variable are entered) and 'sound_filename'. When specifying a sound file,
  be careful not to add a file extension (e.g., .wav, .gsm); the specified sound file
  in /var/lib/asterisk/sounds is then played. The single quotation marks
  are mandatory.

Table 9.4 Time Format Variables in voicemail.conf

A	Weekday (Monday to Sunday).
a	As A.
B	Month (January to December).
b	As B.
h	As B.
d	Day of the month (1 to 31).
e	As d.
Y	Year (e.g., 2007).
I	Hour in 12-hour clock.
l	As I.
H	Hour in 24-hour clock. Single-digit hours are preceded by a 0 ("oh").
k	Hour in 24-hour clock. Single-digit hours have no prefix.
M	Minute.
P	AM or PM.
p	As P.
Q	"Today," "yesterday," or "*weekday, month, day, year.*"
q	Nothing if today, "yesterday," or "*weekday, month, day, year.*"
R	Hour:minute:second.

Here are some sample formats as they would be used in `voicemail.conf`:

```
[zonemessages]
germany=Europe/Berlin | 'vm-received' Q 'digits/at' kM
alberta=Canada/Mountain | 'vm-received' Q 'digits/at' HM
england=Europe/London | 'vm-received' Q 'digits/at' R
military=Zulu | 'vm-received' q 'digits/at' H N 'hours' 'phonetic/z_p'
```

9.4.3 Defined Contexts

As in the dialplan (`extensions.conf`), we can define contexts in `voicemail.conf`. This is useful when there is a need to separate voicemail access by department, branch, or city, and makes it possible to define separate Dial-by-Name directories as well.

The `default` context is mandatory. For most installations, this is all that is required.

9.4.4 Mailboxes

You can configure as many mailboxes as you like in a given context. The proper syntax is as follows:

```
mailbox => password,name[,e-mail[,pager-e-mail[,options]]]
```

The following example sets the voicemail attachment format as WAV and deletes the message once it has been attached to an e-mail and sent:

```
202 => 1234,Gina Smart,gina@widgets-inc.biz,,attachfmt=wav|delete=yes
```

Here are the parameters, explained:

- *mailbox*

  Mailbox number (digits).

- *password*

  Password for the mailbox (may be letters or numbers). Generally, numbers are more practical as they are more likely to be enterable from a standard telephone set.

- *e-mail*

  E-mail address to which to send e-mail notifications.

- *pager-e-mail*

  E-mail address for the pager to which pager notifications should be sent.[6]

---

6. Note that such notifications are still, in essence, e-mail notifications. If you want to send SMS notifications, you can do so with a number of external applications that may be invoked through Asterisk.

- options

  Multiple options must be separated by the pipe character (|); these options can override global options set in [general]. Note that there cannot be any spaces between the parameter name, the = sign, and the value, nor between the options and the pipe character.

  The most important options applied here are as follows:

  - tz=[timezone]

    Sets the time zone previously defined in [zonemessages].

    Example:

    `tz=alberta`

  - attach=[yes|no]

    If defined, sets whether voice message files are attached to e-mail notifications.

  - attachfmt=[gsm|wav|wav49]

    If defined, overrides the format= value set in [general] for message attachments to e-mail notifications.

  - saycid=[yes|no]

    Sets whether the telephone number of the caller is announced when the message is heard.

  - sayduration=[yes|no]

    Sets whether the length of the message will be announced. The default is yes.

  - saydurationm=[length_in_minutes]

    Defines the minimum duration for a message to be before the duration is announced. The default is 0.

  - dialout=[context]

    Specifies the context to be used for dialing out from a message. Calls cannot be made out of voice mail unless this parameter is set.

  - sendvoicemail=[yes|no]

    Sets whether a user may send messages to other voicemail users.

  - callback=[context]

    Specifies the context to be used for dialing out from a message. If defined, this makes it possible to return a call to someone who has left a message.

- `review=[yes|no]`

  Sets whether callers leaving a message may listen to the message before sending it. The default is `no`.

- `operator=[yes|no]`

  Sets whether callers can connect to the operator before, during, or after recording a message by pressing 0 (zero). In Asterisk, the operator is always the o (small letter o) extension in the current context. The default is `no`.

- `envelope=[yes|no]`

  Sets whether message envelope information is announced before the message is heard. Setting this parameter to no overrides other parameters (for example, `tz`). The default is `yes`.

- `delete=[yes|no]`

  Sets whether messages are deleted after they have been attached to an e-mail notification. The default is `no`.

- `nextaftercmd=[yes|no]`

  Sets whether the system jumps to the next message after the current message has been saved or deleted. The default is `no`.

- `forcename=[yes|no]`

  If set to `yes`, forces a system user to record her name when accessing the system for the first time. The default is `no`.

- `forcegreetings=[yes|no]`

  If set to `yes`, forces a system user to record a personal greeting when accessing the system for the first time. The default is `no`.

- `hidefromdir=[yes|no]`

  Sets whether the mailbox for this user will be hidden from the Dial-by-Name directory. The default is `no`.

9.5 Dial-by-Name

Although not part of the voicemail system per se, we discuss the Dial-by-Name directory here because it was developed along with Comedian Mail and gets its configuration from `voicemail.conf`.

The caller is asked to enter the first three letters of the person's last name (if option f is specified, the first name is used) using the telephone keypad. The application searches for the entry in the directory and connects the caller after checking that the entry is the correct one.

Returning to our example for the Robinson family, we make the following small changes:

```
[general]
format=gsm

[default]
; Syntax for new entries looks like this:
; MailboxNumber => password,firstname lastname,e-mail,pager,options
;                            ^^^^^^^^^^^^^^^^^^^^^
200 => 1234,Dave Robinson
201 => 1234,Colleen Robinson
202 => 1234,Matthew Robinson,matt@robinsonfamily.name
203 => 1234,Lisa Robinson,lisa@robinsonfamily.name,,delete=yes
```

From here it is a simple matter to generate a Dial-by-Name directory. We use the Asterisk application **Directory()** in extensions.conf:

```
exten => 800,1,Directory(default,from-internal)
```

The general syntax for **Directory()** is as follows:

```
Directory(vm-context[,dial-context[,options]])
```

▪ *vm-context*

The directory is always generated from a specific context; that is, only the entries in this defined context are entered in the directory. In most cases, the default context is sufficient. In a larger enterprise, you may want to separate directories by department; in this case, you will need additional contexts.

▪ *dial-context*

The directory is used to call to a specific person; dial-context defines which context is used for the call. If undefined, vm-context is used instead.

If the caller presses 0 (zero), the call will be sent to the o (small letter o) extension in this context. If he presses * (asterisk, or star), the call will be sent to the a (small letter a) context.

▪ *options*

Option f sets searching by first name rather than last name.

9.6 Saving Passwords in voicemail.conf

The `voicemail.conf` file need not serve only as a directory source; it can also be used as a general storage file for passwords. You may encounter situations where users must authenticate for reasons other than retrieving voicemail (for example, call agents logging in at a call center).

As a result, it is sometimes necessary to create `voicemail.conf` entries for users, even if they have no need for a mailbox. Don't worry about callers accidentally leaving messages in an unattended mailbox; only mailboxes explicitly referenced through an extension using **VoiceMail()** or **VoiceMailMain()** in `extensions.conf` will be accessible to callers.

For example, you might make two entries in the context `[call-center-agents]` in `voicemail.conf`:

```
1001 => 1234,Annaliese Chips,,,hidefromdir=yes
1002 => 1234,Donald Down,,,hidefromdir=yes
```

You can use these entries to authenticate a user with the dialplan application **VMAuthenticate()** (see Appendix B, "Dialplan Applications"):

```
exten => 988,n,Read(agentno,agent-user,10,,3)
exten => 988,n,VMAuthenticate(${agentno}@call-center-agents)
```

Don't let this example limit your creativity, however. There are many ways to authenticate users, of which this is only one. Saving passwords in AstDB or another database (such as MySQL, for example) is usually preferable.

10

Interactive Voice Response

An interactive voice response (IVR) system lets computer systems interact with telephone callers, who provide input to the system either by pressing the keypad on their telephone set (dual-tone multi-frequency [DTMF] keying, a.k.a. Touch-Tone) or by saying something (natural language speech recognition). Most IVR systems provide selection menus for routing calls without requiring operator intervention, but modern IVR systems can also be very complex applications that handle information or control equipment.

The basic principle common to all IVR systems, however, is that the caller is read a menu and chooses options from that menu to perform actions, or, alternatively, enters information (in numeric format, through pressing the keypad). IVRs can be used to obtain stock quotations, train schedules, and weather reports; they can also be used for automated purchasing systems, such as for concert tickets. The potential applications are limited only by your imagination.

Systems vary in their complexity. The most advanced generate spoken text on-the-fly using text-to-speech (TTS) systems and accept spoken user input with speech recognition. When properly implemented they can provide a high level of user friendliness, but implementation is so complex that they are rarely used, except in larger organizations.

The simplest form of IVR is also the most common. Pre-recorded messages are played to the caller; the caller responds with DTMF keypad input, which Asterisk can recognize easily in the default install.

Public opinion on IVRs is divided. Some people find them a helpful form of automation, whereas others find them exceedingly frustrating. This is usually the result of poor menu design or speech recognition with a high failure rate. A well-functioning IVR can be pleasant for the customer, but a poorly implemented one can scare her away.[1]

Take care when planning an IVR system. Pay special attention to menu design and allow adequate time for a clean development and deployment. Aggressive testing and post-deployment monitoring of premature hangups should be part of your routine. Remember that IVR systems are not an end in themselves, nor are they a panacea. Think of your customer!

10.1 A Simple IVR

The standard Asterisk sound set includes a file called `marryme.gsm`, containing the announcement "Will you marry me? Press 1 for yes or 2 for no."[2]

(Using this IVR for an actual marriage proposal is strongly discouraged.) To build a "marriage proposal" application, the following dialplan will suffice:

```
exten => 30,1,Answer()
exten => 30,2,Background(marryme)
exten => 30,3,Hangup()

exten => 1,1,Playback(thank-you-cooperation)
exten => 1,2,Hangup()

exten => 2,1,Playback(sorry)
exten => 2,2,Hangup()
```

If the caller dials extension 30, Asterisk answers and plays the file `marryme.gsm`. Through use of the **Background()** application, the user is allowed to enter input at any time during playback. The input is interpreted as an extension and the call is passed to that extension. If the caller presses 1, he hears "Thank you for your cooperation," after which Asterisk hangs up.

Note: Playback() (see Appendix B, "Dialplan Applications") only plays back sound files; input is ignored. **Background()** (see Appendix B) plays sound files back while listening for caller input, which is interpreted as an extension as though it had been dialed as one.

---

1. The increasingly multiethnic nature of society everywhere means that speech recognition should be implemented with caution, as accents can be problematic.

2. Allison Smith is a Canadian voice professional who is the "Voice of Asterisk." Its growing popularity has given her a considerable cult following.

10.1.1 Timeouts

If you have multiple extensions beginning with the same digits, you need to allow sufficient time for the IVR to recognize all possible combinations. Let's examine the following example:

```
exten => 30,1,Answer()
exten => 30,2,Background(marryme)
exten => 30,3,Hangup()

exten => 1,1,Playback(thank-you-cooperation)
exten => 1,2,Hangup()

exten => 10,1,NoOp(Test with 10)
exten => 10,2,Hangup()

exten => 100,1,NoOp(Test with 100)
exten => 100,2,Hangup()

exten => 2,1,Playback(sorry)
exten => 2,2,Hangup()
```

Background() waits a set time after each digit in order to distinguish between 1, 10, and 100. Once this time (TIMEOUT) has expired, input is deemed to be complete.

Tip: TIMEOUT lets you set other timeouts. For more information, enter **show function TIMEOUT** in the Asterisk CLI, or see Appendix C, "Dialplan Functions."

TIMEOUT is defined in seconds and may be set in the dialplan as follows:

```
exten => 123,1,Set(TIMEOUT(digit)=3)
```

In this dialplan, Asterisk will proceed immediately if 2 is pressed, but only after the timeout has expired if 1 is pressed. Asterisk intelligently determines whether a digit entered can match multiple extensions and behaves accordingly.

10.1.2 Invalid Input (the i Extension)

An invalid entry (any entry for which no extension in the dialplan matches) can be handled by the i extension. In the following example, the IVR plays an apology to the caller and hangs up. (A real IVR should probably bring the caller back to the main menu in the event of an invalid entry.)

```
exten => 30,1,Answer()
exten => 30,2,Background(marryme)
exten => 30,3,Hangup()
```

```
exten => 1,1,Playback(thank-you-cooperation)
exten => 1,2,Hangup()

exten => 2,1,Playback(sorry)
exten => 2,2,Hangup()

; Any other input is caught by the i extension.
exten => i,1,Background(sorry)
exten => i,2,Hangup()
```

10.1.3 Pauses

The easiest way to create pauses for input is to play back empty sound files. A series of silent sound files of between 1 and 9 seconds in length may be found in /var/lib/asterisk/sounds/silence. If we need to allow 5 seconds following the prompt (a marriage proposal requires careful consideration, after all), here's how we can accomplish that:

```
exten => 30,1,Answer()
exten => 30,2,Background(marryme)
exten => 30,3,Background(silence/5)
exten => 30,4,Hangup()

exten => 1,1,Playback(thank-you-cooperation)
exten => 1,2,Hangup()

exten => 2,1,Playback(sorry)
exten => 2,2,Hangup()

exten => i,1,Background(marryme)
exten => i,2,Hangup()
```

10.2 Multilevel IVR Systems

The problem with multilevel IVRs is that the caller has to enter single digits multiple times (oftentimes the same digit), but gets a different response depending on the menu level. In Asterisk, a number can only be used once in a given context. If we need multiple menus that provide different responses for the same digits, we must place the submenus in different contexts ([cafeteria] in our example). We jump between these contexts using the application **Goto()** (see Appendix B). Let's assume you have some custom sound files stored in /var/lib/asterisk/sounds. Here are the files followed by brief descriptions of what they contain:

- mainmenu.gsm
  "Press 1 for sales, 2 for service, or 3 for the cafeteria."

- `cafeteria.gsm`

  "Press 1 to hear the menu for this week or 2 to hear the menu for next week."

- `cafeteria-menu-this-week.gsm`

  "Monday: Noodles with pesto sauce. Tuesday: Pork chops...."

- `cafeteria-menu-next-week.gsm`

  "Monday: Stew, featuring noodles, basil and, um, pork chops...."

If sales is at extension 100 and the service department is at 150, the dialplan for this IVR would look like this:

```
[example-ivr]
; The menu is repeated until the caller provides input.
;
exten => 30,1,Answer()
exten => 30,2,Background(mainmenu)
exten => 30,3,Background(silence/3)
exten => 30,4,Goto(2)

exten => 1,1,Dial(SIP/100)

exten => 2,1,Dial(SIP/150)

; Goto() jumps to another context ([cafeteria])
;
exten => 3,1,Goto(cafeteria,100,1)

exten => i,1,Goto(30,2)

[cafeteria]
exten => 100,1,Background(cafeteria)
exten => 100,2,Background(silence/3)
exten => 100,3,Goto(1)

exten => 1,1,Playback(cafeteria-menu-this-week)
exten => 1,2,Wait(2)
exten => 1,3,Goto(1)

exten => 2,1,Playback(cafeteria-menu-next-week)
exten => 2,2,Wait(2)
exten => 2,3,Goto(1)

; Invalid input sends the caller back to the main menu
exten => i,1,Goto(example-ivr,30,2)
```

Even though it is technically possible to support an unlimited number of IVR levels, in practice it is advisable to keep the number of menu levels to a maximum of three. Many callers hang up after the third menu level.

10.3 Text-to-Speech

Text-to-speech (TTS) is simply the conversion of written text into a spoken word, using speech synthesis. In our Asterisk system, this means that an external program generates a sound file using a given text file (usually in ASCII format) as the source. The resulting sound file is played back as any other sound file would be, and the caller hears the text spoken out.

Quality of TTS engines varies widely. As a rule of thumb, the open source engines are not as sophisticated as the commercial ones.

Tip: Sometimes you can test high-quality engines through web portals. IBM offers a test portal for its TTS engine at www.ibm.com/software/pervasive/tech/demos/tts.shtml. Like IBM, Cepstral has a demo portal at www.cepstral.com/demos.

The TTS engine Festival (www.cstr.ed.ac.uk/projects/festival) is a widely used open source version, but the voices included with it often lack the quality necessary for professional implementation, particularly if you need voices in languages other than English.

Many Asterisk developers use the engine and voices sold commercially by Cepstral (www.cepstral.com). As of this writing, the pricing was reasonable. The solution described here builds on the Cepstral engine. If you want to work with the open source Festival instead, these instructions are easily modified to work with it. This applies to other TTS engines, too. The implementation model is the same.

10.3.1 Installing Cepstral TTS

Download the voice from www.cepstral.com/downloads. The file (Cepstral_David_i386-linux_4.2.0.tar.gz in this example) is installed with the following commands:

```
tar xvzf Cepstral_David_i386-linux_4.2.0.tar.gz
cd Cepstral_David_i386-linux_4.2.0.tar.gz
./install
```

The engine installs to /opt/swift/bin/swift unless otherwise specified. You can test the installation from the command line as follows:

```
/opt/swift/bin/swift -o /tmp/test.wav -p audio/sampling-
rate=8000,audio/channels=1 "T
```

You can play the resulting file with any audio player, or through Asterisk. To do this, just add a few lines to `extensions.conf`:

```
exten => 1234,1,Answer()
exten => 1234,2,Playback(/tmp/test)
exten => 1234,3,Hangup()
```

To generate some speech output from within Asterisk, we use the **System()** application in the dialplan. Here is an example:

```
exten => 1222,1,Answer()
exten => 1222,2,System(rm -rf /tmp/test.wav)
exten => 1222,3,System(/opt/swift/bin/swift -o /tmp/test.wav -p audio/
sampling-rate=8
exten => 1222,4,Playback(/tmp/test)
exten => 1222,5,Hangup()
```

Cepstral uses SSML (Speech Synthesis Markup Language) in its engine. You can add speech pauses to the output by specifying them as in this example:

```
exten => 1222,1,Answer()
exten => 1222,2,System(rm -rf /tmp/test.wav)
exten => 1222,3,System(/opt/swift/bin/swift -o /tmp/test.wav -p audio/
sampling-rate=8
exten => 1222,4,Playback(/tmp/test.wav)
exten => 1222,5,Hangup()
```

You can learn more about the SSML standard at www.w3.org/TR/speech-synthesis.

11

The Asterisk Database

The problem with using variables in the dialplan is that it is vulnerable to a system crash or a restart of Asterisk. If either of these things happens, the variable is either erased or reset to its default value. This limitation precludes certain implementations entirely. In call forwarding or a calling card application, some values (the remaining balance, for example) need to be stored in a persistent manner so that they can be recalled after a restart. The Asterisk database (AstDB) serves this purpose.

11.1 The Asterisk Database

Those needing a database for Asterisk applications often ask how well the AstDB performs. The answer, to the extent there's a simple one, is that it depends on what you are doing. If you are working with small amounts of data, as you might in a redial or call forwarding scenario, the AstDB performs admirably and you needn't look further. If you are working with large, complex datasets, you should consider whether an external SQL server would be better.

In the overwhelming majority of applications, however, this discussion is purely theoretical. The Asterisk database is built on the Berkeley DB (BDB), which is already a very fast database engine. When handling key/value pairs, BDB is among the fastest of databases. The question of database choice should be raised, therefore, when speed of the database is proven to be the source of problems, or when you want to build a larger system with greater functionality than may be met by a simple key/value database.

In a default installation, Asterisk comes packaged with a database built on BDB. BDB is a simple, high-performance, embedded database library with application programming interfaces (APIs) for C, C++, Java, Perl, Python, Tcl, and many other programming languages. It stores key/value pairs and places no constraints on how data is stored in a record. It does not support SQL or any other query language and allows access only through in-process API calls. This database stores information in the form of key/value pairs, and keys are grouped into families.

Warning: Up to and including Asterisk Version 1.2, these applications were used for database access:

```
DBput(family/key=value)
```

To save a value in the database.

```
DBget(var=family/key)
```

To retrieve a value from the database. If the called key cannot be found, Asterisk jumps to priority 101 in the current extension.

These commands are no longer supported as of Asterisk 1.4. For this reason, only the new method, using the **DB()** function is mentioned here, which also works in 1.2.

Information in this database can be manipulated using the **DB()** function (see Appendix C, "Dialplan Functions.")

11.1.1 Writing Values to the Database

The **DB()** function can be called from within the **Set()** application. To give the apple entry in the fruit family the value 20, write the extension as follows:

```
exten => 1234,1,Set(DB(fruit/apple)=20)
```

11.1.2 Reading Values from the Database

Values may be called from the database using **DB()** in the form ${DB(family/key)}. To print the contents of apple in the fruit family on the console, use the following command:

```
exten => 1234,1,NoOp(fruit/apple has the value
↳ ${DB(fruit/apple)})
```

Warning: In all the code examples, everything is written on one line, without breaks, unless otherwise noted. The limited space for text requires that we wrap the lines in the book.

You can also store the value from this database field in a variable using **Set()**:

```
exten => 1234,1,Set(applequantity=${DB(fruit/apple)})
```

11.1.3 Deleting Values from the Database

The applications **DBdel()** and **DBdeltree()** are used to delete database entries.

11.1.3.1 DBdel() and ${DB_DELETE()}

In Asterisk 1.2, individual entries are deleted using DBdel(). Delete the key apple in the family fruit as follows:

```
exten => 1234,1,DBdel(fruit/apple)
```

DBdel() is deprecated in Asterisk 1.4; use the **DB_DELETE()** function instead. This is called like so:

```
exten => 1234,1,NoOp(${DB_DELETE(fruit/apple)})
```

See also Appendix B, "Dialplan Applications," and Appendix C, "Dialplan Functions."

11.1.3.2 DBdeltree()

If you need to delete an entire key family, use **DBdeltree()**. The fruit family is deleted as follows:

```
exten => 1234,1,DBdeltree(fruit)
```

See Appendix B.

11.2 Database Access from the Command-Line Interface

The system administrator can access the database from the CLI.

Tip: If you have trouble remembering a command or the exact syntax, you can get help at the CLI with **help database**:

```
big-island*CLI> help database
          database del      Removes database key/value
          database deltree  Removes database keytree/values
          database get      Gets database value
          database put      Adds/updates database value
          database show     Shows database contents
          database showkey  Shows database contents
```

For the following examples, remember that if your family, key, or value contain any blank spaces, the entire name must be enclosed in double quotes (" "); for example, **database put test entry "Hello world"**.

11.2.1 Writing Values to the Database

You write values to the database with **database put** *family key value*:

```
big-island*CLI> database put fruit apple 20
Updated database successfully
```

11.2.2 Reading Values from the Database

Read values from the database with database get *family key*:

```
big-island*CLI> database get fruit apple
Value: 20d*CLI>
```

11.2.3 Deleting Values from the Database

Both **database del** *family key* and **database deltree** *family* delete values from the database.

11.2.3.1 database del

To delete only the key apple from the fruit family, use **database del**:

```
big-island*CLI> database del fruit apple
Database entry removed.
```

11.2.3.2 database deltree

You can delete an entire key family with **database deltree**:

```
big-island*CLI> database deltree fruit
Database entries removed.
```

11.2.4 Displaying Database Contents

The commands **database show** and **database showkey** display database contents in the CLI:

```
big-island*CLI> database put shoppinglist eggs 2
Updated database successfully
big-island*CLI> database put shoppinglist butter 250
Updated database successfully
big-island*CLI> database put shoppinglist sugar 500
Updated database successfully
big-island*CLI> database show
/shoppinglist/butter                              : 250
/shoppinglist/eggs                                : 2
/shoppinglist/sugar                               : 500
big-island*CLI> database showkey butter
/shoppinglist/butter                              : 250
big-island*CLI> database deltree shoppinglist
Database entries removed.
```

11.3 Database Access from the System Shell

Using the command **asterisk -rx 'command'**, you can execute any CLI command via a shell script. The following example shows how a database is populated, read, and then finally deleted through external commands:

```
big-island:~# asterisk -rx 'database put test var1 23'
Updated database successfully
big-island:~# asterisk -rx 'database put test var2 42'
Updated database successfully
big-island:~# asterisk -rx 'database show test'
/test/var1                                        : 23
/test/var2                                        : 42
big-island:~# asterisk -rx 'database get test var2'
Value: 42
big-island:~# asterisk -rx 'database deltree test'
Database entries removed.
```

Important: Remember that if the family, key, or value contains any blank spaces, you must use double quotes (") around the string, as follows:

```
big-island:~# asterisk -rx 'database put test entry "Hello World"'
```

11.4 Database Backup

The AstDB is stored in `/var/lib/asterisk/astdb` by default. You can copy that directory as long as Asterisk is stopped.

You can also back up the database from the system shell during operation with the following command:

```
asterisk -rx "database show" > /tmp/backup-asterisk-database.txt
```

Of course, restoring the database from this kind of backup is more difficult.

11.5 Application Example: Call Forwarding

Anybody who wants, when she's not in, to have calls to her local number passed on to her mobile phone needs call forwarding. This can be achieved either through the local SIP phone or centrally through Asterisk. In practice, we prefer that latter approach because it frees us from dependency on end devices, which could be disconnected, fail, or be swapped with others. It also means that the call forwarding setting is not lost in the event of a power failure or system restart.

11.5.1 Simple Call Forwarding

All the staff at Widgets, Inc. can forward calls to their locals to any number they choose. To activate this forwarding, the internal extension 44, followed by the target number, is dialed. To deactivate this forwarding, 44 is dialed again (this time without a target number following it). You can implement this with the following dialplan. (The text wraps in this code example, but remember that every extension must be entered on a single line!)

```
[from-internal]
; call forwarding for a single extension
;
; activate
exten => _44X.,1,Answer()
exten => _44X.,n,Set(DB(CF/${CALLERID(num)})=${EXTEN:2})
exten => _44X.,n,SayDigits(${EXTEN:2})
exten => _44X.,n,NoOp(Forwarding for ${CALLERID(num)} to ${EXTEN:2}
↳activated.)
exten => _44X.,n,Hangup()
; deactivate
exten => 44,1,Answer()
exten => 44,n,DBdel(CF/${CALLERID(num)})
exten => 44,n,Playback(auth-thankyou)
exten => 44,n,NoOp(Forwarding for ${CALLERID(num)} deactivated.)
exten => 44,n,Hangup()
```

```
[from-external]
exten => _X.,1,NoOp(Call from ${CALLERID(num)} for ${EXTEN})
exten => _X.,n,GotoIf($[foo${DB(CF/${EXTEN})} != foo]?normal:forward)
exten => _X.,n(normal),Dial(SIP/${EXTEN})
exten => _X.,n(forward),NoOp(Call for ${EXTEN} is being connected to
↳${DB(CF/${EXTEN})})
exten => _X.,n,Dial(local/${DB(CF/${EXTEN})})
```

11.5.2 Complex Call Forwarding

In this example, all the employees of Widgets, Inc. should be able to activate call forwarding. In addition, however, there must be global forwarding for the whole branch, so that calls can be forwarded to another branch when all the head office staff are away on a team-building exercise. This global forwarding is activated via extension 55 (followed by the destination number). The global forwarding must have a higher priority than the individual forwarding. One way to accomplish this is shown here:

```
[from-internal]
; Call forwarding for a single user
;
; activate
exten => _44X.,1,Answer()
exten => _44X.,n,Set(DB(CF/${CALLERID(num)})=${EXTEN:2})
exten => _44X.,n,SayDigits(${EXTEN:2})
exten => _44X.,n,NoOp(Forwarding for ${CALLERID(num)} to ${EXTEN:2}
↳activated.)
exten => _44X.,n,Hangup()

; deactivate
exten => 44,1,Answer()
exten => 44,n,DBdel(CF/${CALLERID(num)})
exten => 44,n,Playback(auth-thankyou)
exten => 44,n,NoOp(Forwarding for ${CALLERID(num)} deactivated.)
exten => 44,n,Hangup()

; Global forwarding
;
; activate
exten => _55X.,1,Answer()
exten => _55X.,n,Set(DB(CF/system)=${EXTEN:2})
exten => _55X.,n,SayDigits(${EXTEN:2})
exten => _55X.,n,NoOp(System forwarding to ${EXTEN:2} activated.)
exten => _55X.,n,Hangup()

; deactivate
exten => 55,1,Answer()
exten => 55,n,DBdel(CF/system)
exten => 55,n,Playback(auth-thankyou)
exten -> 55,n,NoOp(System forwarding deactivated.)
exten => 55,n,Hangup()
```

```
[from-external]
exten => _X.,1,NoOp(Call from ${CALLERID(num)} for ${EXTEN})
exten => _X.,n,GotoIf($[foo${DB(CF/system)} != foo]?cfsystem:secondrule)
exten => _X.,n(secondrule),GotoIf($[foo${DB(CF/${EXTEN})}
↳!=foo]?cfnormal:normal exten => _X.,n(normal),Dial(SIP/${EXTEN})
exten => _X.,n(cfnormal),NoOp(Call for ${EXTEN} is being connected to
↳${DB(CF/${exten => _X.,n,Dial(local/${DB(CF/${EXTEN})})})
exten => _X.,n(cfsystem),NoOp(Call for ${EXTEN} is being connected to

↳${DB(CF/s exten => _X.,n,Dial(local/${DB(CF/system)})
```

11.6 Application Example: Calling Card

According to Widgets, Inc. policy, personal calls must be made via a virtual calling
card. These cards are charged by the building manager via his phone (by calling
extension 88, followed by the three-digit user extension and then the desired cash
amount in cents).[1]

Such calls are billed at a per-call rate of 1 cent per call. To make a personal call,
the user dials 99, followed by the number being called. Of course, we don't want
anything to be deducted from the account in the event of a busy signal. Also, users
may call up the account balance by dialing 98:

```
[from-building-mgr]
; Charging the virtual calling card
;
exten => _88XXX.,1,Answer()
exten => _88XXX.,2,Set(DB(CallingCard/${EXTEN:2:3})=${EXTEN:5})
exten => _88XXX.,3,SayNumber(${EXTEN:5})
exten => _88XXX.,4,NoOp(CallingCard for ${EXTEN:2:3} has a new balance of
↳${EXTEN:5} cents.)
exten => _88XXX.,5,Hangup()


[from-internal]
; personal calls
;
exten => _99.,1,GotoIf($[${DB(CallingCard/${CALLERID(num)} > 0]?2:200)
exten => _99.,2,Set(DB(CallingCard/${CALLERID(num)})=$[DB(CallingCard/
↳${CALLERID(num)}) - 1])
exten => _99.,3,Dial(local/${EXTEN:2})
exten => _99.,104,Set(DB(CallingCard/${CALLERID(num)})=$[DB(CallingCard/
↳${CALLERID(num)}) + 1])
exten => _99.,200,NoOp(Calling card account ${CALLERID(num) is empty.)
```

---

1. To keep this example as simple as possible, we don't consider the case of a card with an
existing balance to which more money is added (in other words, in which an addition operation is performed). We are also ignoring the very real possibility that an employee could make
personal calls from someone else's extension.

```
exten => _99.,201,Answer()
exten => _99.,202,SayNumber(0)
exten => _99.,203,Hangup()

; The current account balance may be retrieved by dialing 98.
;
exten => 98,1,Answer()
exten => 98,2,SayNumber(${DB(CallingCard/${CALLERID(num)})})
exten => 98,3,Hangup()
```

Queues

A call queue lets you place callers on hold automatically if you don't have some-one available to answer the phone right away. (Most of us call this "being on hold.") Many customers detest queues. Sadly, there is scarcely a business around these days that does not make use of them; perhaps they are better than the alternative, which is to not take the call at all. Whatever your opinion about them, they are part of communications reality, so we will describe the configuration and operation of queues in Asterisk here.[1]

Four files are important for queues:

- queues.conf

  Defines the queues.

- agents.conf

  Defines agents; these are the staff who take calls.

- musiconhold.conf

  Defines the hold music.

- extensions.conf

  The dialplan. Calls are directed to the queue using **Queue()**, and agents are added using **AgentLogin()** or **AgentCallbackLogin()**.

---

1. An example queue setup using AEL (Asterisk Extension Language) is described in doc/ queues-with-callback-members.txt in Asterisk 1.4.

A common cause of misunderstanding is the confusion of "queue members" with callers. Queue members are always and only agents or queue devices, never callers.

Tip: For simplicity's sake, we use the term *agents* primarily. You call queue members agents. For someone who comes from the call center business, that's a bit confusing, because queue members can also be normal SIP extensions.

You are not required to use agents; instead, you can simply assign specific SIP devices to the queue. You do this by writing `member => SIP/1001` to `queues.conf`, for example; alternatively, you can assign the interface dynamically in the dialplan with the command **AddQueueMember()**. See Appendix B, "Dialplan Applications."

Agents may belong to more than one queue, and we want our agents to be able to log in at any station (also called hot-desking).

You configure agents in the `agents.conf` file:

```
[agents]
;   agent_number,password,name
agent => 1001,1234,Ron Popeil
agent => 1002,1234,Don Dinglehopper
```

Configured agents are added to specific queues in the `queues.conf` file:

```
[support-queue]
member => Agent/1001    ; Add agent 1001 to the support-queue
member => Agent/1002    ;  ... 1002 ...
```

The queue itself is invoked through an extension line in `extensions.conf`:

```
exten => 20,1,Queue(support-queue)   ; => queue
exten => 25,1,AgentLogin()           ; login after call
```

This is how it works: Agents log in by dialing extension 25, hear pleasant music, and await callers. Calls to extension 20 are passed into the queue and are answered on a first-in, first-out basis. The agent hears a tone, and the first caller is connected.

12.1 Extension for Testing Hold Music

We add the following extension to `extensions.conf` for testing to see that the music-on-hold is functioning as we expect:

```
exten => 222,1,Answer()
; answer
exten => 222,n,Set(MUSICCLASS()=default)
```

```
; set music-on-hold class to "default"
exten => 222,n,WaitMusicOnHold(20)
; play music-on-hold for 20 seconds
exten => 222,n,Hangup()
; hang-up
```

12.2 musiconhold.conf

Let's begin with the simplest part: setting up music-on-hold.

Keeping Your Hold Music Legal

Note that the music you use for music-on-hold may be subject to licensing fees. If you want to be sure you are using music lawfully, you can obtain use licenses from the performer's rights organization in your country.

In the United States, contact the following:

- The American Society of Composers, Authors, and Publishers (ASCAP), www.ascap.com
- Broadcast Music, Inc. (BMI), www.bmi.com
- SESAC, www.sesac.com

In Canada, contact the following:

- The Society of Composers, Authors, and Music Publishers of Canada (SOCAN) http://www.socan.ca

In the United Kingdom, contact the following:

- PRS for Music, www.prsformusic.com

In Australia and New Zealand, contact the following:

- The Australasian Performing Right Association and Australasian Mechanical Copyright Owners Society, www.apra-amcos.com.au/
- The Australasian Performing Right Association (NZ), www.apra.co.nz

Generally, license fees for use of music in waiting areas and on telephone systems are quite reasonable, so this option is worth considering. Alternatively, you can obtain royalty-free music from a variety of Internet sources. Finally, you may use the three tracks provided with Asterisk.

Allowed configuration settings are described in the sample `musiconhold.conf`. Additional information on MP3 music support through **mpg123** may be found in `doc/README.mp3` (1.2) / `doc/mp3.txt` (1.4). The optimum sample rate for MP3 files is 8000Hz, mono.

Here is the simplest possible configuration, using only one music class (`default`):

```
[default]
mode=quietmp3
directory=/var/lib/asterisk/mohmp3
```

Now we stop and start Asterisk, for example, using **asterisk -rx "restart now"**, to activate the new extension and the audio streams. (A reload is not sufficient.) Now, if you call extension 222 (as defined above), you should hear music-on-hold.

Still, the following is a better configuration, for which we create another music directory `moh-native` in `/var/lib/asterisk`:

```
[default]
mode=files   ; Read music files in an Asterisk native format
directory=/var/lib/asterisk/moh-native
random=yes   ; Play music files in random order
```

For this, it is necessary to convert the source MP3 files into an Asterisk compatible format, which can be accomplished using **sox** and **lame**.

sox (http://sox.sourceforge.net) is available as a Debian package, and is installed as follows:

```
$ apt-get install sox
```

For Mac OS X, if you have installed Darwinports (http://darwinports.opendarwin .org or http://darwinports.com), there is also a package:

```
$ port install sox
```

Many other distributions have **sox** packages; in the absence of packages, you must install it from source.

lame (http://lame.sourceforge.net) is not available as a Debian package, so we must download and install it using the familiar **./configure**, **make**, and **make install** commands.

For Mac OS X, there is a Darwinports package (see **sox**):

```
$ port install lame
```

Or via Fink (http://fink.sourceforge.net/):

```
$ fink install lame
```

Of course, you can do this with the FinkCommander GUI also.

After you've installed the necessary applications, you can create the music files this way:

```
$  lame --decode music.mp3 music.wav
$  sox -V music.wav -r 8000 -c 1 -w music.raw
$  sox -V music.wav -r 8000 -c 1 -w music.gsm
```

This converts our (dummy, only for demonstration) `music.mp3` file into WAV, and then into RAW and GSM formats (with a sample rate of 8kHz). We can also make versions of the file in a-law and u-law formats (by using the `.al` and `.ul` file extensions with **sox**); Asterisk will automatically choose the sound file requiring the least processing. The important common factor is that the sample rate always be 8000Hz.

Important: If you do not specify `autoload` in `modules.conf`, you must ensure that the modules for the formats you wish to use here are loaded before `res_musiconhold`.

For our test, we copied the WAV files provided in `moh/` to `moh-native/`, transcoded them to RAW and GSM formats using **sox**, and then deleted the WAV files (which caused us problems) from the directory:

```
$ cd /var/lib/asterisk/moh-native/
$ for i in *.wav; do \
sox $i -r 8000 -c 1 $(basename $i .wav).raw; \
sox $i -r 8000 -c 1 $(basename $i .wav).gsm; \
done
$ rm *.wav
```

For CPU-critical installations (e.g., large installations), you should take time to read `contrib/utils/README.rawplayer`, which may be found in the Asterisk source directory.

Additional music classes may be defined. For example, in addition to `[default]`, you might choose to classify by genre (e.g., `[rock]`) and provide a different file path (perhaps a subpath) where files of that genre are located. You might also classify by intended audience, particularly if you have custom files with announcements directed at specific types of customer. As always, don't forget to **reload**!

12.3 queues.conf

Queues are defined and configured in `queues.conf`. This file is—as you are already familiar—divided into sections.

Under `[general]`, we always set `persistentmembers=yes`, so that agents are re-added to their respective queues when Asterisk is started.

Every queue goes in its own section. In our example, we are configuring a support queue in its own section `[support]`. The following parameters are available to us:

- `musiconhold`
  Sets the music class of the queue. We choose the above-configured `default`:

  `musiconhold=default`.

- `announce`
  Defines an audio file that is played back to the agent before a call is answered. This is so that an agent working in more than one queue will know which queue the incoming call is coming from and be able to answer appropriately.

  `;announce=queue-support`

  If you use this option, remember to place a corresponding announcement audio file in the `sounds` directory (e.g., `queue-support.gsm`).

- `strategy`
  Sets the call distribution strategy (that is, the method used to decide how incoming calls are distributed to agents):

  - `ringall`
    Ring all agents until one answers (default).

  - `roundrobin`
    Ring agents in order, until one answers.

  - `leastrecent`
    Ring the agent who has been idle the longest.

  - `fewestcalls`
    Ring the agent who has taken the least number of calls in this queue.

  - `random`
    Ring a random agent.

  - `rrmemory`
    Round-robin with memory. Begins the round-robin starting with the agent who is next in line after the last call.

Note: Asterisk 1.6 is expected to do away with the current `roundrobin` method; `rrmemory` will be renamed `roundrobin`.

Agents with a lower penalty will always be preferred in all strategies.

```
strategy=ringall
```

The appropriate setting depends on the circumstances; `ringall` can be extremely annoying for agents, while other values can lead to longer wait times for callers, if agents fail to answer.

- `servicelevel`

  Sets the service level threshold (that is, sets the maximum wait time for callers). This is only really useful for statistical analysis (for example, in asking the question "How many calls were answered within the service level threshold of x seconds?").

  ```
  servicelevel=60
  ```

- `context`

  If set, this is the context we pass the caller to if she presses a single digit while waiting in the queue. This digit is treated as an extension and the call is taken out of the queue and routed to the extension in that context.

  ```
  context=supportqueue-context ; we could set it like this
  ; but we will leave it commented out:
  ;context=supportqueue-Context
  ```

- `timeout`

  Sets the maximum time a telephone may ring before we treat it as unattended.

  ```
  timeout=15
  ```

- `retry`

  Sets the interval (in seconds) before we ring all the agents again.

  ```
  retry=5
  ```

- `weight`

  The relative importance of the queue compared to other queues. If an agent is a member of multiple queues, calls from higher-weight queues are connected first. For example, you might give an emergency queue higher weight.

  ```
  weight=0
  ```

- `wrapuptime`

  The time interval after an agent has finished a call before the system will pass another call to the agent (default: 0).

  ```
  ; we allow our agent a quick pause for throat-clearing
  ; and a drink of water, then it's BACK TO WORK! *whip crack*
  wrapuptime=10
  ```

- `maxlen`

  Maximum number of callers allowed to be waiting in the queue (default: 0, meaning unlimited).

  ```
  maxlen=0
  ```

- `announce-frequency`

  Sets how often a caller will hear an announcement indicating her position in the queue or the estimated wait time. (0 turns this completely off.)

  ```
  announce-frequency=90
  ```

- `announce-holdtime`

  Sets whether the estimated wait time will be announced after the queue position. Possible values are `yes`, `no`, or `once`.

  ```
  announce-holdtime=yes
  ```

- `announce-round-seconds`

  Rounding level for wait-time announcements. If 0, only minutes, not seconds, are announced; other possible values are 0, 1, 5, 10, 15, 20, and 30.[2]
      For example, when set to 30, a wait time of 2:34 will be rounded to 2:30.

  ```
  announce-round-seconds=0
  ```

- `periodic-announce-frequency`

  Sets the interval, in seconds, for periodic announcements (`periodic-announce`; e.g., "We thank you for your patience; all our agents are currently serving other callers. Please continue to hold.").

  ```
  periodic-announce-frequency=60
  ```

- `monitor-format`

  When set, recording (as with the **Monitor()** application) is turned on and the recording format defined. (If not defined, recording is off.) You may specify `gsm`, `wav` (caution: huge files!), or `wav49`.

  ```
  ; if you want recordings, uncomment this line:
  ;monitor-format=gsm
  ```

---

2. Older versions accepted other values, which could result in confusing announcements. See Bug 9514.

By default, files are named (one `-in` and one `-out`) according to `${UNIQUEID}`. You can change this as needed by putting `Set (MONITOR_FILENAME= filename) before Queue()` in `extensions.conf`. We leave it unset in our example.

- `monitor-join`

  Combines the `...-in` and `...-out` files of a recording into one file. Allowed values: `yes` or `no`.

  ```
  monitor-join=yes
  ```

- `joinempty`

  Sets whether callers can be placed in a queue with no agents.

  - `yes`

    Callers may be placed in queues without active agents. (The queue may have no agents logged in, or agents may be logged in but unavailable.)

  - `no`

    Callers may not be placed in queues without active agents.

  - `strict`

    Callers may not be placed in a queue without agents, but *may* be placed in a queue with *unavailable* agents.

    Unavailable should not be confused with busy (that is, on a call). An agent is unavailable if he is a member of the queue but not currently registered on the system.

Warning: Statically defined queue members are always considered available. If you're not mindful, callers could end up in an active queue that is never answered (and that's probably bad for business).

If the caller is not accepted into a queue, **Queue()** ends and execution continues in the next priority in the dialplan.

```
joinempty=no ; we don't want our callers waiting in empty queues.
```

- `leavewhenempty`

  Sets whether waiting callers are removed from the queue when all the agents have logged out. Allowed values are the same as for `joinempty`. After leaving the queue, execution continues in the next priority in the dialplan.

  ```
  leavewhenempty=strict ; don't make 'em wait
  ```

- eventwhencalled

  Sets whether events are created in the Manager interface: `AgentCalled`, `AgentDump`, `AgentConnect`, `AgentComplete`. Possible values are yes and no.

  `eventwhencalled=yes`

- eventmemberstatus

  Sets whether QueueMemberStatus events are created in the Manager interface. (Use with caution; this can generate a lot of events.) Possible values are yes and no.

  `eventmemberstatus=no`

- reportholdtime

  Sets whether the wait time for the incoming caller will be announced to the agent before the call is patched. Some people like to know how long a caller has been waiting, because that caller might be irate. Possible values are yes and no.

  `reportholdtime=no`

- memberdelay

  Sets the interval of silence (in seconds) that the caller hears before she is connected with an agent.

  `memberdelay=1`

- timeoutrestart

  Sets whether the answer-timeout of an agent is reset after a BUSY or CONGESTION signal. This can be useful for agents who are allowed to refuse calls.

  `timeoutrestart=yes`

- autopause

  Sets whether an agent who has failed to answer is automatically paused or not.

  `autopause=no`

- ringinuse

  Sets whether agents who are in a call will be rung. So far, the only channel driver that provides this information is SIP.

  `ringinuse=no`

- member

  It is possible to statically define agents in `queues.conf` like so:

  `member => technology/resource[,penalty]`

An example using a Zap channel would look like this:

```
member => Zap/2
```

This may be used more than once. It should be employed with caution with `joinempty` and `leavewhenempty`, because these queue members are always treated as available, even if the telephone is unattended. The other disadvantage is that an agent is always bound to a specific workstation and cannot log in at another workstation.

In our example, we prefer to use the dynamic agents and add agents in the following format:

```
member => Agent/agentID
```

For two agents `1001` and `1002`, the configuration looks like this:

```
member => Agent/1001
member => Agent/1002
```

These agents must still be defined in `agents.conf` (see below). The agent number is a freely selectable number that uniquely identifies each agent. The number has nothing to do with the extensions that the agent may eventually use.

12.3.1 Announcement Sound Files

The following parameters define which sound files are used for which announcement types. Usually there is no need to change this. For example, *queue-youarenext= sound_filename* sets the filename for the "you are first in line" message to *sound_filename*.

If these parameters are not provided, the default values (shown below) apply:

- `queue-youarenext=queue-youarenext`
  "You are now first in line."
- `queue-thereare=queue-thereare`
  "There are…"
- `queue-callswaiting=queue-callswaiting`
  "Calls waiting."
- `queue-holdtime=queue-holdtime`
  "The current estimated hold time is…"
- `queue-minutes=queue-minutes`
  "Minutes."

- `queue-seconds=queue-seconds`
  "Seconds."
- `queue-thankyou=queue-thankyou`
  "Thank you for your patience."
- `queue-lessthan=queue-less-than`
  "Less than."
- `queue-reporthold=queue-reporthold`
  "Hold time."
- `periodic-announce=queue-periodic-announce`
  "All reps busy, wait for next available."

12.4 agents.conf

Agents (who are also called queue members) are defined in `agents.conf`; additional configurations for the agents are also placed here. Like the other configuration files, it too is divided into sections.

We set `persistentagents=yes` in the `[general]` section, so that agent logins persist (meaning they are stored in the Asterisk database) even if Asterisk is restarted.

We set additional parameters in the `[agents]` section, where we also define the agents. The following parameters are available:

- `autologoff`

  Sets the maximum time a telephone may ring before the agent is automatically logged off.

  Note that agents will not be logged out if the `autologoff` interval is longer than the queue timeout! We set that to 15 seconds, so we set `autologoff` to 14.

  `autologoff=14`

- `ackcall`

  Sets whether agents who have logged in with `AgentCallbackLogin()` must press number sign (#) to accept a call. Be aware that the voice prompt in previous Asterisk versions did not indicate that the agent has to press #. Possible values are `yes` and `no`.

  `ackcall=no`

- endcall

  Sets whether agents may end calls by pressing number sign (#). Possible values are yes and no.

  ```
  endcall=yes
  ```

- wrapuptime

  Sets (same as wrapuptime in queues.conf; why this appears in both files is not clear) the wait time, this time in *milliseconds,* before an agent can be sent another call. Default: 5000.

  ```
  wrapuptime=5000    ; 5 seconds to clear your throat
  ```

- musiconhold

  Sets the music-on-hold class for the agents.

  ```
  musiconhold=default
  ```

- updatecdr

  Adds agent information to the CDR so that we can see which agent answered the call. Possible values are yes and no.

  ```
  updatecdr=yes
  ```

- recordagentcalls

  Record agent calls. Possible values are yes and no. Default: no.

  ```
  recordagentcalls=no
  ```

- recordformat

  Format for call recordings. Possible values are gsm, wav (caution, huge files!), or wav49. Default: wav.

  ```
  recordformat=gsm
  ```

- createlink

  Sets whether the filename of the call recording will be written to the user field of the CDR, so that the call recording can be matched to the CDR later. Possible values are yes and no.

  ```
  createlink=yes
  ```

- urlprefix

  If createlink=yes, the filename can be prefixed with a URL where the recording is made available on a web server.

  ```
  ; an example might look like this:
  ;urlprefix=http://astbox/calls/
  ```

- savecallsin

  Defines the path where recordings are saved. Default: `/var/spool/asterisk/monitor`

  ```
  ; Say we wanted to save recordings in /var/calls/:
  ;savecallsin=/var/calls
  ```

- agent

  Defines individual agents. Can appear more than once.

  ```
  ; Format: agent => agent_id,password,name
  agent => 1001,0000,John Safran
  agent => 1002,0000,Rove McManus
  ```

12.5 Extensions

Both agent and caller access to the queue are configured in the dialplan. Agents log in as queue members via particular extensions, and callers are placed into the queue with the **Queue()** application (see Appendix B, "Dialplan Applications").

12.5.1 Sending Callers to the Queue

A minimal extension, which places callers in the queue, might look like this:

```
exten => 20,1,Answer()
; answer
exten => 20,n,Set(MUSICCLASS()=default)
; set music class "default"
exten => 20,n,Queue(support,t)
; place call in "support" queue
exten => 20,n,Hangup()
; hang-up
```

We prefer something more complete and use this extension:

```
exten => 20,n,Wait(2)
exten => 20,n,Answer()
; answer
exten => 20,n,Set(MUSICCLASS()=default)
; set music class default
exten => 20,n,Queue(support,t)
; place call in "support" queue
; a reminder: the option t allows the agent to
; transfer calls to another extension
; if call can't be placed in queue,
; Queue() sets QUEUESTATUS =
; TIMEOUT | FULL | JOINEMPTY | JOINUNAVAIL |
; LEAVEEMPTY | LEAVEUNAVAIL
```

```
; and exits:
exten => 20,n,Goto(q-${QUEUESTATUS},1)
; jump depending on QUEUESTATUS
; no agents in the queue:
; (you could also route calls to VoiceMail() here)
exten => q-JOINEMPTY,1,Wait(1)
exten => q-JOINEMPTY,n,Playback(vm-nobodyavail,noanswer)
exten => q-JOINEMPTY,n,Playback(vm-goodbye,noanswer)
exten => q-JOINEMPTY,n,Hangup()
; no agents in the queue (or only unavailable agents):
exten => q-JOINUNAVAIL,1,Goto(q-JOINEMPTY,1)
; handle same as JOINEMPTY
; all agents have logged out:
exten => q-LEAVEEMPTY,1,Goto(q-JOINEMPTY,1)
; handle same as JOINEMPTY
; all agents (including unavailable agents) have logged out:
exten => q-LEAVEUNAVAIL,1,Goto(q-JOINEMPTY,1)
; handle same as JOINEMPTY
; no agent is answering:
exten => q-TIMEOUT,1,Goto(q-JOINEMPTY,1)
; handle same as JOINEMPTY
; The number of callers in the queue has hit the maximum:
; (you could also route calls to VoiceMail() here)
exten => q-FULL,1,Busy(5)
exten => q-FULL,n,Hangup()
```

If we call this extension now, we hear the "nobody is available" message because no agents are logged in.

Tip: If you want to pretend you have active agents for testing purposes, set `joinempty=yes` and `leaveempty=no` in `queues.conf`.

Now we need to create an extension through which agents can log in:

```
exten => 25,1,Answer()      ; answer
exten => 25,n,AgentLogin()  ; log the agent in
exten => 25,n,Hangup()      ; hang-up
```

12.5.2 AgentLogin() and AgentCallbackLogin()

`AgentLogin()` establishes an active channel, and incoming calls are simply switched into this active channel. This is analogous to having the phone off the hook and calls simply coming in without the agent actually having to pick up. If the agent hangs up, she is logged out. This also means that an agent cannot be logged in and make outgoing calls at the same time. As a result, `AgentLogin()` is only really appropriate for purely inbound call center groups.

AgentCallbackLogin() has the advantage of letting agents log in, and then having the system call back with calls from the queue. They remain logged in after hanging up and can make outgoing calls. This application is, however, deprecated as of Version 1.4. The example in doc/queues-with-callback-members.txt uses AEL (which can easily be converted into regular dialplan format) to identify agents, **AddQueueMember()** to dynamically add members to the queue and Dial() to call agents back.

If **AgentCallbackLogin()** were to disappear without a replacement, real functionality and ease of configuration would be lost.

Tip: Try out what you've learned. Write a dialplan which brings callers to an IVR menu first; for example, "For sales, press 1; for returns, press 2." And then pass callers to two separate queues, one for each department, depending on the input the caller gives.

12.6 Log File

Events that happen in queues are written in a detailed format to the queue log (usually /var/log/asterisk/queue_log) in addition to the regular call detail recording. Events are recorded in queue_log on a line-by-line basis in the following format:

> *timestamp*|*call_id*|*queue*|*channel*|*event*|*parameter1*[|*parameter2*[|*parameter3*]]

- *timestamp*

  The UNIX time of the event.

- *call_id*

  The unique ID of the call (alphanumeric). Can be NULL or NONE (e.g., in the case of QUEUESTART events).

- *queue*

  The name of the queue (e.g., support). Can be NULL or NONE.

- *channel*

  The name of the bridged channel (e.g., Agent/1001). Can be NULL or NONE.

- *event*

  The name (type) of the event (see next page). The parameters that follow depend on the event type.

Possible events are, among others (see also `doc/queuelog.txt`):

- ABANDON

  The caller abandoned the queue by hanging up. Parameters: position, entry position, wait time.

- AGENTDUMP

  The agent dumped the caller while listening to the queue announcement.

- AGENTLOGIN

  An agent logged in. Parameters: channel (e.g., `SIP/127.0.0.1-0181ac00`).

- AGENTCALLBACKLOGIN

  A callback agent logged in. Parameters: login_extension[@context].

- AGENTLOGOFF

  An agent logged out. Parameters: channel, login time.

- AGENTCALLBACKLOGOFF

  An agent was logged out. Parameters: login_extension[@context], login time, cause (e.g., autologoff).

- COMPLETEAGENT

  Call between caller and agent was ended by agent. Parameters: wait time, call time, queue entry position.

- COMPLETECALLER

  Call between caller and agent was ended by caller. Parameters: wait time, call time, queue entry position.

- CONFIGRELOAD

  Configuration was reloaded (e.g., with **`asterisk -rx "reload"`**).

- CONNECT

  Caller was connected to an agent. Parameters: Wait time.

- ENTERQUEUE

  Caller was placed in the queue. Parameters: URL (if available), caller ID.

- EXITEMPTY

  Caller was exited from the queue because no members were available. Parameters: queue position, entry position, wait time.

- EXITWITHKEY

  Caller left the key by pressing a key. Parameters: key, queue position.

- EXITWITHTIMEOUT

  Caller was in the queue too long and the timeout expired. Parameters: queue position.

- QUEUESTART

  The queue was started. The `call_id`, `queue`, and channel fields are set NULL.

- RINGNOANSWER

  An available agent was called, but did not answer within the timeout. Parameters: ring time (in milliseconds).

- SYSCOMPAT

  The agent accepted a call, but the channels were not compatible and the call was ended.

- TRANSFER

  Caller was transferred to another extension. Parameters: extension, context.

Some commercial queue log analysis tools include QueueMetrics (http://queuemetrics.loway.it) and Easy PABX (http://www.easypabx.com). See also Appendix B.

12.6.1 Importing the Queue Log into MySQL

As yet, Asterisk cannot write the queue log directly into a SQL database. Unfortunately, all the widely used statistical tools assume and expect data to be in a SQL database. A variety of scripts seek to mitigate this problem (some may be found at www.voip-info.org/wiki/view/Asterisk+queue_log+on+MySQL) or are provided with the analysis tools. They all follow the same principle: the queue_log is replaced with a named pipe (FIFO); as soon as Asterisk attempts to write to the log, the log entry is converted and entered into the database.

Warning: This method works in 99% of cases. There remains the risk, however, that the script fails to start before Asterisk or terminates unexpectedly and stops reading from the named pipe, such that Asterisk receives a SIGPIPE when trying to write to the logs and crashes!

Because there is no alternative without native Asterisk SQL support, we provide a sample Perl script from William Lloyd despite the associated risk:[3]

```perl
#!/usr/bin/perl -w
#
# wlloyd at slap.net
# The asterisk version independant way to get queue stats into Mysql,
Postgres
# or whatever is supported by Perl DBI
# It's all about named pipes
# to setup this software
# stop asterisk
# rm /var/log/asterisk/queue_log
# mkfifo /var/log/asterisk/queue_log
# make sure permissions are setup
# chmod 777 /var/log/asterisk/queue_log
# run this program as root or under another user as you see fit.
# should start BEFORE asterisk. Add to /etc/rc.d/rc.local or whatever
# restart asterisk
# requires a DB table like the following..
# CREATE TABLE csr_queue (
# qname varchar(30) default NULL,
# agent varchar(30) default NULL,
# action text,
# info1 text,
# info2 text,
# info3 text,
# timestamp int(11) NOT NULL default '0',
# id tinytext NOT NULL
#) TYPE=MyISAM;
use DBI;
use IO::File;
my $opt_debug = 0;
# if you want postgres change this to "Pg"
my $db_type = "mysql";
my $db_host = "127.0.0.1";
my $db_user_name = 'username';
my $db_password = 'password';
my $db_database = 'asteriskstat';
my $dbh = DBI->connect("DBI:$db_type:dbname=$db_database;host=
$db_host;", $db_user_name, $db_password);
open(FIFO, "< /var/log/asterisk/queue_log")     or die "Can't open
queue_log : $!\n";
while (1) {
    $message = <FIFO>;
    next unless defined $message;  # interrupted or nothing logged
```

---

3. wlloyd at slap.net, released through the Digium `asterisk-users` mailing list. See http://lists.digium.com/pipermail/asterisk-users/2005-July/109892.html.

```perl
    chomp $message;
    # remove chars that will cause DB problems
    $message =~ s/\"\'//g;
    @data = split(/\|/,$message);
    # these messages are almost useless for my purposes
    next if ($data[4] eq "QUEUESTART" );
    next if ($data[4] eq "CONFIGRELOAD" );
    if (!defined($data[5])) {
      $data[5] = '';
    }
    if (!defined($data[6])) {
      $data[6] = '';
    }
    if (!defined($data[7])) {
      $data[7] = '';
    }
    my $sql = "INSERT INTO csr_queue (timestamp, id, qname, agent,
action, info1, info2, info3) VALUES ('$data[0]', '$data[1]', '$data
[2]', '$data[3]', '$data[4]', '$data[5]', '$data[6]', '$data[7]')";
    print "$sql \n\n" if ($opt_debug);
    $dbh->do($sql);
# if you want an actual logfile you might want to uncomment this
#     if ( open(LOG, ">> /var/log/asterisk/queue_log_real") ) {
#         print LOG "$message\n";
#         close(LOG);
#     } else {
#         warn "Couldn't log to /var/log/asterisk_queue_log: $!\n";
#     }
#
}
$dbh->disconnect();
exit 0;
```

QueueMetrics (the free demo version) comes packaged with a Perl script called queueLoader.pl that will also load the queue log into mySQL. (Again, heed the warning above: If you do this in real time, there is a risk that Asterisk will crash.)

13

Features

Asterisk is powerful, but some features routinely used in office environments are either not enabled by default or simply very well hidden. This chapter covers those features, which are configured in the `features.conf` file.

13.1 Call Transfer

If you use the **Dial()** application without any flags or options, you won't be able to transfer a current call to another extension. This feature must be explicitly enabled in **Dial()** via the t, T, or tT options. This is described in detail in Appendix B, "Dialplan Applications." If you want to give both the caller and called party rights to transfer the call, you must use the tT options:

```
exten => _2XXX,1,Dial(SIP/${EXTEN},tT)
```

13.1.1 Blind Transfer

The expression *blind transfer* describes transfers that occur without an interaction between the transferer (the original recipient of the call) and the transferee (the new intended recipient). By default, this is done by entering #1 followed by the new destination number. If you want to transfer the current call to extension 2323, you dial #12323.

You can change the feature code for blind transfers in /etc/asterisk/features.conf in the context [featuremap]:

```
[featuremap]
blindxfer => #1        ; Blind transfer
```

13.1.2 Attended Transfer

In contrast to the blind transfer, the original recipient can discuss the call in private with the intended recipient using attended transfer; the call is only transferred once the original recipient hangs up. By default, attended transfers are initiated with the *2 key combination, followed by the new destination number.

You can change the feature code for attended transfers in /etc/asterisk/features.conf in the context [featuremap]:

```
[featuremap]
atxfer => *21        ; Attended transfer
```

If you are using call limits in your devices, be aware that attended transfers temporarily use two channels.

13.2 Call Parking

To park a call is to place an active call in a virtual parking lot where it can be retrieved from any other extension by dialing the parking position number. For call parking to work in Asterisk, the parkedcalls context must be included (using include) in the current context, as follows:

```
[building-mgr]
exten => _2XXX,1,Dial(SIP/${EXTEN})
include => parkedcalls
```

Only once this context has been included can calls be parked by dialing #700 during an active call. Asterisk will read back the parking position to the caller. (Be sure to note it!) Calling this park position number (e.g.,701) retrieves the call from any telephone in a context that includes the parkedcalls context.

Warning: Parking positions are not secured in any way. If person A parks a call at position 705 and person B dials 705 by accident, she will connect to the parked call.

Feature codes, the number of available parking positions, and the name of the parking context can all be changed in /etc/asterisk/features.conf in the [general] context:

```
[general]
parkext => 7001          ; parking extension
parkpos => 701-7201      ; parking positions
context => parkedcalls   ; context for parked calls
```

14

Conferencing

Asterisk offers an excellent built-in conferencing system. Nevertheless, the same applies here as elsewhere when it comes to Asterisk: There is no perfect solution! You still need to build some functionality in a modular fashion. A good example is three-way calling. Asterisk does not include true three-way calling. To achieve it, you must take a regular conference and limit the number of participants to three.

14.1 Installation and the Timing Device

A default Asterisk installation such as the one we performed in the first chapter of this book isn't able to handle conferencing. The problem is the lack of a timing device. To mix channel audio properly, Asterisk needs a stable and common clock source. For that clock source, there are two options:

- **Telephony interface card and driver**

  Installing a hardware interface card (be it analog or digital ISDN) and its driver automatically provides Asterisk with a clock source.

- **ztdummy module without hardware**

  If you neither have nor need a hardware interface card, you can use the ztdummy kernel module in combination with a 2.6 kernel.[1]

---

1. It is possible to make the ztdummy module work with a 2.4 kernel, but this assumes special USB hardware and is therefore not the best option.

To determine whether your Asterisk installation actually provides conferencing service, go into the Asterisk CLI and enter **show application MeetMe**. If the help text appears, conferencing is installed.

14.2 Defining Conference Rooms

In Asterisk, conferences always take place in virtual conference rooms. These must be defined in /etc/asterisk/meetme.conf, in the [rooms] context, and the syntax is simple and consistent with other Asterisk configuration files: conf => room_number[,PIN][,administrator-PIN].

To create a conference room with the number 1234 that is freely available to any caller, make an entry in meetme.conf as follows:

```
[rooms]
conf => 1234
```

You can secure a conference room with an access number (sometimes called a personal identification number, or PIN). Suppose you want to add rooms 700 to 705, all with the same PIN 5678. The meetme.conf would look like this:

```
[rooms]
conf => 1234
conf => 700,5678
conf => 701,5678
conf => 702,5678
conf => 703,5678
conf => 704,5678
conf => 705,5678
```

14.3 Access from the Dialplan

Conferences are accessed easily using the **MeetMe()** application. Say that any caller dialing 5555 is to be placed into the conference room 1234. The corresponding dialplan entries would look like this:

```
[intern]
exten => 5555,1,Answer()
exten => 5555,n,Wait(1)
exten => 5555,n,MeetMe(1234)
exten => 5555,n,Hangup()
```

14.4 Announcing the Number of Conference Participants

The number of participants in the conference may be announced using the **Meet-MeCount()** application:

```
[internal]
exten => 5555,1,Answer()
exten => 5555,n,Wait(1)
exten => 5555,n,Playback(conf-thereare)
exten => 5555,n,MeetMeCount(1234)
exten => 5555,n,Playback(conf-peopleinconf)
exten => 5555,n,MeetMe(1234)
exten => 5555,n,Hangup()
```

15

External Control of Asterisk

One of Asterisk's biggest advantages is the ability it gives the administrator to control it from the shell or through external applications.

15.1 asterisk -rx "command"

The simplest way to control Asterisk from an external shell or application is to issue the command **asterisk** with the option -rx followed by the CLI command. Any CLI command may be entered from the system shell in this fashion.

Say you want to see the dialplan for extension 23 in the context [my-phones]; you would do this with **asterisk -rx "dialplan show 23@my-phones**" entered in the shell:

```
root@molokai:~>asterisk -rx "dialplan show 23@my-phones"
[ Context 'my-phones' created by 'pbx_config' ]
  '23' =>              1. Wait(1)                           [pbx_config]
                       2. Answer()                          [pbx_config]
                       3. Playback(hello-world)             [pbx_config]
                       4. Wait(1)                           [pbx_config]
                       5. Hangup()                          [pbx_config]

-= 1 extension (5 priorities) in 1 context. =-
root@molokai:~>
```

15.2 Call Files

Call files are like a shell script for Asterisk. A user or application writes a call file into `/var/spool/asterisk/outgoing`, where Asterisk processes it immediately.

Tip: A **mv** (move) is an atomic operation (an operation which does not take effect until it is 100% complete) and as such is ideally suited for `.call` files (provided it is done on the same partition). With **cp** (copy), the file is copied line by line, which could lead to Asterisk processing an incomplete file.

15.2.1 Making a Call Automatically

Let's demonstrate the `.call` file principle with an example. Assume that we have a SIP phone registered with the number 2000 in Asterisk. In addition, we have the following extension in the dialplan:

```
[call-file-test]
exten => 10,1,Answer()
exten => 10,n,Wait(1)
exten => 10,n,Playback(hello-world)
exten => 10,n,Wait(1)
exten => 10,n,Hangup()
```

We create a call file called `a-test.call` in `/tmp` with the following content:

```
Channel: SIP/2000
MaxRetries: 2
RetryTime: 60
WaitTime: 30
Context: call-file-test
Extension: 10
```

Now we move this file with **mv /tmp/a-test.call /var/spool/asterisk/outgoing/**:

```
root@molokai:~>mv /tmp/a-test.call /var/spool/asterisk/outgoing/
```

The following happens:

- Asterisk polls the `/var/spool/asterisk/outgoing` for new call files and processes any it finds.
- Asterisk opens a connection to device `SIP/2000`. If the device is in use or not answered, Asterisk tries two more times (see `MaxRetries`).
- If someone answers `SIP/2000`, Asterisk begins processing extension `10` in the context `[call-file-test]`. In this case, Asterisk plays `hello-world` to the answering party.

When executing a call file, Asterisk compares the change time with the current time. If the change time is in the future, Asterisk ignores the call file. This is an easy way to implement time-based call files.

15.2.2 Parameters

These parameters may be used in call files:

- `Channel: <channel>`
  The channel upon which to initiate the call. Uses the same syntax as the **`Dial()`** command (see Appendix B, "Dialplan Applications").

- `Callerid: <callerid>`
  The caller ID to be used for the call.

- `WaitTime: <number>`
  Number of seconds the system waits for the call to be answered. If not specified, defaults to 45 seconds.

- `MaxRetries: <number>`
  Maximum number of dial retries (if an attempt fails because the device is busy or not reachable). If not specified, defaults to 0 (only one attempt is made).

- `RetryTime: <number>`
  Number of seconds to wait until the next dial attempt. If not specified, defaults to 300 seconds.

- `Account: <account>`
  The account code for the CDR.

- `Context: <context>`
  The destination context.

- `Extension: <exten>`
  The destination extension, in which dialplan execution begins if the device is answered.

- `Priority: <priority>`
  The destination priority. If not specified, defaults to 1.

- `Setvar: <var=value>`
  Lets you set one or more channel variables.

- `Archive: <yes|no>`
  By default, call files are deleted immediately upon execution. If `Archive: yes` is set, they are copied into `/var/spool/asterisk/outgoing_done` instead. Asterisk adds a line to the call file that describes the result:
  ```
  Status: <Expired|Completed|Failed>
  ```

15.2.3 Hotel Wake-Up Call Example

A hotel wants to implement a simple wake-up call system. Clients must be able to set a wake-up call by dialing *77*, whereupon they hear a prompt asking for the date and time of the wake-up call:

```
[hotel-internal]
exten => _*77*XXXXXXXXXXXX,1,Answer()
exten => _*77*XXXXXXXXXXXX,n,Set(year=${EXTEN:4:4})
exten => _*77*XXXXXXXXXXXX,n,Set(month=${EXTEN:8:2})
exten => _*77*XXXXXXXXXXXX,n,Set(day=${EXTEN:10:2})
exten => _*77*XXXXXXXXXXXX,n,Set(hours=${EXTEN:12:2})
exten => _*77*XXXXXXXXXXXX,n,Set(minutes=${EXTEN:14:2})
exten => _*77*XXXXXXXXXXXX,n,NoOp(Wake-up call scheduled for
${CALLERID(num)} at ${hours}:${minutes} on ${day}.${month}.${year}.)
exten => _*77*XXXXXXXXXXXX,n,System(echo -e "Channel:
SIP/${CALLERID(num)}\\nContext: wake-up\\nExtension: 23" >
/tmp/${UNIQUEID}.call)
exten => _*77*XXXXXXXXXXXX,n,System(touch -t
${year}${month}${day}${hours}${minutes} /tmp/${UNIQUEID}.call)
exten => _*77*XXXXXXXXXXXX,n,System(mv /tmp/${UNIQUEID}.call
/var/spool/asterisk/outgoing/)
exten => _*77*XXXXXXXXXXXX,n,Playback(rqsted-wakeup-for)
exten => _*77*XXXXXXXXXXXX,n,SayNumber(${hours})
exten => _*77*XXXXXXXXXXXX,n,SayNumber(${minutes})
exten => _*77*XXXXXXXXXXXX,n,Hangup()

[wake-up]
exten => 23,1,Answer()
exten => 23,n,Wait(1)
exten => 23,n,Playback(this-is-yr-wakeup-call)
exten => 23,n,Wait(1)
exten => 23,n,Hangup()
```

15.3 The Asterisk Manager Interface

The Asterisk Manager Interface (AMI) is a powerful tool that lets external applications communicate with and control Asterisk.

Activate the AMI by setting `enabled=yes` in the `[general]` section in `manager.conf`

Caution: Never do this on a publicly accessible server unless you have taken steps to protect it with packet filters such as **iptables**, **ipfw**, an external firewall, or an SSH tunnel. A malicious person could easily use it to take over your phone system.

15.3.1 Creating an AMI User

To access AMI, we need to create a user account. To do this, we add a user entry called `admin` at the end of `manager.conf`:

```
[admin]
secret = secret5
deny = 0.0.0.0/0.0.0.0
permit = 127.0.0.1/255.255.255.255
read = all,system,call,log,verbose,command,agent,user,config
write = all,system,call,log,verbose,command,agent,user,config
```

The options following `read` and `write` define the allowed command types for this user.[1]

Caution: This generous rights assignment is for test purposes only! The `command` rights level means the user can stop Asterisk. As of Asterisk 1.4, it is even possible to make dialplan changes through the AMI, which also means it is possible to run shell commands with root privileges using **System()**.

15.3.2 Talking to the AMI

After restarting Asterisk, we can connect to the AMI on port 5038 from the system shell using **telnet**:[2]

```
$ telnet 127.0.0.1 5038
Trying 127.0.0.1...
Connected to localhost.
Escape character is '^]'.
Asterisk Call Manager/1.0
```

Now you can enter commands, usually consisting of multiple lines, by hand. For example:

```
Action: Login
ActionID: 1
Username: admin
Secret: secret5
```

Note: All command packets are closed with two carriage returns.

---

1. Learn the rights levels needed for commands by entering **manager show commands** (or **show manager commands** in Asterisk 1.2) in the CLI.

2. Here we use **telnet** as an interface only, and not in the traditional, interactive fashion.

After entering the commands and closing them, you'll receive a response:

```
Response: Success
ActionID: 1
Message: Authentication accepted
```

Of course, we are most interested in automating this interaction with scripts.

Note: The Manager API is not exactly famous for its ability to handle multiple simultaneous connections gracefully (even though this has improved immensely in version 1.4). If you anticipate this kind of load, it is worth considering an AMI proxy, which can handle many connections and bundles them in a single connection. This is completely transparent to the script accessing the AMI. Of course, for the purposes of playing around, it isn't strictly necessary.

Following a successful authentication, packets can be sent in both directions. The packet type is always determined by the first line. The client sends `Action` packets, the server answers with `Response` or can send `Event` packets. Otherwise the order of the lines in a packet is irrelevant. Lines are terminated with a carriage return (decimal ASCII 13) and line feed (decimal ASCII 10) (CR LF) combination. The entire packet is terminated with an additional CR LF combination. An AMI client normally sends a randomized but unique `ActionID` with every `Action`,[3] which the server uses in its response for the purpose of managing overlapping packet streams.

The server sends the client `Event` packets, which can refer to any events; there are also events that occur as the result of a client-initiated `Action`. In this case, the server sends `Response: Follows` followed by the events (which will contain the `ActionID` of the initiating action) and a closing event (usually `actionnameComplete`).

If your client has no need for events, it can turn off these notifications by including `Events: off` in the authentication packet. Once set, the AMI sends only responses to actions initiated by the client.

The events that Asterisk sends are, as of this writing, effectively undocumented. You can find a list with sparse details at www.voip-info.org/wiki/view/asterisk+manager+events. A few additional explanations may be found at http://asterisk-java.sourceforge.net/apidocs/net/sf/asterisk/manager/event/package-frame.html.[4]

---

3. This can be, for example, the name of the script, a timestamp, and a sequence number (e.g., `testscript.php-1169405408-1`).

4 This is primarily Asterisk-Java documentation.

15.3.3 AMI Commands

The list of available commands can be called up in the CLI with **manager show commands** (or **show manager commands**), while information about a specific command can be obtained with **manager show command** *command* (or **show manager command** *command*):

```
mos-eisley*CLI> show manager commands
Action            Privilege           Synopsis
------            ---------           --------
AbsoluteTimeout   call,all            Set Absolute Timeout
AgentCallbackLo   agent,all           Sets an agent as logged in by
                                          callback
AgentLogoff       agnet,all           Sets an agent as no longer
                                          logged in
Agents            agent,all           Lists agents and their status
ChangeMonitor     call,all            Change monitoring filename of
                                          a channel
Command           command,all         Execute Asterisk CLI Command
DBGet             system,all          Get DB Entry
DBPut             system,all          Put DB Entry
Events            <none>              Control Event Flow
ExtensionState    call,all            Check Extension Status
GetConfig         config,all          Retrieve configuration
Getvar            call,all            Gets a Channel Variable
Hangup            call,all            Hangup Channel
IAXnetstats       <none>              Show IAX Netstats
IAXpeers          <none>              List IAX Peers
ListCommands      <none>              List available manager
                                          commands
Logoff            <none>              Logoff Manager
MailboxCount      call,all            Check Mailbox Message Count
MailboxStatus     call,all            Check Mailbox
Monitor           call,all            Monitor a channel
Originate         call,all            Originate Call
Park              call,all            Park a channel
ParkedCalls       <none>              List parked calls
PauseMointor      call,all            Pause monitoring of a channel
Ping              <none>              Keepalive command
PlayDTMF          call,all            Play DTMF signal on a
                                          specific channel.
QueueAdd          agent,all           Add interface to queue.
QueuePause        agent,all           Makes a queue member
                                          temporarily unavailable
QueueRemove       agent, all          Remove interface from queue.
Queues            <none>              Queues
QueueStatus       <none>              Queue Status
Redirect          call,all            Redirect (transfer) a call
SetCDRUserField   call,all            Set the CDR UserField
Setvar            call,all            Set Channel Variable
SIPpeers          system,all          List SIP peers (text format)
SIPshowpeer       system,all          Show SIP peer (text format)
```

```
Status            call,all          Lists channel status
StopMonitor       call,all          Stop monitoring a channel
UnpauseMonitor    call,all          Unpause monitoring of a
                                        channel
UpdateConfig      config,all        Update basic configuration
UserEvent         user,all          Send an arbitrary event
WaitEvent         <none>            Wait for an event to occur
```

These commands are almost always a direct translation of dialplan applications, except in the case of Originate, used to initiate an outgoing call, and Command, which executes a command directly on the CLI. Because our test user admin has all the rights levels (see above), he can execute all commands. The following example shows how we learn how a command is used:

```
mos-eisley*CLI> manager show command command
Action: Command
Synopsis: Execute Asterisk CLI Command
Privilege: command,a
Description: Run a CLI command.
Variables: (Names marked with * are required)
        *Command: Asterisk CLI command to run
        ActionID: Optional Action id for message matching.
```

15.3.4 Getting the Number of Voicemail Messages with Expect

Suppose we want to get the number of messages in a given voice mailbox via the Manager interface. This is easily done using an **expect** script. **expect** is an extended Tcl interpreter used for automating interfaces with interactive shell programs. (See **expect**'s home page at http://expect.nist.gov.)

The following **expect** script connects to the AMI, logs in, then returns the number of new and old messages in the specified mailbox:

```
#!/usr/bin/expect
#
# Usage: ./vmcount.exp 1234@default

# The user account from manager.conf:
set username "admin"
set secret "secret5"
set host "127.0.0.1"
set port "5038"

if {[llength $argv] != 1} {
    send_user "Error: You must specify a mailbox!\n"
    exit 1
}

# First argument is the mailbox:
set mailbox [lindex $argv 0]
send_user "Mailbox: $mailbox\n"
```

```
# Mute output to stdout:
log_user 0

# Open connection to AMI:
spawn telnet $host $port

# Just in case telnet aborts because it cannot connect:
expect_before eof {
    send_user "Failed to connect.\n"
    exit 1
}

# Wait for the text "Manager"; once received, send a login packet:
#
expect "Manager" {
    send_user "Connected.\n"
    send "Action: Login\nUsername: $username\nSecret: $secret\n\n"
    # Please note that telnet automatically converts line feeds
    # (\n) to CR LF (\r\n) - so you must not write \r\n here.
}

# Login successful?:
#
expect {
    -re "Response:\\s*Error" {
        send_user "Login failed.\n"
        exit 1
    }
    -re "Response:\\s*Success" {
        send_user "Logged in.\n"
        # Query the number of messages in the mailbox:
        send "Action: MailboxCount\nMailbox: $mailbox\n\n"
    }
}

expect {
    -re "Response:\\s*Error" {
        send_user "Query of mailbox failed.\n"
        exit 1
    }
    -re "Response:\\s*Success" {}
}
expect {
    -re "NewMessages:\\s*(\[\\d]*)" {
        send_user "New messages: $expect_out(1,string)\n"
    }
}
expect {
    -re "OldMessages:\\s*(\[\\d]*)" {
        send_user "Old messages: $expect_out(1,string)\n"
    }
}

# Log out -- not strictly necessary, but cleaner:
send "Action: Logoff\n\n"
```

We save the script as `vmcount.exp` and set it executable with **chmod   a+x vmcount.exp**. We run it with the mailbox as an option and see how many messages we have:

```
$ ./vmcount.exp 123@default
Mailbox: 123@default
Connected.
Logged in.
New messages: 0
Old messages: 0
```

15.3.5 StarAstAPI for PHP

A disclaimer: Keep your expectations modest. StarAstAPI hasn't been actively maintained for some time.

There are now numerous, more-or-less good APIs for the AMI in a variety of programming languages (PHP, Perl, Python, Ruby, and so on), which we, because of space and time limitations, can't explore here. You can find examples with comments at www.voip-info.org/wiki/view/Asterisk+manager+Examples. If the API for your favorite language doesn't work, we're confident you can figure it out.

In this short example, we test the StarAstAPI (you can obtain this from http://freshmeat.net and other code repositories) in PHP, which assumes a PHP 5 that was compiled with `--enable-sockets`. You can check this from the shell with **php   -m**.

(The API is easily ported to PHP 4, although the code is cluttered and poorly formatted.) When in doubt, just remedy the parse errors.

Unfortunately, the StarAstAPI *files still* contain the obsolete "short open tags" (`<?`). If you encounter them, replace them with the correct syntax (`<?php`). Four demo scripts are included with the API: `sLogin.php` attempts a login, `sCommand.php` executes `reload` on the CLI, `sDial.php` tries a connection to SIP/120, and `sEvents.php` receives events. (If you have followed the examples from voip-info.org, you will need to adapt the user name and password.)

If we connect to Asterisk using **asterisk -vvvr** and simultaneously run **php -q sLogin.php** to open a connection to the AMI using a deliberately incorrect user and password, we will see output like this on the CLI:

```
mos-eisley*CLI>
 == Parsing '/etc/asterisk/manager.conf': Found
[Jan 26 20:08:09] NOTICE[10352]: manager.c:961 authenticate: 127.0.0.1
tried to authenticate with nonexistent user 'mark'
 == Connect attempt from '127.0.0.1' unable to authenticate
mos-eisley*CLI>
```

This failed because the user did not exist, but the demo script still reports success:

```
$ php -q sLogin.php
Login Sucessful
```

The response packet, however, will show the failure:

```
Response: Error
ActionID: 1
Message: Authentication failed
```

The StarAstAPI is, as you can see, not completely clean, but is simple enough that it can be improved easily. If we call **php -q sEvents.php** (this time with the correct user) we see the following:

```
mos-eisley*CLI>
    == Parsing '/etc/asterisk/manager.conf': Found
    == Manager 'admin' logged on from 127.0.0.1
mos-eisley*CLI>
```

As a test, we execute a `reload` in the CLI, which is reflected in the PHP script output:

```
Event: Reload
Privilege: system,all
Message: Reload Requested

Event: ChannelReload
Privilege: system,all
Channel: SIP
ReloadReason: RELOAD (Channel module reload)
Registry_Count: 0
Peer_Count: 0
User_Count: 0
```

Give your creativity free reign! Write a small script that calls all your friends (in the middle of the night, of course).

15.3.6 Getting the Number of Mailbox Messages with PHP

Let's retrieve the number of mailbox messages for an account using PHP and StarAstAPI. Here is the PHP script:

```
#!/usr/bin/php -q
<?php
# option -q turns off the header output when executing CGI-PHP
if ($argc != 2) {
    echo "Error: You must specify a mailbox!\n";
    exit(1);
```

```php
}
# The first argument after the program name is the mailbox:
$mailbox = $argv[1];
echo "Mailbox: $mailbox\n\n";

# Include StarAstAPI:
require_once './StarAstAPI/StarAstAPI.php';

# Connect and log in:
#
$ami = new AstClientConnection();
if ($ami->Login( 'admin', 'secret5', '127.0.0.1', 5038 )) {
    $rp = $ami->GetResponse('1');
    //echo $rp->ToString();
} else {
    exit(1);
}

# Send the following packet:
#    Action: MailboxCount
#    Mailbox: $mailbox
#    ActionID: 2
#
$data = new AstPacketData;
$data->AddKVPair( 'Action' , 'MailboxCount' );
$data->AddKVPair( 'Mailbox' , $mailbox );
$data->AddKVPair( 'ActionID', '2' );
$packet = new AstPacket;
$packet->SetAstPacketType( 'Action' );
$packet->SetAstPacketData( $data );
$ami->SendPacket( $packet );

# Read the response packet bearing ActionID 2:
#
$rPacket = $ami->GetResponse('2');
//echo $rp->ToString();
$rData = $rPacket->GetAstPacketData();
$r = $rData->GetAll();

echo "New messages: ", (int)trim($r['NewMessages:']), "\n";
echo "Old messages: ", (int)trim($r['OldMessages:']), "\n";
echo "\n";

# Log out -- not strictly necessary, but cleaner:
#
$ami->Logoff();
# Unfortunately, StarAstAPI isn't totally discreet.
# It does this:
#echo "Logoff called from somewhere ...";
#socket_close($this->mSocket);

echo "\n";
?>
```

We save this script as vmcount.php and make it executable with **chmod a+x vmcount.exp**, and then invoke it with the mailbox name as an option:

```
$ ./vmcount.php 123@default
Mailbox: 123123123

New messages: 0
Old messages: 0

Logoff called from somewhere ...
```

15.4 The Asynchronous JavaScript Asterisk Manager

As of version 1.4, Asterisk comes packaged with a small web server called Asynchronous JavaScript Asterisk Manager (AJAM), which may be used to access the Asterisk Manager Interface (AMI) via HTTP. The name AJAM derives from AJAX (Asynchronous JavaScript and XML).

Setup assumes the steps from section 15.3 have been carried out, plus some additional parameters. You must set webenabled to yes in the [general] section of manager.conf. Pay attention to httptimeout, which defines the inactivity timeout after which the user is automatically logged out of the web interface. To activate the web server, set these parameters in http.conf:

```
[general]
enabled=yes
enablestatic=yes
bindaddr=127.0.0.1
bindport=8088
prefix=asterisk
```

enablestatic need only be activated if the AJAM will be serving static files from /var/lib/asterisk/static-http. Normally you would set this to no, but it is needed for the purposes of the Asterisk-AJAM demo.

Don't forget to restart!

Warning: Our assessment is that it almost never makes sense to serve other web applications (that is, those intended strictly for administrator access) through the AJAM interface. It is also doubtful that it was intended to, because the rights assignments through read and write (see Appendix E, "AMI Commands") just don't offer sufficient granularity. Always assume that a user can initiate actions other than those you have made available on the web page. It is much better to let your application use a PHP script containing only the specific AMI commands it needs to do its job, and to restrict the AMI rights for the accessing user as extra insurance.

15.4.1 Getting the Number of Voicemail Messages with AJAM

Again, we are solving the problem addressed in the **expect** and PHP examples: We want to find out the number of messages in a specified mailbox. The AJAM offers us a few ways to do this.

15.4.1.1 HTML

The AMI waits for queries at

```
http://localhost:8088/asterisk/manager
```

Packet fields are tacked on the end of the URL. Try these addresses in your web browser:

```
http://localhost:8088/asterisk/manager?action=Login
↳&username=admin&secret=secret

http://localhost:8088/asterisk/manager?action=MailboxCount&mailbox=123
```

The response follows in the form of an HTML page, so it's not really suitable for access via a script.

15.4.1.2 Plain Text

If we replace manager in the URL with rawman, we get plain text output. To log in and get a message count from the mailbox, just enter the following:

```
http://localhost:8088/asterisk/rawman?action=Login
↳&username=admin&secret=secret5

Response: Success
Message: Authentication accepted

http://localhost:8088/asterisk/rawman?action=MailboxCount&mailbox=123

Response: Success
Message: Mailbox Message Count
Mailbox: 123
NewMessages: 0
OldMessages: 0

http://localhost:8088/asterisk/rawman?action=Logoff

Response: Goodbye
Message: Thanks for all the fish.
```

This text output is more script friendly.

15.4.1.3 XML

If we want XML instead, we call `mxml` instead. The XML output is presented formatted for better readability. In practice, AJAM does not put line breaks inside the XML tags. Either way, a compliant XML parser won't care.

```
http://localhost:8088/asterisk/mxml?action=Login
↳&username=admin&secret=secret5

<ajax-response>
    <response type='object' id='unknown'>
        <generic
            response='Success'
            message='Authentication accepted' />
    </response>
</ajax-response>

http://localhost:8088/asterisk/mxml?action=MailboxCount&mailbox=123

<ajax-response>
    <response type='object' id='unknown'>
        <generic
            response='Success'
            message='Mailbox Message Count'
            mailbox='123'
            newmessages='0'
            oldmessages='0' />
    </response>
</ajax-response>

http://localhost:8088/asterisk/mxml?action=Logoff

<ajax-response>
    <response type='object' id='unknown'>
        <generic
            response='Goodbye'
            message='Thanks for all the fish.' />
    </response>
</ajax-response>
```

15.4.2 AJAX and AJAM Considerations

AJAX applications (as the name Asynchronous JavaScript and XML might suggest) use XML as the standard format, even though it is often criticized for its bloated structure. There are alternatives, such as JSON, for example. JSON (JavaScript Object Notation) is (the name gives it away) well suited for JavaScript applications, because the data structure can be converted into an object natively and with little overhead using **eval()**. There are countless implementations for PHP, Perl, and so on, but a JSON implementation for AJAM does not yet exist. One can, however, convert the plain-text output into JSON on the client side, if that turns out to be easier or if it's easily done using available JavaScript libraries. Here's an example to get you thinking:

```
// We assume the received response and
// simulate it here:
var responseText = 'Response: Success\n'
+'Message: Mailbox Message Count\n'
+'Mailbox: 123\n'
+'NewMessages: 0\n'
+'OldMessages: 0\n';

// Escape single quotation marks:
responseText = responseText.replace( /\'/g, "\\'" );
// Wrap fields in quotes:
responseText = responseText.replace( /^([a-z\d]*):\s*(.*)/gmi,
↳ "'$1':'$2'," );
// Convert to object:
eval('var packet = {'+ responseText +'}');

// Now you can access the fields as you would with any object:
alert( packet['NewMessages'] ); // returns "0"
```

15.4.2.1 Ping

When accessing the AJAM with an AJAX application, the **ping** command is particularly useful for keeping authenticated connections alive:

```
http://localhost:8088/asterisk/rawman?action=Ping
Response: Pong
```

15.4.2.2 AJAM Demo

A small sample application demonstrating AJAM access may be run at

```
http://localhost:8088/asterisk/static/ajamdemo.html
```

This uses the highly practical JavaScript library prototype (www.prototypejs.org) for AJAX access and displays to get the Status of the currently active channels. You can use the AJAM demo as a basis for your own AJAX applications.

15.4.2.3 Apache

The Asterisk web server is a minimal implementation and cannot be seen as a wholesale replacement for a "proper" web server that can run PHP scripts or use modules, such as Apache. To unify a system that uses both, you can use Apache as a proxy for AJAM by adding

```
ProxyPass /ajam http://localhost:8088/asterisk
```

in the appropriate place in `httpd.conf`, so that all requests for `/ajam` are passed on to AJAM instead of being served by Apache.

16

Asterisk Gateway Interface

The Asterisk Gateway Interface (AGI) may be compared with CGI (Common Gateway Interface) on a web server. These are external programs called from the within the dialplan (`extensions.conf`). AGI scripts can interact with Asterisk and execute commands. In principle, an AGI script may be written in any programming or script language that can use the UNIX system sockets. The following are simply implementation examples. For a command reference, see Appendix D, "AGI Commands."

16.1 Data Streams

To use AGI scripts, understanding the UNIX system's data stream model is an absolute requirement. You can find more information about this model in Appendix D.

16.1.1 STDIN

The AGI script receives some information from Asterisk via standard input (STDIN). The format looks like this:

```
agi_request: programname.php
agi_channel: Zap/1-2
agi_language: cn
agi_callerid: Joe Bloggs
agi_context: external
agi_extension: 1234
agi_priority: 2
```

This is followed by an empty line, by which Asterisk indicates to the AGI script that the transmission is complete.

Here is a list of all the headers Asterisk can transmit via AGI:

- `agi_request`
  The filename of the called AGI program

- `agi_channel`
  The channel

- `agi_language`
  The language (e.g., en for English or de for German)

- `agi_type`
  The channel type (e.g., sip for SIP or zap for ISDN)

- `agi_uniqueid`
  The unique ID for this call

- `agi_callerid`
  The caller ID (e.g., Charlie Farquharson <2000>)

- `agi_context`
  The originating context

- `agi_extension`
  The originating extension

- `agi_priority`
  The current priority in the extension

- `agi_accountcode`
  Account code

- `agi_calleridname`
  Name field from the caller ID (e.g., Jerry Lewis)

- `agi_callingpres`
  Caller ID in the ZAP channel

16.1.2 STDOUT

Once the AGI script has received information from Asterisk, it issues commands via standard out (STDOUT) back to Asterisk. To monitor this communication, enter **set verbose 5** in the CLI to increase the verbosity level.

16.1.3 STDERR

Standard error (STDERR) is the stream used to bring error or debug messages to the Asterisk console (CLI).

16.2 Examples in Various Programming Languages

By default, Asterisk looks for AGI scripts in `/var/lib/asterisk/agi-bin`.

Ensure that Asterisk is actually allowed to run the script you want it to run. You can guarantee this with the system shell command **chmod 755** *script_name*.

AGI scripts are called by the **AGI()** application, like so:

```
exten => 1234,1,Answer()
exten => 1234,2,AGI(my-agi.php)
exten => 1234,3,Hangup()
```

A few AGI programs in various programming languages are described below.

16.2.1 Perl

The standard Asterisk install comes packaged with a test AGI script called `agi-test.agi`[1] located in the `/var/lib/asterisk/agi-bin` directory. Through this sample, we'll explain how AGI scripts work.

The script is called in `extensions.conf` as follows:

```
exten => 1234,1,Answer()
exten => 1234,2,AGI(agi-test.agi)
exten => 1234,3,Hangup()
```

Here we describe the operation by line or section.

```
#!/usr/bin/perl
use strict;
```

The first line indicates to the shell that this is a Perl script and is to be run by the Perl interpreter located at `/usr/bin/perl`. The directive `use strict` tells the interpreter to force the declaration of variables before use and to handle subs and strings in a way that prevents confusion:

```
$|=1;
```

---

1. The .agi extension is not strictly necessary. You could name the file agi-test.pl if so desired.

This short line tells Perl not to buffer text. This ensures that any output is passed to Asterisk immediately:

```
# Setup some variables
my %AGI; my $tests = 0; my $fail = 0; my $pass = 0;
```

Variables are defined here. The hash %AGI accepts the initial inputs from Asterisk. The remaining variables count the total number of tests, the number of failed tests, and the number of passed tests:

```
while(<STDIN>) {
  chomp;
  last unless length($_);
  if (/^agi_(\w+)\:\s+(.*)$/) {
    $AGI{$1} = $2;
  }
}
```

The values which have just been stored are fed to STDERR for debugging purposes, and then passed to the CLI:

```
print STDERR "AGI Environment Dump:\n";
foreach my $i (sort keys %AGI) {
        print STDERR " -- $i = $AGI{$i}\n";
}
```

Then comes the checkresult subroutine:

```
sub checkresult {
  my ($res) = @_;
  my $retval;
  $tests++;
  chomp $res;
  if ($res =~ /^200/) {
    $res =~ /result=(-?\d+)/;
    if (!length($1)) {
      print STDERR "FAIL ($res)\n";
      $fail++;
    } else {
      print STDERR "PASS ($1)\n";
      $pass++;
    }
  } else {
    print STDERR "FAIL (unexpected result '$res')\n";
    $fail++;
  }
}
```

The checkresult subroutine reads the result of a command to Asterisk and determines whether that test was successful. The variables $fail and $pass are incremented depending on the result. After this basic function has been established, the individual tests can proceed. The `beep.gsm` sound file is played:

```perl
print STDERR "1.  Testing 'sendfile'...";
print "STREAM FILE beep \"\"\n";
my $result = <STDIN>;
&checkresult($result);
```

The text hello world is transmitted to the caller. This works only if the protocol and the device support text:

```perl
print STDERR "2.  Testing 'sendtext'...";
print "SEND TEXT \"hello world\"\n";
my $result = <STDIN>;
&checkresult($result);
```

The image `asterisk-image` is transmitted to the caller. This is also protocol and device dependent:

```perl
print STDERR "3.  Testing 'sendimage'...";
print "SEND IMAGE asterisk-image\n";
my $result = <STDIN>;
&checkresult($result);
```

The number 192837465 is played to the caller:

```perl
print STDERR "4.  Testing 'saynumber'...";
print "SAY NUMBER 192837465 \"\"\n";
my $result = <STDIN>;
&checkresult($result);
```

This command waits for dual-tone multi-frequency (DTMF) input from the caller for 1,000ms:

```perl
print STDERR "5.  Testing 'waitdtmf'...";
print "WAIT FOR DIGIT 1000\n";
my $result = <STDIN>;
&checkresult($result);
```

A 3,000ms GSM file named `testagi.gsm` is recorded. Recording can be interrupted by the caller by pressing 1, 2, 3, or 4:

```perl
print STDERR "6.  Testing 'record'...";
print "RECORD FILE testagi gsm 1234 3000\n";
my $result = <STDIN>;
&checkresult($result);
```

The resulting sound file is played back:

```perl
print STDERR "6a.  Testing 'record' playback...";
print "STREAM FILE testagi \"\"\n";
my $result = <STDIN>;
&checkresult($result);
```

Finally, output is sent to the CLI indicating how many tests succeeded and failed:

```
print STDERR "================== Complete =====================\n";
print STDERR "$tests tests completed, $pass passed, $fail failed\n";
print STDERR "===============================================\n";
```

Warning: In many AGI commands you will see the following construction:

```
fwrite(STDOUT,"BEFEHL $value \"\"\n");
#                          ^^^^^^^
```

The indicated part of this line (between $value and);) is an absolute require-ment, to ensure that the command is executed correctly. This is because we have an empty argument established by two double quotes, which must be escaped. The entire command is closed with a line feed, \n.

16.2.2 PHP

PHP has become one of the most popular programming languages for web applica-tions. Because current versions of PHP support calling applications from the com-mand line, it is also well suited for AGI scripts. To demonstrate this, we'll use a small PHP program (lotto.php), which chooses 6 random numbers from 1 and 49 and plays them back to the caller. A description of the individual steps follows in the program code here:

```
#!/usr/bin/php -q
<?php

# Safety setting. The script won't run for
# longer than 8 seconds.
#################################################
set_time_limit(8);


# Deactivate the output buffer.
# Alternatively, we could execute
# fflush(STDOUT);
# after every output.
#################################################
ob_implicit_flush();


# Deactivate PHP error reporting
#################################################
error_reporting(0);
```

```php
# We need STDIN and STDOUT file handles
# to communicate with Asterisk
#################################################
if (!defined('STDIN'))
  define('STDIN' , fopen('php://stdin' , 'r'));
if (!defined('STDOUT'))
  define('STDOUT', fopen('php://stdout', 'w'));
if (!defined('STDERR'))
  define('STDERR', fopen('php://stderr', 'w'));


# The variables from Asterisk are read and saved
# in the $agi array.
#################################################

$agi = array();

while (!feof(STDIN))
{
  $tmp = trim(fgets(STDIN,4096));
  if (($tmp == '') || ($tmp == "\n"))
    break;
  $var1 = split(':',$tmp);
  $name = str_replace('agi_','',$var1[0]);
  $agi[$name] = trim($var1[1]);
}


# An array with 6 random but not repeating
# numbers from 1 to 49 is generated.
#################################################

$Lottonumbers = array();
do {
  $ZNumber = rand(1,49);
  if (array_search($Number, $Lottonumbers) == FALSE) {
    $Lottonumbers[] = $Number;
  }
} while (count($Lottonumbers) < 6);


# Before the number is announced,
# we wait one second.
#################################################
fwrite(STDOUT,"EXEC Wait 1 \"\"\n");
fflush(STDOUT);


# The numbers are read, one after the other.
# There is a one second pause between numbers.
#################################################
```

```
foreach ($Lottonumber as $value) {
  fwrite(STDOUT,"SAY NUMBER $value \"\"\n");
  fflush(STDOUT);
  fwrite(STDOUT,"EXEC Wait 1 \"\"\n");
  fflush(STDOUT);
}

?>
```

The `lotto.php` program must be saved in `/var/lib/asterisk/agi-bin` and is called from `extensions.conf` as follows:

```
exten => 1234,1,Answer()
exten => 1234,2,AGI(lotto.php)
exten => 1234,3,Hangup()
```

If you use PHP with AGI but don't want to reinvent the wheel, you can use the ready-made phpAGI PHP class. You can find the required files and more information on the project home page at http://phpagi.sourceforge.net.

The following program, which is provided with phpAGI, will give you a quick overview of the function calls.[2]

```php
<?php
/**
 * @package phpAGI_examples
 * @version 2.0
 */

function my_ip(&$agi, $peer)
    {
     $ip = 'unknown';
     $asm = $agi->new_AsteriskManager();
     if($asm->connect())
     {
       $peer = $asm->command("sip show peer $peer");
       $asm->disconnect();

       if(!strpos($peer['data'], ':'))
         echo $peer['data'];
       else
       {
         $data = array();
         foreach(explode("\n", $peer['data']) as $line)
         {
           $a = strpos('z'.$line, ':')- 1;
                if($a >= 0) $data[trim(substr($line, 0, $a))] =
↳trim(substr($line, $a
         }
       }
     }
```

_____

2. To use `text2wav`, you need to have a text-to-speech system (e.g., Festival) installed.

```
    if(isset($data['Addr->IP']))
    {
      $ip = explode(' ', trim($data['Addr->IP']));
      $ip = $ip[0];
    }
  }
  $agi->text2wav("Your IP address is $ip");
}

?>
```

16.2.3 Ruby and Adhearsion

Although Ruby is an older (by IT standards) programming language, it has grown tremendously in popularity in the past few years, mainly as the result of Ruby on Rails (http://rubyonrails.org). Adhearsion (http://adhearsion.com) is a kind of Ruby on Rails for Asterisk. The subject is exciting and deep enough that one could write a dedicated book on it. In the future, we will try to expand and update the coverage of Adhearsion on this book's website.

16.2.3.1 Installation

Like any good open source project, Adhearsion has fragmented and occasionally incomplete documentation. Fortunately, the installation is relatively easy because there is a Ruby *gem* available. (For Ruby Newbies, gem is Ruby's package management system.) Installing Adhearsion is just a matter of entering **gem install adhearsion** in the system shell.

16.2.3.2 Easy Setup

When you use Adhearsion, the Asterisk dialplan becomes short and easy to read. Each inbound context needs only the following line:

```
exten => _.,1,AGI(agi://127.0.0.1)
```

With this line, Adhearsion takes over control of all calls happening in this context. You can, of course, mix a traditional dialplan with Adhearsion if you want.

An Adhearsion application has to be generated first, just as in Ruby on Rails. This is done with the ahn program, called with **ahn create application_name**, as follows:

```
stefan@pbx:~$ ahn create widgets_app
    create
    create  components/simon_game
    create  components/disabled/stomp_gateway create components/
    ami_remote
```

```
      create  components/restful_rpc/spec create config
      create  .ahnrc
      create  components/simon_game/simon_game.rb create
components/ami_remote/ami_remote.rb
      create  components/disabled/stomp_gateway/stomp_gateway.rb create
components/disabled/stomp_gateway/config.yml
      create  components/disabled/stomp_gateway/README.markdown create
components/restful_rpc/restful_rpc.rb
      create  components/restful_rpc/config.yml
      create  components/restful_rpc/README.markdown create
components/restful_rpc/example-client.rb
      create  components/restful_rpc/spec/restful_rpc_spec.rb
      create  config/startup.rb
      create  dialplan.rb
      create  events.rb
      create  README
      create  Rakefile
stefan@pbx:~$ cd widgets_app
stefan@pbx:~/widgets_app$
```

The application is then started with **ahn start** . (And, yes, this command ends with the blank space, as shown here and in the following code snippet.)

```
stefan@pbx:~/widgets_app$ ahn start .
 INFO ahn: Adhearsion initialized!
```

From this point forward, Adhearsion can control calls. You can monitor progress on the screen.

16.2.3.3 Sample Program

The Adhearsion dialplan, or rather the Ruby program, is defined in `dialplan.rb`. Here is a simple example:

```
internal {
  case extension
  when 22
    play "hello-world"
    hangup
  else
    dial "SIP/#{extension}"
  end
}
```

`internal` is, in this example, the context from which Adhearsion was called.

16.2.4 Other Programming Languages

Remember, you can write AGI programs in any language. There are already libraries for many of the most popular languages, including the following:

- Java
- Perl
- PHP
- Python
- Ruby
- C

The easiest way to start is with a web search using the keywords AGI and the name of the language you want to use.

17

Fax Server

Many attempts have been made to cleanly integrate Asterisk with faxing, with inconsistent results. For a long time, app_rxfax, available at www.soft-switch .org was the accepted convention. Unfortunately it is error prone and unreliable. Faxes often arrive piecemeal, or worse, not at all.

For that reason, we describe a more robust solution using IAXmodem here (see http://iaxmodem.sourceforge.net).

The IAXmodem application emulates a fax modem, which may be operated by a fax application of the administrator's choosing. We'll use the popular Hylafax. For simplicity and consistency, the installation platform will be the same Debian Linux and Asterisk 1.4 we have used for the other examples (see Appendix A, "Installation Instructions for Asterisk 1.4").

17.1 Installing IAXmodem

IAXmodem simulates a fax modem and makes it available to Asterisk via IAX2. All the steps in this chapter must be performed as the root user.

To install IAXmodem, we need some additional Debian packages, which may be installed with the command **apt-get -y install g++ libtiff-tools libtiff4 libtiff4-dev**.

```
debian:~# apt-get -y install g++ libtiff-tools libtiff4 libtiff4-dev
Reading Package Lists... Done
```

```
Building Dependency Tree... Done
The following extra packages will be installed:
  g++ libjpeg62 libjpeg62-dev libtiffxx0 zlib1g-dev

[...]

Setting up zlib1g-dev (1.2.2-4.sarge.2) ...
Setting up libtiff4-dev (3.7.2-7) ...

debian:~#
```

We switch into the appropriate directory with **cd /usr/src** to install the IAX-modem source code:

```
debian:~# cd /usr/src
debian:/usr/src#
```

The sources for IAXmodem can be downloaded with any typical web browser from http://iaxmodem.sourceforge.net (the version used in this example is 0.3.0). After downloading the archive, copy it to /usr/src and unpack it with **tar -xvzf iaxmodem-0.3.0.tar.gz**:

```
debian:/usr/src# tar -xvzf iaxmodem-0.3.0.tar.gz
iaxmodem-0.3.0/
iaxmodem-0.3.0/iaxmodem.c
iaxmodem-0.3.0/iaxmodem.init.debian
iaxmodem-0.3.0/Makefile.in
iaxmodem-0.3.0/CHANGES
iaxmodem-0.3.0/lib/
iaxmodem-0.3.0/lib/spandsp/
iaxmodem-0.3.0/lib/spandsp/Makefile.am

[...]

iaxmodem-0.3.0/TODO
iaxmodem-0.3.0/FAQ
iaxmodem-0.3.0/build
iaxmodem-0.3.0/iaxmodem.init.fedora
debian:/usr/src#
```

Change into the unpacked directory with **cd iaxmodem-0.3.0**:

```
debian:/usr/src# cd iaxmodem-0.3.0
debian:/usr/src/iaxmodem-0.3.0#
```

Now compile the sources with **./configure && make**:

```
debian:/usr/src/iaxmodem-0.3.0# ./configure && make
checking for a BSD-compatible install... /usr/bin/install -c
checking whether build environment is sane... yes
checking for gawk... no
```

```
checking for mawk... mawk
checking whether make sets $(MAKE)... yes
checking for gcc... gcc

[...]

cc -DMODEMVER=\"0.3.0\" -DDSPVER=\"spandsp-0.0.3-snapshot-20070223+\" -
DIAXVER=\"libiax2-0.2.3-CVS-20060222+\" -Wall -g -DSTATICLIBS -
DUSE_UNIX98_PTY -std=c99 -Ilib/libiax2/src -Ilib/spandsp/src -c iaxmodem.c
cc -DMODEMVER=\"0.3.0\" -DDSPVER=\"spandsp-0.0.3-snapshot-20070223+\" -
DIAXVER=\"libiax2-0.2.3-CVS-20060222+\" -Wall -g -DSTATICLIBS -
DUSE_UNIX98_PTY -std=c99 -Ilib/libiax2/src -Ilib/spandsp/src iaxmodem.o
lib/spandsp/src/.libs/libspandsp.a lib/libiax2/src/.libs/libiax.a -o
iaxmodem -lm -lutil -ltiff
[...]

debian:/usr/src/iaxmodem-0.3.0#
```

Copy the resulting binary into /usr/bin with **cp iaxmodem /usr/bin/**:

```
debian:/usr/src/iaxmodem-0.3.0# cp iaxmodem /usr/bin/
debian:/usr/src/iaxmodem-0.3.0#
```

17.1.1 Configuring IAXmodem

Now we can configure the modem. IAXmodem expects to find configuration files in /etc/iaxmodem. Create it with **mkdir /etc/iaxmodem**:

```
debian:/usr/src/iaxmodem-0.3.0# mkdir /etc/iaxmodem
debian:/usr/src/iaxmodem-0.3.0#
```

Create the configuration file with **touch /etc/iaxmodem/ttyIAX0**:

```
debian:/usr/src/iaxmodem-0.3.0# touch /etc/iaxmodem/ttyIAX0
debian:/usr/src/iaxmodem-0.3.0#
```

This file must contain the following parameters:

- device

  The device node to be created in /dev. This is the device Hylafax uses to connect to IAXmodem. You can choose any name you like, but we prefer to adhere to the convention and so choose a device name appropriate for a serial interface, ttyIAX0.

- owner

  This is the owner of the device (in the form *user:group*). It is best to use the same user and group under which Hylafax runs.

- port

  The port that IAXmodem listens on. Asterisk uses 4569 to listen for IAX2 connections, so you must choose something else (e.g., 4570).

- refresh

  This sets how long IAXmodem waits between registrations with Asterisk. If this number is 0, the modem does not register at all.

- server

  IP address of the server running Asterisk. If this is on the same machine as IAXmodem, use the localhost address 127.0.0.1.

- peername

  The name under which IAXmodem registers with Asterisk.

- secret

  The password used for Asterisk registration.

- codec

  The codec used by IAXmodem. Allowed codecs are alaw, ulaw, and slinear. Compressed codecs are not appropriate for faxing; fax transmissions are themselves compressed and don't tolerate further compression; moreover, most compressed codecs are lossy and a fax transmission will not tolerate losses. This is one of the major reasons why faxing over VoIP remains problematic.

Using an appropriate editor (e.g., vi), we write the following configuration in the file /etc/iaxmodem/ttyIAX0:

```
device          /dev/ttyIAX0
owner           uucp:uucp
mode            660
port            4570
refresh         50
server          127.0.0.1
peername        iaxmodem
secret          password
codec           alaw
```

IAXmodem is now configured and can be started. The best way to do this is with **init**. Add a line to start IAXmodem to /etc/inittab with **echo "IA00:23: respawn:/usr/bin/iaxmodem ttyIAX0" >> /etc/inittab**:

```
debian:/usr/src/iaxmodem-0.3.0# echo "IA00:23:respawn:
↳/usr/bin/iaxmodem ttyIAX0" >> /etc/inittab
debian:/usr/src/iaxmodem-0.3.0#
```

The device name ttyIAX0 is the same device name as specified in /etc/iaxmodem.

17.1.2 Configuring faxgetty and Logging

To receive faxes, we need a getty that listens for connections on the IAXmodem. This is accomplished through an additional entry in /etc/inittab. Add it with **echo "mo00:23:respawn:/usr/sbin/faxgetty ttyIAX0" >> /etc/inittab**:

```
mo00:23:respawn:/usr/local/sbin/faxgetty ttyIAX0
```

Create a log directory for IAXmodem with **mkdir /var/log/iaxmodem/** and the log files with **touch /var/log/iaxmodem/ttyIAX0** and **touch /var/log/iaxmodem/iaxmodem**:

```
debian:/usr/src/iaxmodem-0.3.0# mkdir /var/log/iaxmodem/
debian:/usr/src/iaxmodem-0.3.0# touch /var/log/iaxmodem/ttyIAX0
↳debian:/usr/src/iaxmodem-0.3.0# touch /var/log/iaxmodem/iaxmodem
debian:/usr/src/iaxmodem-0.3.0#
```

To make sure everything will start as expected at boot time, reboot the system with **shutdown -r now**:

```
debian:/usr/src/iaxmodem-0.3.0# shutdown -r now

Broadcast message from root@debian (pts/1) (Sat May 5 00:15:49 2007):
The system is going down for reboot NOW!
debian:/usr/src/iaxmodem-0.3.0#
```

17.2 Installing Hylafax

We'll install Hylafax from the Debian Repository to simplify installation. Do this with **apt-get -y install hylafax-server.** Dependencies are automatically resolved:

```
debian:~# apt-get install -y hylafax-server
Reading Package Lists... Done
Building Dependency Tree... Done
The following extra packages will be installed:
  enscript gs-common gs-esp hylafax-client libcupsimage2 libcupsys2 mailx
  metamail ps
Suggested packages:
  gv postscript-viewer lpr gs-pdfencrypt gs-cjk-resource mgetty-viewfax
hylafax-doc m
Recommended packages:
  psfontmgr netpbm transfig
The following NEW packages will be installed:
  enscript gs-common gs-esp hylafax-client hylafax-server libcupsimage2
  libcupsys2 ma
```

```
[...]

Update /var/spool/hylafax/status/any.info.

        HylaFAX configuration parameters are:

        [1] Init script starts faxq:          yes
        [2] Init script starts hfaxd          yes
        [3] Start old protocol:               no
        [4] Start paging protocol:            no

Are these ok [yes]?
Modem support functions written to /var/spool/hylafax/etc/setup.modem.
Configuration parameters written to /var/spool/hylafax/etc/setup.cache.

Restarting HylaFAX server processes.

Should I restart the HylaFAX server processes [yes]?
You do not appear to have any modems configured for use. Modems are
configured for use with HylaFAX with the faxaddmodem(8) command.
Do you want to run faxaddmodem to configure a modem [yes]?
Done verifying system setup.
Updating /etc/hylafax/setup.cache from /var/spool/hylafax/etc/setup.cache.
Updating /etc/hylafax/setup.modem from /var/spool/hylafax/etc/
setup.modem.apt-get -y install hylafax-server
/var/spool/hylafax
Starting HylaFAX: faxq hfaxd faxgetty.

debian:~#
```

The next step is the configuration of the fax server. Do this with **faxsetup**:

```
debian:/usr/src/hylafax-4.3.4# faxsetup

[...]

Update /var/spool/hylafax/status/any.info.

        HylaFAX configuration parameters are:

        [1] Init script starts faxq:          yes
        [2] Init script starts hfaxd          yes
        [3] Start old protocol:               no
        [4] Start paging protocol:            no

Are these ok [yes]?
```

Simply press Enter after the following 2 or 3 questions:

```
You have a HylaFAX scheduler process running.  faxq will be
restarted shortly, as soon as some other work has been completed.
Can I terminate this faxq process (4048) [yes]?
Should I restart the HylaFAX server processes [yes]?
```

```
/etc/init.d/hylafax start
Not starting HylaFAX daemons since they are already running.

[...]

Modems are configured for use with HylaFAX with the faxaddmodem(8)
    command.
Do you want to run faxaddmodem to configure a modem [yes]?
```

We confirm restart of the server processes with yes and are asked if we want to install a modem. Our IAXmodem is already set up so we can proceed and confirm again with yes.

Specify the modem and confirm with Enter:

```
Serial port that modem is connected to [ttyS0]? ttyIAX0

Ok, time to setup a configuration file for the modem. The manual
page config(5) may be useful during this process. Also be aware
that at any time you can safely interrupt this procedure.

Reading scheduler config file /var/spool/hylafax/etc/config.
```

Many questions follow, but only a few of them are really important. This is where you set international dialing codes, the fax number, country and area code, and the CSID (call subscriber ID), which is printed on the top line of the fax page on the receiver's end. Confirm with yes.

```
No existing configuration, let's do this from scratch.

Country code [1]? 1
Area code []? 403
Phone number of fax modem [+1.999.555.1212]? +1 888 555 4091
Local identification string (for TSI/CIG) ["NothingSetup"]?
Long distance dialing prefix [1]? 1
International dialing prefix [011]? 011
Dial string rules file (relative to /var/spool/hylafax) [etc/dialrules]?
Tracing during normal server operation [1]?
Tracing during send and receive sessions [11]?
Protection mode for received facsimile [0600]?
Protection mode for session logs [0600]?
Protection mode for ttyIAX0 [0600]?
Rings to wait before answering [1]?
Modem speaker volume [off]?
Command line arguments to getty program ["-h %l dx_%s"]?
Pathname of TSI access control list file (relative to /var/spool/hylafax)
[""]?
Pathname of Caller-ID access control list file (relative to
/var/spool/hylafax) [""]?
Tag line font file (relative to /var/spool/hylafax) [etc/lutRS18.pcf]?
Tag line format string ["From %%l|%c|Page %%P of %%T"]?
Time before purging a stale UUCP lock file (secs) [30]?
```

```
Hold UUCP lockfile during inbound data calls [Yes]?
Hold UUCP lockfile during inbound voice calls [Yes]?
Percent good lines to accept during copy quality checking [95]?
Max consecutive bad lines to accept during copy quality checking [5]?
Max number of pages to accept in a received facsimile [25]?
Syslog facility name for ServerTracing messages [daemon]?
Set UID to 0 to manipulate CLOCAL [""]?
Use available priority job scheduling mechanism [""]?
```

A confirmation page follows where you can double-check your entries:

```
The non-default server configuration parameters are:

CountryCode:             1
AreaCode:                403
FAXNumber:               +1 888 555 4091
LongDistancePrefix:      0
InternationalPrefix:     00
DialStringRules:         etc/dialrules
SessionTracing:          11
RingsBeforeAnswer:       1
SpeakerVolume:           off
GettyArgs:               "-h %l dx_%s"
LocalIdentifier:         "NothingSetup"
TagLineFont:             etc/lutRS18.pcf
TagLineFormat:           "From %%l|%c|Page %%P of %%T"
MaxRecvPages:             25


Are these ok [yes]?
```

Answering yes brings us to modem detection:

```
Now we are going to probe the tty port to figure out the type
of modem that is attached. This takes a few seconds, so be patient.
Note that if you do not have the modem cabled to the port, or the
modem is turned off, this may hang (just go and cable up the modem
or turn it on, or whatever).

Probing for best speed to talk to modem: 38400 OK.

About fax classes:

The difference between fax classes has to do with how HylaFAX interacts
with the modem and the fax protocol features that are used when sending
or receiving faxes. One class isn't inherently better than another;
however, one probably will suit a user's needs better than others.

Class 1 relies on HylaFAX to perform the bulk of the fax protocol.
Class 2 relies on the modem to perform the bulk of the fax protocol.
Class 2.0 is similar to Class 2 but may include more features.
Class 1.0 is similar to Class 1 but may add V.34-fax capability.
Class 2.1 is similar to Class 2.0 but adds V.34-fax capability.
```

```
HylaFAX generally will have more features when using Class 1/1.0 than
when using most modems' Class 2 or Class 2.0 implementations. Generally
any problems encountered in Class 1/1.0 can be resolved by modifications
to HylaFAX, but usually any problems encountered in Class 2/2.0/2.1 will
require the modem manufacturer to resolve it.

If you're unsure and your modem supports it, use Class 1.

This modem looks to have support for Class 1 and 1.0.
How should it be configured [1]?

Hmm, this looks like a Class 1 modem.
Product code (ATI0) is "spandsp".
Other information (ATI3) is "www.soft-switch.org".
DTE-DCE flow control scheme [default]?
Modem manufacturer is "spandsp".
Modem model is "IAXmodem".

Using prototype configuration file iaxmodem...

The modem configuration parameters are:

ModemResetCmds:          "ATH1\nAT+VCID=1"

Are these ok [yes]?
```

The modem was detected and we are asked if it is a Class 1 modem, and we confirm this because it is exactly what we want. The default reset commands are also acceptable. Confirm with `yes`.

Answer the first question In the next dialog with `no`, because we don't need to configure any further modems. The second question is confirmed with by pressing Enter, which starts the fax server:

```
Creating new configuration file /var/spool/hylafax/etc/config.ttyIAX0...
Creating fifo /var/spool/hylafax/FIFO.ttyIAX0 for faxgetty... done.
Done setting up the modem configuration.

[...]

Do you want to run faxaddmodem to configure another modem [yes]? no

[...]

Should I run fax modem for each configured modem [yes]?
/usr/sbin/fax modem ttyIAX0

Done verifying system setup.
/var/spool/hylafax

debian:~#
```

Hylafax is now configured for sending faxes.

17.3 Receiving Faxes

Our fax solution still has to be integrated into Asterisk. To do this, we configure the
IAXmodem as an IAX2 peer by adding a section to /etc/asterisk/iax.conf:

```
[general]
bindport = 4569
bindaddr = 0.0.0.0
disallow=all
allow=ulaw
allow=alaw

[iaxmodem]
type=friend
secret=password
port=4570
host=dynamic
context=fax-out
disallow=all
allow=alaw
```

Global settings are defined in the general section. In this example we are binding the standard IAX2 port of 4569. The bindaddr defines the IP address (and
thereby the interface) on which the IAX2 channel driver listens for connections; in
this case, it is set to listen on all interfaces.

The IAXmodem is set to type friend, which allows both incoming and outgoing
connections. The secret and port parameters match those in the IAXmodem configuration we did above, and context defines the entry context for outgoing connections.

Enter **iax2 show peers** in the Asterisk console to see our new IAXmodem:

```
*CLI> iax2 show peers
Name/Username    Host                 Mask              Port        Status
Iaxmodem         127.0.0.1
   (D)   255.255.255.255    4570          Unmonitored
1 iax2 peers [0 onlinc, 0 offline, 1 unmonitored]
*CLI>
```

We are, of course, not done yet. Asterisk still needs an extension so that it knows
what to do with an incoming fax. Our objective is to ensure that any incoming faxes
are passed on to Hylafax. In this example, we assume that all faxes come in
through a SIP provider. A real configuration will have to reflect the installation and
account settings of the SIP provider you use; for the sake of example, a configuration in sip.conf might look like this:

```
[...]

[123456]
type=friend
```

```
insecure=very;
nat=yes
username=123456
fromuser=12345
fromdomain=my-voip-provider.com
secret=secret
host=my-voip-provider.com
qualify=yes
context=fax-in

[...]
```

The corresponding context in `extensions.conf` would look like this:

```
[fax-in]
exten => _X.,1,Dial(IAX2/iaxmodem)
```

Any faxes coming in will now be routed to Hylafax via IAXmodem and ultimately e-mailed to the user address defined in the `faxmaster` alias.

Note: By default, Hylafax saves incoming faxes in `/var/spool/hylafax/recvq`.

17.4 Sending Faxes

The next obvious step is configuring our system to send faxes. Here, too, we need a context (this time it is `[fax-out]`) in `extensions.conf`. If IAXmodem wants to send a fax, it will automatically land in this context. If the faxes are to go out our hypothetical SIP connection 123456, the entry in `extensions.conf` will look like this:

```
[fax-out]
exten => _X.,1,Dial(SIP/123456/${EXTEN})
```

We can test sending of faxes with **sendfax -n -d <faxnumber> <file.txt>:**

```
debian:~# sendfax -n -d 6045557977 /etc/issue.net
```

We should see this in the CLI:

```
-- Accepting AUTHENTICATED call from 127.0.0.1:
      > requested format = alaw,
      > requested prefs = (),
      > actual format = alaw,
      > host prefs = (alaw),
      > priority = mine
   -- Executing Answer("IAX2/iaxmodem-3", "") in new stack
      Executing Dial("IAX2/iaxmodem-3", "SIP/123456/6045557977")
      in new stack
```

```
-- Called 123456/6045557977
-- SIP/123456-0818f630 is making progress passing it to
   IAX2/iaxmodem-3
-- SIP/123456-0818f630 answered IAX2/iaxmodem-3
-- parse_srv: SRV mapped to host my-voip-provider.com, port 5060
== Spawn extension (fax-out, 6045557977, 2) exited non-zero on 'IAX2/
iaxmodem-3'
-- Executing Hangup("IAX2/iaxmodem-3", "") in new stack
== Spawn extension (fax-out, h, 1) exited non-zero on 'IAX2/iaxmodem-3'
-- Hungup 'IAX2/iaxmodem-3'
```

If we issue the command **faxstat -s** during the transmission, we will see the following:

```
debian:~# faxstat -s
HylaFAX scheduler on w077.example.com: Running
Modem ttyIAX0 (123456): Sending job 7

JID  Pri S  Owner Number       Pages Dials     TTS   Status
7    127 R   root 06912345678   0:1   0:12
debian:~#
```

Done! Now you can send and receive faxes via Asterisk using Hylafax.

The Hylafax website, www.hylafax.org, has numerous examples and how-tos that will help you integrate your Hylafax installation with your existing office infrastructure effectively.

17.5 Sending Received Faxes as E-Mail

The following steps illustrate how we can configure Hylafax to transmit incoming faxes to a predefined e-mail address.[1]

The recipient will receive the fax as an e-mail attachment.

To do this, the configuration file /var/spool/hylafax/etc/FaxDispatch must contain the following parameters:

- SENDTO

  The destination e-mail address for incoming faxes.

- FILETYPE

  The format of the attachment. In addition to pdf, tiff (Tagged Image File Format) and ps (Postscript) are also acceptable options.

---

1. Our example assumes a properly configured MTA (e.g., Sendmail, Postfix, or a lightweight SMTP engine such as ssmtp).

The parameters are set with an equal sign, just as in Asterisk configuration files:

```
SENDTO=fax-incoming@company.com
FILETYPE=pdf
```

After the file has been saved, you must restart the fax server with **/etc/ init.d/hylafax restart**:

```
debian:~# /etc/init.d/hylafax restart
Starting HylaFAX: faxq hfaxd.
debian:~#
```

We can test e-mail transmission by sending ourselves a fax with **sendfax -n -d <faxnumber> <file.txt>**:

```
debian:~# sendfax -n -d 6045557977 /etc/issue.net
```

After a short time, your target e-mail address should receive an e-mail in the following format:

```
recvq/fax000000016.tif (ftp://debian:4559/recvq/fax000000016.tif):

             Sender: IAXmodem
              Pages: 4
            Quality: Normal
               Size: North American Letter
           Received: 2007:06:02 02:49:45
    Time To Receive: 1:58
        Signal Rate: 9600 bit/s
        Data Format: 2-D MMR
      Error Correct: Yes
            CallID1: 2007
            CallID2: IAXmodem 1
        Received On: ttyIAX0
             CommID: 000000033 (ftp://debian:4559/log/c000000033)

  [...]


Jun 02 02:51:46.99: [ 3320]: RECV FAX: bin/faxrcvd
                    "recvq/fax000000016.tif" "ttyIAX0"
Jun 02 02:51:47.00: [ 3320]: RECV FAX: end
Jun 02 02:51:47.00: [ 3320]: SESSION END
Jun 02 02:51:47.01: [ 3320]: RECV FAX (000000033):
                          recvq/fax000000016.tif from IAXmod
```

The attachment will be a PDF file. In this example, the PDF is named `fax000000016.pdf`.

Now you can not only send and receive faxes, but received faxes are also received as e-mail attachments.

18

Busy Lamp Field, Hints, Pickup

Philipp Kempgen
Stephen Bosch

The Busy Lamp Field, or BLF, is a common feature of many modern telephone systems. The BLF can be colored LED lamps or LCD indicators that allow a user to monitor the status of other users on the system, almost always next to a button that performs an operation on the monitored station depending on the status of the indicator. Although it is most frequently seen at reception switchboards, it can also be found on regular desktop sets.

The indicator tells the operator the status of the line. A line can have many states:

- Ringing
- Busy or "in use" (which usually includes active "Do not disturb" states)
- On hold
- Available

The states can be indicated with LED lights (which may change color or blink, depending on the state) or LCD icons (which change appearance; for example, by providing a "do not enter" sign for a line that is in use). Not all models of telephone provide all the states.

If the system is so configured, the button can be configured to "pick up" a call when a station is ringing. (For some sets, this is done by entering a feature code.)

We cover the configuration of these features for Asterisk and a selection of SIP phones here.

18.1 Controlling Indicators

The BLF is implemented through hints and subscriptions. A hint tells Asterisk to monitor the state of a SIP channel. SIP devices can subscribe to those hints, and when the state changes, Asterisk sends an event package to the subscribed devices to inform them of the state change.

18.1.1 sip.conf

To make BLF work, some additional entries must be added to the [general] section of sip.conf:

```
[general]
allowsubscribe = yes
notifyringing = yes
notifyhold = yes
limitonpeers = yes
```

Individual SIP accounts need the entries subscribecontext and call-limit. The first, subscribecontext, tells Asterisk which context to check whether this user wants to subscribe to the hint of another user. Most often, this is the same as the context for all the other internal users. The second, call-limit, limits the number of simultaneous calls that this user is allowed (when this is greater than zero). This is important because it tells Asterisk how to set the state of the station in question. For example, if the call limit is set to 2, and the station has an active call, Asterisk will treat the station as available. If there are already two active calls, then Asterisk treats the station as busy. A setting of at least 2 is best; lower settings can cause problems with call transfers.

For call pick-up functions, the additional parameters, callgroup and pickupgroup are important. For example, a SIP user with pickupgroup=2 can only pick up calls from other users with callgroup=2. This can be used to manage rights; because there are only 64 possible groups, however (0 to 63), this works in small to medium environments only. For larger installations, all the users are set to the same group and rights are handled through another mechanism. (We hint at how later.)

```
[21]
type = friend
context = internal-users
secret = 9847825134
host = dynamic
mailbox = 2000
subscribecontext = internal-users
call-limit = 10
callgroup = 2
pickupgroup = 2
```

18.1.2 Hints

Asterisk does not know how to correlate dialplan extensions to specific users or devices unless we tell it how. This is done in the dialplan using *hints*. If you want to think in terms of dialplan priorities, a hint is a special kind of priority that precedes priority 1. The hint for a device must be in the same context as that specified in `sip.conf` using `subscribecontext`, or the status information will not be relayed. This is to prevent a station from monitoring all the other stations without permission.

The following examples show how to define hints. Here is a simple example in `.conf` format:

```
[internal-users]

exten => 21,hint,SIP/21
exten => 21,1,Dial(SIP/${EXTEN},40)
exten => 21,n,VoiceMail(${EXTEN},u)

exten => 22,hint,SIP/22
exten => 22,1,Dial(SIP/${EXTEN},40)
exten => 22,n,VoiceMail(${EXTEN},u)
```

The same is accomplished in AEL like this:

```
context internal-users {

    hint(SIP/21) 21 => {
       Dial(SIP/${EXTEN},40);
       VoiceMail(${EXTEN},u);
    }
    hint(SIP/22) 22 => {
       Dial(SIP/${EXTEN},40);
       VoiceMail(${EXTEN},u);
    }
}
```

Let's take a look at the typical example of a group of extensions beginning with a common digit, here in `extensions.conf`:

```
[internal-users]
exten => 21,hint,SIP/21
exten => 22,hint,SIP/22

exten => _2X,1,Dial(SIP/${EXTEN},40)
exten => _2X,n,VoiceMail(${EXTEN},u)
```

In `extensions.ael`, the same dialplan section would look like this:

```
context internal-users {

    hint(SIP/21) 21 => {}
    hint(SIP/22) 22 => {}
```

```
  _2X => {
    Dial(SIP/${EXTEN},40);
    VoiceMail(${EXTEN},u);
  }
}
```

You may want to try using patterns to reduce the maintenance overhead for your dialplan. Unfortunately, in Asterisk 1.4, this doesn't work.

Warning: In Asterisk 1.4, hints may not use patterns, so these hints would not be allowed (`extensions.conf`, `extensions.ael` respectively):

```
exten => _2X,hint,SIP/${EXTEN}
hint(SIP/${EXTEN}) _2X => {...}
```

As of Asterisk 1.6, you can use patterns in this way, although the early implementations were buggy. Be careful, however: When hints are processed, there is no channel yet, so no channel variables will be available for the hint definition to use (with the exception of `${EXTEN}`).

You can also combine multiple users in a single hint, as in this example:

```
[internal-users]

exten => 20,hint,SIP/21&SIP/22

exten => 20,1,Dial(SIP/21&SIP/22,40)
```

To accomplish the same thing in `extensions.ael`, a similar pattern is followed:

```
context internal-users {

  hint(SIP/21&SIP/22) 20 => {}

  20 => {
    Dial(SIP/21&SIP/22,40);
  }
}
```

Tip: You can also place all your hints in a dedicated context and then `include` it if you need it in another context.

You must reload the dialplan in order to apply hints (**asterisk -rx 'module reload'**).

18.2 Subscriptions

This section provides the technical background for subscriptions for those who are interested. It's not really required for configuration purposes. If you just want to configure your phones, proceed to the next section.

SIP provides more than one method for subscribing to a station's status. The customary method is described in RFC 4235, "Dialog Event Package," which is based on RFC 3265, "SIP Event Notification." With this method, a monitoring device sends a separate SUBSCRIBE message to the phone system for each station it wants to monitor. The system replies with a NOTIFY message whenever the status of subscribed station changes. (Asterisk does not yet support PUBLISH.)

The main difference between the subscription standards is the XML they use in the body of the notification. The standards supported by Asterisk are described in Table 18.1.

The xpidf+xml and cpim-pidf+xml formats are very similar.

18.3 Configuring Telephones for BLF

To actually use BLF, you have to configure BLF on the phone you want to use for monitoring. This will vary for every model of phone you use. We briefly cover two commonly used phones here, to get you started. Be warned that the firmware, and thus the configuration parameters, can change at any time. For the last word, consult the documentation for the phone you are configuring.

Table 18.1 Subscription Methods Supported by Asterisk

Specification	MIME Type
SIP Dialog Event Package (RFC 4235) (http://tools.ietf.org/html/rfc4235)	application/dialog-info+xml
Presence Information Data Format (PIDF) (RFC 3863) (http://tools.ietf.org/html/rfc3863)	application/pidf+xml
Pre-version of RFC 3863 (see PIDF above), XPIDF, CPIM with Microsoft extensions (http://msdn.microsoft.com/en-us/library/ cc246193(PROT.10).aspx)	application/xpidf+xml
Pre-version of RFC 3863 (see PIDF above), XPIDF, CPIM	application/cpim-pidf+xml

---

Note: *Central provisioning* is the term used for configuration of telephones through configuration files distributed through a central server, whether by TFTP, FTP, HTTP, or some other means. The setup for central provisioning is beyond the scope of this chapter. Because central provisioning is a very common way to configure phones and an absolute must for large installations, most manufacturers provide instructions for this in the administration guide (as opposed to the user's guide) for the phone.

18.3.1 BLF on the Aastra 480i

Aastra telephones, like most IP telephones, can be configured via a web interface or through a configuration file, which the phone loads from a configuration server (nearly always the Asterisk server itself).

18.3.1.1 Configuring via a Web Browser

Find the IP address for the phone. You can do this on the phone by pressing the Options key and scrolling to Network Setup. (The default password is 22222.)

Enter the IP address into a web browser on a computer on the same network. When you do this, you'll be prompted for a username and password. By default, the username is admin, and the password is 22222. Once you are logged in, click Softkeys and XML in the Operation section of the configuration menu on the left. Figure 18.1 shows the Softkeys Configuration page.

Choose the key you would like to configure and set the type to BLF. The Aastra phones automatically condense softkeys on the idle screen. For example, if you only configure key 5, it will appear as key 1 on your phone.

The Label field sets the display name for the BLF, and the value is the number of the extension you want to monitor. The check boxes on the right (Idle, Connected, Incoming, and Outgoing) determine in which states the key will display. (This is because the keys can be set to perform a variety of functions and you may not need them in every state.) The defaults usually suffice.

Once you have entered all the BLF settings you want, you have to save the settings before restarting the phone. The Save Settings button is in the lower-left corner of the window. (It's not visible in the figure above because you have to scroll down to see it.) Restart the phone by going to Operation on the left and clicking Reset and then Restart.

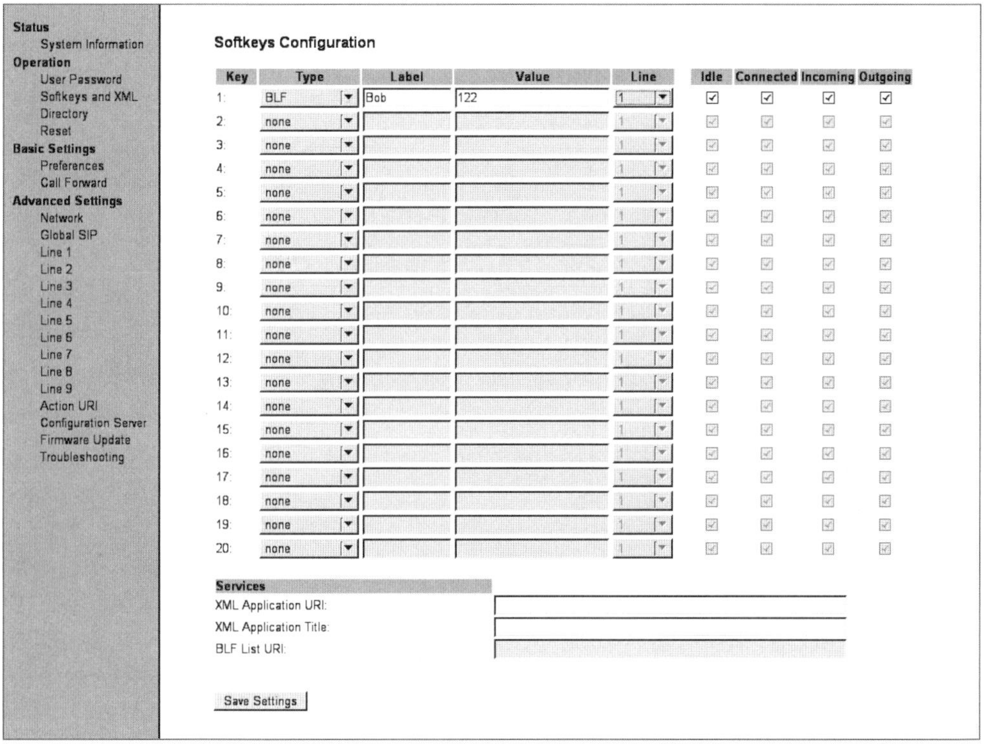

Status
 System Information
Operation
 User Password
 Softkeys and XML
 Directory
 Reset
Basic Settings
 Preferences
 Call Forward
Advanced Settings
 Network
 Global SIP
 Line 1
 Line 2
 Line 3
 Line 4
 Line 5
 Line 6
 Line 7
 Line 8
 Line 9
 Action URI
 Configuration Server
 Firmware Update
 Troubleshooting

Softkeys Configuration

Key	Type	Label	Value	Line	Idle	Connected	Incoming	Outgoing
1:	BLF	Bob	122	1	✓	✓	✓	✓
2:	none			1	✓	✓	✓	✓
3:	none			1	✓	✓	✓	✓
4:	none			1	✓	✓	✓	✓
5:	none			1	✓	✓	✓	✓
6:	none			1	✓	✓	✓	✓
7:	none			1	✓	✓	✓	✓
8:	none			1	✓	✓	✓	✓
9:	none			1	✓	✓	✓	✓
10:	none			1	✓	✓	✓	✓
11:	none			1	✓	✓	✓	✓
12:	none			1	✓	✓	✓	✓
13:	none			1	✓	✓	✓	✓
14:	none			1	✓	✓	✓	✓
15:	none			1	✓	✓	✓	✓
16:	none			1	✓	✓	✓	✓
17:	none			1	✓	✓	✓	✓
18:	none			1	✓	✓	✓	✓
19:	none			1	✓	✓	✓	✓
20:	none			1	✓	✓	✓	✓

Services
XML Application URI:
XML Application Title:
BLF List URI:

Save Settings

Figure 18.1 Aastra 480i Busy Lamp Field web configuration.

18.3.1.2 Configuring with a Configuration File

Another way to configure BLF is by provisioning the phone through a central server.

Aastra phones can be configured globally through the `aastra.cfg` file, or on a per-phone basis through a configuration file with a filename containing the phone's MAC address, as in, for example, `00085D1840D1.cfg`.

An example BLF configuration for an Aastra 480i would look like this:

```
softkey1 type: blf
softkey1 label: "Station 2"
softkey1 value: 119
softkey1 line: 2
```

Note that this is not the entire configuration file. Although you can put these lines wherever you want in the file, it makes sense to keep them together for clarity.

You need at least three parameters in the configuration file to activate BLF; four are listed here. All the parameters are preceded by *softkeyN*, where *N* is the number of the softkey:

- `type`

  This sets the key type. Aastra sets have highly configurable softkeys. You need to specify what type of function you want the key to perform. In this case, we want the type `blf`, as in this example:

  ```
  softkey1 type: blf
  ```

- `label`

  This sets the display name for the softkey. You can set it to whatever you like, but generally this is set to the name of the user whose phone you want to monitor. This is a short field which can only contain up to 9 characters. You must enclose it in quotes:

  ```
  softkey1 label: "S. Smart"
  ```

- `value`

  Sets the extension you want to monitor, usually by number:

  ```
  softkey1 value: 112
  ```

- `line`

  You can specify a line number for the softkey; softkeys configured in this way will automatically dial the monitored extension when pressed.

  ```
  softkey1 line: 2
  ```

18.3.2 Polycom Soundpoint IP 501

You can only configure Busy Lamp Field on the Soundpoint IP 501 through central provisioning. On Polycom phones, BLF is called Buddy Watch.

Warning: The Polycom phones use XML configuration files. You must be extremely careful with spacing and formatting when editing these files. All tags must be closed; typographical errors, extra lines, and extra spaces have been known to cause boot problems.

The BLF is a phone-specific setting made in the directory file for the phone you want to configure. The filename for the directory file follows the convention `mac_address-directory.xml`. For example, `0004f2047ff5-directory.xml` is a valid directory filename. A directory file with presence configured for two extensions might look like this:

```
<?xml version="1.0" encoding="UTF-8" standalone="yes"?>
<!-- $Revision: 1.2 $ $Date: 2004/12/21 18:28:05 $ -->
<directory>
        <item_list>
              <item>
                      <ln>Smart</ln>
                      <fn>Sally</fn>
                      <ct>201</ct>
                      <sd>1</sd>
                      <rt>3</rt>
                      <dc/>
                      <ad>0</ad>
                      <ar>0</ar>
                      <bw>1</bw>
                      <bb>0</bb>
              </tem>
              <item>
                      <ln>Duff</ln>
                      <fn>Don</fn>
                      <ct>202</ct>
                      <sd>2</sd>
                      <rt>3</rt>
                      <dc/>
                      <ad>0</ad>
                      <ar>0</ar>
                      <bw>1</bw>
                      <bb>0</bb>
              </item>
        </item_list>
</directory>
```

All the fields for a given entry must be enclosed in the `<item></item>` tags. These are the fields you'll need to set:

- `<ct>extension</ct>`

  This sets the extension you want to monitor and is usually a number:

  `<ct>201</ct>`

- `<sd>number</sd>`

  This sets the speed dial number for the entry. The Soundpoint IP 501 has three line buttons, which can be line appearances, speed dial buttons, or BLF indicators. Line appearances always take priority. For example, if you configure three line appearances, no buttons or fields will remain on the idle screen for BLF (although you can still check the states of monitored extensions by pressing the Buddies softkey). If you have a single line appearance, it will occupy the first button. If you set <sd> to 1, the entry will occupy the next available line button.

- <bw>[0,1]</bw>

  This is the Buddy Watch parameter (essentially, another Polycom name for BLF). Allowed values are 0 and 1. If this is set to 1, the extension specified in <ct></ct> will be monitored. When this is set, the Buddy icon appears next to the button if the line is available; the Do Not Enter icon appears when the line is in use for any other reason.

- <bb>[0,1]</bb>

  This is the Buddy Block parameter. Allowed values are 0 and 1. If this is set to 1, it blocks the extension defined in this entry from monitoring this phone.

Configuring BLF for SIP phones is generally straightforward. The different terminology used by different phone manufacturers can be a source of confusion. Equivalent terms include *presence, buddy watch, extension monitoring,* and *subscription.* As with all the hardware-specific examples, consult with the manufacturer's documentation if you are not sure. Things change frequently.

18.4 Testing the Hints

Now we can test to see that things are working.

Note: Output in the following examples has been truncated to save space.

Enter **core show hints** in the Asterisk CLI:

```
*CLI> core show hints
    =- Registered Asterisk Dial Plan Hints
    21          : SIP/21         State:Unavailable      Watchers    0
    22          : SIP/22         State:Unavailable      Watchers    0
    ----------------
    - 2 hints registered
```

We can see that extensions we want to monitor (SIP/21 and SIP/22) have not yet contacted the Asterisk server ("Status: Unavailable"), nor have the monitoring extensions subscribed to the hints for the unavailable extensions ("Watchers 0"). We can confirm this with an additional command, **sip show subscriptions**:

```
*CLI> sip show subscriptions
Peer           User      Call ID     Extension    Last state    Type
0 active SIP subscriptions
```

Once the monitoring extension has contacted the server, we run the command
again:

```
*CLI> sip show subscriptions
Peer            User    Call ID       Extension    Last state    Type
192.168.0.2  21        815d944554e  22            Unavailable   dialog-info+xml
1 active SIP subscription
```

Now we can see that extension 21 is monitoring extension 22. Once the extension
22 contacts the Asterisk server, we will see this in the CLI:

```
Extension Changed 22 new state Idle for Notify User 21
```

Now we enter **core show hints** again:

```
*CLI> core show hints
   -= Registered Asterisk Dial Plan Hints =-
   21          : SIP/21        State:Idle          Watchers    0
   22          : SIP/22        State:Idle          Watchers    1
-----------------
- 2 hints registered
```

In contrast to the first output of this command, we can see that the phones have a
defined setting (Idle) and at least one phone is monitoring extension 22 (Watchers 1).
Configuration is now complete, and Asterisk will send an event to the monitoring
extensions as soon as the monitored extension changes state. How this is manifested will depend on the phone. Some phones have an icon on an LED display,
others have LED indicators that blink, change color, or both. Status changes are
also reported in the CLI:

```
Extension Changed 22 new state Ringing for Notify User 21
Extension Changed 22 new state InUse for Notify User 21
Extension Changed 22 new state Idle for Notify User 21
```

18.5 Configuring Pickup

The Busy Lamp Field only tells us the status of a monitored extension, but it
doesn't let us do anything about it. If we want to be able to pick up a call intended
for another extension, we have to do some additional configuration.

There are two basic kinds of call pickup. Pickup of calls for a single user is sometimes called *directed pickup;* pick up of calls for a call group is known as *group
pickup*. Let's start with the directed pickup, which is simpler.

18.5.1 extensions.conf for Pickup

We need to add additional entries to `extensions.conf` to make pickup work:

```
[internal-users]

; ...

; Pickup
exten => _*8X.,1,Set(nst=${EXTEN:2})
exten => _*8X.,n,Verbose(1,${CALLERID(num)} wants to pick up call from
↳ ${nst})
exten => _*8X.,n,Pickup(${nst}@internal-users)
```

To accomplish the same thing in `extensions.ael`, we make the following entries:

```
context internal-users {

  // ...

  // Pickup
  _*8X. => {
    Set(nst=${EXTEN:2});
    Verbose(1,${CALLERID(num)} wants to pick up call from ${nst});
    // An authentication step could be put here
    Pickup(${nst}@internal-users);
  }
}
```

Now we have defined an Asterisk extension that picks up a call to *<extension>* in the `internal-users` context if `*8`*<extension>* (e.g. `*821`) is dialed, provided that *<extension>* (e.g., 21) is ringing but has not yet been answered. (See also Appendix B, "Dialplan Applications.")

Warning: The `Pickup()` application should not be confused with similar applications packaged with bristuff (such as `PickUp()`, `DPickup()`, `PickUp2()`). You might find these in the form of patches when doing Internet searches on the topic.

18.5.2 Configuring the Phones for Pickup

Some phones support BLF + directed call pickup. In this configuration, when a monitored phone is ringing, the user can pick up the call simply by pressing the adjacent associated button. We will set this up here. Because Polycom sets do not support BLF + directed call pickup with Asterisk, we provide only the Aastra instructions.

Tip: This behavior can be replicated on Polycom phones by having Asterisk do the state detection. In such a scenario, if the device is idle, dialing the extension will ring the extension. If the extension is already ringing, and a phone in the correct context dials it, that phone will pick up the extension. This way, the button adjacent to the "watched buddy" can be used to do call pickup, even though it is simply a speed dial button.

Early on, Aastra sets relied on the SIP subscription event package to determine what to dial in order to do call pickup. This meant that, in theory, it was only necessary to enable directed call pickup in the web configuration to make this feature work.

Unfortunately, Asterisk does not transmit this information by default. To do this, Asterisk must be patched using the patch provided at https://issues.asterisk.org/view.php?id=5014. This is still an option for you if recompiling and installing Asterisk is not too bothersome; then you need only to activate directed call pickup in the Aastra web configuration. Log in to the phone, find Basic Settings on the left, and click Preferences. Scroll to the bottom of the screen and find the Directed Call Pickup Settings section. Check the box next to Directed Call Pickup to enable it, as shown in Figure 18.2.

Click Save Settings, then restart the phone by going to Operation on the left, clicking Reset and then Restart.

Another way to do this, which does not require patching and recompiling Asterisk, is to tell the Aastra phone what the pickup prefix is. Currently, this can be done only via the Aastra configuration file, not through the web interface. For the purposes of our example, add the following lines to the configuration file (as mentioned before, in the form `mac_address.cfg`):

```
directed call pickup: 1
directed call pickup prefix: *8
```

Force the phone to restart so that it loads the configuration. Pressing the softkey associated with the BLF indicator of a ringing extension should now pick up the call.

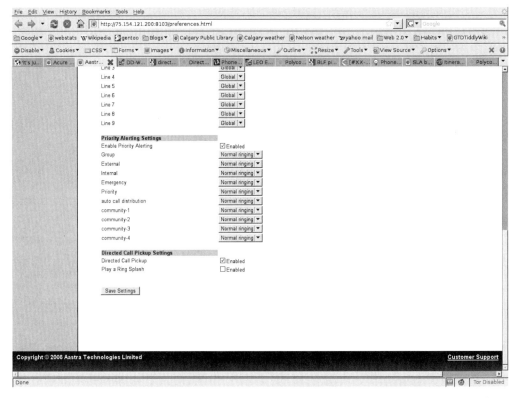

Figure 18.2 Activating directed call pickup on the Aastra 480i.

18.5.3 Configuring Group Pickup

Group pickup is very similar in operation to directed call pickup. One way is to make pickup extensions, such as in this example:

```
context internal-users {

    // Sales
    hint(SIP/21&SIP/22&SIP/23) 20 => {
      Dial(SIP/21&SIP/22&SIP/23,40);
    }

    // Pickup calls to sales
    _*820 => {
      Verbose(1,${CALLERID(num)} wants to pick up call to sales);
      Pickup(21@internal-users&22@internal-users&23@internal-users);
    }
}
```

To monitor this extension, you set 20 as the value in the Softkeys and XML configuration page of the Aastra.

The PICKUPMARK functionality of the **Pickup()** application (Appendix B) provides a bit more flexibility:

```
context internal-users {

    // Sales staff
    _2X => {
      Set(__PICKUPMARK=sales);
      Dial(SIP/${EXTEN},40);
    }
    hint(SIP/21) 21 => {}
    hint(SIP/22) 22 => {}
    hint(SIP/23) 23 => {}

    // Sales department
    hint(SIP/21&SIP/22&SIP/23) 20 => {
      Set(__PICKUPMARK=sales);
      Dial(SIP/21&SIP/22&SIP/23,40);
    }

    // Pickup of calls to sales
    _*820 => {
      Verbose(1,${CALLERID(num)} wants to pick up a call to sales);
      Pickup(sales@PICKUPMARK);
    }
}
```

Advanced users who want to avoid using up an extra extension (20 in the example above) for the pickup group can try this configuration:

```
context internal-users {

    // Sales staff
    _2X => {
      Set(__PICKUPMARK=200);
      // To define which users belong to which pickup group,
      // you can use an external AGI script
      Dial(SIP/${EXTEN}&Local/**${PICKUPMARK},40);
    }
    hint(SIP/21) 21 => {}
    hint(SIP/22) 22 => {}
    hint(SIP/23) 23 => {}

    // Pickup groups
    _**X. => {
      Verbose(1,Pickup group ${EXTEN:2} is ringing);
    }
    hint(SIP/21&SIP/22&SIP/23) **200 => {}
```

```
// Directed call pickup
_*8X. => {
  Set(nst=${EXTEN:2});
  Verbose(1,${CALLERID(num)} wants to pick up call to ${nst});
  // You could put an authorization test here
  Pickup(${nst}@internal-users);
}

// Group call pickup
_*8**X. => {
  Set(pmark=${EXTEN:4});
  Verbose(1,${CALLERID(num)} wants to pick up call to ${pmark});
  // You could put an authorization test here
  Pickup(${pmark}@PICKUPMARK);
}
}
```

We mentioned earlier that pickup rights could be handled through another mechanism; the comment lines above show where you would perform an authentication test, which could be done either in the dialplan or through an external AGI script.

On the monitoring extension, you set the monitored extension to `**200`.

18.6 Advanced BLF Indicators in Asterisk 1.6

`DEVICE_STATE()`

You already know how to monitor the status of SIP users. Here is a quick review for the sake of comparison. In `extensions.conf`, a user hint looks like this:

```
exten => 21,hint,SIP/21
```

In `extensions.ael`, the same hint looks like this:

```
hint(SIP/21) 21 => {}
```

As of Asterisk 1.6, you can monitor the state of MeetMe conferences using hints. Assume a MeetMe conference room of 885, which you want to tie to Asterisk extension 881. In `extensions.conf`, this would look like this:

```
exten => 881,hint,MeetMe:885
exten => 881,1,MeetMe(885)
```

In `extensions.ael`, it would look like this:

```
hint(MeetMe:885) 881 => {
  MeetMe(885);
}
```

The hint in the preceding example ties the status of extension 881 to the status of MeetMe room 885. The numbers are deliberately different, for the sake of clarity; in practice, they are often the same. You can now configure a BLF softkey with the extension 881, and it will indicate the status of the conference (that is, active or not).

Even more interesting, though, is that Asterisk 1.6 lets you tie hints to virtual devices. In `extensions.conf`, this looks like this:

```
exten => 99,hint,Custom:my-status
```

In `extensions.ael`, the same construction looks like this:

```
hint(Custom:my-status) 99 => {}
```

The status of such a custom device can be set to anything you like using the **DEVICE_STATE()** function from elsewhere in the dialplan.

Warning: For a time, this function was called **DEVSTATE()**, which is the same as **DEVICE_STATE()**. This function is not to be confused with the application **DevState()** or other applications and functions with similar names, even if they work similarly. You might see these alternates mentioned in list archives and Internet forums.

The function uses this construction:

```
Set (DEVICE_STATE (Custom:my-status)=INUSE);
```

The status is then transmitted to any phones that have subscribed to the hint. The possible values are as follows:

- UNKNOWN
  Unknown; the device cannot be contacted.
- UNAVAILABLE
  The device is unavailable.
- NOT_INUSE
  The device is available.
- INUSE
  The device is in a call.
- BUSY
  The device is in a call and will not take additional calls. (The device is busy.)

- RINGING

  The device is ringing.

- RINGINUSE

  The device is in use, but is also ringing. (An additional caller is calling and the device is receiving call waiting notification.)

- ONHOLD

  The device has a call on hold.

- INVALID

  The status is invalid. (You might use this if the device status cannot be determined because it doesn't exist; for example, for a dynamic conference room.)

How this status is actually manifested on a monitoring extension depends on the device manufacturer. Not all devices distinguish between all the status types.

The default status is UNKNOWN, which would be the case after an Asterisk restart before the status is set.

Once a phone has subscribed to the status of Custom:my-status at extension 99 in the above example, you can test whether it works by adding these statements to the dialplan (shown here in AEL format):

```
hint(Custom:my-status) 99 => {}

*91 => {
  Set(DEVICE_STATE(Custom:my-status)=RINGING);
}
*92 => {
  Set(DEVICE_STATE(Custom:my-status)=INUSE);
}
*93 => {
  Set(DEVICE_STATE(Custom:my-status)=NOT_INUSE);
}
*94 => {
  if ("${DEVICE_STATE(Custom:my-status)}" = "INUSE") {
    Set(DEVICE_STATE(Custom:my-status)=NOT_INUSE);
  }
  else {
    Set(DEVICE_STATE(Custom:my-status)=INUSE);
  }
}
```

You can now set the device status by dialing the configured extensions accordingly. (Again, how each model of phone indicates those states will vary.) Most of the extensions are self-explanatory; *94 toggles between INUSE and NOT_INUSE. After you've figured out the basic operation, there are no limits on what you can do with indicators of this type. For example, you could use it to indicate whether you are logged into a queue, or even if your sprinkler system is currently watering the lawn.

Of course, this hint will appear in the output of **core show hints** along with all the other more conventional hints:

```
*CLI> core show hints
   -= Registered Asterisk Dial Plan Hints =-
   99@test            : Custom:my-status   State:InUse        Watchers   0
   21@internal-users : SIP/21              State:Unavailable  Watchers   0
   22@internal-users : SIP/22              State:Unavailable  Watchers   0
   ----------------
- 1 hints registered
```

The CLI command **devstate list** lets you find the status of all your Custom hints:

```
*CLI> devstate list
--------------------------------------------------------------------
--- Custom Device States -------------------------------------------
--------------------------------------------------------------------
---
--- Name: 'Custom:my-status' State: 'INUSE'
---
--------------------------------------------------------------------
--------------------------------------------------------------------
```

Finally, **devstate change device state** lets you set the status from the CLI:

```
*CLI> devstate change Custom:my-status RINGING
Changing my-status to RINGING
*CLI> devstate list
--------------------------------------------------------------------
--- Custom Device States -------------------------------------------
--------------------------------------------------------------------
---
--- Name: 'Custom:my-status' State: 'RINGING'
---
--------------------------------------------------------------------
--------------------------------------------------------------------
```

A

Installation Instructions for Asterisk 1.4

These instructions install the components necessary to use all the features covered in this book (except where otherwise indicated). As a result, it is a little more comprehensive than a typical installation but offers the advantage that you won't have to install additional components in the future.

> **Tip:** If you don't want to install a new operating system on the computer, you can try installing using a KNOPPIX Live System.

We assume a freshly installed Debian GNU/Linux 5.0 (Lenny). You can find an ISO image for the installation CD at www.debian.org/releases/lenny/debian-installer. We recommend the 150MB network install image. (You will most likely need the one for the i386 architecture.) You can find installation instructions for Debian GNU/Linux at www.debian.org/releases/lenny/i386.

Before continuing, please log in as root.

> **Tip:** Some users will undoubtedly want to install an SSH server immediately, in order to work on the server from their regular workstation. The installation command to do this is **aptitude install ssh**. We will assume you are comfortable with SSH or will be working directly at the server console.

Begin by making sure that your package lists are current:

```
debian:~# aptitude update
Get http://ftp.debian.org lenny Release.gpg
[...]
1820kB downloaded in 13s (138kB/s)
Reading package lists... Done

debian:~#
```

Also, make sure that any already installed packages are the most current:

```
debian:~# aptitude -y upgrade
Reading package lists... Done
Building dependency tree
[...]
No packages will be installed, upgraded, or removed.
0 packages upgraded, 0 newly installed, 0 to remove and 0 not upgraded.
Need to get 0B of archives. After unpacking 0B will be used.
debian:~#
```

Just in case the updates installed a new kernel, we restart the system with **shutdown -r now** (or reboot):

```
debian:~# shutdown -r now

Broadcast message from root@debian (pts/0) (Mon Mar 23 17:42:04 2009):

The system is going down for reboot NOW!
```

After boot, log in again as root.

Tip: It is a good idea to install NTP (Network Time Protocol); this keeps the system time accurate and current:

```
debian:~# aptitude -y install ntp ntpdate
```

First, we must install a working build environment (i.e., compiler, linker, autoconf, and so on and any necessary dependencies) so that we can compile and install Asterisk:

```
debian:~# aptitude -y install build-essential
[...]
The following NEW packages will be installed:
  binutils{a} build-essential bzip2{a} cpp{a} cpp-4.3{a}
  dpkg-dev{a} g++{a} g++-4.3{a} gcc{a} gcc-4.3{a}
  libc6-dev{a} libgmp3c2{a} libgomp1{a} libmpfr1ldbl{a}
  libstdc++6-4.3-dev{a} libtimedate-perl{a}
  linux-libc-dev{a} make{a} patch{a}
```

```
0 packages upgraded, 19 newly installed, 0 to remove and 0 not upgraded.
Need to get 19.4MB of archives. After unpacking 60.4MB will be used.
[...]
Setting up libstdc++6-4.3-dev (4.3.2-1.1) ...
Setting up g++-4.3 (4.3.2-1.1) ...
Setting up g++ (4:4.3.2-2) ...
Setting up build-essential (11.4) ...
[...]
```

The Linux kernel headers are also required:

```
debian:~# aptitude -y install linux-headers-`uname -r`
The following NEW packages will be installed:
  cpp-4.1{a} gcc-4.1{a} gcc-4.1-base{a}
  linux-headers-2.6.26-1-686
  linux-headers-2.6.26-1-common{a}
  linux-kbuild-2.6.26{a}
0 packages upgraded, 19 newly installed, 0 to remove and 0 not upgraded.
Need to get 7326kB of archives. After unpacking 41.3MB will be used.
[...]
```

A.1 LibPRI

The LibPRI library contains the PRI stack, which includes some functions used for BRI connections. This is not required for all types of installations, but won't do any harm, either. Obtain the package this way:

```
debian:~# cd /usr/src/
debian:/usr/src# wget http://downloads.digium.com/pub/libpri/libpri-1.4-current.tar.gz
[...]
2009-03-23 10:24:55 (108 KB/s) - »libpri-1.4-current.tar.gz« saved [94075/94075]
```

Then, unpack it with tar:

```
debian:/usr/src# tar xvzf libpri-1.4-current.tar.gz
libpri-1.4.9/
libpri-1.4.9/build_tools/
[...]
debian:/usr/src# rm libpri-1.4-current.tar.gz
```

Finally, compile and install it:

```
debian:/usr/src# cd libpri-1.4.9/
debian:/usr/src/libpri-1.4.9# make
[...]
debian:/usr/src/libpri-1.4.9# make install
[...]
```

A.2 DAHDI

DAHDI handles telephony adapter card (both analog and digital) communications, and until Asterisk 1.4.21 was known as Zaptel (see Appendix H, "From Zaptel to DAHDI"). Even if you have no such cards installed, you should install this package because it performs other important functions, such as providing a time source for conferences.

Get the DAHDI package like so:

```
debian:~# cd /usr/src/
debian:/usr/src# wget http://downloads.digium.com/pub/telephony/
dahdi-linux/dahdi-linux-current.tar.gz
[...]
2009-03-23 10:33:37 (165 KB/s) - »dahdi-linux-current.tar.gz« saved
[1251822/1251822]

debian:/usr/src# wget http://downloads.digium.com/pub/telephony/
dahdi-tools/dahdi-tools-current.tar.gz
[...]
2009-03-23 10:37:49 (147 KB/s) - »dahdi-tools-current.tar.gz« saved
[413598/413598]

debian:/usr/src#
```

Unpack the tar.gz archives:

```
debian:/usr/src# tar xvzf dahdi-linux-current.tar.gz
dahdi-linux-2.1.0.4/
dahdi-linux-2.1.0.4/include/
[...]
debian:/usr/src# tar xvzf dahdi-tools-current.tar.gz
dahdi-tools-2.1.0.2/
dahdi-tools-2.1.0.2/hdlctest.c
[...]
debian:/usr/src# rm dahdi-linux-current.tar.gz
debian:/usr/src# rm dahdi-tools-current.tar.gz
```

Now, install the dependencies:

```
debian:/usr/src# aptitude -y install libusb-dev libnewt-dev
[...]
The following NEW packages will be installed:
  libnewt-dev libpng12-0{a} libpng12-dev{a}
  libslang2-dev{a} libusb-dev zlib1g-dev{a}
0 packages upgraded, 6 newly installed, 0 to remove and 0 not upgraded.
Need to get 1176kB of archives. After unpacking 3113kB will be used.
[...]
Setting up libnewt-dev (0.52.2-11.3) ...
Setting up libusb-dev (2:0.1.12-13) ...
[...]
```

Finally, compile and install DAHDI and the tools:

```
debian:/usr/src# cd dahdi-linux-2.1.0.4/
debian:/usr/src/dahdi-linux-2.1.0.4# make
[...]
make[1]: Leaving directory '/usr/src/linux-headers-2.6.26-1-686'
debian:/usr/src/dahdi-linux-2.1.0.4# make install
[...]
  DEPMOD    2.6.26-1-686
make[1]: Leaving directory '/usr/src/linux-headers-2.6.26-1-686'
[...]
###################################################
###
### DAHDI installed successfully.
### If you have not done so before, install the package
### dahdi-tools.
###
###################################################

debian:/usr/src/dahdi-linux-2.1.0.4# cd ..
debian:/usr/src# cd dahdi-tools-2.1.0.2/
debian:/usr/src/dahdi-tools-2.1.0.2# ./configure
[...]
checking for newtBell in -lnewt... yes
checking newt.h usability... yes
checking newt.h presence... yes
checking for newt.h... yes
checking for usb_init in -lusb... yes
checking usb.h usability... yes
checking usb.h presence... yes
checking for usb.h... yes
[...]
debian:/usr/src/dahdi-tools-2.1.0.2# make
[...]
make[1]: Leaving directory '/usr/src/dahdi-tools-2.1.0.2'
debian:/usr/src/dahdi-tools-2.1.0.2# make install
[...]
###################################################
###
### DAHDI tools installed successfully.
### If you have not done so before, install init scripts with:
###
###   make config
###
###################################################
debian:/usr/src/dahdi-tools-2.1.0.2#
```

To install init scripts and configuration files, run **make config**:

```
debian:/usr/src/dahdi-tools-2.1.0.2# make config
install -D dahdi.init /etc/init.d/dahdi
/usr/bin/install -c -D -m 644 init.conf.sample /etc/dahdi/init.conf
/usr/bin/install -c -D -m 644 modules.sample /etc/dahdi/modules
```

```
/usr/bin/install -c -D -m 644 modprobe.conf.sample /etc/modprobe.d/dahdi
/usr/bin/install -c -D -m 644 blacklist.sample /etc/modprobe.d/dahdi.blacklist
/usr/sbin/update-rc.d dahdi defaults 15 30
 Adding system startup for /etc/init.d/dahdi ...
   /etc/rc0.d/K30dahdi -> ../init.d/dahdi
   /etc/rc1.d/K30dahdi -> ../init.d/dahdi
   /etc/rc6.d/K30dahdi -> ../init.d/dahdi
   /etc/rc2.d/S15dahdi -> ../init.d/dahdi
   /etc/rc3.d/S15dahdi -> ../init.d/dahdi
   /etc/rc4.d/S15dahdi -> ../init.d/dahdi
   /etc/rc5.d/S15dahdi -> ../init.d/dahdi
DAHDI has been configured.

If you have any DAHDI hardware it is now recommended you
edit /etc/dahdi/modules in order to load support for only
the DAHDI hardware installed in this system. By default
support for all DAHDI hardware is loaded at DAHDI start.

I think that the DAHDI hardware you have on your system is:
debian:/usr/src/dahdi-tools-2.1.0.2#
```

There is no harm in leaving the defaults in place. If you have no cards installed, you can safely comment out all the modules in /etc/dahdi/modules. To avoid a reboot, we simply restart DAHDI:

```
debian:~# /etc/init.d/dahdi restart
Unloading DAHDI hardware modules: done
Loading DAHDI hardware modules:

No hardware timing source found in /proc/dahdi, loading dahdi_dummy
Running dahdi_cfg: done.
debian:~#
```

As there are no built-in cards in our example, dahdi_dummy (formerly ztdummy) is loaded so that we have a timing source.

A.3 Asterisk

Now it's time to install Asterisk itself. You can obtain the necessary source files at the Asterisk home page, www.asterisk.org. Make sure to get a stable (not a development) version:

```
debian:~# cd /usr/src/
debian:/usr/src# wget http://downloads.digium.com/pub/asterisk/asterisk-
1.4-current.tar.gz
[...]
2009-03-23 10:45:32 (172 KB/s) - »asterisk-1.4-current.tar.gz« saved
[11858197/11858197]
```

Now, unpack the sources:

```
debian:/usr/src# tar xvzf asterisk-1.4-current.tar.gz
asterisk-1.4.24/
[...]
debian:/usr/src# rm asterisk-1.4-current.tar.gz
```

Next, install the required dependencies:

```
debian:/usr/src# aptitude -y install libncurses5-dev libnewt-dev
[...]
The following NEW packages will be installed:
  libncurses5-dev
0 packages upgraded, 1 newly installed, 0 to remove and 0 not upgraded.
Need to get 1546kB of archives. After unpacking 6599kB will be used.
[...]

debian:/usr/src# aptitude -y install libcurl4-openssl-dev
[...]
The following NEW packages will be installed:
  ca-certificates{a} comerr-dev{a} libcurl3{a}
  libcurl4-openssl-dev libglib2.0-0{a}
  libglib2.0-data{a} libidn11{a} libidn11-dev{a}
  libkadm55{a} libkrb5-dev{a} libldap2-dev{a}
  libpcre3{a} libssh2-1{a} libssh2-1-dev{a}
  libssl-dev{a} openssl{a} pkg-config{a}
0 packages upgraded, 17 newly installed, 0 to remove and 0 not upgraded.
Need to get 8171kB of archives. After unpacking 23.2MB will be used.
[...]

debian:/usr/src# aptitude -y install libspeex-dev libspeexdsp-dev
[...]
The following NEW packages will be installed:
  libspeex-dev libspeex1{a}
  libspeexdsp-dev libspeexdsp1{a}
0 packages upgraded, 2 newly installed, 0 to remove and 0 not upgraded.
Need to get 320kB of archives. After unpacking 717kB will be used.
[...]

debian:/usr/src# aptitude -y install libiksemel-dev
[...]
The following NEW packages will be installed:
  libiksemel-dev libiksemel3{a}
0 packages upgraded, 2 newly installed, 0 to remove and 0 not upgraded.
Need to get 91.9kB of archives. After unpacking 299kB will be used.
[...]

debian:/usr/src# aptitude -y install unixodbc-dev
[...]
The following NEW packages will be installed:
  autotools-dev{a} defoma{a} file{a} fontconfig{a}
  fontconfig-config{a} libaudio2{a} libdrm2{a}
  libexpat1{a} libfontconfig1{a} libfreetype6{a}
  libgl1-mesa-glx{a} libglu1-mesa{a} libice6{a}
```

```
   libjpeg62{a} liblcms1{a} libltdl3{a} libltdl3-dev{a}
   libmagic1{a} libmng1{a} libodbcinstq1c2{a}
   libqt3-mt{a} libsm6{a} libtool{a} libxcursor1{a}
   libxdamage1{a} libxfixes3{a} libxft2{a} libxi6{a}
   libxinerama1{a} libxmu6{a} libxrandr2{a}
   libxrender1{a} libxt6{a} libxxf86vm1{a}
   odbcinst1debian1{a} ttf-dejavu{a} ttf-dejavu-core{a}
   ttf-dejavu-extra{a} ucf{a}unixodbc{a} unixodbc-dev
0 packages upgraded, 41 newly installed, 0 to remove and 0 not upgraded.
Need to get 12.8MB of archives. After unpacking 32.2MB will be used.
[...]
```

To prepare for compilation, run the configure script with the Asterisk sources:

```
debian:/usr/src# cd asterisk-1.4.24/
debian:/usr/src/asterisk-1.4.24# ./configure
checking build system type... i686-pc-linux-gnu
checking host system type... i686-pc-linux-gnu
checking for gcc... gcc
[...]
configure: Package configured for:
configure: OS type  : linux-gnu
configure: Host CPU : i686
debian:/usr/src/asterisk-1.4.24#
```

Now you can begin compiling:

Tip: Experienced users can interactively select specific modules by invoking **make menuselect**.

```
debian:/usr/src/asterisk-1.4.24# make
[...]
   [CC] astman.c -> astman.o
   [CC] md5.c -> md5.o
   [LD] astman.o md5.o -> astman
   [CC] stereorize.c -> stereorize.o
[...]
make[1]: Leaving directory '/usr/src/asterisk-1.4.24/main'
+--------- Asterisk Build Complete ---------+
+ Asterisk has successfully been built, and +
+ can be installed by running:              +
+                                           +
+               make install                +
+-------------------------------------------+
debian:/usr/src/asterisk-1.4.24#
```

Finally, install the compiled sources by invoking **make install**:

```
debian:/usr/src/asterisk-1.4.24# make install
[...]
```

```
+---- Asterisk Installation Complete -------+
+                                           +
+     YOU MUST READ THE SECURITY DOCUMENT   +
+                                           +
+ Asterisk has successfully been installed. +
+ If you would like to install the sample   +
+ configuration files (overwriting any      +
+ existing config files), run:              +
+                                           +
+                make samples               +
+                                           +
+---------------- or --------------------+  .
+                                           +
+ You can go ahead and install the asterisk +
+ program documentation now or later run:   +
+                                           +
+                make progdocs              +
+                                           +
+ **Note** This requires that you have      +
+ doxygen installed on your local system    +
+-------------------------------------------+
debian:/usr/src/asterisk-1.4.24#
```

All the binaries and libraries are now installed, but the configuration directory
/etc/asterisk is completely empty. Because we don't want to start completely
from scratch, we want the sample configuration files. Get those by invoking **make
samples**:

```
debian:/usr/src/asterisk-1.4.24# make samples
[...]
debian:/usr/src/asterisk-1.4.24#
```

To set up Asterisk so that it starts and stops automatically at boot time and shutdown, we need the init scripts. We get those with **make config**:

```
debian:/usr/src/asterisk-1.4.24# make config
update-rc.d: warning: /etc/init.d/asterisk missing LSB information
update-rc.d: see <http://wiki.debian.org/LSBInitScripts>
 Adding system startup for /etc/init.d/asterisk ...
   /etc/rc2.d/K91asterisk -> ../init.d/asterisk
   /etc/rc3.d/K91asterisk -> ../init.d/asterisk
   /etc/rc4.d/K91asterisk -> ../init.d/asterisk
   /etc/rc5.d/K91asterisk -> ../init.d/asterisk
   /etc/rc2.d/S50asterisk -> ../init.d/asterisk
   /etc/rc3.d/S50asterisk -> ../init.d/asterisk
   /etc/rc4.d/S50asterisk -> ../init.d/asterisk
   /etc/rc5.d/S50asterisk -> ../init.d/asterisk
debian:/usr/src/asterisk-1.4.24#
```

We're almost done. Unfortunately, the Asterisk installer sets the init sequence incorrectly for a Debian Linux system. We correct that as follows:

```
debian:/usr/src/asterisk-1.4.24# cd
debian:~# update-rc.d -f asterisk remove
 Removing any system startup links for /etc/init.d/asterisk ...
   /etc/rc2.d/S50asterisk
   /etc/rc2.d/K91asterisk
   /etc/rc3.d/S50asterisk
   /etc/rc3.d/K91asterisk
   /etc/rc4.d/S50asterisk
   /etc/rc4.d/K91asterisk
   /etc/rc5.d/S50asterisk
   /etc/rc5.d/K91asterisk
debian:~# update-rc.d asterisk defaults 50 15
update-rc.d: warning: /etc/init.d/asterisk missing LSB information update-
rc.d: see <http://wiki.debian.org/LSBInitScripts>
 Adding system startup for /etc/init.d/asterisk ...
   /etc/rc0.d/K15asterisk -> ../init.d/asterisk
   /etc/rc1.d/K15asterisk -> ../init.d/asterisk
   /etc/rc6.d/K15asterisk -> ../init.d/asterisk
   /etc/rc2.d/S50asterisk -> ../init.d/asterisk
   /etc/rc3.d/S50asterisk -> ../init.d/asterisk
   /etc/rc4.d/S50asterisk -> ../init.d/asterisk
   /etc/rc5.d/S50asterisk -> ../init.d/asterisk
debian:~#
```

Done! Asterisk is now installed. You can determine the installed version by running **asterisk -V** (uppercase V):

```
debian:~# asterisk -V
Asterisk 1.4.24
debian:~#
```

B

Dialplan Applications

This appendix describes the use of dialplan applications (in other words, /etc/ asterisk/extensions.conf) in detail. Applications are contained within modules, so only the applications contained in the modules you've actually loaded into Asterisk will be available to your dialplan. You can define which modules will be loaded through the [modules] section of the /etc/asterisk/modules.conf configuration file, either with the blanket autoload=yes statement or by explicitly loading specific modules with load => app_*application_name*.so. You can determine which applications are available to you by invoking the command **core show applications** in the Asterisk CLI. Detailed information about a specific application can be viewed by invoking **core application *application_name***.

Tip: The commands **core application xyz** and **core applications** work only beginning with Asterisk version 1.4. If you're using Asterisk 1.2, you can get the same result with the commands **show application xyz** and **show applications**. The old commands will work in Asterisk 1.4, but are deprecated.

Be careful not to confuse applications with functions. Functions are called as parameters to applications.

Note: Asterisk configuration files use the poorly defined INI format made famous by Microsoft, for which there is no consistently applied grammar, let alone a published one. Asterisk's config file parser also doesn't follow the conventional process of lexical analysis, tokenization, and syntactical analysis. This is why the developers of the Asterisk fork OpenPBX switched to the Mac OS X "property list" (`.plist`) configuration file format.

This is just to say that, due to the lack of a specification, it's not always clear where and whether spaces are allowed, or where quotation marks are expected. Usually, multiple formats are accepted. If a syntax isn't working for your specific Asterisk version, the only way to find out for sure is to make changes and test it.

As always, we welcome feedback. Should you find any errors in the book, please let the authors know.

In many cases, parameters may be omitted; you will still have to include the commas to indicate the empty fields (which usually means Asterisk will assume a default value), as follows:

```
exten => s,1,Dial(IAX2/User:password@example.com/123,,tT)
```

In general, failure results in a return code of –1; success returns 0. A return code of –1 means that Asterisk hangs up the channel and stops processing the dialplan for that call.

Note: The parameter delimiter is a comma (,) or pipe (|) depending on the Asterisk version. This book uses the comma.

Experienced Asterisk users may wonder why some applications have been omitted. The reason is simple: Some were already deprecated in Asterisk 1.2 and don't exist at all in 1.4. Those applications are not described here; you can find the corresponding functions that replaced them in **Appendix C, "Dialplan Functions."** The file compares ("diffs") of the internal help always compare the newer 1.4 to the older 1.2.

The examples use the hypothetical and arbitrarily chosen extension 123 and the priority 1; in practice this will not always be practical.

Prior to Asterisk 1.2, many applications jump to priority n+101 (if it exists and where n is the current priority). This old behavior (known as "priority jumping") can be enabled with the option j (jump) with some commands or by setting priorityjumping=yes in the [general] section of extensions.conf. This behavior is deprecated, however. The accepted method is to query channel variables.

The applications are listed in alphabetic order, under their functional purpose:

Call management (answering, patching, hanging up, and so on)

answer—Answering
busy—Indicate busy
chanisavail—Test to see whether a channel is available
channelredirect—Redirect a channel to another extension
congestion—Indicate congestion
dial—Initiate a call, patch a call, or connect a channel
disa—DISA (Direct Inward System Access)
followme—"Follow me" functionality
hangup—Hang up the channel
page—Page (one-way communications) to a group of devices
park—Park the call
pickup—Pick up a call in a pickup group
retrydial—`Dial()` with retry
ringing—Indicate ringing

Flow control and timeouts

continuewhile—Return to the beginning of a `while` loop
endwhile—End a `while` loop
exec—Execute an application
execif—Conditionally execute an application
execiftime—Execute an application if a time condition is met
exitwhile—Exit a `while` loop
gosub—Jump to a subroutine
gosubif—Conditionally jump to a subroutine
goto—Jump to the given priority, extension, or context
gotoif—Conditional `Goto()`
gotoiftime—Time conditional `Gosub()`
random—Jump to a random point in the dialplan
return—Return from a `Gosub()` or `GosubIf()`
tryexec—Check to see whether an application will run
while—Begin a `while` loop

Macros

macro—Call a macro
macroexclusive—Call a macro; allow only one instance of that macro to run
macroexit—Exit a macro
macroif—Conditionally call a macro

Caller identification

lookupblacklist—Look up caller ID in the blacklist

lookupcidname—Look up caller ID name in the database

privacymanager—Request input of the caller's number if caller ID cannot be obtained

setcallerpres—Set caller ID presentation flags

softhangup—Hang up the channel but return 0

zapateller—Block telemarketers

Call detail records (CDRs)

appendcdruserfield—Append a value to the CDR user field

forkcdr—Break the CDR into two entries

nocdr—Deactivate CDR for this call

resetcdr—Reset CDR

setamaflags—Set AMA flags

setcdruserfield—Set CDR user field

Voicemail

directory—Provide dial-by-name directory

mailboxexists—Check if mailbox exists

voicemail—Voicemail

voicemailmain—Administer voicemail

vmauthenticate—Authenticate the caller using records in `voicemail.conf`

Conferencing

meetme—Conference call

meetmeadmin—Administer conference

meetmecount—Count participants in a conference

Change variables

importvar—Import variables from a channel

read—Read digits dialed by caller into a variable

readfile—Read a file into a variable

realtime—Read data from the real-time system into a variable

realtimeupdate—Change variables in the real-time system

set—Set a channel variable

setglobalvar—Set a global variable

Music and sound output

background—Begin playing a sound and advance to the next priority immediately

backgrounddetect—`Background()` with speech detection

controlplayback—`Playback()` with shuttle controls (forward, reverse, exit)

datetime—Say date and time ("speaking clock")

echo—Echo audio back to caller

festival—Say text with Festival text-to-speech engine

milliwatt—Generate milliwatt test tone

mp3player—Play MP3 file or stream

musiconhold—Play hold music

nbscat—Play Network Broadcast Sound stream

playback—Play sound file

playtones—Play tones

progress—Indicate call progress in-band

sayalpha—Spell out text alphanumerically

saydigits—Say digits

saynumber—Say number

sayphonetic—Spell out text using NATO alphabet

sayunixtime—Say time in UNIX format

setmusiconhold—Set music-on-hold class

stopplaytones—Stop `Playtones()`

Recording and monitoring

agentmonitoroutgoing—Record outgoing calls of an agent

changemonitor—Change the recording file for `Monitor()`

chanspy—Eavesdrop on a channel

dial—Enable in-call recording with w or W parameters

dictate—Take and play back dictation

extenspy—Eavesdrop on an extension

mixmonitor—Like `Monitor()`, but mixes into a single file

monitor—Records a call

pausemonitor—Pauses recording of a call

record—Records incoming audio

stopmonitor—Stop `Monitor()`

unpausemonitor—Unpause recording of a call

zapbarge—Eavesdrop on a ZAP channel

zapscan—Scan through ZAP channels for eavesdropping

Database

   dbdel—Delete a database entry
   dbdeltree—Delete a database branch

General

   authenticate—Authenticate a caller
   senddtmf—Send dual-tone multi-frequency (DTMF) tones
   sendimage—Send an image
   sendtext—Send a text
   sendurl—Send a URL
   transfer—Transfer a call
   vmauthenticate—Authenticate a caller using account information
      from `voicemail.conf`
   wait—Wait a specified time
   waitexten—Wait for the caller to dial an extension
   waitforring—Wait for ring
   waitforsilence—Wait for silence
   waitmusiconhold—Wait and play music-on-hold

Scripts

   agi—Run an AGI application
   deadagi—Run `AGI()` on a hung-up channel
   dumpchan—Dump channel information to the CLI
   eagi—See `AGI()`
   externalivr—Run an external IVR
   log—Log an event at the specified verbosity level
   macro—Call a macro
   noop—Do nothing, but write the event to the CLI or logs
   read—Read caller input (digits) into a variable
   system—Execute a command in the system shell
   trysystem—Like `System()`, but always returns 0
   userevent—Send an event to the Manager interface
   verbose—Write a message to the CLI at the specified verbosity level

SIP

   sipdtmfmode—Change DTMF mode during SIP connection
   sipaddheader—Add a SIP header to an outgoing call

ZAP

flash—Perform a switchhook "flash" on a ZAP trunk

zapbarge—Eavesdrop on a ZAP channel

zapras—Starts the Zaptel ISDN RAS (Remote Access Server) on a ZAP channel

zapscan—Scan through ZAP channels for eavesdropping

Queues, call center functions

addqueuemember—Dynamically add an interface to the queue

agentcallbacklogin—Log in a call agent (with callback)

agentlogin—Log out a call agent

agentmonitoroutgoing—Record outgoing calls of an agent

parkandannounce—Park the call and announce

parkedcall—Take a parked call

pausequeuemember—Pause a call agent

queue—Queue the incoming call

queuelog—Write a message to the queue log

removequeuemember—Remove an interface from the queue

unpausequeuemember—Reactivate a paused call agent

ADSI

adsiprog—Load an ADSI script into the phone

getcpeid—Query an ADSI device for ADSI CPE ID

Miscellaneous

amd—Answering machine detection

alarmreceiver—Emulate an alarm or fire panel receiver

iax2provision—Provision an IAXy device

morsecode—Send text in Morse code

settransfercapability—Set ISDN transfer capability

sms—Send or receive SMS (Short Message System) messages

B.1 AbsoluteTimeout()

Sets the absolute maximum time allowed for a call:

```
AbsoluteTimeout(seconds)
```

> **Warning:** AbsoluteTimeout() is removed as of Asterisk 1.4 and has been
> replaced with the TIMEOUT() (see Appendix C) function:
>
>   Set(TIMEOUT(absolute)=seconds)

Asterisk versions:

```
--------| 1.2 |             |      |           |         |

    Internal help for this application in Asterisk 1.2:
      -= Info about application 'AbsoluteTimeout' =-

    [Synopsis]
    Set absolute maximum time of call

    [Description]
      AbsoluteTimeout(seconds): This application will set the absolute
      maximum amount of time permitted for a call. A setting of 0 disables
      the timeout.
      AbsoluteTimeout has been deprecated in favor of
      Set(TIMEOUT(absolute)=timeout)
```

Diff of the internal help from Asterisk 1.2 to 1.4:

Not available in Asterisk 1.4

See also **digittimeout**, **responsetimeout**, and Appendix C, "Dialplan Functions."

B.2 AddQueueMember()

Dynamically adds an interface into the queue:

```
AddQueueMember(queue[,interface[,penalty[|,options]]])
```

Dynamically adds the specified interface to the specified queue, which is configured
in queues.conf. The penalty setting, if provided, will influence the priority assigned
to the interface in the queue. Agents with lower penalty values will receive calls
before agents with higher penalty values.

If the specified interface is already in the queue and the n+101 priority exists
(where n is the current priority), the call jumps to that priority; otherwise an error
code (–1) is returned. (Depending on the version of Asterisk, you may need to provide the j option to enable priority jumping.)

If **AddQueueMember()** is called without the interface parameter, the current
user's active interface is used.

Some versions of Asterisk allow commas as an option separator.

This application sets the channel variable ${AQMSTATUS} to ADDED, MEMBERALREADY (member exists in the queue) or NOSUCHQUEUE (queue does not exist) depending on circumstance:

```
; add SIP/3000 to "supportqueue":
exten => 123,1,AddQueueMember(supportqueue,SIP/3000)

; add the active interface with a penalty of 2:
exten => 123,1,AddQueueMember(supportqueue,,2)
```

Asterisk versions:

```
--------| 1.2 |--------| 1.4 |--------| 1.6 |--------
```

Internal help for this application in Asterisk 1.4:

```
    -= Info about application 'AddQueueMember' =-

[Synopsis]
Dynamically adds queue members

[Description]
   AddQueueMember(queuename[|interface[|penalty[|options[|membername]]]]
):
Dynamically adds interface to an existing queue.
If the interface is already in the queue and there exists an n+101
priority then it will then jump to this priority. Otherwise it will
return an error.
The option string may contain zero or more of the following characters:
        'j' -- jump to +101 priority when appropriate.
  This application sets the following channel variable upon completion:
      AQMSTATUS    The status of the attempt to add a queue member as a
                     text string, one of
           ADDED | MEMBERALREADY | NOSUCHQUEUE
Example: AddQueueMember(techsupport|SIP/3000)
```

Diff of the internal help from Asterisk 1.2 to 1.4:

```
--- in Asterisk 1.2
+++ in Asterisk 1.4
@@ -4,7 +4,7 @@
   Dynamically adds queue members

   [Description]
-      AddQueueMember(queuename[|interface[|penalty[|options]]]):
+      AddQueueMember(queuename[|interface[|penalty[|options
      [|membername]]]]):
   Dynamically adds interface to an existing queue.
   If the interface is already in the queue and there exists an n+101
   priority then it will then jump to this priority. Otherwise it will
   return an error
```

Diff of the internal help from Asterisk 1.4 to 1.6:

```
--- in Asterisk 1.4
+++ in Asterisk 1.6
@@ -4,14 +4,11 @@
   Dynamically adds queue members

   [Description]
-     AddQueueMember(queuename[|interface[|penalty[|options[|membername]]]]):
+     AddQueueMember(queuename[,interface[,penalty[,options[,membername
[,stateinterface]]]]]):
   Dynamically adds interface to an existing queue.
-  If the interface is already in the queue and there exists an n+101 priority
-  then it will then jump to this priority.  Otherwise it will return an error
-  The option string may contain zero or more of the following characters:
-        'j' -- jump to +101 priority when appropriate.
+  If the interface is already in the queue it will return an error.
     This application sets the following channel variable upon completion:
        AQMSTATUS    The status of the attempt to add a queue member as
 a
                        text string, one of
             ADDED | MEMBERALREADY | NOSUCHQUEUE
-  Example: AddQueueMember(techsupport|SIP/3000)
+  Example: AddQueueMember(techsupport,SIP/3000)
```

See also `queue`, `removequeuemember`, `queues.conf`.

B.3 ADSIProg()

Loads an ADSI script into an ADSI-capable phone:

```
ADSIProg([script])
```

Programs an ADSI (Analog Display Services Interface) phone with the provided script. If no script is provided, the default `asterisk.adsi` is used. The pathname for the script is relative to the default Asterisk configuration directory, which is usually `/etc/asterisk`. The absolute path is also accepted.

Use **GetCPEID()** to obtain the CPE (customer premises equipment) ID and other information about the ADSI device.

```
; Program the ADSI phone with the telcordia-1.adsi script:
exten => 123,1,ADSIProg(telcordia-1.adsi)
```

Asterisk versions:

```
--------| 1.2 |--------| 1.4 |--------| 1.6 |--------
```

Internal help for this application in Asterisk 1.4:

```
    -= Info about application 'ADSIProg' =-

[Synopsis]
Load Asterisk ADSI Scripts into phone

[Description]
  ADSIProg(script): This application programs an ADSI Phone with the
given
script. If nothing is specified, the default script (asterisk.adsi) is
used.
```

Diff of the internal help from Asterisk 1.2 to 1.4:

None

Diff of the internal help from Asterisk 1.4 to 1.6:

None

See also `getcpeid`, `adsi.conf`.

B.4 AgentCallbackLogin()

Allows call agent login with callback:

```
AgentCallbackLogin([agentid][,options[,extension@context]])
```

Allows an agent identified through the agent ID to log in to the queue. A call in the queue will cause the agent's phone to ring. (This is in contrast to **AgentLogin()**, in which the agent's phone is off-hook and new calls are indicated by a tone.)

For an incoming call for the specified agent, the specified extension (at the specified context, if provided) is called.

The option s makes the login silent; the agent login is not reported:

```
; logs in Agent 33 silently. Calls for this agent go to SIP/300:
exten => 123,1,AgentCallbackLogin(33,s,${CALLERID(num)})

; Assuming that the agent number is the same as the agent extension,
we can do:
exten => 123,1,AgentCallbackLogin(${CALLERID(num)},s,${CALLERID(num)})
```

Numerous examples are available at www.voip-info.org/wiki/index.php?page=
Asterisk+cmd+AgentCallbackLogin.

Asterisk versions:

```
--------| 1.2 |--------| 1.4 |           |     |
```

Internal help for this application in Asterisk 1.4:

```
    -= Info about application 'AgentCallbackLogin' =-

[Synopsis]
Call agent callback login

[Description]
  AgentCallbackLogin([AgentNo][|[options][|[exten]@context]]):
Asks the agent to login to the system with callback.
The agent's callback extension is called (optionally with the specified
context).
The option string may contain zero or more of the following characters:
      's' -- silent login - do not announce the login ok segment agent
logged in/off
```

Diff of the internal help from Asterisk 1.2 to 1.4:

None

Diff of the internal help from Asterisk 1.4 to 1.6:

Not available in Asterisk 1.6

See also agentlogin.

B.5 AgentLogin()

Allows call agent login:

```
AgentLogin([agentid][,options])
```

Logs the current caller (optionally identified through agentid) into the queue as a
call agent. Once logged in, the agent can take calls with the phone off-hook; each
call is preceded by a warning tone. Calls are ended by pressing the star (*) key.

The option s makes the login silent:

```
; logs in Agent 33 silently.
exten => 123,1,AgentLogin(33,s)
```

Asterisk versions:

```
--------| 1.2 |--------| 1.4 |--------| 1.6 |--------
```

Internal help for this application in Asterisk 1.4:

```
    -= Info about application 'AgentLogin' =-

[Synopsis]
Call agent login

[Description]
  AgentLogin([AgentNo][|options]):
Asks the agent to login to the system.  Always returns -1.  While
logged in, the agent can receive calls and will hear a 'beep'
when a new call comes in.  The agent can dump the call by pressing
the star key.
The option string may contain zero or more of the following characters:
      's' -- silent login - do not announce the login ok segment after
agent logged in/off
```

Diff of the internal help from Asterisk 1.2 to 1.4:

None

Diff of the internal help from Asterisk 1.4 to 1.6:

```
--- in Asterisk 1.4
+++ in Asterisk 1.6
@@ -4,10 +4,10 @@
   Call agent login

   [Description]
-    AgentLogin([AgentNo][|options]):
+    AgentLogin([AgentNo][,options]):
  Asks the agent to login to the system.  Always returns -1.  While
  logged in, the agent can receive calls and will hear a 'beep'
  when a new call comes in. The agent can dump the call by pressing
  the star key.
  The option string may contain zero or more of the following characters:
-        's' -- silent login - do not announce the login ok segment
after agent logged in/off
+        's' -- silent login - do not announce the login ok segment
after agent logged in/off
```

B.6 AgentMonitorOutgoing()

Records all the outgoing calls of an agent.

```
AgentMonitorOutgoing([options])
```

This application attempts to determine the ID of an agent making an outgoing call by comparing the caller ID of the agent with a global variable set by the **AgentCallbackLogin()** application. As such, it should be used with **AgentCallbackLogin()**, and always in a later priority. This application uses monitoring functions in chan_agent rather than **Monitor()**, so call recording must be configured in agents.conf.

By default, recordings are saved in /var/spool/asterisk/monitor. You can override this behavior with the parameter savecallsin in agents.conf.

Warning: Be aware that recording of calls may be subject to freedom of information and privacy legislation in your jurisdiction. As a matter of professional practice, you should know the terms under which it is lawful to record telephone calls. In most jurisdictions, it is illegal to record a call without the knowledge of the participants.

If the caller ID and/or agent ID for the agent cannot be determined, the call jumps to priority n+101, if it exists.

Unless the options specify otherwise, the application returns 0.

The following options may be used:

- **d**

  forces the return of −1 in the event of error if there is no n+101 priority.

- **c**

  changes the call detail record so that the source of the call is agent/agentid rather than the caller ID.

- **n**

  suppresses error messages if the caller/agent ID cannot be determined. This is useful if a common context for agent and nonagent calls is desired.

  ```
  ; record outgoing calls of this agent and adjust the CDR accordingly
  exten => 123,1,AgentMonitorOutgoing(c)
  ```

Asterisk versions:

```
--------| 1.2 |--------| 1.4 |--------| 1.6 |--------
```

Internal help for this application in Asterisk 1.4:

```
    -= Info about application 'AgentMonitorOutgoing' =-

[Synopsis]
Record agent's outgoing call

[Description]
  AgentMonitorOutgoing([options]):
Tries to figure out the id of the agent who is placing outgoing
call based on comparison of the callerid of the current interface
and the global variable placed by the AgentCallbackLogin application.
That's why it should be used only with the AgentCallbackLogin app.
Uses the monitoring functions in chan_agent instead of Monitor
application. That have to be configured in the
agents.conf file.

Return value:
Normally the app returns 0 unless the options are passed. Also if the
callerid or the agentid are not specified it'll look for n+101 priority.

Options:
        'd' - make the app return -1 if there is an error condition and
there is
                no extension n+101
        'c' - change the CDR so that the source of the call is
'Agent/agent_id'
        'n' - don't generate the warnings when there is no callerid or
the
                agentid is not known.
             It's handy if you want to have one context for agent and
non-agent calls.
```

Diff of the internal help from Asterisk 1.2 to 1.4:

None

Diff of the internal help from Asterisk 1.4 to 1.6:

```
--- in Asterisk 1.4
+++ in Asterisk 1.6
@@ -9,15 +9,13 @@
   comparison of the callerid of the current interface and the global
variable
   placed by the AgentCallbackLogin application. That's why it should
be used only
   with the AgentCallbackLogin app. Uses the monitoring functions in
chan_agent
-  instead of Monitor application. That have to be configured in the
agents.conf file.
```

```
+   instead of Monitor application. That has to be configured in the
agents.conf file.

    Return value:
-   Normally the app returns 0 unless the options are passed. Also if
the callerid or
-   the agentid are not specified it'll look for n+101 priority.
+   Normally the app returns 0 unless the options are passed.

    Options:
-               'd' - make the app return -1 if there is an error condition
and there is
-                       no extension n+101
+               'd' - make the app return -1 if there is an error condition
                'c' - change the CDR so that the source of the call is
'Agent/agent_id'
                'n' - don't generate the warnings when there is no callerid
or the
                        agentid is not known.
```

B.7 AGI()

Runs an AGI-compliant application (similar to **EAGI()**, **FastAGI()**, **DeadAGI()**):

```
AGI(program[,arguments])
```

It also runs an AGI-compliant program called `program` on the current channel. AGI scripts or programs can be implemented in almost any conceivable language (e.g., Perl, PHP) and may be used to manipulate the channel, play sound files, interpret DTMF tones, and so on. Asterisk communicates with the AGI program over `stdin` and `stdout`. The arguments are passed directly to the AGI program at execution time.

The AGI program must be flagged as executable in the file system. The path is relative to the Asterisk AGI directory, which is at `/var/lib/asterisk/agi-bin` by default, or may be specified as an absolute path.

To run a program or script on another server, invoke it using a URL in the form `agi://host[:port[/program]]`. This way the script, program, or daemon can be running at all times and be ready to accept connections, which improves the performance of your applications. This method, known as FastAGI, is similar in function and purpose to FastCGI as used in web servers. The default port is 4573 if no port number is provided. If `program` is specified, it is provided as an environment variable `agi_network_script` to the FastAGI program. There is a sample FastAGI Perl script in the `agi` folder (`fastagi-test`). Use this as a starting point for your own FastAGI scripts.

For a list of commands an AGI script can send to Asterisk, see **agi-commands**, or enter **agi show** in the Asterisk CLI.

To run AGI programs on inactive channels (as in the case of an h extension, where the channel is on-hook), used **DeadAGI()** instead. Should your AGI program need access to the incoming audio stream, use **EAGI()** rather than **AGI()**. The incoming audio stream is provided on file descriptor 3.[1]

It returns −1 on hangup or if the program requests a hangup; returns 0 if not.

```
; call an AGI script:
exten => 123,1,AGI(my-agi-script.agi,hello,world)
exten => 123,n,Verbose(1,AGISTATUS: ${AGISTATUS})

; call another AGI script:
exten => 124,1,AGI(my-other-agi-script.agi,--x=hello,--y=world)
exten => 124,n,Verbose(1,AGISTATUS: ${AGISTATUS})

; call a FastAGI script on another server:
exten => 125,1,AGI(agi://192.168.1.130/test,--x=hello,--y=world)
exten => 125,n,Verbose(1,AGISTATUS: ${AGISTATUS})
```

Asterisk versions:

```
--------| 1.2 |--------| 1.4 |--------| 1.6 |--------
```

Internal help for this application in Asterisk 1.4:

```
    -= Info about application 'AGI' =-

[Synopsis]
Executes an AGI compliant application

[Description]
    [E|Dead]AGI(command|args): Executes an Asterisk Gateway Interface
compliant
program on a channel. AGI allows Asterisk to launch external programs
written in any language to control a telephony channel, play audio,
read DTMF digits, etc. by communicating with the AGI protocol on stdin
and stdout.
    This channel will stop dialplan execution on hangup inside of this
application, except when using DeadAGI. Otherwise, dialplan execution
will continue normally.
    A locally executed AGI script will receive SIGHUP on hangup from the
channel
except when using DeadAGI. This can be disabled by setting the
AGISIGHUP channel
variable to "no" before executing the AGI application.
```

---

1. A reminder: 0: stdin, 1: stdout, 2: stderr. File descriptor 3 is freely assignable.

Using 'EAGI' provides enhanced AGI, with incoming audio available out
of band
on file descriptor 3

  Use the CLI command 'agi show' to list available agi commands
  This application sets the following channel variable upon completion:
      AGISTATUS   The status of the attempt to the run the AGI script
                      text string, one of SUCCESS | FAILURE | HANGUP

Diff of the internal help from Asterisk 1.2 to 1.4:

```
--- in Asterisk 1.2
+++ in Asterisk 1.4
@@ -9,9 +9,16 @@
   written in any language to control a telephony channel, play audio,
   read DTMF digits, etc. by communicating with the AGI protocol on
stdin
   and stdout.
-  Returns -1 on hangup (except for DeadAGI) or if application
requested
-   hangup, or 0 on non-hangup exit.
-  Using 'EAGI' provides enhanced AGI, with incoming audio available
out of band
+    This channel will stop dialplan execution on hangup inside of this
+  application, except when using DeadAGI.  Otherwise, dialplan
execution
+  will continue normally.
+    A locally executed AGI script will receive SIGHUP on hangup from
the channel
+  except when using DeadAGI. This can be disabled by setting the
AGISIGHUP channel
+  variable to "no" before executing the AGI application.
+    Using 'EAGI' provides enhanced AGI, with incoming audio available
out of band
   on file descriptor 3

-  Use the CLI command 'show agi' to list available agi commands
+    Use the CLI command 'agi show' to list available agi commands
+    This application sets the following channel variable upon
completion.
+        AGISTATUS       The status of the attempt to the run the AGI
script
+                        text string, one of SUCCESS | FAILURE | HANGUP
```

Diff of the internal help from Asterisk 1.4 to 1.6:

```
--- in Asterisk 1.4
+++ in Asterisk 1.6
@@ -4,21 +4,22 @@
   Executes an AGI compliant application

   [Description]
-    [E|Dead]AGI(command|args): Executes an Asterisk Gateway Interface
compliant
```

- program on a channel. AGI allows Asterisk to launch external
programs
- written in any language to control a telephony channel, play audio,
- read DTMF digits, etc. by communicating with the AGI protocol on
stdin
- and stdout.
-    This channel will stop dialplan execution on hangup inside of this
- application, except when using DeadAGI. Otherwise, dialplan
execution
- will continue normally.
+   [E|Dead]AGI(command,args): Executes an Asterisk Gateway Interface
compliant
+ program on a channel. AGI allows Asterisk to launch external
programs written
+ in any language to control a telephony channel, play audio, read
DTMF digits,
+ etc. by communicating with the AGI protocol on stdin and stdout.
+   As of 1.6.0, this channel will not stop dialplan execution on
hangup inside
+ of this application. Dialplan execution will continue normally, even
upon
+ hangup until the AGI application signals a desire to stop (either by
exiting
+ or, in the case of a net script, by closing the connection).
     A locally executed AGI script will receive SIGHUP on hangup from
the channel
- except when using DeadAGI. This can be disabled by setting the
AGISIGHUP channel
- variable to "no" before executing the AGI application.
+ except when using DeadAGI. A fast AGI server will correspondingly
receive a
+ HANGUP in OOB data. Both of these signals may be disabled by setting
the
+ AGISIGHUP channel variable to "no" before executing the AGI
application.
     Using 'EAGI' provides enhanced AGI, with incoming audio available
out of band
- on file descriptor 3
+ on file descriptor 3.

-    Use the CLI command 'agi show' to list available agi commands
+   Use the CLI command 'agi show' to list available agi commands.
     This application sets the following channel variable upon
completion:
        AGISTATUS   The status of the attempt to the run the AGI
script
-                        text string, one of SUCCESS | FAILURE | HANGUP
+   text string, one of
SUCCESS | FAILURE | NOTFOUND | HANGUP

See also deadagi, fastagi.

B.8 AlarmReceiver()

Receives alarm reports from a burglar or fire alarm panel:

```
AlarmReceiver()
```

It emulates an alarm receiver and allows Asterisk to receive and process alarm reports in proprietary alarm panel signaling formats from burglar and fire alarm panels. Only Ademco Contact ID formatted alarm reports are supported at this time.

When **AlarmReceiver()** is called, Asterisk performs a handshake with the connected alarm panel, waits for it to transmit events, then validates and stores them. When the panel has hung up, **AlarmReceiver()** runs the system command specified in the eventcmd of alarmreceiver.conf. The alarmreceiver.conf also contains DTMF timing settings and acknowledgment tone volume.

This application has not been certified for use in critical environments where it is the only means of polling alarm events. Use it at your own risk! Before implementing, be sure to test it thoroughly.

It always returns 0.

```
; Process alarm events:
exten => s,1,AlarmReceiver()
```

Asterisk versions:

```
--------| 1.2 |--------| 1.4 |--------| 1.6 |--------
```

Internal help for this application in Asterisk 1.4:

```
    -= Info about application 'AlarmReceiver' =-
[Synopsis]
Provide support for receiving alarm reports from a burglar or fire alarm panel

[Description]
  AlarmReceiver(): Only 1 signalling format is supported at this time:
Ademco
Contact ID. This application should be called whenever there is an
alarm
panel calling in to dump its events. The application will handshake
with the
alarm panel, and receive events, validate them, handshake them, and
store them
until the panel hangs up. Once the panel hangs up, the application will
run the
system command specified by the eventcmd setting in alarmreceiver.conf
and pipe
the events to the standard input of the application. The configuration
```

```
file also
contains settings for DTMF timing, and for the loudness of the
acknowledgement
tones.
```

Diff of the internal help from Asterisk 1.2 to 1.4:

```
--- in Asterisk 1.2
+++ in Asterisk 1.4
@@ -1,7 +1,7 @@
    -= Info about application 'AlarmReceiver' =-

  [Synopsis]
-  Provide support for receving alarm reports from a burglar or fire
alarm panel
+  Provide support for receving alarm reports from a burglar or fire
alarm panel

  [Description]
     AlarmReceiver(): Only 1 signalling format is supported at this
time: Ademco
```

Diff of the internal help from Asterisk 1.4 to 1.6:

None

See also `alarmreceiver.conf`.

B.9 AMD()

Answering machine detection. Attempts to detect an answering machine at the remote end of a call.

```
AMD([initialSilence[,greeting[,afterGreetingSilence[,totalAnalysisTime[,
minWordLength[,betweenWordsSilence[,maxNumberOfWords[,silenceThreshold]]]]
]]]])
```

If a call is initiated through a `.call` file, you can use **AMD()** to sense an answering machine at the remote end. Defaults are set in `amd.conf`:

- **`initialSilence`**

  Maximum duration of silence preceding the remote announcement. If this is exceeded, sets `${AMDSTATUS}` to MACHINE.

- **`greeting`**

  Maximum duration of an announcement. If this is exceeded, sets `${AMDSTATUS}` to MACHINE.

- **afterGreetingSilence**

  Maximum duration of silence following the remote announcement. If this is exceeded, sets ${AMDSTATUS} to HUMAN.

- **totalAnalysisTime**

  Maximum duration **AMD()** is allowed to determine whether remote end is HUMAN or MACHINE.

- **minWordsSilence**

  Minimum allowed duration of a sound for it to be considered a word.

- **betweenWordsSilence**

  Minimum allowed duration of silence between words.

- **maxNumberOfWords**

  Maximum number of words in the announcement. If this is exceeded, sets ${AMDSTATUS} to MACHINE.

- **silenceThreshold**

  The silence threshold.

This application delivers its output in the channel variables AMDSTATUS and AMDCAUSE.

AMDSTATUS can be assigned the following values:

- MACHINE

  The remote end is a machine.

- HUMAN

  The remote end is a human.

- NOTSURE

  Threshold cases are indicated with NOTSURE.

- HANGUP

  The remote end has hung up.

AMDCAUSE can be assigned the following values:

- TOOLONG-<%d total_time>
- INITIALSILENCE-<%d silenceDuration>-<%d initialSilence>
- HUMAN-<%d silenceDuration>-<%d afterGreetingSilence>

- MAXWORDS-*<%d wordsCount>*-*<%d maximumNumberOfWords>*

- LONGGREETING-*<%d voiceDuration>*-*<%d greeting>*

```
; This extension is called through a .call file:
exten => 10,1,AMD()
exten => 10,n,Goto(Status-${AMDSTATUS})
exten => 10,n(Status-HUMAN),Playback(message)
exten => 10,n,Hangup()
exten => 10,n(Status-MACHINE),Hangup()
exten => 10,n(Status-NOTSURE),Hangup()
exten => 10,n(Status-HANGUP),Hangup()
```

Asterisk versions:

```
|       |          | 1.4 |--------| 1.6 |--------
```

Internal help for this application in Asterisk 1.4:

```
   -= Info about application 'AMD' =-

[Synopsis]
Attempts to detect answering machines

[Description]
  AMD([initialSilence][|greeting][|afterGreetingSilence][|totalAnalysisTime]
     [|minimumWordLength][|betweenWordsSilence][|maximumNumberOfWords]
     [|silenceThreshold])
  This application attempts to detect answering machines at the
beginning
  of outbound calls.  Simply call this application after the call
  has been answered (outbound only, of course).
  When loaded, AMD reads amd.conf and uses the parameters specified as
  default values. Those default values get overwritten when calling AMD
  with parameters.
- 'initialSilence' is the maximum silence duration before the greeting.
If
  exceeded then MACHINE.
- 'greeting' is the maximum length of a greeting. If exceeded then
MACHINE.
- 'afterGreetingSilence' is the silence after detecting a greeting.
   If exceeded then HUMAN.
- 'totalAnalysisTime' is the maximum time allowed for the algorithm to
decide
   on a HUMAN or MACHINE.
- 'minimumWordLength'is the minimum duration of Voice to considered as
a word.
- 'betweenWordsSilence' is the minimum duration of silence after a word
to
   consider the audio that follows as a new word.
```

```
    - 'maximumNumberOfWords'is the maximum number of words in the greeting.
      If exceeded then MACHINE.
    - 'silenceThreshold' is the silence threshold.
    This application sets the following channel variable upon completion:
        AMDSTATUS - This is the status of the answering machine detection.
                    Possible values are:
                    MACHINE | HUMAN | NOTSURE | HANGUP
        AMDCAUSE - Indicates the cause that led to the conclusion.
                   Possible values are:
                   TOOLONG-<%%%%d total_time>
                   INITIALSILENCE-<%%%%d silenceDuration>-<%%%%d
initialSilence>
                   HUMAN-<%%%%d silenceDuration>-<%%%%d afterGreetingSilence>
                   MAXWORDS-<%%%%d wordsCount>-<%%%%d maximumNumberOfWords>
                   LONGGREETING-<%%%%d voiceDuration>-<%%%%d greeting>
```

Diff of the internal help from Asterisk 1.2 to 1.4:

Not available in Asterisk 1.2

Diff of the internal help from Asterisk 1.4 to 1.6:

```
--- in Asterisk 1.4
+++ in Asterisk 1.6
@@ -4,9 +4,9 @@
   Attempts to detect answering machines

   [Description]
-    AMD([initialSilence][|greeting][|afterGreetingSilence]
[|totalAnalysisTime]
-        [|minimumWordLength][|betweenWordsSilence]
[|maximumNumberOfWords]
-        [|silenceThreshold])
+    AMD([initialSilence],[greeting],[afterGreetingSilence],
[totalAnalysisTime]
+        ,[minimumWordLength],[betweenWordsSilence],
[maximumNumberOfWords]
+        ,[silenceThreshold],[|maximumWordLength])
   This application attempts to detect answering machines at the
beginning
   of outbound calls.  Simply call this application after the call
   has been answered (outbound only, of course).
@@ -26,7 +26,8 @@
   - 'maximumNumberOfWords'is the maximum number of words in the
greeting.
      If exceeded then MACHINE.
   - 'silenceThreshold' is the silence threshold.
-  This application sets the following channel variable upon
completion:
+  - 'maximumWordLength' is the maximum duration of a word to accept.
If exceeded then MACHINE
+  This application sets the following channel variables upon
```

```
completion:
        AMDSTATUS - This is the status of the answering machine
detection.
                     Possible values are:
                     MACHINE | HUMAN | NOTSURE | HANGUP
@@ -37,3 +38,4 @@
                     HUMAN-<%%%%d silenceDuration>-
<%%%%d afterGreetingSilence>
                     MAXWORDS-<%%%%d wordsCount>-
<%%%%d maximumNumberOfWords>
                     LONGGREETING-<%%%%d voiceDuration>-<%%%%d greeting>
+                    MAXWORDLENGTH-<%%%%d consecutiveVoiceDuration>
```

See also **call-file.**

B.10 Answer()

Answers a ringing channel.

```
Answer([delay])
```

Instructs Asterisk to answer the channel if it is ringing. If the channel is not ring-ing, this application has no effect.

It is generally recommended that the channel be answered before other applica-tions are called, unless there is a specific reason for not doing so. Most applications require that the channel be answered before they are run; if this is not done, the behavior may be unexpected.

The optional delay parameter specifies how long Asterisk should wait, in milli-seconds, before answering the channel.

Returns 0 upon success.

```
exten => 123,1,Answer()
exten => 123,n,Wait(1)
exten => 123,n,Playback(hello)
exten => 123,n,Hangup()
```

Asterisk versions:

```
--------| 1.2 |--------| 1.4 |--------| 1.6 |--------
```

Internal help for this application in Asterisk 1.4:

```
    -= Info about application 'Answer' =-

[Synopsis]
Answer a channel if ringing
```

```
[Description]
  Answer([delay]): If the call has not been answered, this application
will
answer it. Otherwise, it has no effect on the call. If a delay is
specified,
Asterisk will wait this number of milliseconds before returning to
the dialplan after answering the call.
```

Diff of the internal help from Asterisk 1.2 to 1.4:

None

Diff of the internal help from Asterisk 1.4 to 1.6:

None

See also hangup.

B.11 AppendCDRUserField()

Appends a string to the user field in the CDR.

```
AppendCDRUserField(string)
```

This application has been deprecated in favor of the **CDR(userfield)** function.

Asterisk versions:

```
--------| 1.2 |--------| 1.4 |            |      |
```

Internal help for this application in Asterisk 1.4:

```
     -= Info about application 'AppendCDRUserField' =-

[Synopsis]
Append to the CDR user field

[Description]
[Synopsis]
AppendCDRUserField(value)

[Description]
AppendCDRUserField(value): Append value to the CDR user field
        The Call Data Record (CDR) user field is an extra field you
        can use for data not stored anywhere else in the record.
        CDR records can be used for billing or storing other arbitrary
data
        (I.E. telephone survey responses)
        Also see SetCDRUserField().

This application is deprecated in favor of Set(CDR(userfield)=...)
```

Diff of the internal help from Asterisk 1.2 to 1.4:

```
--- in Asterisk 1.2
+++ in Asterisk 1.4
@@ -14,3 +14,5 @@
            CDR records can be used for billing or storing other
arbitrary data
            (I.E. telephone survey responses)
             Also see SetCDRUserField().
+
+   This application is deprecated in favor of Set(CDR(userfield)=...)
```

Diff of the internal help from Asterisk 1.4 to 1.6:

Not available in Asterisk 1.6

See also Appendix C.

B.12 Authenticate()

Requires that the caller enter a password before proceeding to the next priority.

```
Authenticate(password[,options[,maxDigits]])
```

Requires that the caller enters the specified password correctly before proceeding to the next priority. Allows the caller three chances to enter the password correctly before hanging up.

If `password` begins with forward slash (/), Asterisk will assume it is a filename to a file containing a list of valid passwords (exactly one per line). Passwords may also be stored in the Asterisk database (AstDB); see option **d** below.

The following options are allowed (also in combination):

- **a**

  (accountcode) Sets the CDR `accountcode` field and the channel variable `ACCOUNTCODE` to the entered password.

- **d**

  (database) Interprets the entered password as a key in the Asterisk database. If a database key is used, the value in the associated record is ignored and can be arbitrary.

- **r**

  (remove) Removes the database key after successful password entry (valid only with option **d**).

- **j**

  (jump) In the event of three failed attempts, jump to priority n+101 (if it exists) instead of hanging up.

Warning: When using option d, note that Asterisk looks for a *key* with a name
equivalent to the password: as in */passwords/1234*. The value in the record itself
is irrelevant. A more logical implementation is to place the password as a value in
the record, as in /passwords/type => 1234.

If maxDigits is set, input is ended as soon as the user has entered enough digits; this saves having to enter pound sign (#). (Default: 0; no limits on input.)

Returns 0 if the user enters the correct password within three attempts, otherwise hangs up the channel and returns –1.

```
; Request password 1234:
exten => 123,1,Answer()
exten => 123,2,Authenticate(1234,j,4)          ; an exceptional use of
priority jumping because we want to tell the caller
                                               ; that she has entered
the wrong password
exten => 123,3,Playback(pin-accepted)
exten => 123,103,Playback(pin-rejected)
```

Asterisk versions:

--------| 1.2 |--------| 1.4 |--------| 1.6 |--------

Internal help for this application in Asterisk 1.4:

```
    -= Info about application 'Authenticate' =-

[Synopsis]
Authenticate a user

[Description]
  Authenticate(password[|options[|maxdigits]]): This application asks
the caller
to enter a given password in order to continue dialplan execution. If
the password
begins with the '/' character, it is interpreted as a file which
contains a list of
valid passwords, listed 1 password per line in the file.
  When using a database key, the value associated with the key can be
anything.
Users have three attempts to authenticate before the channel is hung
up. If the
passsword is invalid, the 'j' option is specified, and priority n+101
exists,
dialplan execution will continnue at this location.
```

```
   Options:
       a - Set the channels' account code to the password that is entered
       d - Interpret the given path as database key, not a literal file
       j - Support jumping to n+101 if authentication fails
       m - Interpret the given path as a file which contains a list of
account
           codes and password hashes delimited with ':', listed one per
line in
           the file. When one of the passwords is matched, the channel
will have
           its account code set to the corresponding account code in the
file.
       r - Remove the database key upon successful entry (valid with 'd'
only)
       maxdigits   - maximum acceptable number of digits. Stops reading
after
           maxdigits have been entered (without requiring the user to
           press the '#' key).
           Defaults to 0 - no limit - wait for the user press the '#'
key.
```

Diff of the internal help from Asterisk 1.2 to 1.4:

```
--- in Asterisk 1.2
+++ in Asterisk 1.4
@@ -4,9 +4,9 @@
   Authenticate a user

   [Description]
-    Authenticate(password[|options]): This application asks the caller
to enter a
-  given password in order to continue dialplan execution. If the
password begins
-  with the '/' character, it is interpreted as a file which contains a
list of
+    Authenticate(password[|options[|maxdigits]]): This application
asks the caller
+  to enter a given password in order to continue dialplan execution.
If the password
+  begins with the '/' character, it is interpreted as a file which
contains a list of
   valid passwords, listed 1 password per line in the file.
     When using a database key, the value associated with the key can
be anything.
   Users have three attempts to authenticate before the channel is hung
up. If the
@@ -21,3 +21,7 @@
           the file. When one of the passwords is matched, the channel
will have
           its account code set to the corresponding account code in
the file.
```

```
          r - Remove the database key upon successful entry (valid with
'd' only)
+      maxdigits              - maximum acceptable number of digits.
Stops reading after
+           maxdigits have been entered (without requiring the user to
+           press the '#' key).
+           Defaults to 0 - no limit - wait for the user press the '#'
key.
```

Diff of the internal help from Asterisk 1.4 to 1.6:

```
--- in Asterisk 1.4
+++ in Asterisk 1.6
@@ -4,18 +4,15 @@
   Authenticate a user

   [Description]
-    Authenticate(password[|options[|maxdigits]]): This application
asks the caller
+    Authenticate(password[,options[,maxdigits]]): This application
asks the caller
   to enter a given password in order to continue dialplan execution.
If the password
   begins with the '/' character, it is interpreted as a file which
contains a list of
   valid passwords, listed 1 password per line in the file.
      When using a database key, the value associated with the key can
be anything.
-  Users have three attempts to authenticate before the channel is hung
up. If the
-  passsword is invalid, the 'j' option is specified, and priority
n+101 exists,
-  dialplan execution will continnue at this location.
+  Users have three attempts to authenticate before the channel is hung
up.
      Options:
          a - Set the channels' account code to the password that is
entered
          d - Interpret the given path as database key, not a literal
file
-          j - Support jumping to n+101 if authentication fails
          m - Interpret the given path as a file which contains a list of
account
          codes and password hashes delimited with ':', listed one
per line in
          the file. When one of the passwords is matched, the channel
will have
```

See also vmauthenticate.

B.13 Background()

Plays a sound file while listening for DTMF input from the caller.

```
Background(soundfile1[&soundfile2...][,options[,language]])
```

Plays the specified sound files while waiting for the caller to dial an extension. Playback stops the moment the first digit is pressed. Filenames must be provided without file extensions; Asterisk chooses the file format with the minimum transcoding cost.

Allowed options may not be combined:

- **skip**

  Playback is skipped if the channel is not in the up state when the application is run. If skip is specified, the application ends immediately when the channel is hung up.

- **noanswer**

  The channel is answered only after the specified sound file has been played. The default behavior is to answer the channel automatically before playing the sound file. Note that not all channel types allow playback of a message before being answered.

The language parameter can be used to specify a language for the sound files played, in the event this should be different than the language currently specified for the channel.

Returns −1 if hung up or if the specified sound file does not exist, otherwise returns 0.

```
exten => 123,1,Answer()
exten => 123,n,Background(please-enter-extension)
```

Asterisk versions:

```
--------| 1.2 |--------| 1.4 |--------| 1.6 |--------
```

Internal help for this application in Asterisk 1.4:

```
   -= Info about application 'BackGround' =-

[Synopsis]
Play an audio file while waiting for digits of an extension to go to.

[Description]
  Background(filename1[&filename2...][|options[|langoverride][|context]]):
This application will play the given list of files (do not put extension)
```

while waiting for an extension to be dialed by the calling channel. To
continue waiting for digits after this application has finished playing
files, the WaitExten application should be used. The 'langoverride'
option
explicitly specifies which language to attempt to use for the requested
sound
files. If a 'context' is specified, this is the dialplan context that
this
application will use when exiting to a dialed extension. If one of the
requested sound files does not exist, call processing will be
terminated.
  Options:
    s - Causes the playback of the message to be skipped
        if the channel is not in the 'up' state (i.e. it
        hasn't been answered yet). If this happens, the
        application will return immediately.
    n - Don't answer the channel before playing the files.
    m - Only break if a digit hit matches a one digit
        extension in the destination context.
See Also: Playback (application) -- Play sound file(s) to the channel,
                          that cannot be interrupted

Diff of the internal help from Asterisk 1.2 to 1.4:

```
--- in Asterisk 1.2
+++ in Asterisk 1.4
@@ -1,23 +1,25 @@
    -= Info about application 'BackGround' =-

  [Synopsis]
- Play a file while awaiting extension
+ Play an audio file while waiting for digits of an extension to go
to.

  [Description]
    Background(filename1[&filename2...][|options[|langoverride]
[|context]]):
- This application will play the given list of files while waiting for
an
- extension to be dialed by the calling channel. To continue waiting
for digits
- after this application has finished playing files, the WaitExten
application
- should be used. The 'langoverride' option explicity specifies which
language
- to attempt to use for the requested sound files. If a 'context' is
specified,
- this is the dialplan context that this application will use when
exiting to a
- dialed extension.  If one of the requested sound files does not
exist, call processing will be
+ This application will play the given list of files (do not put
extension)
```

```
+   while waiting for an extension to be dialed by the calling channel.
To
+   continue waiting for digits after this application has finished
playing
+   files, the WaitExten application should be used. The 'langoverride'
option
+   explicitly specifies which language to attempt to use for the
requested sound
+   files. If a 'context' is specified, this is the dialplan context
that this
+   application will use when exiting to a dialed extension.  If one of
the requested sound files does not exist, call processing will be
    terminated.
     Options:
-       s - causes the playback of the message to be skipped
+       s - Causes the playback of the message to be skipped
             if the channel is not in the 'up' state (i.e. it
-             hasn't been answered yet.) If this happens, the
+             hasn't been answered yet). If this happens, the
             application will return immediately.
-       n - don't answer the channel before playing the files
-       m - only break if a digit hit matches a one digit
-             extension in the destination context
+       n - Don't answer the channel before playing the files.
+       m - Only break if a digit hit matches a one digit
+             extension in the destination context.
+   See Also: Playback (application) -- Play sound file(s) to the
channel,
+                                       that cannot be interrupted
```

Diff of the internal help from Asterisk 1.4 to 1.6:

```
--- in Asterisk 1.4
+++ in Asterisk 1.6
@@ -4,7 +4,7 @@
   Play an audio file while waiting for digits of an extension to go
to.

   [Description]
-     Background(filename1[&filename2...][|options[|langoverride]
[|context]]):
+     Background(filename1[&filename2...][,options[,langoverride]
[,context]]):
   This application will play the given list of files (do not put
extension)
   while waiting for an extension to be dialed by the calling channel.
To
   continue waiting for digits after this application has finished
playing
@@ -21,5 +21,8 @@
       n - Don't answer the channel before playing the files.
       m - Only break if a digit hit matches a one digit
             extension in the destination context.
```

```
+  This application sets the following channel variable upon
completion:
+   BACKGROUNDSTATUS    The status of the background attempt as a text
string, one of
+                  SUCCESS | FAILED
   See Also: Playback (application) -- Play sound file(s) to the
channel,
                                that cannot be interrupted
```

See also playback, backgrounddetect, show translation.

B.14 BackgroundDetect()

Plays a sound file while listening for sound from the caller.

```
BackgroundDetect(soundfile[,silence[,min[,max]]])
```

Similar to **Background()**, but listens for sound also.

During playback of the sound file, the application monitors audio on the incoming audio channel. If it detects a sound longer than *min* milliseconds in duration but shorter than *max* milliseconds, followed by a period of silence of at least *silence* milliseconds, it stops playback and passes the call to the talk extension, if it exists.

If *silence*, *min*, and *max* are not specified, the defaults are used: 1,000ms, 100ms, and unlimited, respectively.

Returns –1 on hangup, otherwise returns 0 (such as when playback is interrupted due to input).

```
exten => 123,1,BackgroundDetect(symphony)
exten => 123,n,Playback(vm-sorry)
exten => talk,1,Playback(yes-please)
```

Asterisk versions:

```
--------| 1.2 |--------| 1.4 |--------| 1.6 |--------
```

Internal help for this application in Asterisk 1.4:

```
    -= Info about application 'BackgroundDetect' =-

[Synopsis]
Background a file with talk detect

[Description]
  BackgroundDetect(filename[|sil[|min|[max]]]): Plays back a given
filename, waiting for interruption from a given digit (the digit must
start the beginning of a valid extension, or it will be ignored).
During the playback of the file, audio is monitored in the receive
direction, and if a period of non-silence which is greater than 'min'
ms
```

yet less than 'max' ms is followed by silence for at least 'sil' ms
then
the audio playback is aborted and processing jumps to the 'talk'
extension
if available. If unspecified, sil, min, and max default to 1000, 100,
and
infinity respectively.

Diff of the internal help from Asterisk 1.2 to 1.4:

None

Diff of the internal help from Asterisk 1.4 to 1.6:

```
--- in Asterisk 1.4
+++ in Asterisk 1.6
@@ -4,12 +4,12 @@
   Background a file with talk detect

   [Description]
-    BackgroundDetect(filename[|sil[|min|[max]]]):  Plays   back   a   given
-  filename, waiting for interruption from a given digit (the digit must
-  start the beginning of a valid extension, or it will be ignored).
-  During the playback of the file, audio is monitored in the receive
-  direction, and if a period of non-silence which is greater than
'min' ms
-  yet less than 'max' ms is followed by silence for at least 'sil' ms
then
-  the audio playback is aborted and processing jumps to the 'talk'
extension
-  if available.  If unspecified, sil, min, and max default to 1000,
100, and
-  infinity respectively.
+    BackgroundDetect(<filename>[,<sil>[,<min>[,<max>
[,<analysistime>]]]]):
+  Plays back <filename>, waiting for interruption from a given digit
(the digit
+  must start the beginning of a valid extension, or it will be
ignored).  During
+  the playback of the file, audio is monitored in the receive
direction, and if
+  a period of non-silence which is greater than <min> ms yet less than
<max> ms
+  is followed by silence for at least <sil> ms, which occurs during
the first
+  <analysistime> ms, then the audio playback is aborted and processing
jumps to
+  the <talk> extension, if available.  If unspecified, <sil>, <min>,
<max>, and
+  <analysistime> default to 1000, 100, infinity, and infinity
respectively.
```

See also **playback, background.**

B.15 Bridge()

Connects the active channel with a specified channel.

```
Bridge(channel,options)
```

Success or failure is signaled through the BRIDGERESULT variable.

- **SUCCESS**

  Connection successful.

- **LOOP**

  You have attempted to connect a channel to itself.

- **NONEXISTENT**

  The target channel does not exist.

- **INCOMPATIBLE**

  The channel uses an incompatible technology.

- **FAILURE**

  Other error.

Asterisk versions:

```
    |        |          |          |          | 1.6 |--------
```

Internal help for this application in Asterisk 1.6:

```
        -= Info about application 'Bridge' =-

    [Synopsis]
    Bridge two channels

    [Description]
    Usage: Bridge(channel[,options])
            Allows the ability to bridge two channels via the dialplan.
    The current channel is bridged to the specified 'channel'.
      Options:
        p - Play a courtesy tone to 'channel'.
    This application sets the following channel variable upon completion:
      BRIDGERESULT    The result of the bridge attempt as a text string, one
    of
                SUCCESS | FAILURE | LOOP | NONEXISTENT | INCOMPATIBLE
```

Diff of the internal help from Asterisk 1.4 to 1.6:

Not available in Asterisk 1.4

See also dial.

B.16 Busy()

Sets the channel as "busy."

```
Busy([timeout])
```

Instructs the channel to indicate busy and waits until the caller hangs up or the timeout expires (*timeout*, in seconds).

This application indicates a busy state only on the bridged channel. Every channel type has its own way of indicating that a device is busy. To play an actual busy tone, use **Playtones (busy)**.

Always returns −1.

```
exten => 123,1,Playback(vm-sorry)
exten => 123,n,Playtones(busy)
exten => 123,n,Busy()
```

Asterisk versions:

```
--------| 1.2 |--------| 1.4 |--------| 1.6 |--------
```

Internal help for this application in Asterisk 1.4:

```
   -= Info about application 'Busy' =-
[Synopsis]
Indicate the Busy condition

[Description]
  Busy([timeout]): This application will indicate the busy condition to
the calling channel. If the optional timeout is specified, the calling
channel
will be hung up after the specified number of seconds. Otherwise, this
application will wait until the calling channel hangs up.
```

Diff of the internal help from Asterisk 1.2 to 1.4:

None

Diff of the internal help from Asterisk 1.4 to 1.6:

None

See also **congestion, progress, playtones.**

B.17 ChangeMonitor()

Changes the monitoring filename of a channel.

```
ChangeMonitor(filename-prefix)
```

Changes the filename prefix for sound files written while recording the channel
with **Monitor()**. This application has no effect if the affected channel is not being
monitored.

```
; Monitor channel with a filename prefix of 'audioclip'
exten => 123,1,Monitor(audioclip)
; Change filename prefix to 'audioclip2'
exten => 123,n,ChangeMonitor(audioclip2)
```

Asterisk versions:

```
--------| 1.2 |--------| 1.4 |--------| 1.6 |--------
```

Internal help for this application in Asterisk 1.4:

```
    -= Info about application 'ChangeMonitor' =-

[Synopsis]
Change monitoring filename of a channel

[Description]
ChangeMonitor(filename_base)
Changes monitoring filename of a channel. Has no effect if the channel
is not monitored
The argument is the new filename base to use for monitoring this
channel.
```

Diff of the internal help from Asterisk 1.2 to 1.4:

None

Diff of the internal help from Asterisk 1.4 to 1.6:

```
--- in Asterisk 1.4
+++ in Asterisk 1.6
@@ -4,6 +4,6 @@
   Change monitoring filename of a channel

   [Description]
-  ChangeMonitor(filename_base)
-  Changes monitoring filename of a channel. Has no effect if the
channel is not monitored
+    ChangeMonitor(filename_base):
+  Changes monitoring filename of a channel. Has no effect if the
channel is not monitored.
   The argument is the new filename base to use for monitoring this
channel.
```

See also monitor, stopmonitor.

B.18 ChanIsAvail()

Indicates whether the specified channel is available.

```
ChanIsAvail(technology1/resource1[&technology2/resource2...][,options])
```

Verifies that the one or more of the queried channels is available, in the order specified. Returns 0 on success or –1 on failure.

If the s (state) option is given, Asterisk will treat the channel as unavailable if it is in use, even if it is capable of taking another call. Option j sets priority jumping to n+101 if the channel is unavailable.

Note: The mere fact of a channel being available does not automatically mean that it is free for use or that the device on the channel will accept a call. That is determined using a **Dial()** to the channel.

ChanIsAvail() sets the following channel variables:

- **${AVAILCHAN}**

  The name of the accessible channel, including the session number of the call.

- **${AVAILORIGCHAN}**

  The canonical channel name (i.e., the channel name without session number).

- **${AVAILSTATUS}**

  Status code of the channel:

- **AST_DEVICE_UNKNOWN (0)**

  Status of the channel is unknown. It is a valid channel, but we don't know about its state.

- **AST_DEVICE_NOT_INUSE (1)**

  The channel is not in use.

- **AST_DEVICE_IN_USE (2)**

  The channel is in use.

- **AST_DEVICE_BUSY (3)**

  The channel is busy.

- **AST_DEVICE_INVALID (4)**
  The channel is unknown.

- **AST_DEVICE_UNAVAILABLE (5)**
  The channel is not available and not registered.

- **AST_DEVICE_RINGING (6)**
  The channel is ringing.

Note: This application does not behave as expected on MGCP channels.

```
; Check the availability of Zap/1 and Zap/2:
exten => 123,1,ChanIsAvail(Zap/1&Zap/2,j)
; As an exception, using priority jumping, because we want to announce
; something to the caller if no channel is available

; at least one channel is available - dial this channel:
exten => 123,2,NoOp(${AVAILORIGCHAN} is available)
exten => 123,3,Dial(${AVAILORIGCHAN}/123456)

; if the call goes to priority 102, neither Zap/1 nor Zap/2 is
available
exten => 123,102,Playback(all-channels-busy)
```

Asterisk versions:

--------| **1.2** |--------| **1.4** |--------| **1.6** |--------

Internal help for this application in Asterisk 1.4:

```
    -= Info about application 'ChanIsAvail' =-

[Synopsis]
Check channel availability

[Description]   ChanIsAvail(Technology/resource[&Technology2/
resource2...][|options]):
This application will check to see if any of the specified channels are
available. The following variables will be set by this application:
  ${AVAILCHAN}     - the name of the available channel, if one exists
  ${AVAILORIGCHAN} - the canonical channel name that was used to create
the channel
  ${AVAILSTATUS}   - the status code for the available channel
  Options:
    s - Consider the channel unavailable if the channel is in use at
all
    j - Support jumping to priority n+101 if no channel is available
```

Diff of the internal help from Asterisk 1.2 to 1.4:

None

Diff of the internal help from Asterisk 1.4 to 1.6:

```
--- in Asterisk 1.4
+++ in Asterisk 1.6
@@ -4,12 +4,15 @@
   Check channel availability

   [Description]
-    ChanIsAvail(Technology/resource[&Technology2/resource2...]
[|options]):
+    ChanIsAvail(Technology/resource[&Technology2/resource2...]
[,options]):
   This application will check to see if any of the specified channels
are
-  available. The following variables will be set by this application:
-    ${AVAILCHAN}    - the name of the available channel, if one
exists
-    ${AVAILORIGCHAN} - the canonical channel name that was used to
create the channel
-    ${AVAILSTATUS}  - the status code for the available channel
+  available.
     Options:
-     s - Consider the channel unavailable if the channel is in use at
all
-     j - Support jumping to priority n+101 if no channel is available
+     a - Check for all available channels, not only the first one.
+     s - Consider the channel unavailable if the channel is in use at
all.
+     t - Simply checks if specified channels exist in the channel
list
+           (implies option s).
+  This application sets the following channel variable upon
completion:
+    AVAILCHAN    - the name of the available channel, if one exists
+    AVAILORIGCHAN - the canonical channel name that was used to create
the channel
+    AVAILSTATUS  - the status code for the available channel
```

B.19 ChannelRedirect()

Redirects a channel to another extension and priority.

```
ChannelRedirect(channel,[context,]extension,priority)
```

Redirects the specified channel to another extension and priority.

- **Channel**
  Name of the channel to be redirected

- **Context**
  Context to which the channel should be redirected

- **Extension**
  Extension to which the channel should be redirected

- **priority**
  Priority in the new extension

Asterisk versions:

```
  |      |           | 1.4 |--------| 1.6 |--------
```

Internal help for this application in Asterisk 1.4:

```
    -= Info about application 'ChannelRedirect' =-

[Synopsis]
Redirects given channel to a dialplan target.

[Description]
ChannelRedirect(channel|[[context|]extension|]priority):
   Sends the specified channel to the specified extension priority
```

Diff of the internal help from Asterisk 1.2 to 1.4:

Not available in Asterisk 1.2

Diff of the internal help from Asterisk 1.4 to 1.6:

```
--- in Asterisk 1.4
+++ in Asterisk 1.6
@@ -4,5 +4,8 @@
   Redirects given channel to a dialplan target.

   [Description]
-  ChannelRedirect(channel|[[context|]extension|]priority):
+  ChannelRedirect(channel,[[context,]extension,]priority)
     Sends the specified channel to the specified extension priority
+  This application sets the following channel variables upon
completion:
+    CHANNELREDIRECT_STATUS - Are set to the result of the redirection
+                             either NOCHANNEL or SUCCESS
```

See also goto, transfer.

B.20 ChanSpy()

Enables eavesdropping on a channel.

```
ChanSpy([channelprefix[,options]])
```

Allows eavesdropping on a conversation on any specified channel. (This is different from **ZapBarge()**/**ZapScan()**, which are bound to Zap channels only.) Note that this application only listens on single channels, rather than the conversation per se, even though it does capture incoming and outgoing audio on the channel.

Warning: Be aware that listening to calls may be subject to freedom of information and privacy legislation in your jurisdiction. As a matter of professional practice, you should know the terms under which it is lawful to eavesdrop on telephone calls. In most jurisdictions it is illegal to eavesdrop on a call without the knowledge of the participants.

If *channelprefix* is specified, only channels with a name beginning with that string are available for listening.

Options may be combined:

- **b**

  (bridged) Restrict to bridged (i.e., connected) channels.

- **g(grp)**

  (group) Restrict to channels whose ${SPYGROUP} channel variable contains the value *grp* in a colon-delimited (:) list.

- **q**

  (quiet) Do not play beep tones (or announce the channel name) when switching channels.

- **r([name])**

  (record) Record the spying session in a file in the /var/spool/asterisk/ monitor directory. The default filename prefix is chanspy.

- **v[(value)]**

  (volume) Set the initial volume. The range of allowed values is from −4 (quiet) to 4 (loud).

- **w**

  (whisper) Activate whisper mode. Allows the user on the spying channel to speak to the channel user on the spied-on channel, without the remote user hearing. (This option is available as of Asterisk 1.4.)

- **W**

    (private whisper) Private whisper mode. Like w, except that the spying chan-
    nel cannot hear the spied-on channel. (It's not immediately clear where this
    would be useful, but Asterisk has found myriad applications!)

While listening, the following keypad input is accepted:

- **#**

    Increases volume stepwise (from –4 to 4).

- *****

    Switches to another channel.

- ***X..X#***

    A set of digits of arbitrary length, ended with #, is attached to *channelpre-
    fix*. For example, if the extension invokes **ChanSpy(Agent)** and the user on
    the spying channel dials 1234#, the spying channel will begin spying on the
    channel Agent/1234.

    ```
    ; Eavesdrop on an agent:
    exten => 123,1,ChanSpy(Agent)
    exten => 123,n,Hangup()

    ; Example using g:
    ; for calls to 0, set SPYGROUP 10005:
    exten => _0.,1,Set(SPYGROUP=10005)
    ;...
    ; Listen to channels in SPYGROUP 10005:
    exten => 123,1,ChanSpy(,g(10005))
    exten => 123,n,Hangup()
    ```

Asterisk versions:

```
--------| 1.2 |--------| 1.4 |--------| 1.6 |--------
```

Internal help for this application in Asterisk 1.4:

```
    -= Info about application 'ChanSpy' =-

[Synopsis]
Listen to a channel, and optionally whisper into it

[Description]
  ChanSpy([chanprefix][|options]): This application is used to listen
to the
audio from an Asterisk channel. This includes the audio coming in and
out of the channel being spied on. If the 'chanprefix' parameter is
specified,
only channels beginning with this string will be spied upon.
```

```
   While spying, the following actions may be performed:
     - Dialing # cycles the volume level.
     - Dialing * will stop spying and look for another channel to spy
on.
     - Dialing a series of digits followed by # builds a channel name to
append
        to 'chanprefix'. For example, executing ChanSpy(Agent) and then
dialing
        the digits '1234#' while spying will begin spying on the channel
        'Agent/1234'.
   Options:
       b               - Only spy on channels involved in a bridged call.
       g(grp)          - Match only channels where their ${SPYGROUP}
variable is set to
                         contain 'grp' in an optional : delimited list.
       q               - Don't play a beep when beginning to spy on a
channel, or speak the
                         selected channel name.
       r[(basename)]   - Record the session to the monitor spool directory.
An
                         optional base for the filename may be specified.
The
                         default is 'chanspy'.
       v([value])      - Adjust the initial volume in the range from -4 to
4. A
                         negative value refers to a quieter setting.
       w               - Enable 'whisper' mode, so the spying channel can
talk to
                         the spied-on channel.
       W               - Enable 'private whisper' mode, so the spying
channel can
                         talk to the spied-on channel but cannot listen to
that
                         channel.
```

Diff of the internal help from Asterisk 1.2 to 1.4:

```
--- in Asterisk 1.2
+++ in Asterisk 1.4
@@ -1,27 +1,33 @@
     -= Info about application 'ChanSpy' =-

   [Synopsis]
-  Listen to the audio of an active channel
+  Listen to a channel, and optionally whisper into it

   [Description]
     ChanSpy([chanprefix][|options]): This application is used to
listen to the
-  audio from an active Asterisk channel. This includes the audio
coming in and
```

```
+   audio from an Asterisk channel. This includes the audio coming in
and
    out of the channel being spied on. If the 'chanprefix' parameter is
specified,
    only channels beginning with this string will be spied upon.
-   While Spying, the following actions may be performed:
+   While spying, the following actions may be performed:
       - Dialing # cycles the volume level.
       - Dialing * will stop spying and look for another channel to spy
on.
       - Dialing a series of digits followed by # builds a channel name
to append
          to 'chanprefix'. For example, executing ChanSpy(Agent) and
then dialing
-         the digits '1234#' while spying will begin spying on the
channel,
+         the digits '1234#' while spying will begin spying on the
channel
          'Agent/1234'.

    Options:
-       b - Only spy on channels involved in a bridged call.
-       g(grp) - Match only channels where their ${SPYGROUP} variable is
set to
-               'grp'.
-       q - Don't play a beep when beginning to spy on a channel.
+       b               - Only spy on channels involved in a bridged call.
+       g(grp)          - Match only channels where their ${SPYGROUP}
variable is set to
+                         contain 'grp' in an optional : delimited list.
+       q               - Don't play a beep when beginning to spy on a
channel, or speak the
+                         selected channel name.
        r[(basename)] - Record the session to the monitor spool
directory. An
                         optional base for the filename may be specified.
The
                         default is 'chanspy'.
        v([value])    Adjust the initial volume in the range from  4 to
4. A
-                         negative value refers to a quieter setting.
+       v([value])    - Adjust the initial volume in the range from -4
to 4. A
+                         negative value refers to a quieter setting.
+       w             - Enable 'whisper' mode, so the spying channel can
talk to
+                         the spied-on channel.
+       W             - Enable 'private whisper' mode, so the spying
channel can
+                         talk to the spied-on channel but cannot listen
to that
+                         channel.
```

Diff of the internal help from Asterisk 1.4 to 1.6:

```
--- in Asterisk 1.4
+++ in Asterisk 1.6
@@ -4,7 +4,7 @@
   Listen to a channel, and optionally whisper into it
   [Description]
-    ChanSpy([chanprefix][|options]): This application is used to
listen to the
+    ChanSpy([chanprefix][,options]): This application is used to
listen to the
   audio from an Asterisk channel. This includes the audio coming in
and
   out of the channel being spied on. If the 'chanprefix' parameter is
specified,
   only channels beginning with this string will be spied upon.
@@ -14,20 +14,54 @@
       - Dialing a series of digits followed by # builds a channel name
to append
         to 'chanprefix'. For example, executing ChanSpy(Agent) and
then dialing
         the digits '1234#' while spying will begin spying on the
channel
-         'Agent/1234'.
+         'Agent/1234'. Note that this feature will be overriden if the
'd' option
+            is used
+    Note: The X option supersedes the three features above in that if
a valid
+            single digit extension exists in the correct context ChanSpy
will
+            exit to it. This also disables choosing a channel based on
'chanprefix'
+            and a digit sequence.
    Options:
-       b               - Only spy on channels involved in a bridged call.
-       g(grp)          - Match only channels where their ${SPYGROUP}
variable is set to
-                         contain 'grp' in an optional : delimited list.
-       q               - Don't play a beep when beginning to spy on a
channel, or speak the
-                         selected channel name.
-       r[(basename)] - Record the session to the monitor spool
directory. An
-                         optional base for the filename may be specified.
The
-                         default is 'chanspy'.
-       v([value])      - Adjust the initial volume in the range from -4
to 4. A
-                         negative value refers to a quieter setting.
-       w               - Enable 'whisper' mode, so the spying channel can
talk to
-                         the spied-on channel.
```

```
-      W               - Enable 'private whisper' mode, so the spying
channel can
-                        talk to the spied-on channel but cannot listen
to that
-                          channel.
+      b                       - Only spy on channels involved in a
bridged call.
+      B                       - Instead of whispering on a single
channel barge in on both
+                              channels involved in the call.
+      d                       - Override the typical numeric DTMF
functionality and instead
+                              use DTMF to switch between spy modes.
+                                  4 = spy mode
+                                  5 = whisper mode
+                                  6 = barge mode
+      g(grp)                  - Only spy on channels in which one or
more of the groups
+                              listed in 'grp' matches one or more
groups from the
+                              SPYGROUP variable set on the channel to
be spied upon.
+                              Note that both 'grp' and SPYGROUP can
contain either a
+                              single group or a colon-delimited list
of groups, such
+                              as 'sales:support:accounting'.
+      n([mailbox][@context]) - Say the name of the person being spied
on if that person has recorded
+                              his/her name. If a context is
specified, then that voicemail context will
+                              be searched when retrieving the name,
otherwise the "default" context
+                              will be searched. If no mailbox is
specified, then the channel name will
+                              be used when searching for the name
(i.e. if SIP/1000 is the channel being
+                              spied on and no mailbox is specified,
then "1000" will be used when searching
+                              for the name).
+      q                       - Don't play a beep when beginning to spy
on a channel, or speak the
+                              selected channel name.
+      r[(basename)]           - Record the session to the monitor spool
directory. An
+                              optional base for the filename may be
specified. The
+                              default is 'chanspy'.
+      s                       - Skip the playback of the channel type
(i.e. SIP, IAX, etc) when
+                              speaking the selected channel name.
+      v([value])              - Adjust the initial volume in the range
from -4 to 4. A
```

```
+                              negative value refers to a quieter
setting.
+      w                     - Enable 'whisper' mode, so the spying
channel can talk to
+                             the spied-on channel.
+      W                     - Enable 'private whisper' mode, so the
spying channel can
+                             talk to the spied-on channel but cannot
listen to that
+                             channel.
+      o                     - Only listen to audio coming from this
channel.
+      X                     - Allow the user to exit ChanSpy to a
valid single digit
+                             numeric extension in the current
context or the context
+                             specified by the SPY_EXIT_CONTEXT
channel variable. The
+                             name of the last channel that was spied
on will be stored
+                             in the SPY_CHANNEL variable.
+      e(ext)               - Enable 'enforced' mode, so the spying
channel can
+                             only monitor extensions whose name is
in the 'ext' :
+                             delimited list.
```

See also extenspy, zapbarge, zapscan, monitor.

B.21 CheckGroup()

Limits the number of channels in a group.

```
CheckGroup(maximum[@category][|options])
```

Warning: CheckGroup() is removed as of Asterisk 1.4 and has been replaced by the **GROUP_COUNT()** function.

Asterisk versions:

```
--------| 1.2 |            |       |           |       |
```

Internal help for this application in Asterisk 1.2:

```
    -= Info about application 'CheckGroup' =-

[Synopsis]
Check the channel count of a group against a limit
```

```
[Description]
Usage: CheckGroup(max[@category][|options])
  Checks that the current number of total channels in the
current channel's group does not exceed 'max'.  If the number
does not exceed 'max', we continue to the next step.
 The option string may contain zero of the following character:
        'j' -- jump to n+101 priority if the number does in fact exceed
max,
                and priority n+101 exists. Execuation then continues at
that
                step, otherwise -1 is returned.
 This application sets the following channel variable upon successful
completion:
        CHECKGROUPSTATUS  The status of the check that the current
channel's
                        group does not exceed 'max'. It's value is
one of
                OK | OVERMAX
```

Diff of the internal help from Asterisk 1.2 to 1.4:

Not available in Asterisk 1.4

See also `setgroup`; and in Appendix C, see `group_count`, `getgroupcount`, `getgroupmatchcount`.

B.22 ClearHash()

Clears all the key/value pairs from the specified hashname.

```
ClearHash(hashname)
```

For more information, see Appendix C.

Asterisk versions:

```
|       |       |       |       |       | 1.0 |========
```

Internal help for this application in Asterisk 1.6:

```
    -= Info about application 'ClearHash' =-

[Synopsis]
Clear the keys from a specified hashname

[Description]
ClearHash(<hashname>)
  Clears all keys out of the specified hashname
```

Diff of the internal help from Asterisk 1.4 to 1.6:

Not available in Asterisk 1.4

See also Appendix C.

B.23 Congestion()

Indicates congestion (insufficient resources available) on the channel.

```
Congestion([timeout])
```

Indicates congestion on the channel and waits until the caller hangs up or until the specified timeout *timeout* has expired.

   This application indicates congestion in the system but does not indicate this to the caller. Should you wish to notify the caller, use **Playtones(congestion)**.

   Returns −1.

```
; for Caller ID is 888-555-8701, always signal congestion:
exten => 123,1,GotoIf($[${CALLERID(num)} = 8885558701]?10)
exten => 123,n,Playtones(congestion)
exten => 123,n,Congestion(5)
exten => 123,n,Hangup()
exten => 123,10,Dial(Zap/1)
```

Asterisk versions:

```
--------| 1.2 |--------| 1.4 |--------| 1.6 |--------
```

Internal help for this application in Asterisk 1.4:

```
    -= Info about application 'Congestion' =-

[Synopsis]
Indicate the Congestion condition

[Description]
  Congestion([timeout]): This application will indicate the congestion
condition to the calling channel. If the optional timeout is specified,
the
calling channel will be hung up after the specified number of seconds.
Otherwise, this application will wait until the calling channel hangs
up.
```

Diff of the internal help from Asterisk 1.2 to 1.4:

```
--- in Asterisk 1.2
+++ in Asterisk 1.4
```

```
@@ -4,7 +4,7 @@
   Indicate the Congestion condition

   [Description]
-    Congestion([timeout]): This application will indicate the
congenstion
+    Congestion([timeout]): This application will indicate the
congestion
   condition to the calling channel. If the optional timeout is
specified, the
   calling channel will be hung up after the specified number of
seconds.
   Otherwise, this application will wait until the calling channel
hangs up.
```

Diff of the internal help from Asterisk 1.4 to 1.6:

None

See also busy, progress, playtones.

B.24 ContinueWhile()

Returns to the beginning of a while loop.

```
ContinueWhile()
```

The **ContinueWhile()** application can interrupt a while loop while in progress. Asterisk returns to the beginning of the loop and evaluates the condition.

Asterisk versions:

```
    |      |          | 1.4 |--------| 1.6 |--------
```

Internal help for this application in Asterisk 1.4:

```
   -= Info about application 'ContinueWhile' =-

[Synopsis]
Restart a While loop

[Description]
Usage:  ContinueWhile()
Returns to the top of the while loop and re-evaluates the conditional.
```

Diff of the internal help from Asterisk 1.2 to 1.4:

Not available in Asterisk 1.2

Diff of the internal help from Asterisk 1.4 to 1.6:

```
--- in Asterisk 1.4
+++ in Asterisk 1.6
@@ -4,5 +4,4 @@
   Restart a While loop

   [Description]
-  Usage:  ContinueWhile()
-  Returns to the top of the while loop and re-evaluates the conditional.
+    ContinueWhile(): Returns to the top of the while loop and re-
evaluates the conditional.
```

See also while, exitwhile, ael.

B.25 ControlPlayback()

Plays a sound file with fast forward and rewind controls.

```
ControlPlayback(soundfile[,skipms[,ffchar[,rewchar[,stopchar[,pausechar]]]]])
```

Plays the specified file; the caller can manipulate playback by pressing the defined keys *ffchar* and *rewchar*. The defaults are # (forward) and * (backward). Play-back is stopped when *stopchar* is pressed (if it is defined). If the file does not exist, the application jumps to priority n+101, if it exists.

The *pausechar* option is similar in behavior to *stopchar* except that playback can be resumed by pressing *pausechar* a second time.

The *skipms* defines how far forward or backward **ControlPlayback()** will skip in the file when *ffchar* or *rewchar* is pressed.

Returns −1 if the caller hangs up during playback.

```
; play "symphony" to the caller with playback control:
exten => 123,1,ControlPlayback(symphony,5000,#,*,5,0)
```

Asterisk versions:

```
--------| 1.2 |--------| 1.4 |--------| 1.6 |--------
```

Internal help for this application in Asterisk 1.4:

```
   -= Info about application 'ControlPlayback' =-

[Synopsis]
Play a file with fast forward and rewind

[Description]
  ControlPlayback(file[|skipms[|ff[|rew[|stop[|pause
```

```
[|restart|options]]]]]]]):
This application will play back the given filename. By default, the '*'
key
can be used to rewind, and the '#' key can be used to fast-forward.
Parameters:
  skipms   - This is number of milliseconds to skip when rewinding or
             fast-forwarding.
  ff       - Fast-forward when this DTMF digit is received.
  rew      - Rewind when this DTMF digit is received.
  stop     - Stop playback when this DTMF digit is received.
  pause    - Pause playback when this DTMF digit is received.
  restart - Restart playback when this DTMF digit is received.
Options:
  j - Jump to priority n+101 if the requested file is not found.
This application sets the following channel variable upon completion:
  CPLAYBACKSTATUS -  This variable contains the status of the attempt
as a text
                    string, one of: SUCCESS | USERSTOPPED | ERROR
```

Diff of the internal help from Asterisk 1.2 to 1.4:

None

Diff of the internal help from Asterisk 1.4 to 1.6:

```
--- in Asterisk 1.4
+++ in Asterisk 1.6
@@ -4,7 +4,7 @@
   Play a file with fast forward and rewind
   [Description]
-    ControlPlayback(file[|skipms[|ff[|rew[|stop
[|pause[|restart|options]]]]]]]):
+    ControlPlayback(file[,skipms[,ff[,rew[,stop[,pause
[,restart,options]]]]]]]):
   This application will play back the given filename. By default, the
'*' key
   can be used to rewind, and the '#' key can be used to fast-forward.
   Parameters:
@@ -16,7 +16,11 @@
     pause    - Pause playback when this DTMF digit is received.
     restart - Restart playback when this DTMF digit is received.
   Options:
-    j - Jump to priority n+101 if the requested file is not found.
-  This application sets the following channel variable upon
completion:
+    o(#) - Start at # ms from the beginning of the file.
+  This application sets the following channel variables upon
completion:
     CPLAYBACKSTATUS -  This variable contains the status of the
attempt as a text
                       string, one of: SUCCESS | USERSTOPPED | ERROR
+    CPLAYBACKOFFSET - This contains the offset in ms into the file
where
```

```
+                          playback was at when it stopped.  -1 is end of
file.
+    CPLAYBACKSTOPKEY - If the playback is stopped by the user this
variable contains
+                          the key that was pressed.
```

See also **playback, background.**

B.26 Curl()

Loads a URL.

```
Curl(URL|postdata)
```

Warning: **Curl()** is removed as of Asterisk 1.4 and was replaced by the **CURL()**
function.

Asterisk versions:

```
--------| 1.2 |            |       |          |       |
```

Internal help for this application in Asterisk 1.2:

```
    -= Info about application 'Curl' =-

[Synopsis]
Load an external URL

[Description]
  Curl(URL[|postdata]): This application will request the specified
URL.
It is mainly used for signalling external applications of an event.
Parameters:
  URL      - This is the external URL to request.
  postdata - This information will be treated as POST data.
This application will set the following variable:
  CURL - This variable will contain the resulting page.
This application has been deprecated in favor of the CURL function.
```

Diff of the internal help from Asterisk 1.2 to 1.4:

Not available in Asterisk 1.4

See also Appendix C.

B.27 Cut()

Cuts a variables contents using the specified delimiter.

```
Cut(newvariable=variable,delimiter,fieldnumber)
```

Warning: `Cut()` is removed as of Asterisk 1.4 and has been replaced by the `CUT()` function.

Asterisk versions:

```
--------| 1.2 |          |      |         |      |
```

Internal help for this application in Asterisk 1.2:

```
     -= Info about application 'Cut' =-

[Synopsis]
Splits a variable's contents using the specified delimiter

[Description]
  Cut(newvar=varname,delimiter,fieldspec): This application will split
the
contents of a variable based on the given delimiter and store the
result in
a new variable.
Parameters:
  newvar    - new variable created from result string
  varname   - variable you want cut
  delimiter - defaults to '-'
  fieldspec - number of the field you want (1-based offset)
              may also be specified as a range (with -)
              or group of ranges and fields (with &)
This application has been deprecated in favor of the CUT function.
```

Diff of the internal help from Asterisk 1.2 to 1.4:

Not available in Asterisk 1.4

See also Appendix C.

B.28 DAHDIBarge()

Monitor a DAHDI channel.

```
DAHDIBarge([channel])
```

Note: This application came out of the renaming of Zaptel to DAHDI.[2]

Asterisk versions:

```
--------| 1.2*|--------| 1.4*|--------| 1.6 |--------
```

(* different name)

Internal help for this application in Asterisk 1.6:

```
    -= Info about application 'DAHDIBarge' =-

[Synopsis]
Barge in (monitor) DAHDI channel

[Description]
  DAHDIBarge([channel]): Barges in on a specified DAHDI
channel or prompts if one is not specified.  Returns
-1 when caller user hangs up and is independent of the
state of the channel being monitored.
```

Diff of the internal help from Asterisk 1.2 (zapbarge) to 1.4 (zapbarge):

None

Diff of the internal help from Asterisk 1.4 (zapbarge) to 1.6:

```
--- in Asterisk 1.4
+++ in Asterisk 1.6
@@ -1,10 +1,10 @@
-    -= Info about application 'ZapBarge' =-
+    -= Info about application 'DAHDIBarge' =-

   [Synopsis]
-  Barge in (monitor) Zap channel
+  Barge in (monitor) DAHDI channel

   [Description]
-   ZapBarge([channel]): Barges in on a specified zap
+   DAHDIBarge([channel]): Barges in on a specified DAHDI
   channel or prompts if one is not specified.  Returns
   -1 when caller user hangs up and is independent of the
   state of the channel being monitored.
```

See also zapbarge.

---

2. For a complete description, see zapbarge.

B.29 DAHDIRAS()

Executes a DAHDI ISDN Remote Access Server.

```
DAHDIRAS(arguments)
```

Note: This application came out of the renaming of Zaptel to DAHDI.[3]

Asterisk versions:

```
--------| 1.2*|--------| 1.4*|--------| 1.6 |--------
```

(* different name)

Internal help for this application in Asterisk 1.6:

```
    -= Info about application 'DAHDIRAS' =-

[Synopsis]
Executes DAHDI ISDN RAS application

[Description]
  DAHDIRAS(args): Executes a RAS server using pppd on the given
channel.
The channel must be a clear channel (i.e. PRI source) and a DAHDI
channel to be able to use this function (No modem emulation is
included).
Your pppd must be patched to be DAHDI aware. Arguments should be
separated by , characters.
```

Diff of the internal help from Asterisk 1.2 (zapras) to 1.4 (zapras):

None

Diff of the internal help from Asterisk 1.4 (zapras) to 1.6:

```
--- in Asterisk 1.4
+++ in Asterisk 1.6
@@ -1,11 +1,11 @@
-    -= Info about application 'ZapRAS' =-
+    -= Info about application 'DAHDIRAS' =-

   [Synopsis]
-  Executes Zaptel ISDN RAS application
+  Executes DAHDI ISDN RAS application
```

---

3. For a complete description, see zapras.

```
     [Description]
-     ZapRAS(args): Executes a RAS server using pppd on the given
channel.
-   The channel must be a clear channel (i.e. PRI source) and a Zaptel
+     DAHDIRAS(args): Executes a RAS server using pppd on the given
channel.
+   The channel must be a clear channel (i.e. PRI source) and a DAHDI
     channel to be able to use this function (No modem emulation is
included).
-   Your pppd must be patched to be zaptel aware. Arguments should be
-   separated by | characters.
+   Your pppd must be patched to be DAHDI aware. Arguments should be
+   separated by , characters.
```

See also zapras.

B.30 DAHDIScan()

Scans DAHDI channels for monitoring purposes.

```
DAHDIScan([channel_group])
```

Note: This application came out of the renaming of Zaptel to DAHDI.[4]

Asterisk versions:

```
--------| 1.2*|--------| 1.4*|--------| 1.6 |--------
```

(* different name)

Internal help for this application in Asterisk 1.6:

```
     -= Info about application 'DAHDIScan' =-

[Synopsis]
Scan DAHDI channels to monitor calls

[Description]
  DAHDIScan([group]) allows a call center manager to monitor DAHDI
channels in
a convenient way.  Use '#' to select the next channel and use '*' to
exit
Limit scanning to a channel GROUP by setting the option group argument.
```

---

4. For a complete description, see zapscan.

Diff of the internal help from Asterisk 1.2 (zapscan) to 1.4 (zapscan):

None

Diff of the internal help from Asterisk 1.4 (zapscan) to 1.6:

```
--- in Asterisk 1.4
+++ in Asterisk 1.6
@@ -1,9 +1,9 @@
-      -= Info about application 'ZapScan' =-
+      -= Info about application 'DAHDIScan' =-

    [Synopsis]
-   Scan Zap channels to monitor calls
+   Scan DAHDI channels to monitor calls

    [Description]
-    ZapScan([group]) allows a call center manager to monitor Zap
channels in
+    DAHDIScan([group]) allows a call center manager to monitor DAHDI
channels in
    a convenient way.  Use '#' to select the next channel and use '*' to
exit
    Limit scanning to a channel GROUP by setting the option group
argument.
```

See also zapscan.

B.31 DAHDISendKeypadFacility()

Sends digits out-of-band on an ISDN PRI connection.

```
DAHDISendKeypadFacility(digits)
```

Note. This application came out of the renaming of Zaptel to DAHDI.[5]

Asterisk versions:

```
    |     |          | 1.4*|--------| 1.6 |--------
```

(* different name)

---

5. For a complete description, see `zapsendkeypadfacility`.

Internal help for this application in Asterisk 1.6:

```
    -= Info about application 'DAHDISendKeypadFacility' =-

[Synopsis]
Send digits out of band over a PRI

[Description]
  DAHDISendKeypadFacility(): This application will send the given
string of digits in a Keypad Facility
  IE over the current channel.
```

Diff of the internal help from Asterisk 1.2 to 1.4:

Not available in Asterisk 1.2

Diff of the internal help from Asterisk 1.4 (zapsendkeypadfacility) to 1.6:

```
--- in Asterisk 1.4
+++ in Asterisk 1.6
@@ -1,8 +1,8 @@
-     -= Info about application 'ZapSendKeypadFacility' =-
+     -= Info about application 'DAHDISendKeypadFacility' =-

  [Synopsis]
  Send digits out of band over a PRI

  [Description]
-     ZapSendKeypadFacility(): This application will send the given
string of digits in a Keypad Facility
+     DAHDISendKeypadFacility(): This application will send the given
string of digits in a Keypad Facility
      IE over the current channel.
```

See also **zapsendkeypadfacility**.

B.32 DateTime()

Say the current time.

```
DateTime([unixtime[,timezone[,format]]])
```

Says the current time. It is not yet deprecated but is now only an alias to **SayUnixTime()**; see the description there for use instructions.

Asterisk versions:

```
--------|1.2|--------|1.4|--------|1.6|--------
```

Internal help for this application in Asterisk 1.4:

```
    -= Info about application 'DateTime' =-

[Synopsis]
Says a specified time in a custom format

[Description]
DateTime([unixtime][|[timezone][|format]])
  unixtime: time, in seconds since Jan 1, 1970.  May be negative.
            defaults to now.
  timezone: timezone, see /usr/share/zoneinfo for a list.
            defaults to machine default.
  format:   a format the time is to be said in.  See voicemail.conf.
            defaults to "ABdY 'digits/at' IMp"
```

Diff of the internal help from Asterisk 1.2 to 1.4:

None

Diff of the internal help from Asterisk 1.4 to 1.6:

```
--- in Asterisk 1.4
+++ in Asterisk 1.6
@@ -4,10 +4,10 @@
   Says a specified time in a custom format

   [Description]
-  DateTime([unixtime][|[timezone][|format]])
-    unixtime: time, in seconds since Jan 1, 1970.  May be negative.
+  DateTime([unixtime][,[timezone][,format]])
+    unixtime  - time, in seconds since Jan 1, 1970.  May be negative.
                 defaults to now.
-    timezone: timezone, see /usr/share/zoneinfo for a list.
+    timezone  - timezone, see /usr/share/zoneinfo for a list.
                 defaults to machine default.
-    format:   a format the time is to be said in.  See voicemail.conf.
+    format:   - a format the time is to be said in.  See
  voicemail.conf.
                 defaults to "ABdY 'digits/at' IMp"
```

See also sayunixtime.

B.33 DBdel()

Deletes a key from the Asterisk database (AstDB).

```
DBdel(family/key)
```

Deletes the specified key from the Asterisk database.
  Returns 0.

```
exten => 123,1,Set(DB(test/name)=Richard)     ; save key in AstDB
exten => 123,n,Set(name=${DB(test/name)})     ; retrieve key
exten => 123,n,DBdel(test/name)               ; delete key
```

Warning: DBdel() is deprecated as of Asterisk 1.4; use the **DB_DELETE()** function instead.

Asterisk versions:

```
--------|1.2|--------|1.4|--------|1.6|--------
```

Internal help for this application in Asterisk 1.4:

```
    -= Info about application 'DBdel' =-

[Synopsis]
Delete a key from the database

[Description]
  DBdel(family/key): This application will delete a key from the
Asterisk
database.
  This application has been DEPRECATED in favor of the DB_DELETE
function.
```

Diff of the internal help from Asterisk 1.2 to 1.4:

```
--- in Asterisk 1.2
+++ in Asterisk 1.4
@@ -6,3 +6,4 @@
   [Description]
     DBdel(family/key): This application will delete a key from the
Asterisk
    database.
+    This application has been DEPRECATED in favor of the DB_DELETE
function.
```

Diff of the internal help from Asterisk 1.4 to 1.6:

None

See also **dbdeltree**, and in Appendix C, see **db**, **db_delete**.

B.34 DBdeltree()

Deletes a family or branch from the Asterisk database.

```
DBdeltree(family[/branch])
```

Deletes the specified branch from the Asterisk database.
Returns 0.

```
; save entries in AstDB:
exten => 123,1,Set(DB(colors/one)=red)
exten => 123,n,Set(DB(colors/two)=blue)
; now delete the key family named test
exten => 123,n,DBdeltree(colors)
```

Asterisk versions:

```
--------|1.2|--------|1.4|--------|1.6|--------
```

Internal help for this application in Asterisk 1.4:

```
    -= Info about application 'DBdeltree' =-

[Synopsis]
Delete a family or keytree from the database

[Description]
  DBdeltree(family[/keytree]): This application will delete a family or
keytree
from the Asterisk database
```

Diff of the internal help from Asterisk 1.2 to 1.4:

None

Diff of the internal help from Asterisk 1.4 to 1.6:

None

See also dbdel, and in Appendix C, see db.

B.35 DBget()

Reads a value from the Asterisk database (AstDB).

```
DBget(variable=family/key|options)
```

Warning: DBget() is removed as of Asterisk 1.4 and has been replaced by the
DB() function.

Asterisk versions:

```
--------|1.2|        |   |        |   |   |
```

Internal help for this application in Asterisk 1.2:

```
     -= Info about application 'DBget' =-

[Synopsis]
Retrieve a value from the database

[Description]
   DBget(varname=family/key[|options]): This application will retrieve
a value
from the Asterisk database and store it in the given variable.
   Options:
      j - Jump to priority n+101 if the requested family/key isn't
found.
   This application sets the following channel variable upon
completion:
      DBGETSTATUS - This variable will contain the status of the attempt
                  FOUND | NOTFOUND
   This application has been deprecated in favor of the DB function.
```

Diff of the internal help from Asterisk 1.2 to 1.4:

Not available in Asterisk 1.4

See also dbput, and in Appendix C, see db.

B.36 DBput()

Stores a value in the Asterisk database (AstDB).

```
DBput(family/key=value)
```

Warning: DBput() is removed as of Asterisk 1.4 and has been replaced by the **DB()** function.

Asterisk versions:

```
--------|1.2|          |   |        |    |
```

Internal help for this application in Asterisk 1.2:

```
     -= Info about application 'DBput' =-

[Synopsis]
Store a value in the database
```

```
[Description]
    DBput(family/key=value): This application will store the given value
in the
specified location in the Asterisk database.
    This application has been deprecated in favor of the DB function.
```

Diff of the internal help from Asterisk 1.2 to 1.4:

Not available in Asterisk 1.4

See also dbget, and in Appendix C, see db.

B.37 DeadAGI()

Runs an AGI compliant program on an inactive channel.

```
DeadAGI(program,arguments)
DeadAGI(program[,argument1[,argument2[,...]]])
```

Runs an AGI-compliant program on an inactive (on hook) channel. With AGI (Asterisk Gateway Interface), you can manipulate channels with programs written in practically any conceivable language. AGI programs can control a channel, play audio, interpret and store DTMF tones; AGI programs exchange data with Asterisk on stdin and stdout. The specified arguments are passed unadulterated to the AGI program.

This application was developed for use on inactive (on hook) channels, because the standard AGI interface will not work on a channel after it is hung up. It is not necessary for the channel to be "dead" at the time of execution, however!

Returns −1 if the application causes a hangup, or returns 0 on exit without hangup.

```
; run AGI on a hung-up channel:
exten => h,1,DeadAGI(agi-test)
```

Warning: The channel will be treated as active as long as the AGI program is running. This can have implications for CDRs.

Note also that DeadAGI applications receive a SIGHUP signal when the channel is hung up, and it may need to be explicitly ignored in your AGI program:

Perl	$SIG{HUP} = "IGNORE";
PHP	pcntl_signal(SIGHUP, SIG_IGN);
	(PHP must be compiled with process control enabled using --enable-pcntl; this is not always the case!)
Ruby	trap('SIGHUP','IGNORE')

It is also important for the AGI program to stop communicating after the hangup, or it will receive a `SIGPIPE` signal and end (unless the signal is explicitly ignored as in the example above).

Asterisk versions:

```
--------|1.2|--------|1.4|--------|1.6|--------
```

Internal help for this application in Asterisk 1.4:

```
    -= Info about application 'DeadAGI' =-

[Synopsis]
Executes AGI on a hung-up channel

[Description]
  [E|Dead]AGI(command|args): Executes an Asterisk Gateway Interface
compliant
program on a channel. AGI allows Asterisk to launch external programs
written in any language to control a telephony channel, play audio,
read DTMF digits, etc. by communicating with the AGI protocol on stdin
and stdout.
  This channel will stop dialplan execution on hangup inside of this
application, except when using DeadAGI.  Otherwise, dialplan execution
will continue normally.
  A locally executed AGI script will receive SIGHUP on hangup from the
channel
except when using DeadAGI. This can be disabled by setting the
AGISIGHUP channel
variable to "no" before executing the AGI application.
  Using 'EAGI' provides enhanced AGI, with incoming audio available out
of band
on file descriptor 3

  Use the CLI command 'agi show' to list available agi commands
  This application sets the following channel variable upon completion:
      AGISTATUS       The status of the attempt to the run the AGI script
                      text string, one of SUCCESS | FAILURE | HANGUP
```

Diff of the internal help from Asterisk 1.2 to 1.4:

```
--- in Asterisk 1.2
+++ in Asterisk 1.4
@@ -9,9 +9,16 @@
   written in any language to control a telephony channel, play audio,
   read DTMF digits, etc. by communicating with the AGI protocol on
stdin
   and stdout.
- Returns -1 on hangup (except for DeadAGI) or if application
requested
-   hangup, or 0 on non-hangup exit.
```

```
-  Using 'EAGI' provides enhanced AGI, with incoming audio available
out of band
+   This channel will stop dialplan execution on hangup inside of this
+  application, except when using DeadAGI.  Otherwise, dialplan
execution
+  will continue normally.
+   A locally executed AGI script will receive SIGHUP on hangup from
the channel
+  except when using DeadAGI. This can be disabled by setting the
AGISIGHUP channel
+  variable to "no" before executing the AGI application.
+   Using 'EAGI' provides enhanced AGI, with incoming audio available
out of band
   on file descriptor 3

-  Use the CLI command 'show agi' to list available agi commands
+   Use the CLI command 'agi show' to list available agi commands
+   This application sets the following channel variable upon
completion:
+       AGISTATUS      The status of the attempt to the run the AGI
script
+                      text string, one of SUCCESS | FAILURE | HANGUP
```

Diff of the internal help from Asterisk 1.4 to 1.6:

```
--- in Asterisk 1.4
+++ in Asterisk 1.6
@@ -4,21 +4,22 @@
   Executes AGI on a hung-up channel

   [Description]
-   [E|Dead]AGI(command|args): Executes an Asterisk Gateway Interface
compliant
-  program on a channel. AGI allows Asterisk to launch external
programs
-  written in any language to control a telephony channel, play audio,
-  read DTMF digits, etc. by communicating with the AGI protocol on
stdin
-  and stdout.
-   This channel will stop dialplan execution on hangup inside of this
-  application, except when using DeadAGI.  Otherwise, dialplan
execution
-  will continue normally.
+   [E|Dead]AGI(command,args): Executes an Asterisk Gateway Interface
compliant
+  program on a channel. AGI allows Asterisk to launch external
programs written
+  in any language to control a telephony channel, play audio, read
DTMF digits,
+  etc. by communicating with the AGI protocol on stdin and stdout.
+   As of 1.6.0, this channel will not stop dialplan execution on
hangup inside
```

+   of this application. Dialplan execution will continue normally, even upon
+   hangup until the AGI application signals a desire to stop (either by exiting
+   or, in the case of a net script, by closing the connection).
    A locally executed AGI script will receive SIGHUP on hangup from the channel
-   except when using DeadAGI. This can be disabled by setting the AGISIGHUP channel
-   variable to "no" before executing the AGI application.
+   except when using DeadAGI. A fast AGI server will correspondingly receive a
+   HANGUP in OOB data. Both of these signals may be disabled by setting the
+   AGISIGHUP channel variable to "no" before executing the AGI application.
    Using 'EAGI' provides enhanced AGI, with incoming audio available out of band
-   on file descriptor 3
+   on file descriptor 3.

-    Use the CLI command 'agi show' to list available agi commands
+   Use the CLI command 'agi show' to list available agi commands.
     This application sets the following channel variable upon completion:
         AGISTATUS   The status of the attempt to the run the AGI script
-                        text string, one of SUCCESS | FAILURE | HANGUP
+   text string, one of

SUCCESS | FAILURE | NOTFOUND | HANGUP

See also **agi, fastagi.**

B.38 Dial()

Connects channels.

```
Dial(technology/resource,timeout,options,URL)
Dial(technology1/resource1[&tech2/resource2[&...]],timeout,options,URL)
Dial(technology/user:password@host/extension,timeout,options)
```

Connects two channels together.[6]

Dial() is perhaps the most important application in Asterisk. We recommend you read this section carefully and more than once if necessary.

---

6. Generally, channels of any type supported by Asterisk may be connected (for example, IAX, SIP, H.323, Skinny, PRI, FXO, FXS, Local).

Dial() accepts every valid channel type (e.g., SIP, IAX2, H.323, MGCP, Local, or Zap) but the allowable parameters are channel-specific; i.e., what parameters a channel requires or will accept depends on the nature of the channel technology. For example, a SIP channel will require an IP address and user information, whereas a ZAP channel requires a telephone number.

When a network-based channel type is specified, the parameters (such as IP address, user name, password, and remote extension) can be supplied as options to **Dial()** or, alternatively, be included in a host configuration section in the appropriate .conf file. If this second approach is used, all the required configuration information must be present.

Here's an example:

```
exten => s,1,Dial(SIP/richard:secret@widgets.biz)
```

This extension would accomplish the same thing:

```
exten => s,1,Dial(SIP/a_SIP_friend)
```

... as long as a_SIP_friend is defined as a channel in sip.conf:

```
[a_SIP_friend]
fromuser=richard
password=secret
host=widgets.biz
```

Sometimes an extension is attached to the address information, as in this example:

```
exten => s,1,Dial(IAX2/user:secret@widgets.biz/500)
```

The remote system is asked to connect the call to extension 500 in the incoming channel. This extension is not required because the channel configuration on the remote system is used, or, alternatively, the call is passed to the default s extension in the incoming context.

In the end, the remote host decides how the call will be processed; all you can do is request special call handling.

If no timeout is specified, the channel will ring indefinitely. This behavior is not necessarily undesirable and so it's not automatically necessary to set this parameter. Just be aware that "indefinite" can end up being a very long time.

The timeout is specified in seconds. It always follows the device information:

```
exten => s,1,Dial(IAX2/user:secret@widgets.biz/500,20)
```

With **Dial()**, you can ring multiple channels simultaneously. The call is handled on a "first come, first served" basis; the first extension to pick up answers the call, and all the other extensions stop ringing and become available:

```
exten => s,1,Dial(SIP/2000&SIP/2001&SIP/2303)
```

A big part of the power in the **Dial()** application is in the options, which always follow the device and timeout information, like so:

```
exten => s,1,Dial(IAX2/user:secret@widgets.biz/500,60,options)
```

Warning: If you want to provide options, you still need to provide a comma-delimited space for the timeout value even if it is empty:

```
exten => s,1,Dial(IAX2/user:secret@widgets.biz/500,,options)
```

Here are the valid options to the **Dial()** application:

- **d**

  Allows the caller to dial another single-digit extension while waiting for the current extension to answer. (For example, a caller dials 4 while the phone is ringing and the call is immediately passed to the 4 extension. The extension is in the current context unless $\{\texttt{EXITCONTEXT}\}$ is set.)

- **t**

  Blind transfer initiated by the *called* party. Allows the called party to transfer the call by pressing the blind transfer key (normally #). Reinvites are not possible when this option is selected because Asterisk must monitor the connection to detect when the called party presses the # key.

- **T**

  Blind transfer initiated by the *calling* party. Allows the calling party to transfer the call by pressing the blind transfer key (normally #). Reinvites are not possible when this option is selected because Asterisk must monitor the connection to detect when the called party presses the # key.

- **w**

  Allows the *called* party to start recording the call by pressing the `automon` key sequence (as defined in `features.conf`). If the `TOUCH_MONITOR` variable is set, its value is passed to **Monitor()** as a parameter when recording starts. If it is not set, `WAV,,m` is passed to **Monitor()**.

- **W**

  Allows the *calling* party to start recording the call by pressing the `automon` key sequence (as defined in `features.conf`).

- **f**

  Sets the caller ID as the number of the line making or redirecting the outgoing call. Some PSTNs don't allow IDs from extensions other than those assigned to you. For example, if you have a PRI, you would use f to overwrite the caller

ID provided by a SIP extension to that belonging to the outgoing Zap channel on the PRI.

▪ **o**

Uses the caller ID received on the incoming leg of a call as the caller ID for the outgoing leg. This is useful if a call is accepted and then transferred; in the normal case, the caller ID of the initial recipient is used for the outgoing leg, which can be confusing to the ultimate recipient. For example, say Joe calls Mary; Mary decides that Joe really needs to speak to Don and transfers the call. If option o is set, Don will see Joe's number on his display when Mary transfers him, instead of Mary's number.

▪ **r**

Generate a ringing tone for the calling party. Normally Asterisk will generate a ringing tone when it is appropriate. Option **r** forces it to do so no matter the circumstance. Sometimes called devices don't provide useful call progress information (or none at all) and **r** is needed; however, this can also lead to strange behavior, such as initial ringing interrupted by a busy signal.

▪ **m[class]**

Plays music to the caller until the call is answered. Optionally you can provide the Music-on-Hold class (as defined in `musiconhold.conf`).

▪ **M(x[^arg])**

Runs the macro x when the call is answered, optionally passing ^ (caret) separated arguments. The macro may set the MACRO_RESULT channel variable to one of the following values:

  ▪ **ABORT**

  Hangs up both ends of the call.

  ▪ **CONGESTION**

  Indicates congestion on the line.

  ▪ **BUSY**

  Indicates that the line is busy (and jumps to n+101).

  ▪ **CONTINUE**

  Hangs the called end up and continues in the dialplan.

▪ **GOTO:<context>^<extension>^<priority>**

Jumps to the specified point in the dialplan.

▪ **h**

Allows the *called* party to hang up by pressing *.

- **H**

  Allows the *calling* party to hang up by pressing *.

- **C**

  Resets the Call Detail Record (CDR) for this call. Normally, the CDR clock is reset from the moment the call is answered by Asterisk; if CDRs are being used for billing purposes, sometimes it is appropriate to reset the timer when the connection between two parties is actually established.

- **P[(x)]**

  Uses the Privacy Manager if no caller ID is present, where the optional variable x is a family in the AstDB. The Privacy Manager asks the caller to enter a ten-digit telephone number if no caller ID is provided, providing a simple way to screen for telemarketers and solicitors blocking their caller ID. See also `LookupBlacklist()`.

- **g**

  Proceeds in the context when the target channel has been hung up.

- **G(context^extension^priority)**

  Drops both channels into the specified *context*, *extension*, and *priority* when the call is answered.

- **A(x)**

  Plays an announcement to the called party, where x is the sound file prefix. For example, A(confirm) would play the most efficient version of confirm (such as `confirm.gsm`, or `confirm.wav`) that can be found in the `/var/lib/asterisk/sounds` directory.

- **D([called][:*calling*])**

  Sends DTMF digits after the call is answered but before it is bridged. The `called` digits are transmitted to the called party, the `calling` digits to the calling party. One or both parameters may be set.

- **L(x[:y][:z])**

  Limits call duration to x milliseconds. At y ms before the maximum allowed duration, and thereafter every z ms until the end of the call, a warning is given. The x must be defined, y and z are optional. The behavior can be further controlled with the following variables:

  - **LIMIT_PLAYAUDIO_CALLER=*yes*/*no***

    Sets whether the calling party should hear announcements.

  - **LIMIT_PLAYAUDIO_CALLEE=*yes*/*no***

    Sets whether the called party should hear announcements.

- **LIMIT_TIMEOUT_FILE=*filename***

  Specifies the sound file to be played after the maximum duration is reached and the call is ended.

- **LIMIT_CONNECT_FILE=*filename***

  Specifies the sound file to be played when the call is connected.

- **LIMIT_WARNING_FILE=*filename***

  Specifies the sound file to be played for the warning signal when *y* is set.

- **j**

  Turns priority jumping on. That is, the call jumps to priority n+101 (where n is the current priority) if all the channels respond busy.

  A call may be parked instead of transferred (which is the case when t or T is used). By default, calls are parked by transferring them to extension 700, but this behavior may be configured in features.conf.

- **n**

  Privacy Manager setting. Caller introductions are not to be saved in the priv-callerintros directory.

- **N**

  Privacy Manager setting. Calls are not screened if caller ID information is provided.

When **Dial()** completes, the following variables are set:

- **DIALEDTIME**

  The total elapsed time from the time the Dial() command is executed until its completion.

- **ANSWEREDTIME**

  The time elapsed during conversation.

- **NOANSWER**

  The channel was not answered before the ring timeout had expired.

- **BUSY**

  The called channel is currently busy.

- **ANSWER**

  The called channel was answered.

- **CANCEL**

  The call was interrupted before it could be completed.

```
; dial a number on Zap channel 2, let it ring a maximum of 10 seconds:
exten => 123,1,Dial(Zap/2/1234567,10,tTm)
; otherwise proceed in the dialplan:
exten => 123,n,Playback(sorry)
exten => 123,n,Hangup()

; dial extension 500 over IAX on host widgets.biz:
exten => 123,1,Dial(IAX/username:password@widgets.biz/500)
```

Asterisk versions:

```
--------|1.2|--------|1.4|--------|1.6|--------
```

Internal help for this application in Asterisk 1.4:

```
    -= Info about application 'Dial' =-

[Synopsis]
Place a call and connect to the current channel

[Description]
  Dial(Technology/resource[&Tech2/resource2...][|timeout][|options]
[|URL]):
This application will place calls to one or more specified channels. As
soon
as one of the requested channels answers, the originating channel will
be
answered, if it has not already been answered. These two channels will
then
be active in a bridged call. All other channels that were requested
will then
be hung up.
  Unless there is a timeout specified, the Dial application will wait
indefinitely until one of the called channels answers, the user hangs
up, or
if all of the called channels are busy or unavailable. Dialplan
executing will
continue if no requested channels can be called, or if the timeout
expires.

  This application sets the following channel variables upon
completion:
    DIALEDTIME   - This is the time from dialing a channel until when
it
                 is disconnected.
    ANSWEREDTIME - This is the amount of time for actual call.
    DIALSTATUS   - This is the status of the call:
                   CHANUNAVAIL | CONGESTION | NOANSWER | BUSY |
  ANSWER | CANCEL
                   DONTCALL | TORTURE | INVALIDARGS
  For the Privacy and Screening Modes, the DIALSTATUS variable will be
set to
```

DONTCALL if the called party chooses to send the calling party to the
'Go Away'
script. The DIALSTATUS variable will be set to TORTURE if the called
party
wants to send the caller to the 'torture' script.
   This application will report normal termination if the originating
channel
hangs up, or if the call is bridged and either of the parties in the
bridge
ends the call.
   The optional URL will be sent to the called party if the channel
supports it.
   If the OUTBOUND_GROUP variable is set, all peer channels created by
this
application will be put into that group (as in Set(GROUP()=...).
   If the OUTBOUND_GROUP_ONCE variable is set, all peer channels created
by this
application will be put into that group (as in Set(GROUP()=...). Unlike
OUTBOUND_GROUP,
however, the variable will be unset after use.

   Options:
     A(x) - Play an announcement to the called party, using 'x' as the
file.
     C   - Reset the CDR for this call.
     d   - Allow the calling user to dial a 1 digit extension while
waiting for
          a call to be answered. Exit to that extension if it exists
in the
          current context, or the context defined in the EXITCONTEXT
variable,
          if it exists.
     D([called][:calling]) - Send the specified DTMF strings *after* the
called
          party has answered, but before the call gets bridged. The
'called'
          DTMF string is sent to the called party, and the 'calling'
DTMF
          string is sent to the calling party. Both parameters can be
used
          alone.
     f   - Force the callerid of the *calling* channel to be set as the
          extension associated with the channel using a dialplan
'hint'.
          For example, some PSTNs do not allow CallerID to be set to
anything
          other than the number assigned to the caller.
     g   - Proceed with dialplan execution at the current extension if
the
          destination channel hangs up.
     G(context^exten^pri) - If the call is answered, transfer the

calling party to
          the specified priority and the called party to the specified
priority+1.
          Optionally, an extension, or extension and context may be
specified.
          Otherwise, the current extension is used. You cannot use any
additional
          action post answer options in conjunction with this option.
    h   - Allow the called party to hang up by sending the '*' DTMF
digit.
    H   - Allow the calling party to hang up by hitting the '*' DTMF
digit.
    i   - Asterisk will ignore any forwarding requests it may receive
on this
          dial attempt.
    j   - Jump to priority n+101 if all of the requested channels were
busy.
    k   - Allow the called party to enable parking of the call by
sending
          the DTMF sequence defined for call parking in features.conf.
    K   - Allow the calling party to enable parking of the call by
sending
          the DTMF sequence defined for call parking in features.conf.
    L(x[:y][:z]) - Limit the call to 'x' ms. Play a warning when 'y' ms
are
          left. Repeat the warning every 'z' ms. The following special
          variables can be used with this option:
          * LIMIT_PLAYAUDIO_CALLER   yes|no (default yes)
                                     Play sounds to the caller.
          * LIMIT_PLAYAUDIO_CALLEE   yes|no
                                     Play sounds to the callee.
          * LIMIT_TIMEOUT_FILE   File to play when time is up.
          * LIMIT_CONNECT_FILE   File to play when call begins.
          * LIMIT_WARNING_FILE   File to play as warning if 'y' is
defined.
                                     The default is to say the time
remaining.
    m([class]) - Provide hold music to the calling party until a
requested
          channel answers. A specific MusicOnHold class can be
          specified.
    M(x[^arg]) - Execute the Macro for the *called* channel before
connecting
          to the calling channel. Arguments can be specified to the
Macro
          using '^' as a delimeter. The Macro can set the variable
          MACRO_RESULT to specify the following actions after the
Macro is
          finished executing.
          * ABORT   Hangup both legs of the call.
          * CONGESTION   Behave as if line congestion was encountered.
          * BUSY   Behave as if a busy signal was encountered.

```
This will also
                          have the application jump to priority n+101
if the
                          'j' option is set.
              * CONTINUE    Hangup the called party and allow the calling
party
                          to continue dialplan execution at the next
priority.
              * GOTO:<context>^<exten>^<priority> - Transfer the call to
the
                          specified priority. Optionally, an extension,
or
                          extension and priority can be specified.
              You cannot use any additional action post answer options in
conjunction
                with this option. Also, pbx services are not run on the peer
(called) channel,
                so you will not be able to set timeouts via the TIMEOUT()
function in this macro.
   n     - This option is a modifier for the screen/privacy mode. It
specifies
                that no introductions are to be saved in the
priv-callerintros
                directory.
   N     - This option is a modifier for the screen/privacy mode. It
specifies
                that if callerID is present, do not screen the call.
   o     - Specify that the CallerID that was present on the *calling*
channel
                be set as the CallerID on the *called* channel. This was the
                behavior of Asterisk 1.0 and earlier.
   O([x]) - "Operator Services" mode (Zaptel channel to Zaptel channel
                only, if specified on non-Zaptel interface, it will be
ignored).
                When the destination answers (presumably an operator
services
                station), the originator no longer has control of their
line.
                They may hang up, but the switch will not release their
line
                until the destination party hangs up (the operator).
Specified
                without an arg, or with 1 as an arg, the originator
hanging up
                will cause the phone to ring back immediately. With a 2
specified,
                when the "operator" flashes the trunk, it will ring their
phone
                back.
   p     - This option enables screening mode. This is basically
Privacy mode
                without memory.
```

```
     P([x]) - Enable privacy mode. Use 'x' as the family/key in the
database if
          it is provided. The current extension is used if a database
          family/key is not specified.
     r    - Indicate ringing to the calling party. Pass no audio to the
calling
          party until the called channel has answered.
     S(x) - Hang up the call after 'x' seconds *after* the called party
has
          answered the call.
     t    - Allow the called party to transfer the calling party by
sending the
          DTMF sequence defined in features.conf.
     T    - Allow the calling party to transfer the called party by
sending the
          DTMF sequence defined in features.conf.
     w    - Allow the called party to enable recording of the call by
sending
          the DTMF sequence defined for one-touch recording in
features.conf.
     W    - Allow the calling party to enable recording of the call by
sending
          the DTMF sequence defined for one-touch recording in
features.conf.
```

Diff of the internal help from Asterisk 1.2 to 1.4:

```
--- in Asterisk 1.2
+++ in Asterisk 1.4
@@ -61,7 +61,13 @@
          action post answer options in conjunction with this
option.
     h    - Allow the called party to hang up by sending the '*' DTMF
digit.
     H    - Allow the calling party to hang up by hitting the '*'
DTMF digit.
+    i    - Asterisk will ignore any forwarding requests it may
receive on this
+          dial attempt.
     j    - Jump to priority n+101 if all of the requested channels
were busy.
+    k    - Allow the called party to enable parking of the call by
sending
+          the DTMF sequence defined for call parking in
features.conf.
+    K    - Allow the calling party to enable parking of the call by
sending
+          the DTMF sequence defined for call parking in
features.conf.
```

         L(x[:y][:z]) - Limit the call to 'x' ms. Play a warning when 'y'
ms are
                left. Repeat the warning every 'z' ms. The following
special
                variables can be used with this option:
@@ -92,7 +98,8 @@
                              specified priority. Optionally, an
extension, or
                              extension and priority can be specified.
                You cannot use any additional action post answer options
in conjunction
-                with this option.
+   with this option. Also, pbx services are not run on the
peer (called) channel,
+   so you will not be able to set timeouts via the TIMEOUT()
function in this macro.
         n   - This option is a modifier for the screen/privacy mode. It
specifies
                that no introductions are to be saved in the
priv-callerintros
                directory.
@@ -101,6 +108,16 @@
         o   - Specify that the CallerID that was present on the
calling channel
                be set as the CallerID on the *called* channel. This was
the
                behavior of Asterisk 1.0 and earlier.
+   O([x]) - "Operator Services" mode (Zaptel channel to Zaptel
channel
+   only, if specified on non-Zaptel interface, it will be
ignored).
+   When the destination answers (presumably an operator
services
+   station), the originator no longer has control of their
line.
+   They may hang up, but the switch will not release their
line
+   until the destination party hangs up (the operator).
Specified
+   without an arg, or with 1 as an arg, the originator
hanging up
+   will cause the phone to ring back immediately. With a 2
specified,
+   when the "operator" flashes the trunk, it will ring
their phone
+   back.
         p   - This option enables screening mode. This is basically
Privacy mode
                without memory.
         P([x]) - Enable privacy mode. Use 'x' as the family/key in the
database if

Diff of the internal help from Asterisk 1.4 to 1.6:

```
--- in Asterisk 1.4
+++ in Asterisk 1.6
@@ -4,7 +4,7 @@
   Place a call and connect to the current channel

   [Description]
-    Dial(Technology/resource[&Tech2/resource2...][|timeout][|options]
[|URL]):
+    Dial(Technology/resource[&Tech2/resource2...][,timeout][,options]
[,URL]):
   This application will place calls to one or more specified channels.
As soon
   as one of the requested channels answers, the originating channel
will be
   answered, if it has not already been answered. These two channels
will then
@@ -39,6 +39,8 @@
     Options:
        A(x) - Play an announcement to the called party, using 'x' as
the file.
        C    - Reset the CDR for this call.
+       c    - If DIAL cancels this call, always set the flag to tell
the channel
+            driver that the call is answered elsewhere.
        d    - Allow the calling user to dial a 1 digit extension while
waiting for
             a call to be answered. Exit to that extension if it
exists in the
             current context, or the context defined in the
EXITCONTEXT variable,
@@ -48,10 +50,14 @@
             DTMF string is sent to the called party, and the
'calling' DTMF
             string is sent to the calling party. Both parameters can
be used
             alone.
+       e    - execute the 'h' extension for peer after the call ends.
This
+            operation will not be performed if the peer was parked
        f    - Force the callerid of the *calling* channel to be set as
the
             extension associated with the channel using a dialplan
'hint'.
             For example, some PSTNs do not allow CallerID to be set
to anything
             other than the number assigned to the caller.
+       F(context^exten^pri) - When the caller hangs up, transfer the
called party
```

```
+            to the specified context and extension and continue
execution.
       g    - Proceed with dialplan execution at the current extension
if the
             destination channel hangs up.
       G(context^exten^pri) - If the call is answered, transfer the
calling party to
```
@@ -59,15 +65,18 @@
```
             Optionally, an extension, or extension and context may be
specified.
             Otherwise, the current extension is used. You cannot use
any additional
             action post answer options in conjunction with this
option.
-      h    - Allow the called party to hang up by sending the '*' DTMF
digit.
-      H    - Allow the calling party to hang up by hitting the '*'
DTMF digit.
+      h    - Allow the called party to hang up by sending the '*' DTMF
digit, or
+            whatever sequence was defined in the featuremap section
for
+            'disconnect' in features.conf
+      H    - Allow the calling party to hang up by hitting the '*'
DTMF digit, or
+            whatever sequence was defined in the featuremap section
for
+            'disconnect' in features.conf
       i    - Asterisk will ignore any forwarding requests it may
receive on this
             dial attempt.
-      j    - Jump to priority n+101 if all of the requested channels
were busy.
       k    - Allow the called party to enable parking of the call by
sending
-            the DTMF sequence defined for call parking in
features.conf.
+            the DTMF sequence defined for call parking in the
featuremap section of features.conf.
       K    - Allow the calling party to enable parking of the call by
sending
-            the DTMF sequence defined for call parking in
features.conf.
+            the DTMF sequence defined for call parking in the
featuremap section of features.conf.
       L(x[:y][:z]) - Limit the call to 'x' ms. Play a warning when 'y'
ms are
             left. Repeat the warning every 'z' ms. The following
special
             variables can be used with this option:
```
@@ -84,14 +93,12 @@
```
             specified.
```

```
        M(x[^arg]) - Execute the Macro for the *called* channel before
connecting
                to the calling channel. Arguments can be specified to the
Macro
-               using '^' as a delimeter. The Macro can set the variable
+               using '^' as a delimiter. The Macro can set the variable
                MACRO_RESULT to specify the following actions after the
Macro is
                finished executing.
                * ABORT        Hangup both legs of the call.
                * CONGESTION   Behave as if line congestion was
encountered.
-               * BUSY         Behave as if a busy signal was
encountered. This will also
-                              have the application jump to priority
n+101 if the
-                              'j' option is set.
+               * BUSY         Behave as if a busy signal was
encountered.
                * CONTINUE     Hangup the called party and allow the
calling party
                               to continue dialplan execution at the next
priority.
                * GOTO:<context>^<exten>^<priority> - Transfer the call
to the
@@ -108,8 +115,8 @@
        o      - Specify that the CallerID that was present on the
*calling* channel
                be set as the CallerID on the *called* channel. This was
the
                behavior of Asterisk 1.0 and earlier.
-       O([x]) - "Operator Services" mode (Zaptel channel to Zaptel
channel
-              only, if specified on non-Zaptel interface, it will be
ignored).
+       O([x]) - "Operator Services" mode (DAHDI channel to DAHDI
channel
+              only, if specified on non-DAHDI interface, it will be
ignored).
                When the destination answers (presumably an operator
services
                station), the originator no longer has control of their
line.
                They may hang up, but the switch will not release their
line
@@ -128,10 +135,35 @@
        S(x) - Hang up the call after 'x' seconds *after* the called
party has
                answered the call.
        t      - Allow the called party to transfer the calling party by
sending the
-                DTMF sequence defined in features.conf.
```

```
+               DTMF sequence defined in the blindxfer setting in the
featuremap section
+               of features.conf.
     T     - Allow the calling party to transfer the called party by
sending the
-               DTMF sequence defined in features.conf.
+               DTMF sequence defined in the blindxfer setting in the
featuremap section
+               of features.conf.
+     U(x[^arg]) - Execute via Gosub the routine 'x' for the *called*
channel before connecting
+               to the calling channel. Arguments can be specified to the
Gosub
+               using '^' as a delimiter. The Gosub routine can set the
variable
+               GOSUB_RESULT to specify the following actions after the
Gosub returns.
+               * ABORT        Hangup both legs of the call.
+               * CONGESTION    Behave as if line congestion was
encountered.
+               * BUSY         Behave as if a busy signal was
encountered.
+               * CONTINUE     Hangup the called party and allow the
calling party
+                              to continue dialplan execution at the next
priority.
+               * GOTO:<context>^<exten>^<priority> - Transfer the call
to the
+                              specified priority. Optionally, an
extension, or
+                              extension and priority can be specified.
+               You cannot use any additional action post answer options
in conjunction
+               with this option. Also, pbx services are not run on the
peer (called) channel,
+               so you will not be able to set timeouts via the TIMEOUT()
function in this routine.
     w     - Allow the called party to enable recording of the call by
sending
-               the DTMF sequence defined for one-touch recording in
features.conf.
+               the DTMF sequence defined in the automon setting in the
featuremap section
+               of features.conf.
     W     - Allow the calling party to enable recording of the call
by sending
-               the DTMF sequence defined for one-touch recording in
features.conf.
+               the DTMF sequence defined in the automon setting in the
featuremap section
+               of features.conf.
```

```
+       x    - Allow the called party to enable recording of the call by
sending
+                the DTMF sequence defined in the automixmon setting in
the featuremap section
+                of features.conf.
+       X    - Allow the calling party to enable recording of the call
by sending
+                the DTMF sequence defined in the automixmon setting in
the featuremap section
+                of features.conf.
```

See also `retrydial`.

B.39 Dictate()

Virtual dictation machine.

```
Dictate([path[,filename]])
```

Starts a virtual dictation machine. The options define the base directory (default: `/var/spool/asterisk/dictate/`) and (numerical) filename. The files are recorded in RAW format.

The user can control dictation with these keys:

- **0**

  Help

- **1**

  Switches between playback and record modes

- *****

  Pause / continue

- **#**

  Select file / enter new filename (e.g., 1234#)

In playback mode:

- **2**

  Switches the playback speed by increments (1x, 2x, 3x, 4x)

- **7**

  Jumps back a set time interval

- **8**

  Jumps forward a set time interval

In record mode:

- **8**

  Erase recording and start over

  ```
  ; Take dictation:
  exten => 123,1,Dictate()
  ```

To provide every system user her own dictation machine, you might set the path to /var/spool/asterisk/dictate/${EXTEN}.

Asterisk versions:

```
--------|1.2|--------|1.4|--------|1.6|--------
```

Internal help for this application in Asterisk 1.4:

```
    -= Info about application 'Dictate' =-

[Synopsis]
Virtual Dictation Machine

[Description]
  Dictate([<base_dir>[|<filename>]])
Start dictation machine using optional base dir for files.
```

Diff of the internal help from Asterisk 1.2 to 1.4:

```
--- in Asterisk 1.2
+++ in Asterisk 1.4
@@ -4,5 +4,5 @@
   Virtual Dictation Machine

   [Description]
-    Dictate([<base_dir>])
+    Dictate([<base_dir>[|<filename>]])
   Start dictation machine using optional base dir for files.
```

Diff of the internal help from Asterisk 1.4 to 1.6:

```
--- in Asterisk 1.4
+++ in Asterisk 1.6
@@ -4,5 +4,5 @@
   Virtual Dictation Machine

   [Description]
-    Dictate([<base_dir>[|<filename>]])
+    Dictate([<base_dir>[,<filename>]])
   Start dictation machine using optional base dir for files.
```

See also record.

B.40 DigitTimeout()

Sets the maximum timeout between digits when the caller is dialing.

```
DigitTimeout(seconds)
```

Warning: `DigitTimeout()` is removed as of Asterisk 1.4 and has been replaced by the `TIMEOUT()` function:

```
Set(TIMEOUT(digit)=seconds)
```

Asterisk versions:

```
--------|1.2|        |    |        |    |
```

Internal help for this application in Asterisk 1.2:

```
    -= Info about application 'DigitTimeout' =-

[Synopsis]
Set maximum timeout between digits

[Description]
  DigitTimeout(seconds): Set the maximum amount of time permitted
between
digits when the user is typing in an extension. When this timeout
expires,
after the user has started to type in an extension, the extension will
be
considered complete, and will be interpreted. Note that if an extension
typed in is valid, it will not have to timeout to be tested, so
typically
at the expiry of this timeout, the extension will be considered invalid
(and thus control would be passed to the 'i' extension, or if it
doesn't
exist the call would be terminated). The default timeout is 5 seconds.
  DigitTimeout has been deprecated in favor of
Set(TIMEOUT(digit)=timeout)
```

Diff of the internal help from Asterisk 1.2 to 1.4:

Not available in Asterisk 1.4

See also **absolutetimeout, responsetimeout,** and in Appendix C, see **timeout.**

B.41 Directory()

Provides a directory of users or user voicemail box numbers. (For more on system directories and Dial-by-Name, see Chapter 9, "Voicemail.")

```
Directory(voicemail-context[,dialplan-context[,options]])
```

Lets callers search a directory by the user's name. The list of names and extensions is configured in `voicemail.conf`. The *voicemail-context* parameter is required.

The dialplan-context defines the context to be used when dialing the user's extension. If this is not provided, *voicemail-context* is assumed. The only option currently accepted is f, which allows dialing by first name.

If the user dials 0 (zero) and the extension o exists in the current context, the call goes to this extension. Likewise, pressing * sends the call to the extension a, if it exists. This behavior is similar to that found in **Voicemail()**.

Returns 0 unless the caller hangs up.

```
exten => *,1,Directory(default,incoming)
exten => #,1,Directory(default,incoming,f)
```

Asterisk versions:

```
--------|1.2|--------|1.4|--------|1.6|--------
```

Internal help for this application in Asterisk 1.4:

```
    -= Info about application 'Directory' =-

[Synopsis]
Provide directory of voicemail extensions

[Description]
  Directory(vm-context[|dial-context[|options]]): This application will
present
the calling channel with a directory of extensions from which they can
search
by name. The list of names and corresponding extensions is retrieved
from the
voicemail configuration file, voicemail.conf.
  This application will immediately exit if one of the following DTMF
digits are
received and the extension to jump to exists:
    0 - Jump to the 'o' extension, if it exists.
    * - Jump to the 'a' extension, if it exists.

  Parameters:
    vm-context   - This is the context within voicemail.conf to use for
the
                  Directory.
```

```
        dial-context - This is the dialplan context to use when looking for
an
                        extension that the user has selected, or when
jumping to the
                        'o' or 'a' extension.

  Options:
      e - In addition to the name, also read the extension number to the
          caller before presenting dialing options.
      f - Allow the caller to enter the first name of a user in the
directory
          instead of using the last name.
```

Diff of the internal help from Asterisk 1.2 to 1.4:

```
--- in Asterisk 1.2
+++ in Asterisk 1.4
@@ -21,5 +21,7 @@
                        'o' or 'a' extension.

     Options:
+       e - In addition to the name, also read the extension number to
the
+           caller before presenting dialing options.
       f - Allow the caller to enter the first name of a user in the
directory
           instead of using the last name.
```

Diff of the internal help from Asterisk 1.4 to 1.6:

```
--- in Asterisk 1.4
+++ in Asterisk 1.6
@@ -4,7 +4,7 @@
   Provide directory of voicemail extensions

   [Description]
-    Directory(vm-context[|dial-context[|options]]): This application
will present
+    Directory(vm-context[,dial-context[,options]]): This application
will present
   the calling channel with a directory of extensions from which they
can search
   by name. The list of names and corresponding extensions is retrieved
from the
   voicemail configuration file, voicemail.conf.
@@ -21,7 +21,26 @@
                        'o' or 'a' extension.

     Options:
-       e - In addition to the name, also read the extension number to
the
-           caller before presenting dialing options.
-       f - Allow the caller to enter the first name of a user in the
directory
```

```
-               instead of using the last name.
+        e              In addition to the name, also read the extension
number to the
+                      caller before presenting dialing options.
+        f[(<n>)]    Allow the caller to enter the first name of a user
in the
+                      directory instead of using the last name.   If
specified, the
+                      optional number argument will be used for the number
of
+                      characters the user should enter.
+        l[(<n>)]    Allow the caller to enter the last name of a user in
the
+                      directory.  This is the default.  If specified, the
+                      optional number argument will be used for the number
of
+                      characters the user should enter.
+        b[(<n>)]    Allow the caller to enter either the first or the
last name
+                      of a user in the directory.  If specified, the
optional number
+                      argument will be used for the number of characters the
user
+                      should enter.
+        m              Instead of reading each name sequentially and asking
for
+                      confirmation, create a menu of up to 8 names.
+        p(<n>)      Pause for n milliseconds after the digits are
typed.  This is
+                      helpful for people with cellphones, who are not
holding the
+                      receiver to their ear while entering DTMF.
+
+     Only one of the f, l, or b options may be specified.  If more
than one is
+       specified, then Directory will act as if 'b' was specified.  The
number
+       of characters for the user to type defaults to 3.
```

See also `voicemail.conf`.

B.42 DISA()

Direct Inward System Access lets outside callers enter the system and provides them with an internal dial tone.

```
DISA(password[,context[,callerid[,mailbox[@voicemail-context]]]])

DISA(password-file[,callerid[,mailbox[@voicemail-context]]])
```

Provides an internal dial tone to outside callers such that they can make calls as though calling from an internal extension. Upon hearing the dial tone, an access code must be entered followed by the # key. If it is correct, the caller hears another dial tone; this is the system dial tone and the caller can now dial and initiate calls.

Caution: This type of access represents a serious and real security risk and should be planned and considered carefully before use, if it must be used at all!

The *password* option is a numeric access code that must be entered for the caller to be able to make calls out. Following this particular syntax, all the users that call in will use the same access code. If you want to allow unsecured access, enter the string `no-password` instead of an actual password.

The *context* option specifies the context in which the initiated call will be placed. If it is not provided, **DISA()** assumes the context named disa.

The *callerid* option sets the mailbox number (and the optional *voicemail-context*) of a mailbox. If the mailbox contains new messages the caller will hear a stuttered dial tone to indicate this.

Alternatively, you may use password-file to define multiple access passwords. Each line of this file can contain either an access code or a combination of an access code and a context, separated by the pipe (|) character. If no context is specified, `disa` is assumed.

If the caller successfully authenticates with a valid access code, **DISA()** will start the call in the specified context.

```
; Allow outside callers to dial 800 numbers, provided they know the
; password (1234). Set the caller ID so that the call appears to be
; coming from inside the company:
[incoming]
exten => 123,1,DISA(1234,disa,Widgets Inc <212-555-3412>)
[disa]
exten => _0800XXXXXXXX,1,Dial(Zap/4/${EXTEN})
```

Asterisk versions:

--------| 1.2 |--------| 1.4 |--------| 1.6 |--------

Internal help for this application in Asterisk 1.4:

```
    -= Info about application 'DISA' =-

[Synopsis]
DISA (Direct Inward System Access)

[Description]
DISA(<numeric passcode>[|<context>]) or DISA(<filename>)
```

The DISA, Direct Inward System Access, application allows someone from
outside the telephone switch (PBX) to obtain an "internal" system
dialtone and to place calls from it as if they were placing a call from
within the switch.
DISA plays a dialtone. The user enters their numeric passcode, followed by
the pound sign (#). If the passcode is correct, the user is then given
system dialtone on which a call may be placed. Obviously, this type
of access has SERIOUS security implications, and GREAT care must be
taken NOT to compromise your security.

There is a possibility of accessing DISA without password. Simply
exchange your password with "no-password".

    Example: exten => s,1,DISA(no-password|local)

Be aware that using this compromises the security of your PBX.

The arguments to this application (in extensions.conf) allow either
specification of a single global passcode (that everyone uses), or
individual passcodes contained in a file. It also allows specification
of the context on which the user will be dialing. If no context is
specified, the DISA application defaults the context to "disa".
Presumably a normal system will have a special context set up
for DISA use with some or a lot of restrictions.

The file that contains the passcodes (if used) allows specification
of either just a passcode (defaulting to the "disa" context, or
passcode|context on each line of the file. The file may contain blank
lines, or comments starting with "#" or ";". In addition, the
above arguments may have |new-callerid-string appended to them, to
specify a new (different) callerid to be used for this call, for
example: numeric-passcode|context|"My Phone" <(234) 123-4567> or
full-pathname-of-passcode-file|"My Phone" <(234) 123-4567>. Last
but not least, |mailbox[@context] may be appended, which will cause
a stutter-dialtone (indication "dialrecall") to be used, if the
specified mailbox contains any new messages, for example:
numeric-passcode|context||1234 (w/a changing callerid). Note that
in the case of specifying the numeric-passcode, the context must be
specified if the callerid is specified also.

If login is successful, the application looks up the dialed number in
the specified (or default) context, and executes it if found.
If the user enters an invalid extension and extension "i" (invalid)
exists in the context, it will be used. Also, if you set the 5th
argument
to 'NOANSWER', the DISA application will not answer initially.

Diff of the internal help from Asterisk 1.2 to 1.4:

```
--- in Asterisk 1.2
+++ in Asterisk 1.4
@@ -4,7 +4,7 @@
   DISA (Direct Inward System Access)
```

```
   [Description]
-  DISA(<numeric passcode>[|<context>]) or disa(<filename>)
+  DISA(<numeric passcode>[|<context>]) or DISA(<filename>)
   The DISA, Direct Inward System Access, application allows someone
from
   outside the telephone switch (PBX) to obtain an "internal" system
   dialtone and to place calls from it as if they were placing a call
from
@@ -24,7 +24,7 @@

   The arguments to this application (in extensions.conf) allow either
   specification of a single global passcode (that everyone uses), or
-  individual passcodes contained in a file. It also allow
specification
+  individual passcodes contained in a file. It also allows
specification
   of the context on which the user will be dialing. If no context is
   specified, the DISA application defaults the context to "disa".
   Presumably a normal system will have a special context set up
@@ -48,4 +48,5 @@
   If login is successful, the application looks up the dialed number
in
   the specified (or default) context, and executes it if found.

   If the user enters an invalid extension and extension "i" (invalid)
-  exists in the context, it will be used.
+  exists in the context, it will be used. Also, if you set the 5th
argument
+  to 'NOANSWER', the DISA application will not answer initially.
```

Diff of the internal help from Asterisk 1.4 to 1.6:

```
--- in Asterisk 1.4
+++ in Asterisk 1.6
@@ -4,49 +4,42 @@
   DISA (Direct Inward System Access)

   [Description]
-  DISA(<numeric passcode>[|<context>]) or DISA(<filename>)
+  DISA(<numeric passcode>[,<context>[,<cid>[,mailbox[,options]]]]) or
+  DISA(<filename>[,,,,options])
   The DISA, Direct Inward System Access, application allows someone
from
   outside the telephone switch (PBX) to obtain an "internal" system
   dialtone and to place calls from it as if they were placing a call
from
   within the switch.
   DISA plays a dialtone. The user enters their numeric passcode,
followed by
   the pound sign (#). If the passcode is correct, the user is then
given
-  system dialtone on which a call may be placed. Obviously, this type
-  of access has SERIOUS security implications, and GREAT care must be
```

```
 -  taken NOT to compromise your security.
 -
 -  There is a possibility of accessing DISA without password. Simply
 -  exchange your password with "no-password".
 +  system dialtone within <context> on which a call may be placed. If
the user
 +  enters an invalid extension and extension "i" exists in the
specified
 +  context, it will be used.

 -      Example: exten => s,1,DISA(no-password|local)
 +  If you need to present a DISA dialtone without entering a password,
simply
 +  set <passcode> to "no-password".

 -  Be aware that using this compromises the security of your PBX.
 +  Be aware that using this may compromise the security of your PBX.

    The arguments to this application (in extensions.conf) allow either
    specification of a single global passcode (that everyone uses), or
 -  individual passcodes contained in a file. It also allows
specification
 -  of the context on which the user will be dialing. If no context is
 -  specified, the DISA application defaults the context to "disa".
 -  Presumably a normal system will have a special context set up
 -  for DISA use with some or a lot of restrictions.
 +  individual passcodes contained in a file.

 -  The file that contains the passcodes (if used) allows specification
 -  of either just a passcode (defaulting to the "disa" context, or
 -  passcode|context on each line of the file. The file may contain
blank
 -  lines, or comments starting with "#" or ";". In addition, the
 -  above arguments may have |new-callerid-string appended to them, to
 -  specify a new (different) callerid to be used for this call, for
 -  example: numeric-passcode|context|"My Phone" <(234) 123-4567> or
 -  full-pathname-of-passcode-file|"My Phone" <(234) 123-4567>.  Last
 -  but not least, |mailbox[@context] may be appended, which will cause
 ▪  a stutter-dialtone (indication "dialrecall") to be used, if the
 -  specified mailbox contains any new messages, for example:
 -  numeric-passcode|context||1234 (w/a changing callerid).  Note that
 -  in the case of specifying the numeric-passcode, the context must be
 -  specified if the callerid is specified also.
 +  The file that contains the passcodes (if used) allows a complete
 +  specification of all of the same arguments available on the command
 +  line, with the sole exception of the options. The file may contain
blank
 +  lines, or comments starting with "#" or ";".

 -  If login is successful, the application looks up the dialed number
in
 -  the specified (or default) context, and executes it if found.
 -  If the user enters an invalid extension and extension "i" (invalid)
```

```
    -    exists in the context, it will be used. Also, if you set the 5th
    argument
    -    to 'NOANSWER', the DISA application will not answer initially.
    +    <context> specifies the dialplan context in which the user-entered
    extension
    +    will be matched. If no context is specified, the DISA application
    defaults
    +    the context to "disa". Presumably a normal system will have a
    special
    +    context set up for DISA use with some or a lot of restrictions.
    +
    +    <cid> specifies a new (different) callerid to be used for this call.
    +
    +    <mailbox[@context]> will cause a stutter-dialtone (indication
    "dialrecall")
    +    to be used, if the specified mailbox contains any new messages.
    +
    +    The following options are available:
    +      n - the DISA application will not answer initially.
    +      p - the extension entered will be considered complete when a '#' is
    entered.
```

B.43 DumpChan()

Prints information about the calling channel on the console.

```
DumpChan([min_verbose_level])
```

Shows information about the calling channel and the contents of all the channel variables. If *min_verbose_level* is set, only messages at the same or higher verbosity level are printed.
   Returns 0.

```
exten => 123,1,Answer()
exten => 123,n,DumpChan()
exten => 123,n,Background(enter-ext-of-person)
```

Asterisk versions:

```
--------|1.2|--------|1.4|--------|1.6|--------
```

Internal help for this application in Asterisk 1.4:

```
    -= Info about application 'DumpChan' =-

[Synopsis]
Dump Info About The Calling Channel

[Description]
   DumpChan([<min_verbose_level>])
```

```
Displays information on channel and listing of all channel
variables. If min_verbose_level is specified, output is only
displayed when the verbose level is currently set to that number
or greater.
```

Diff of the internal help from Asterisk 1.2 to 1.4:

None

Diff of the internal help from Asterisk 1.4 to 1.6:

None

See also noop, log, verbose.

B.44 DUNDiLookup()

Looks up a number using DUNDi.

```
DUNDiLookup(number[|DUNDi-context[|options]])
```

Warning: DUNDiLookup() is removed as of Asterisk 1.4 and has been replaced by the DUNDILOOKUP() function.

Asterisk versions:

```
--------|1.2|          |      |         |      |
```

Internal help for this application in Asterisk 1.2:

```
    -= Info about application 'DUNDiLookup' =-

[Synopsis]
Look up a number with DUNDi

[Description]
DUNDiLookup(number[|context[|options]])
     Looks up a given number in the global context specified or in
the reserved 'e164' context if not specified.  Returns -1 if the
channel
is hungup during the lookup or 0 otherwise.  On completion, the
variable
${DUNDTECH} and ${DUNDDEST} will contain the technology and destination
of the appropriate technology and destination to access the number. If
no
answer was found, and the priority n + 101 exists, execution will
continue
```

at that location. Note that this will only occur if the global priority
jumping option is enabled in extensions.conf. If the 'b' option is
specified,
the internal DUNDi cache will by bypassed.

Diff of the internal help from Asterisk 1.2 to 1.4:

Not available in Asterisk 1.4

See also Appendix C.

B.45 EAGI()

Calls an AGI-compliant application.

```
EAGI(program[,argument1[,argument2[,...]]])
```

(Similar to **AGI()**, **FastAGI()**, **DeadAGI()**)

EAGI() (Extended AGI) works just like regular **AGI()** (agi), but also provides
access to the incoming audio stream. The incoming audio stream is provided on file
descriptor 3.[7]

AGI scripts or programs can be implemented in almost any conceivable language
(e.g., Perl, PHP) and may be used to manipulate the channel, play sound files, interpret DTMF tones, and so on. Asterisk communicates with the AGI program over
stdin and stdout. The arguments are passed directly to the AGI program at execution time.

The AGI program must be flagged as executable in the filesystem. The path is
relative to the Asterisk AGI directory, which is at /var/lib/asterisk/agi-bin
by default, or it may be stated as an absolute path.

To run AGI programs on inactive channels (as in the case of an h extension,
where the channel is on-hook), use **DeadAGI()** instead.

Returns −1 on hangup or if the AGI program requests a hangup; returns 0 if no
hangup is requested.

```
; call my AGI script:
exten => 123,1,AGI(agi-script.agi)
exten => 123,n,EAGI(eagi-script.agi)
```

Asterisk versions:

```
--------|1.2|--------|1.4|--------|1.6|--------
```

---

7. A reminder: 0: stdin, 1: stdout, 2:stderr. File descriptor 3 is freely assignable.

Internal help for this application in Asterisk 1.4:

```
   -= Info about application 'EAGI' =-

[Synopsis]
Executes an EAGI compliant application

[Description]
   [E|Dead]AGI(command|args): Executes an Asterisk Gateway Interface
compliant
program on a channel. AGI allows Asterisk to launch external programs
written in any language to control a telephony channel, play audio,
read DTMF digits, etc. by communicating with the AGI protocol on stdin
and stdout.
   This channel will stop dialplan execution on hangup inside of this
application, except when using DeadAGI.  Otherwise, dialplan execution
will continue normally.
   A locally executed AGI script will receive SIGHUP on hangup from the
channel
except when using DeadAGI. This can be disabled by setting the
AGISIGHUP channel
variable to "no" before executing the AGI application.
   Using 'EAGI' provides enhanced AGI, with incoming audio available out
of band
on file descriptor 3

   Use the CLI command 'agi show' to list available agi commands
   This application sets the following channel variable upon completion:
      AGISTATUS      The status of the attempt to the run the AGI script
                     text string, one of SUCCESS | FAILURE | HANGUP
```

Diff of the internal help from Asterisk 1.2 to 1.4:

```
--- in Asterisk 1.2
+++ in Asterisk 1.4
@@ -9,9 +9,16 @@
    written in any language to control a telephony channel, play audio,
    read DTMF digits, etc. by communicating with the AGI protocol on
stdin
    and stdout.
-  Returns -1 on hangup (except for DeadAGI) or if application
requested
-   hangup, or 0 on non-hangup exit.
-  Using 'EAGI' provides enhanced AGI, with incoming audio available
out of band
+   This channel will stop dialplan execution on hangup inside of this
+  application, except when using DeadAGI.  Otherwise, dialplan
execution
+  will continue normally.
+   A locally executed AGI script will receive SIGHUP on hangup from
the channel
+  except when using DeadAGI. This can be disabled by setting the
AGISIGHUP channel
```

```
+   variable to "no" before executing the AGI application.
+     Using 'EAGI' provides enhanced AGI, with incoming audio available
out of band
    on file descriptor 3

-   Use the CLI command 'show agi' to list available agi commands
+     Use the CLI command 'agi show' to list available agi commands
+     This application sets the following channel variable upon
completion:
+          AGISTATUS        The status of the attempt to the run the AGI
script
+                                text string, one of SUCCESS | FAILURE | HANGUP
```

Diff of the internal help from Asterisk 1.4 to 1.6:

```
--- in Asterisk 1.4
+++ in Asterisk 1.6
@@ -4,21 +4,22 @@
    Executes an EAGI compliant application

    [Description]
-     [E|Dead]AGI(command|args): Executes an Asterisk Gateway Interface
compliant
-   program on a channel. AGI allows Asterisk to launch external
programs
-   written in any language to control a telephony channel, play audio,
-   read DTMF digits, etc. by communicating with the AGI protocol on
stdin
-   and stdout.
-     This channel will stop dialplan execution on hangup inside of this
-   application, except when using DeadAGI.  Otherwise, dialplan
execution
-   will continue normally.
+     [E|Dead]AGI(command,args): Executes an Asterisk Gateway Interface
compliant
+   program on a channel. AGI allows Asterisk to launch external
programs written
+   in any language to control a telephony channel, play audio, read
DTMF digits,
+   etc. by communicating with the AGI protocol on stdin and stdout.
+     As of 1.6.0, this channel will not stop dialplan execution on
hangup inside
+   of this application. Dialplan execution will continue normally, even
upon
+   hangup until the AGI application signals a desire to stop (either by
exiting
+   or, in the case of a net script, by closing the connection).
+     A locally executed AGI script will receive SIGHUP on hangup from
the channel
-   except when using DeadAGI. This can be disabled by setting the
AGISIGHUP channel
-   variable to "no" before executing the AGI application.
```

```
+   except when using DeadAGI. A fast AGI server will correspondingly
receive a
+   HANGUP in OOB data. Both of these signals may be disabled by setting
the
+   AGISIGHUP channel variable to "no" before executing the AGI
application.
        Using 'EAGI' provides enhanced AGI, with incoming audio available
out of band
-   on file descriptor 3
+   on file descriptor 3.

-       Use the CLI command 'agi show' to list available agi commands
+       Use the CLI command 'agi show' to list available agi commands.
        This application sets the following channel variable upon
completion:
            AGISTATUS      The status of the attempt to the run the AGI
script
-                          text string, one of SUCCESS | FAILURE | HANGUP
+                          text string, one of
SUCCESS | FAILURE | NOTFOUND | HANGUP
```

See also agi.

B.46 Echo()

Repeats incoming audio to the caller.

```
Echo()
```

Takes any incoming audio and returns it on the same channel. This application is used primarily for troubleshooting and testing of delay (latency) and sound quality on VoIP connections. The caller can end the call by pressing #.

Returns 0 if the caller ends the call with # or –1 if the caller hangs up.

```
exten => 123,1,Echo()
exten => 123,n,Playback(vm-goodbye)
```

Asterisk versions:

```
--------|1.2|--------|1.4|--------|1.6|--------
```

Internal help for this application in Asterisk 1.4:

```
    -= Info about application 'Echo' =-

[Synopsis]
Echo audio, video, or DTMF back to the calling party
```

```
[Description]
  Echo(): This application will echo any audio, video, or DTMF frames
read from
the calling channel back to itself. If the DTMF digit '#' is received,
the
application will exit.
```

Diff of the internal help from Asterisk 1.2 to 1.4:

```
--- in Asterisk 1.2
+++ in Asterisk 1.4
@@ -1,9 +1,9 @@
    -= Info about application 'Echo' =-

   [Synopsis]
-  Echo audio read back to the user
+  Echo audio, video, or DTMF back to the calling party

   [Description]
-    Echo():  Echo audio read from channel back to the channel.
-  User can exit the application by either pressing the '#' key,
-  or hanging up.
+    Echo(): This application will echo any audio, video, or DTMF
frames read from
+  the calling channel back to itself. If the DTMF digit '#' is
received, the
+  application will exit.
```

Diff of the internal help from Asterisk 1.4 to 1.6:

None

See also milliwatt.

B.47 EndWhile()

Ends a while loop.

```
EndWhile()
```

Returns to the previously called **While()** statement. For a complete description of while loops in Asterisk, see while.

```
exten => 123,1,Answer()
exten => 123,n,Set(i=1)
exten => 123,n,While($[${i} < 5])
exten => 123,n,SayNumber(${i})
exten => 123,n,Set(i=$[${i} + 1])
exten => 123,n,EndWhile()
exten => 123,n,Hangup()
```

Asterisk versions:

```
--------|1.2|--------|1.4|--------|1.6|--------
```

Internal help for this application in Asterisk 1.4:

```
    -= Info about application 'EndWhile' =-

[Synopsis]
End a while loop

[Description]
Usage:  EndWhile()
Return to the previous called While
```

Diff of the internal help from Asterisk 1.2 to 1.4:

```
--- in Asterisk 1.2
+++ in Asterisk 1.4
@@ -1,7 +1,7 @@
    -= Info about application 'EndWhile' =-

  [Synopsis]
- End A While Loop
+ End a while loop

  [Description]
  Usage:  EndWhile()
```

Diff of the internal help from Asterisk 1.4 to 1.6:

```
--- in Asterisk 1.4
+++ in Asterisk 1.6
@@ -4,5 +4,4 @@
  End a while loop

  [Description]
- Usage:  EndWhile()
- Return to the previous called While
+    EndWhile(): Return to the previous called While()
```

See also while, gotoif.

B.48 EnumLookup()

Looks up a number using ENUM.

```
EnumLookup(number)
```

Warning: `EnumLookup()` is removed as of Asterisk 1.4 and was replaced by the
`ENUMLOOKUP()` function.

Asterisk versions:

```
--------|1.2|            |     |         |     |
```

Internal help for this application in Asterisk 1.2:

```
        -= Info about application 'EnumLookup' =-

    [Synopsis]
    Lookup number in ENUM

    [Description]
      EnumLookup(exten[|option]):  Looks up an extension via ENUM and sets
    the variable 'ENUM'. For VoIP URIs this variable will
    look like 'TECHNOLOGY/URI' with the appropriate technology.
    Currently, the enumservices SIP, H323, IAX, IAX2 and TEL are
    recognized.

    Returns status in the ENUMSTATUS channel variable:
        ERROR         Failed to do a lookup
        <tech>        Technology of the successful lookup: SIP, H323, IAX,
    IAX2 or TEL
        BADURI        Got URI Asterisk does not understand.
      The option string may contain zero or the following character:
            'j' -- jump to +101 priority if the lookup isn't successful.
                    and jump to +51 priority on a TEL entry.
```

Diff of the internal help from Asterisk 1.2 to 1.4:

Not available in Asterisk 1.4

See also Appendix C.

B.49 Eval()

Evaluates a variable twice.

```
    Eval(result_variable=expression)
```

Warning: `Eval()` is removed as of Asterisk 1.4 and has been replaced by the
`EVAL()` function.

Asterisk versions:

```
--------|1.2|            |     |         |     |
```

Internal help for this application in Asterisk 1.2:

```
    -= Info about application 'Eval' =-

[Synopsis]
Evaluates a string

[Description]
Usage: Eval(newvar=somestring)
  Normally Asterisk evaluates variables inline.  But what if you want
to
store variable offsets in a database, to be evaluated later?  Eval is
the answer, by allowing a string to be evaluated twice in the dialplan,
the first time as part of the normal dialplan, and the second using
Eval.
```

Diff of the internal help from Asterisk 1.2 to 1.4:

Not available in Asterisk 1.4

See also Appendix C.

B.50 Exec()

Executes an Asterisk application dynamically.

```
Exec(application(arguments))
```

Allows the execution of an arbitrary dialplan application, even if this application is not hardcoded into the dialplan. Returns the value returned by the application or –2 if the application cannot be found. The *arguments* are passed to the called application.

This application enables the calling of applications out of a database or other external source.

```
exten => 123,1,Set(app=SayDigits(12345))
exten => 123,2,Exec(${app})
```

Warning: A negative return value will mean that execution of the dialplan ends. If this is not desired, use **TryExec()**.

Asterisk versions:

```
--------|1.2|--------|1.4|--------|1.6|--------
```

Internal help for this application in Asterisk 1.4:

```
    -= Info about application 'Exec' =-

[Synopsis]
Executes dialplan application

[Description]
Usage: Exec(appname(arguments))
  Allows an arbitrary application to be invoked even when not
hardcoded into the dialplan.  If the underlying application
terminates the dialplan, or if the application cannot be found,
Exec will terminate the dialplan.
  To invoke external applications, see the application System.
  If you would like to catch any error instead, see TryExec.
```

Diff of the internal help from Asterisk 1.2 to 1.4:

```
--- in Asterisk 1.2
+++ in Asterisk 1.4
@@ -1,11 +1,13 @@
    -= Info about application 'Exec' =-

  [Synopsis]
-  Executes internal application
+  Executes dialplan application

  [Description]
  Usage: Exec(appname(arguments))
    Allows an arbitrary application to be invoked even when not
-  hardcoded into the dialplan. To invoke external applications
-  see the application System. Returns whatever value the
-  app returns or a non-zero value if the app cannot be found.
+  hardcoded into the dialplan.  If the underlying application
+  terminates the dialplan, or if the application cannot be found,
+  Exec will terminate the dialplan.
+    To invoke external applications, see the application System.
+    If you would like to catch any error instead, see TryExec.
```

Diff of the internal help from Asterisk 1.4 to 1.6:

```
--- in Asterisk 1.4
+++ in Asterisk 1.6
@@ -4,8 +4,8 @@
   Executes dialplan application

   [Description]
-  Usage: Exec(appname(arguments))
-    Allows an arbitrary application to be invoked even when not
+    Exec(appname(arguments)):
```

> + Allows an arbitrary application to be invoked even when not
>   hardcoded into the dialplan. If the underlying application
>   terminates the dialplan, or if the application cannot be found,
>   Exec will terminate the dialplan.

See also execif, tryexec, system.

B.51 ExecIf()

Executes an Asterisk application under specific conditions.

```
ExecIf(expression,application,arguments)
```

If expression evaluates to true, the defined application is executed with the provided arguments, and the return value is returned. See doc/README.variables (1.2) /doc/channelvariables.txt (1.4) for more information about standard expressions for Asterisk.

If expression evaluates to false, execution moves to the next priority in the extension.

```
exten => 123,1,ExecIf($[${CALLERID(num)} = 101],SayDigits,123)
exten => 123,n,SayDigits(678)
```

Asterisk versions:

```
--------|1.2|--------|1.4|--------|1.6|--------
```

Internal help for this application in Asterisk 1.4:

```
    -= Info about application 'ExecIf' =-

[Synopsis]
Executes dialplan application, conditionally

[Description]
Usage:  ExecIF (<expr>|<app>|<data>)
If <expr> is true, execute and return the result of <app>(<data>).
If <expr> is true, but <app> is not found, then the application
will return a non-zero value.
```

Diff of the internal help from Asterisk 1.2 to 1.4:

```
--- in Asterisk 1.2
+++ in Asterisk 1.4
@@ -1,7 +1,7 @@
    -= Info about application 'ExecIf' =-

 [Synopsis]
-  Conditional exec
+  Executes dialplan application, conditionally
```

```
    [Description]
    Usage:  ExecIF (<expr>|<app>|<data>)
```

Diff of the internal help from Asterisk 1.4 to 1.6:

```
--- in Asterisk 1.4
+++ in Asterisk 1.6
@@ -4,7 +4,7 @@
    Executes dialplan application, conditionally

    [Description]
-   Usage:  ExecIF (<expr>|<app>|<data>)
-   If <expr> is true, execute and return the result of <app>(<data>).
-   If <expr> is true, but <app> is not found, then the application
+     ExecIF (<expr>?<appiftrue>(<args>)[:<appiffalse>(<args>)])
+   If <expr> is true, execute and return the result of
<appiftrue>(<args>).
+   If <expr> is true, but <appiftrue> is not found, then the
application
    will return a non-zero value.
```

See also **exec.**

B.52 ExecIfTime()

Executes an application based on the current time.

```
ExecIf(times|daysofweek|daysofmonth|months?application[,arguments])
```

If the current time is in the time window defined, *application* is executed with *arguments* and the result returned. The time window is defined the same way as it is for include, **GotoIfTime()** (see gotoiftime) or **IFTIME()** (see Appendix C).

If the current time is not in the time window defined, execution continues in the next priority.

Asterisk versions:

```
--------|1.2|--------|1.4|--------|1.6|--------
```

Internal help for this application in Asterisk 1.4:

```
    -= Info about application 'ExecIfTime' =-

[Synopsis]
Conditional application execution based on the current time

[Description]
    ExecIfTime(<times>|<weekdays>|<mdays>|<months>?appname[|appargs]):
This application will execute the specified dialplan application, with
optional
arguments, if the current time matches the given time specification.
```

Diff of the internal help from Asterisk 1.2 to 1.4:

```
--- in Asterisk 1.2
+++ in Asterisk 1.4
@@ -6,6 +6,4 @@
   [Description]
     ExecIfTime(<times>|<weekdays>|<mdays>|<months>?appname[|appargs]):
   This application will execute the specified dialplan application,
with optional
-   arguments, if the current time matches the given time specification.
Further
-   information on the time speicification can be found in examples
illustrating
-   how to do time-based context includes in the dialplan.
+   arguments, if the current time matches the given time specification.
```

Diff of the internal help from Asterisk 1.4 to 1.6:

```
--- in Asterisk 1.4
+++ in Asterisk 1.6
@@ -4,6 +4,6 @@
   Conditional application execution based on the current time

   [Description]
-     ExecIfTime(<times>|<weekdays>|<mdays>|<months>
?appname[|appargs]):
+     ExecIfTime(<times>,<weekdays>,<mdays>,<months>
?appname[(appargs)]):
   This application will execute the specified dialplan application,
with optional
     arguments, if the current time matches the given time specification.
```

See also exec, execif, gotoiftime, and in Appendix C, see iftime.

B.53 ExitWhile()

Exits a while loop, irrespective of whether its condition has been satisfied.

```
ExitWhile()
```

With **ExitWhile()**, you can interrupt further execution whether or not the while condition has been satisfied.

Asterisk versions:

```
  |      |            |1.4|--------|1.6|--------
```

Internal help for this application in Asterisk 1.4:

```
    -= Info about application 'ExitWhile' =-

[Synopsis]
End a While loop

[Description]
Usage:  ExitWhile()
Exits a While loop, whether or not the conditional has been satisfied.
```

Diff of the internal help from Asterisk 1.2 to 1.4:

Not available in Asterisk 1.2

Diff of the internal help from Asterisk 1.4 to 1.6:

```
--- in Asterisk 1.4
+++ in Asterisk 1.6
@@ -4,5 +4,4 @@
   End a While loop

   [Description]
-  Usage:  ExitWhile()
-  Exits a While loop, whether or not the conditional has been
satisfied.
+    ExitWhile(): Exits a While() loop, whether or not the conditional
has been satisfied.
```

See also `while`, `endwhile`.

B.54 ExtenSpy()

Eavesdrop on a channel attached to a specific extension and whisper to it if desired.

```
ExtenSpy(extension[@context][,options])
```

ExtenSpy() can listen to incoming and outgoing audio on channels used by the specified extension. The options:

- **b**

  Only listens to channels that belong to a bridged call.

- **g(grp)**

  Only listens to channels where the channel variable ${SPYGROUP} is set to grp. ${SPYGROUP} can contain a colon-separated list of values.

- **q**

  Do not play a tone or say the channel name when listening starts on a
  channel.

- **r([name])**

  Records the listening session to the spool directory. A filename may be specified if desired; chanspy is the default.

- **v([value])**

  Sets the initial volume. The value may be between –4 and 4.

- **w**

  Enables "whisper" mode. Lets the spying channel talk to the spied-on channel.

- **W**

  Enables "private whisper mode." The "spying" channel can whisper to the
  spied-on channel, but cannot listen.

The following key controls are available while listening:

- **#**

  Stepwise volume adjustment (–4 to 4)

- *****

  Switch to another channel

Asterisk versions:

```
|     |              |1.4|--------|1.6|--------
```

Internal help for this application in Asterisk 1.4:

```
    -= Info about application 'ExtenSpy' =-

[Synopsis]
Listen to a channel, and optionally whisper into it

[Description]
  ExtenSpy(exten[@context][|options]): This application is used to
listen to the
audio from an Asterisk channel. This includes the audio coming in and
out of the channel being spied on. Only channels created by outgoing
calls for the
specified extension will be selected for spying. If the optional
context is not
supplied, the current channel's context will be used.
  While spying, the following actions may be performed:
    - Dialing # cycles the volume level.
    - Dialing * will stop spying and look for another channel to spy
on.
```

```
Options:
    b               - Only spy on channels involved in a bridged call.
      g(grp)        - Match only channels where their ${SPYGROUP}
variable is set to
                      contain 'grp' in an optional : delimited list.
    q               - Don't play a beep when beginning to spy on a
channel, or speak the
                      selected channel name.
      r[(basename)] - Record the session to the monitor spool directory.
An
                      optional base for the filename may be specified.
The
                      default is 'chanspy'.
      v([value])    - Adjust the initial volume in the range from -4 to
4. A
                      negative value refers to a quieter setting.
      w             - Enable 'whisper' mode, so the spying channel can
talk to
                      the spied-on channel.
      W             - Enable 'private whisper' mode, so the spying
channel can
                      talk to the spied-on channel but cannot listen to
that
                      channel.
```

Diff of the internal help from Asterisk 1.2 to 1.4:

Not available in Asterisk 1.2

Diff of the internal help from Asterisk 1.4 to 1.6:

```
--- in Asterisk 1.4
+++ in Asterisk 1.6
@@ -4,7 +4,7 @@
   Listen to a channel, and optionally whisper into it

   [Description]
-     ExtenSpy(exten[@context][|options]): This application is used to
listen to the
+     ExtenSpy(exten[@context][,options]): This application is used to
listen to the
   audio from an Asterisk channel. This includes the audio coming in
and
   out of the channel being spied on. Only channels created by outgoing
calls for the
   specified extension will be selected for spying. If the optional
context is not
@@ -12,19 +12,48 @@
     While spying, the following actions may be performed:
       - Dialing # cycles the volume level.
       - Dialing * will stop spying and look for another channel to spy
on.
+     Note: The X option superseeds the two features above in that if a
valid
```

```
+           single digit extension exists in the correct context it
ChanSpy will
+           exit to it.
    Options:
-      b              - Only spy on channels involved in a bridged call.
-      g(grp)         - Match only channels where their ${SPYGROUP}
variable is set to
-                       contain 'grp' in an optional : delimited list.
-      q              - Don't play a beep when beginning to spy on a
channel, or speak the
-                       selected channel name.
-      r[(basename)]  - Record the session to the monitor spool
directory. An
-                       optional base for the filename may be specified.
The
-                       default is 'chanspy'.
-      v([value])     - Adjust the initial volume in the range from -4
to 4. A
-                       negative value refers to a quieter setting.
-      w              - Enable 'whisper' mode, so the spying channel can
talk to
-                       the spied-on channel.
-      W              - Enable 'private whisper' mode, so the spying
channel can
-                       talk to the spied-on channel but cannot listen
to that
-                       channel.
+      b                  - Only spy on channels involved in a
bridged call.
+      B                  - Instead of whispering on a single
channel barge in on both
+                           channels involved in the call.
+      d                  - Override the typical numeric DTMF
functionality and instead
+                           use DTMF to switch between spy modes.
+                                  4 = spy mode
+                                  5 = whisper mode
+                                  6 = barge mode
+      g(grp)             - Only spy on channels in which one or
more of the groups
+                           listed in 'grp' matches one or more
groups from the
+                           SPYGROUP variable set on the channel to
be spied upon.
+                           Note that both 'grp' and SPYGROUP can
contain either a
+                           single group or a colon-delimited list
of groups, such
+                           as 'sales:support:accounting'.
+      n([mailbox][@context]) - Say the name of the person being spied
on if that person has recorded
+                           his/her name. If a context is
specified, then that voicemail context will
```

```
+                                be searched when retrieving the name,
otherwise the "default" context
+                                will be searched. If no mailbox is
specified, then the channel name will
+                                be used when searching for the name
(i.e. if SIP/1000 is the channel being
+                                spied on and no mailbox is specified,
then "1000" will be used when searching
+                                for the name).
+       q                        - Don't play a beep when beginning to spy
on a channel, or speak the
+                                selected channel name.
+       r[(basename)]            - Record the session to the monitor spool
directory. An
+                                optional base for the filename may be
specified. The
+                                default is 'chanspy'.
+       s                        - Skip the playback of the channel type
(i.e. SIP, IAX, etc) when
+                                speaking the selected channel name.
+       v([value])               - Adjust the initial volume in the range
from -4 to 4. A
+                                negative value refers to a quieter
setting.
+       w                        - Enable 'whisper' mode, so the spying
channel can talk to
+                                the spied-on channel.
+       W                        - Enable 'private whisper' mode, so the
spying channel can
+                                talk to the spied-on channel but cannot
listen to that
+                                channel.
+       o                        - Only listen to audio coming from this
channel.
+       X                        - Allow the user to exit ChanSpy to a
valid single digit
+                                numeric extension in the current
context or the context
+                                specified by the SPY_EXIT_CONTEXT
channel variable. The
+                                name of the last channel that was spied
on will be stored
+                                in the SPY_CHANNEL variable.
```

See also chanspy, zapbarge, zapscan, monitor.

B.55 ExternalIVR()

Start an external IVR application.

```
ExternalIVR(shell-command[,arg1[,arg2[,...]]])
```

Forks a process and starts an external IVR[8] application. This application then receives all DTMF events and responds accordingly. The application receives notification if the channel is hung up but must shutdown on its own. The protocol for this interface is described in `doc/externalivr.txt`.

Asterisk versions:

```
--------|1.2|--------|1.4|--------|1.6|--------
```

Internal help for this application in Asterisk 1.4:

```
    -= Info about application 'ExternalIVR' =-

[Synopsis]
Interfaces with an external IVR application

[Description]
  ExternalIVR(command[|arg[|arg...]]): Forks a process to run the
supplied command,
and starts a generator on the channel. The generator's play list is
controlled by the external application, which can add and clear entries
via simple commands issued over its stdout. The external application
will receive all DTMF events received on the channel, and notification
if the channel is hung up. The application will not be forcibly
terminated
when the channel is hung up.
See doc/externalivr.txt for a protocol specification.
```

Diff of the internal help from Asterisk 1.2 to 1.4:

```
--- in Asterisk 1.2
+++ in Asterisk 1.4
@@ -4,11 +4,11 @@
    Interfaces with an external IVR application

    [Description]
-    ExternalIVR(command[|arg[|arg...]]): Forks an process to run the
supplied command,
+    ExternalIVR(command[|arg[|arg...]]): Forks a process to run the
supplied command,
    and starts a generator on the channel. The generator's play list is
    controlled by the external application, which can add and clear
entries
    via simple commands issued over its stdout. The external application
    will receive all DTMF events received on the channel, and
notification
    if the channel is hung up. The application will not be forcibly
```

---

8. Interactive voice response.

terminated
   when the channel is hung up.
-   See doc/README.externalivr for a protocol specification.
+   See doc/externalivr.txt for a protocol specification.

Diff of the internal help from Asterisk 1.4 to 1.6:

```
--- in Asterisk 1.4
+++ in Asterisk 1.6
@@ -4,11 +4,18 @@
    Interfaces with an external IVR application

    [Description]
-    ExternalIVR(command[|arg[|arg...]]): Forks a process to run the
supplied command,
-   and starts a generator on the channel. The generator's play list is
-   controlled by the external application, which can add and clear
entries
-   via simple commands issued over its stdout. The external application
-   will receive all DTMF events received on the channel, and
notification
-   if the channel is hung up. The application will not be forcibly
terminated
-   when the channel is hung up.
+    ExternalIVR(command|ivr://ivrhosti([,arg[,arg...]])[,options]):
Either forks a process
+   to run given command or makes a socket to connect to given host and
starts
+   a generator on the channel. The generator's play list is controlled
by the
+   external application, which can add and clear entries via simple
commands
+   issued over its stdout. The external application will receive all
DTMF events
+   received on the channel, and notification if the channel is hung up.
The
+   application will not be forcibly terminated when the channel is hung
up.
    See doc/externalivr.txt for a protocol specification.
+   The 'n' option tells ExternalIVR() not to answer the channel.
+   The 'i' option tells ExternalIVR() not to send a hangup and exit
when the
+      channel receives a hangup, instead it sends an 'I' informative
message
+      meaning that the external application MUST hang up the call with
an H command
+   The 'd' option tells ExternalIVR() to run on a channel that has been
hung up
+      and will not look for hangups.  The external application must exit
with
+      an 'E' command.
```

B.56 FastAGI()

Calls an AGI-compliant application over a network connection.

```
FastAGI(agi://hostname[:port][/script],arguments)
```

(Similar to **AGI()**, **DeadAGI()**, **EAGI()**.)

Runs an AGI-compliant program on the current channel, but calls the application from another host on the network. The intent is to help distribute the load of processor-intensive AGI scripts or programs to remote servers and reduce startup latency of those programs. (A FastAGI script can be started before it is actually needed, much like a FastCGI script on a web server.)

FastAGI() attempts to connect directly to a running FastAGI program listening for connections on the specified port on the server hostname. The default port is 4573 if it is not specified. If script is defined, it is used to populate the variable agi_network_script and passed to the FastAGI program. The arguments are also passed to the program.

A sample FastAGI script can be found at agi/fastagi-test. Use this as a starting point for writing your own FastAGI applications.

Returns −1, if the application ends and requests a hangup; returns 0 if it ends without requesting a hangup.

```
; Connect to the sample FastAGI program "fastagi-test",
; which must nevertheless be running on the local machine:
exten => 123,1,Answer()
exten => 123,n,FastAGI(agi://localhost/fastagi-test)

; Connect to the FastAGI script "test" on the host "testbox"
; at port 9000 and pass parameter "123":
exten => 124,1,Answer()
exten => 124,n,FastAGI(agi://testbox:9000/test,123)
```

See also agi, deadagi.

B.57 Festival()

Uses the free Festival text-to-speech (TTS) engine to read text to the caller.

```
Festival(text[,keys])
```

Connects to the locally running Festival server (which must be installed separately), sends it the specified text and plays the resulting audio back to the caller. If *keys* are defined, pressing of the defined keys will interrupt playback and return the value of the key depressed; if any is provided as the value to keys, all *keys* will be recognized and the call will be passed to the appropriate extension in the dialplan.

Festival must be started *before* Asterisk, and the channel must be answered with **Answer()**, in order for this application to work.

```
exten => 123,1,Answer()
exten => 123,n,Festival('Hello World',#)
```

As an alternative to the application **Festival()**, you may also use the **System()** command to call Festival's command-line program **text2wave** and play back the resulting audio stream with **Background()** or **Playback()**, like so (for example only; pay attention to pathnames!):

```
exten => 123,1,Answer()
exten => 123,n,System(echo 'Hello World' | text2wave
↳-o sound.wav -otype wav -)
exten => 123,n,Background(sound)
```

Asterisk versions:

```
--------|1.2|--------|1.4|--------|1.6|--------
```

Internal help for this application in Asterisk 1.4:

```
    -= Info about application 'Festival' =-

[Synopsis]
Say text to the user

[Description]
  Festival(text[|intkeys]):  Connect to Festival, send the argument,
get back the waveform,play it to the user, allowing any given interrupt
keys to immediately terminate and return
the value, or 'any' to allow any number back (useful in dialplan)
```

Diff of the internal help from Asterisk 1.2 to 1.4:

None

Diff of the internal help from Asterisk 1.4 to 1.6:

```
--- in Asterisk 1.4
+++ in Asterisk 1.6
@@ -4,5 +4,6 @@
    Say text to the user

    [Description]
-    Festival(text[|intkeys]):  Connect to Festival, send the argument,
get back the waveform,play it to the user, allowing any given interrupt
keys to immediately terminate and return
```

```
+    Festival(text[,intkeys]):  Connect to Festival, send the argument,
get back the waveform,
+  play it to the user, allowing any given interrupt keys to
immediately terminate and return
     the value, or 'any' to allow any number back (useful in dialplan)
```

See also **contrib/README.festival.**

B.58 Flash()

Performs a "flash hook" on a Zap (DAHDi) channel.

```
Flash()
```

Performs a flash on a Zap (DAHDi) channel.

Note: A flash (also called a switchhook-flash, flash hook, or link) is simply a short depressing of the hook switch on an analog telephone for between 80 and 500 milliseconds (depending on the carrier), used primarily as a signaling method to provide feature control for simple analog telephone sets (such as for call waiting, three-way calling, call transfer, and similar services).

Returns 0 upon success, or –1 if the channel is not a Zap channel.

```
exten => 123,1,Flash()
```

If an outgoing line supports flash-transfer (usually an extra service), you might use it on a Zap channel like so:

```
[macro-flash-transfer]
exten => s,1,Playback(transfer)
exten => s,n,Flash()
exten => s,n,Wait(1)
exten => s,n,SendDTMF(${ARG1})
exten => s,n,Wait(1)
exten => s,n,Hangup()

[outside-extensions]
; Transfer incoming calls on extension 6001 to the outside number
; (514)5554138:
exten => 6001,1,Macro(flash-transfer,5145554138)
```

Sometimes it is necessary to adjust the flash duration; ask your carrier for the specification in your area. This can be done in `zapata.conf` with a parameter; e.g., `flash=200` (the value is in milliseconds).

Asterisk versions:

```
--------|1.2|--------|1.4|--------|1.6|--------
```

Internal help for this application in Asterisk 1.4:

```
   -= Info about application 'Flash' =-

[Synopsis]
Flashes a Zap Trunk

[Description]
Performs a flash on a zap trunk.  This can be used
to access features provided on an incoming analogue circuit
such as conference and call waiting. Use with SendDTMF() to
perform external transfers
```

Diff of the internal help from Asterisk 1.2 to 1.4:

```
--- in Asterisk 1.2
+++ in Asterisk 1.4
@@ -4,6 +4,7 @@
   Flashes a Zap Trunk

   [Description]
-    Flash(): Sends a flash on a zap trunk.  This is only a hack for
-  people who want to perform transfers and such via AGI and is
generally
-  quite useless oths application will only work on Zap trunks.
+  Performs a flash on a zap trunk.  This can be used
+  to access features provided on an incoming analogue circuit
+  such as conference and call waiting. Use with SendDTMF() to
+  perform external transfers
```

Diff of the internal help from Asterisk 1.4 to 1.6:

```
--- in Asterisk 1.4
+++ in Asterisk 1.6
@@ -1,10 +1,10 @@
     -= Info about application 'Flash' =-

   [Synopsis]
-  Flashes a Zap Trunk
+  Flashes a DAHDI Trunk

   [Description]
-  Performs a flash on a zap trunk.  This can be used
+  Performs a flash on a DAHDI trunk.  This can be used
   to access features provided on an incoming analogue circuit
   such as conference and call waiting. Use with SendDTMF() to
   perform external transfers
```

B.59 FollowMe()

Follow-Me/Find-Me functionality.

        FollowMe(*followMeID,options*)

Read the configuration file `followme.conf` for a complete explanation.

Asterisk versions:

            | | |1.4|--------|1.6|--------

Internal help for this application in Asterisk 1.4:

```
    -= Info about application 'FollowMe' =-

[Synopsis]
Find-Me/Follow-Me application

[Description]
  FollowMe(followmeid|options):
This application performs Find-Me/Follow-Me functionality for the
caller
as defined in the profile matching the <followmeid> parameter in
followme.conf. If the specified <followmeid> profile doesn't exist in
followme.conf, execution will be returned to the dialplan and call
execution will continue at the next priority.

  Options:
    s    - Playback the incoming status message prior to starting the
follow-me step(s)
    a    - Record the caller's name so it can be announced to the
callee on each step
    n    - Playback the unreachable status message if we've run out of
steps to reach the
           or the callee has elected not to be reachable.
Returns -1 on hangup
```

Diff of the internal help from Asterisk 1.2 to 1.4:

Not available in Asterisk 1.2

Diff of the internal help from Asterisk 1.4 to 1.6:

```
--- in Asterisk 1.4
+++ in Asterisk 1.6
@@ -4,7 +4,7 @@
   Find-Me/Follow-Me application

   [Description]
-    FollowMe(followmeid|options):
```

```
+     FollowMe(followmeid[,options]):
   This application performs Find-Me/Follow-Me functionality for the
caller
   as defined in the profile matching the <followmeid> parameter in
   followme.conf. If the specified <followmeid> profile doesn't exist
in
```

B.60 ForkCDR()

Generates an additional CDR in the current call.

```
ForkCDR()
```

Generates an additional CDR beginning from the moment the **ForkCDR()** command is called. Used for the purpose of distinguishing between total call duration and actual conversation duration for billing purposes. Option v will pass all the CDR variables to the new CDR.

```
exten => 123,n,ForkCDR()
```

Asterisk versions:

--------|1.2|--------|1.4|--------|1.6|--------

Internal help for this application in Asterisk 1.4:

```
    -= Info about application 'ForkCDR' =-

[Synopsis]
Forks the Call Data Record

[Description]
  ForkCDR([options]):  Causes the Call Data Record to fork an
additional
cdr record starting from the time of the fork call
  Options:
     a  - update the answer time on the NEW CDR just after it's been
initiated..
         The new CDR may have been answered already, the reset that
forkcdr.
         does will erase the answer time. This will bring it back, but.
         the answer time will be a copy of the fork/start time. It
will.
         only do this if the initial cdr was indeed already answered..
     D  - Copy the disposition forward from the old cdr, after the .
         init..
     d  - Clear the dstchannel on the new CDR after reset..
     e  - end the original CDR. Do this after all the necc. data.
         is copied from the original CDR to the new forked CDR..
```

```
    R -  do NOT reset the new cdr..
    s(name=val) - Set the CDR var 'name' in the original CDR, with
value.
                    'val'.
    v  - When the new CDR is forked, it gets a copy of the vars
attached
        to the current CDR. The vars attached to the original CDR are
removed
        unless this option is specified.
```

Diff of the internal help from Asterisk 1.2 to 1.4:

```
--- in Asterisk 1.2
+++ in Asterisk 1.4
@@ -6,4 +6,20 @@
   [Description]
     ForkCDR([options]):  Causes the Call Data Record to fork an
additional
   cdr record starting from the time of the fork call
-  If the option 'v' is passed all cdr variables will be passed along
also.
+    Options:
+      a  - update the answer time on the NEW CDR just after it's been
inited..
+          The new CDR may have been answered already, the reset that
forkcdr.
+          does will erase the answer time. This will bring it back,
but.
+          the answer time will be a copy of the fork/start time. It
will.
+          only do this if the initial cdr was indeed already
answered..
+       D - Copy the disposition forward from the old cdr, after the .
+           init..
+       d - Clear the dstchannel on the new CDR after reset..
+       e - end the original CDR. Do this after all the necc. data.
+           is copied from the original CDR to the new forked CDR..
+       R - do NOT reset the new cdr..
+      s(name=val) - Set the CDR var 'name' in the original CDR, with
value.
+                    'val'.
+      v  - When the new CDR is forked, it gets a copy of the vars
attached
+          to the current CDR. The vars attached to the original CDR
are removed
+          unless this option is specified.
```

Diff of the internal help from Asterisk 1.4 to 1.6:

```
--- in Asterisk 1.4
+++ in Asterisk 1.6
@@ -5,21 +5,82 @@
```

[Description]
    ForkCDR([options]): Causes the Call Data Record to fork an
additional
-   cdr record starting from the time of the fork call
+   cdr record starting from the time of the fork call. This new cdr
record will
+   be linked to end of the list of cdr records attached to the channel.
The original CDR is
+   has a LOCKED flag set, which forces most cdr operations to skip it,
except
+   for the functions that set the answer and end times, which ignore
the LOCKED
+   flag. This allows all the cdr records in the channel to be 'ended'
together
+   when the channel is closed.
+   The CDR() func (when setting CDR values) normally ignores the LOCKED
flag also,
+   but has options to vary its behavior. The 'T' option (described
below), can
+   override this behavior, but beware the risks.
+
+   Detailed Behavior Description:
+   First, this app finds the last cdr record in the list, and makes
+   a copy of it. This new copy will be the newly forked cdr record.
+   Next, this new record is linked to the end of the cdr record list.
+   Next, The new cdr record is RESET (unless you use an option to
prevent this)
+   This means that:
+   1. All flags are unset on the cdr record
+   2. the start, end, and answer times are all set to zero.
+   3. the billsec and duration fields are set to zero.
+   4. the start time is set to the current time.
+   5. the disposition is set to NULL.
+   Next, unless you specified the 'v' option, all variables will be
+   removed from the original cdr record. Thus, the 'v' option allows
+   any CDR variables to be replicated to all new forked cdr records.
+   Without the 'v' option, the variables on the original are
effectively
+   moved to the new forked cdr record.
+   Next, if the 's' option is set, the provided variable and value
+   are set on the original cdr record.
+   Next, if the 'a' option is given, and the original cdr record has an
+   answer time set, then the new forked cdr record will have its answer
+   time set to its start time. If the old answer time were carried
forward,
+   the answer time would be earlier than the start time, giving strange
+   duration and billsec times.
+   Next, if the 'd' option was specified, the disposition is copied
from
+   the original cdr record to the new forked cdr.
+   Next, if the 'D' option was specified, the destination channel field
+   in the new forked CDR is erased.

+ Next, if the 'e' option was specified, the 'end' time for the
original
+ cdr record is set to the current time. Future hang-up or ending
events
+ will not override this time stamp.
+ Next, If the 'A' option is specified, the original cdr record will
have
+ it ANS_LOCKED flag set, which prevent future answer events
+ from updating the original cdr record's disposition. Normally, an
+ 'ANSWERED' event would mark all cdr records in the chain as
'ANSWERED'.
+ Next, if the 'T' option is specified, the original cdr record will
have
+ its 'DONT_TOUCH' flag set, which will force the cdr_answer, cdr_end,
and
+ cdr_setvar functions to leave that cdr record alone.
+ And, last but not least, the original cdr record has its LOCKED flag
+ set. Almost all internal CDR functions (except for the funcs that
set
+ the end, and answer times, and set a variable) will honor this flag
+ and leave a LOCKED cdr record alone.
+ This means that the newly created forked cdr record will affected
+ by events transpiring within Asterisk, with the previously noted
+ exceptions.
     Options:
-       a - update the answer time on the NEW CDR just after it's been
inited..
+   a - update the answer time on the NEW CDR just after it's been
inited..
          The new CDR may have been answered already, the reset that
forkcdr.
-          does will erase the answer time. This will bring it back,
but.
+   does will erase the answer time. This will bring it back,
but
          the answer time will be a copy of the fork/start time. It
will.
          only do this if the initial cdr was indeed already
answered,,
-       D - Copy the disposition forward from the old cdr, after the .
+   A - Lock the original CDR against the answer time being updated.
+   This will allow the disposition on the original CDR to
remain the same.
+   d - Copy the disposition forward from the old cdr, after the .
          init..
-       d - Clear the dstchannel on the new CDR after reset..
-       e - end the original CDR. Do this after all the necc. data.
+   D - Clear the dstchannel on the new CDR after reset..
+   e - end the original CDR. Do this after all the necc. data.
          is copied from the original CDR to the new forked CDR..
       R - do NOT reset the new cdr..

```
        s(name=val) - Set the CDR var 'name' in the original CDR, with
value.
                        'val'.
+       T - Mark the original CDR with a DONT_TOUCH flag. setvar,
answer, and end
+           cdr funcs will obey this flag; normally they don't honor
the LOCKED
+           flag set on the original CDR record.
+           Beware-- using this flag may cause CDR's not to have their
end times
+           updated! It is suggested that if you specify this flag,
you might
+           wish to use the 'e' flag as well!
        v - When the new CDR is forked, it gets a copy of the vars
attached
            to the current CDR. The vars attached to the original CDR
are removed
            unless this option is specified.
```

See also nocdr, resetcdr.

B.61 GetCPEID()

Retrieve the CPE-ID (customer premises equipment ID) of an ADSI-capable telephone.

```
GetCPEID()
```

Retrieves the CPE-ID and additional information, displaying it on the Asterisk CLI. This information is often needed to properly configure zapata.conf to support ADSI features.

Returns –1 on hangup.

```
exten => 123,1,GetCPEID()
```

Asterisk versions:

`--------|1.2|--------|1.4|--------|1.6|--------`

Internal help for this application in Asterisk 1.4:

```
   -= Info about application 'GetCPEID' =-

[Synopsis]
Get ADSI CPE ID

[Description]
  GetCPEID: Obtains and displays ADSI CPE ID and other information in
order
to properly setup zapata.conf for on-hook operations.
```

Diff of the internal help from Asterisk 1.2 to 1.4:

None

Diff of the internal help from Asterisk 1.4 to 1.6:

```
--- in Asterisk 1.4
+++ in Asterisk 1.6
@@ -4,5 +4,5 @@
   Get ADSI CPE ID

   [Description]
-    GetCPEID: Obtains and displays ADSI CPE ID and other information
in order
-  to properly setup zapata.conf for on-hook operations.
+    GetCPEID(): Obtains and displays ADSI CPE ID and other information
in order
+  to properly setup dahdi.conf for on-hook operations.
```

See also adsiprog, adsi.conf, zapata.conf.

B.62 GetGroupCount()

Returns the number of channels in a channel group.

```
GetGroupCount(group[@category])
```

Warning: **GetGroupCount()** is removed as of Asterisk 1.4 and has been replaced by the **GROUP_COUNT()** function.

Asterisk versions:

```
--------|1.2|          |     |        |     |
```

Internal help for this application in Asterisk 1.2:

```
   -= Info about application 'GetGroupCount' =-

[Synopsis]
Get the channel count of a group

[Description]
Usage: GetGroupCount([groupname][@category])
  Calculates the group count for the specified group, or uses
the current channel's group if not specifed (and non-empty).
Stores result in GROUPCOUNT.
Note: This application has been deprecated, please use the function
GROUP_COUNT.
```

Diff of the internal help from Asterisk 1.2 to 1.4:

Not available in Asterisk 1.4

See also `setgroup`, `checkgroup`, `getgroupmatchcount`, and in Appendix C, see `group_count`.

B.63 GetGroupMatchCount()

Returns the number of channels in all the groups that match the supplied pattern.

```
GetGroupMatchCount(pattern[@category])
```

Warning: `GetGroupMatchCount()` is removed as of Asterisk 1.4 and has been replaced by the **GROUP_MATCH_COUNT()** function.

Asterisk versions:

```
--------| 1.2 |        |      |       |     |
```

Internal help for this application in Asterisk 1.2:

```
    -= Info about application 'GetGroupMatchCount' =-

[Synopsis]
Get the channel count of all groups that match a pattern

[Description]
Usage: GetGroupMatchCount(groupmatch[@category])
  Calculates the group count for all groups that match the specified
pattern. Uses standard regular expression matching (see regex(7)).
Stores result in GROUPCOUNT.  Always returns 0.
Note: This application has been deprecated, please use the function
GROUP_MATCH_COUNT.
```

Diff of the internal help from Asterisk 1.2 to 1.4:

Not available in Asterisk 1.4

See also `setgroup`, `checkgroup`, `getgroupcount`, and in Appendix C, see `group_match_count`.

B.64 Gosub()

Jumps to the specified priority, extension, and context and allows return.

```
Gosub([[context,]extension,]priority)
Gosub(named_priority)
```

Like **Goto()** but allows the dialplan subroutine to return with **Return()**.
Returns 0, or –1 if the target is invalid.

```
exten => 123,1,Gosub(set-cid)
exten => 123,n,Dial(SIP/${EXTEN})

exten => 123,10(set-cid),Set(CALLERID(all)=Widgets Inc <212-555-3412>)
exten => 123,n,Return()
```

Asterisk versions:

```
--------|1.2|--------|1.4|--------|1.6|--------
```

Internal help for this application in Asterisk 1.4:

```
    -= Info about application 'Gosub' =-

[Synopsis]
Jump to label, saving return address

[Description]
Gosub([[context|]exten|]priority)
   Jumps to the label specified, saving the return address.
```

Diff of the internal help from Asterisk 1.2 to 1.4:

None

Diff of the internal help from Asterisk 1.4 to 1.6:

```
--- in Asterisk 1.4
ιιι in Asterisk 1.6
@@ -4,5 +4,5 @@
   Jump to label, saving return address

   [Description]
-  Gosub([[context|]exten|]priority)
-    Jumps to the label specified, saving the return address.
+    Gosub([[context,]exten,]priority[(arg1[,...][,argN])]):
+  Jumps to the label specified, saving the return address.
```

See also **gosubif, goto, gotoif, return, macro.**

B.65 GosubIf()

Jumps to the specified priority if a condition is satisfied and allows return.

```
GosubIf(condition?labeliftrue:labeliffalse)
```

Jumps to the specified priority if a condition is satisfied (similar to **GotoIf()**) but allows the subroutine to return with **Return()**.

Returns 0 or –1 if the target is invalid.

```
exten => telcid,1,Set(CALLERID(all)=Widgets <212-555-3412>)
exten => telcid,n,Return()
exten => faxcid,1,Set(CALLERID(all)=Widgets <212-555-3412>)
exten => faxcid,n,Return()

exten => _0.,1,GosubIf($[${CHANNEL:4:2} = 43]?faxcid,1:telcid,1)
exten => _0.,n,Dial(${TRUNK}/${EXTEN:1},,T)
```

Asterisk versions:

```
--------| 1.2 |--------| 1.4 |--------| 1.6 |--------
```

Internal help for this application in Asterisk 1.4:

```
    -= Info about application 'GosubIf' =-

[Synopsis]
Conditionally jump to label, saving return address

[Description]
GosubIf(condition?labeliftrue[:labeliffalse])
  If the condition is true, then jump to labeliftrue.  If false, jumps
to
labeliffalse, if specified.  In either case, a jump saves the return
point
in the dialplan, to be returned to with a Return.
```

Diff of the internal help from Asterisk 1.2 to 1.4:

```
--- in Asterisk 1.2
+++ in Asterisk 1.4
@@ -1,7 +1,7 @@
     -= Info about application 'GosubIf' =-

  [Synopsis]
- Jump to label, saving return address
+ Conditionally jump to label, saving return address

  [Description]
  GosubIf(condition?labeliftrue[:labeliffalse])
```

Diff of the internal help from Asterisk 1.4 to 1.6:

```
--- in Asterisk 1.4
+++ in Asterisk 1.6
@@ -4,7 +4,7 @@
   Conditionally jump to label, saving return address

   [Description]
-  GosubIf(condition?labeliftrue[:labeliffalse])
-    If the condition is true, then jump to labeliftrue.  If false,
jumps to
+    GosubIf(condition?labeliftrue[(arg1[,...])]
[:labeliffalse[(arg1[,...])]]]):
+  If the condition is true, then jump to labeliftrue.  If false, jumps
to
   labeliffalse, if specified.  In either case, a jump saves the return
point
   in the dialplan, to be returned to with a Return.
```

See also gosub, goto, gotoif, return, macro.

B.66 Goto()

Jumps to a specified priority, extension, and context.

```
Goto([[context,]extension,]priority)
Goto(named_priority)
```

Hands the currently active channel to the specified priority (and optionally, extension and context).

Optionally, a named priority may be specified to access a labeled priority. Named priorities work only in the current extension.

Always returns 0, even if the target is invalid.

```
exten => 123,1,Answer()
exten => 123,2,Set(COUNT=1)
exten => 123,3,SayNumber(${COUNT})
exten => 123,4,Set(COUNT=$[ ${COUNT} + 1 ])
exten => 123,5,Goto(3)

; as above but with a named priority
exten => 124,1,Answer()
exten => 124,2,Set(COUNT=1)
exten => 124,3(announcement),SayNumber(${COUNT})
exten => 124,4,Set(COUNT=$[ ${COUNT} + 1 ])
exten => 124,5,Goto(announcement)
```

Asterisk versions:

```
--------| 1.2 |--------| 1.4 |--------| 1.6 |--------
```

Internal help for this application in Asterisk 1.4:

```
    -= Info about application 'Goto' =-

[Synopsis]
Jump to a particular priority, extension, or context

[Description]
  Goto([[context|]extension|]priority): This application will set the
current
context, extension, and priority in the channel structure. After it
completes, the
pbx engine will continue dialplan execution at the specified location.
If no specific extension, or extension and context, are specified, then
this
application will just set the specified priority of the current
extension.
  At least a priority is required as an argument, or the goto will
return a -1,
and the channel and call will be terminated.
  If the location that is put into the channel information is bogus,
and asterisk cannot
find that location in the dialplan,
then the execution engine will try to find and execute the code in the
'i' (invalid)
extension in the current context. If that does not exist, it will try
to execute the
'h' extension. If either or neither the 'h' or 'i' extensions have been
defined, the
channel is hung up, and the execution of instructions on the channel is
terminated.
What this means is that, for example, you specify a context that does
not exist, then
it will not be possible to find the 'h' or 'i' extensions, and the call

will terminate!
```

Diff of the internal help from Asterisk 1.2 to 1.4:

None

Diff of the internal help from Asterisk 1.4 to 1.6:

```
--- in Asterisk 1.4
+++ in Asterisk 1.6
@@ -4,7 +4,7 @@
   Jump to a particular priority, extension, or context
```

```
    [Description]
-     Goto([[context|]extension|]priority): This application will set
the current
+     Goto([[context,]extension,]priority): This application will set
the current
    context, extension, and priority in the channel structure. After it
completes, the
    pbx engine will continue dialplan execution at the specified
location.
    If no specific extension, or extension and context, are specified,
then this
```

See also gotoif, gotoiftime, gosub, gosubif, macro.

B.67 GotoIf()

Jumps to a specified priority, extension, and context if a condition is satisfied.

```
GotoIf(condition?labeliftrue:labeliffalse)
```

Hands the currently active channel to the priority specified by *labeliftrue* if the condition is true, or to the priority specified in *labeliffalse* if the condition is false. Either *labeliftrue* or *labeliffalse* may be omitted (in which case execution continues with the next priority) but not both! You must include the colon delimiter (:).

In this case, a priority can be

- A single priority (e.g., 10)
- an extension and a priority (e.g., 123,10)
- a context, extension, and priority (e.g., incoming,123,10)
- a named priority in the same extension (e.g., ok)

```
exten => 123,1,GotoIf($[ ${test} = 5 ]?ok:no)
exten => 123,10(ok),Playback(tt-monkeys)
exten => 123,20(no),Playback(tt-weasels)
```

Asterisk versions:

```
--------| 1.2 |--------| 1.4 |--------| 1.6 |--------
```

Internal help for this application in Asterisk 1.4:

```
    -= Info about application 'GotoIf' =-

[Synopsis]
Conditional goto

[Description]
  GotoIf(condition?[labeliftrue]:[labeliffalse]): This application will
```

```
set the current
context, extension, and priority in the channel structure based on the
evaluation of
the given condition. After this application completes, the
pbx engine will continue dialplan execution at the specified location
in the dialplan.
The channel will continue at
'labeliftrue' if the condition is true, or 'labeliffalse' if the
condition is
false. The labels are specified with the same syntax as used within the
Goto
application.  If the label chosen by the condition is omitted, no jump
is
performed, and the execution passes to the next instruction.
If the target location is bogus, and does not exist, the execution
engine will try
to find and execute the code in the 'i' (invalid)
extension in the current context. If that does not exist, it will try
to execute the
'h' extension. If either or neither the 'h' or 'i' extensions have been
defined, the
channel is hung up, and the execution of instructions on the channel is
terminated.
Remember that this command can set the current context, and if the
context specified
does not exist, then it will not be able to find any 'h' or 'i'
extensions there, and
the channel and call will both be terminated!
```

Diff of the internal help from Asterisk 1.2 to 1.4:

```
--- in Asterisk 1.2
+++ in Asterisk 1.4
@@ -19,5 +19,5 @@
   'h' extension. If either or neither the 'h' or 'i' extensions have
been defined, the
   channel is hung up, and the execution of instructions on the channel
is terminated.
   Remember that this command can set the current context, and if the
context specified
-  does not exist, then it will not be able to find any 'h' or 'i'
extensions there,
-  and the channel and call will both be terminated!
+  does not exist, then it will not be able to find any 'h' or 'i'
extensions there, and
+  the channel and call will both be terminated!
```

Diff of the internal help from Asterisk 1.4 to 1.6:

None

See also goto, gotoiftime, gosub, gosubif, macro.

B.68 GotoIfTime()

Jumps to the specified priority if the time condition is met.

```
GotoIfTime(times,daysofweek,daysofmonth,months
↳?[[context,]extension,]priority)
```

Jumps to the specified priority if the current time falls within the specified time window. Each element may be defined with a * (always) or a – (range).

The arguments for this application are as follows:

- **times**
  Time interval, in 24-hour format with hours and minutes (e.g., 9:00-17:00)

- **daysofweek**
  Weekdays (mon, tue, wed, thu, fri, sat, sun) (e.g., mon-fri)

- **daysofmonth**
  Days of the month (1-31) (e.g., 1-15

- **months**
  Months of the year (jan, feb, mar, apr, mai, jun, jul, aug, sep, oct, nov, dec) (e.g., apr-oct)

```
; During business hours, jump to incoming-open context.
; We are open Monday to Friday from 9:00 to 18:00 (9 a.m. to 6 p.m.):
exten => s,1,GotoIfTime(09:00-17:59,mon-fri,*,*?incoming-open,s,1)
; Also Saturdays from 9 to 12:
exten => s,n,GotoIfTime(09:00-11:59,sat,*,*?incoming-open,s,1)
; After hours go to incoming-closed:
exten => s,n,Goto(incoming-closed,s,1)
```

Asterisk versions:

```
--------| 1.2 |--------| 1.4 |--------| 1.6 |--------
```

Internal help for this application in Asterisk 1.4:

```
    -= Info about application 'GotoIfTime' =-

[Synopsis]
Conditional Goto based on the current time

[Description]
  GotoIfTime(<times>|<weekdays>|<mdays>|<months>
?[[context|]exten|]priority):
This application will set the context, extension, and priority in the
channel structure
if the current time matches the given time specification. Otherwise,
nothing is done.
Further information on the time specification can be found in examples
illustrating how to do time-based context includes in the dialplan.
```

If the target jump location is bogus, the same actions would be taken
as for Goto.

Diff of the internal help from Asterisk 1.2 to 1.4:

```
--- in Asterisk 1.2
+++ in Asterisk 1.4
@@ -9,4 +9,4 @@
   if the current time matches the given time specification. Otherwise,
nothing is done.
   Further information on the time specification can be found in
examples
   illustrating how to do time-based context includes in the dialplan.
-  If the target jump location is bogus, Asterisk will respond as
outlined in Goto.
+  If the target jump location is bogus, the same actions would be
taken as for Goto.
```

Diff of the internal help from Asterisk 1.4 to 1.6:

```
--- in Asterisk 1.4
+++ in Asterisk 1.6
@@ -4,9 +4,14 @@
   Conditional Goto based on the current time

   [Description]
-     GotoIfTime(<times>|<weekdays>|<mdays>|<months>
?[[context|]exten|]priority):
+     GotoIfTime(<times>,<weekdays>,<mdays>,<months>
?[labeliftrue]:[labeliffalse]):
   This application will set the context, extension, and priority in
the channel structure
-  if the current time matches the given time specification. Otherwise,
nothing is done.
+  based on the evaluation of the given time specification. After this
application completes,
+  the pbx engine will continue dialplan execution at the specified
location in the dialplan.
+  If the current time is within the given time specification, the
channel will continue at
+  'labeliftrue'. Otherwise the channel will continue at
'labeliffalse'. If the label chosen
+  by the condition is omitted, no jump is performed, and execution
passes to the next
+  instruction. If the target jump location is bogus, the same actions
would be taken as for
+  Goto.
   Further information on the time specification can be found in
examples
   illustrating how to do time-based context includes in the dialplan.
-  If the target jump location is bogus, the same actions would be
taken as for Goto.
```

See also gotoif, execiftime, and in Appendix C, see iftime.

B.69 Hangup()

Hangs up the active channel.

```
Hangup()
```

Hangs up the active channel unconditionally.
Returns −1.

```
exten => 123,1,Answer()
exten => 123,n,Playback(vm-goodbye)
exten => 123,n,Hangup()
```

Asterisk versions:

```
--------| 1.2 |--------| 1.4 |--------| 1.6 |--------
```

Internal help for this application in Asterisk 1.4:

```
    -= Info about application 'Hangup' =-

[Synopsis]
Hang up the calling channel

[Description]
  Hangup([causecode]): This application will hang up the calling
channel.
If a causecode is given the channel's hangup cause will be set to the
given
value.
```

Diff of the internal help from Asterisk 1.2 to 1.4:

```
--- in Asterisk 1.2
+++ in Asterisk 1.4
@@ -4,4 +4,6 @@
   Hang up the calling channel

   [Description]
-    Hangup(): This application will hang up the calling channel.
+    Hangup([causecode]): This application will hang up the calling
channel.
+  If a causecode is given the channel's hangup cause will be set to
the given
+  value.
```

Diff of the internal help from Asterisk 1.4 to 1.6:

None

See also answer.

B.70 HasNewVoicemail()

Checks to see whether there are new messages in the specified voice mailbox.

```
HasNewVoicemail(mailbox[/folder][@context][,variable,[options]])
```

Warning: **HasVoicemail()** is removed as of Asterisk 1.6 and was already super-fluous in 1.2. The same thing can be achieved with the **VMCOUNT()** function:

```
Set(number=${VMCOUNT(4000|0)})
```

New messages reside in folder 0.

Asterisk versions:

```
--------| 1.2 |--------| 1.4 |          |        |
```

Internal help for this application in Asterisk 1.4:

```
    -= Info about application 'HasNewVoicemail' =-

[Synopsis]
Conditionally branches to priority + 101 with the right options set

[Description]
HasNewVoicemail(vmbox[/folder][@context][|varname[|options]])
Assumes folder 'INBOX' if folder is not specified. Optionally sets
<varname> to the number of messages
in that folder.
  The option string may contain zero of the following character:
        'j' -- jump to priority n+101, if there is new voicemail in
folder 'folder' or INBOX
  This application sets the following channel variable upon completion:
        HASVMSTATUS             The result of the new voicemail check
returned as a text string as follows
                <# of messages in the folder, 0 for NONE>

This application has been deprecated in favor of the VMCOUNT() function
```

Diff of the internal help from Asterisk 1.2 to 1.4:

```
--- in Asterisk 1.2
+++ in Asterisk 1.4
@@ -12,3 +12,5 @@
    This application sets the following channel variable upon completion:
         HASVMSTATUS             The result of the new voicemail
check returned as a text string as follows
                <# of messages in the folder, 0 for NONE>
+
+  This application has been deprecated in favor of the VMCOUNT()
function
```

Diff of the internal help from Asterisk 1.4 to 1.6:

Not available in Asterisk 1.6

See also hasvoicemail, and in Appendix C, see vmcount.

B.71 HasVoicemail()

Checks to see whether there are messages in the specified voice mailbox.

```
HasVoicemail(mailbox[/folder][@context][,variable,[options]])
```

Warning: **HasVoicemail()** is removed as of Asterisk 1.6 and was already superfluous as of 1.2. The same thing can be achieved using the **VMCOUNT()** function:

```
Set(number=${VMCOUNT(4000)})
```

Asterisk versions:

```
--------| 1.2 |--------| 1.4 |          |       |
```

Internal help for this application in Asterisk 1.4:

```
     -= Info about application 'HasVoicemail' =-

[Synopsis]
Conditionally branches to priority + 101 with the right options set

[Description]
HasVoicemail(vmbox[/folder][@context][|varname[|options]])
   Optionally sets <varname> to the number of messages in that
folder.  Assumes folder of INBOX if not specified.
   The option string may contain zero or the following character:
          'j' -- jump to priority n+101, if there is voicemail in the
folder indicated.
   This application sets the following channel variable upon completion:
        HASVMSTATUS               The result of the voicemail check
returned as a text string as follows
                  <# of messages in the folder, 0 for NONE>

This application has been deprecated in favor of the VMCOUNT() function
```

Diff of the internal help from Asterisk 1.2 to 1.4:

```
     --- in Asterisk 1.2
     +++ in Asterisk 1.4
     @@ -11,3 +11,5 @@
          This application sets the following channel variable upon
```

```
completion:
            HASVMSTATUS                The result of the voicemail check
returned as a text string as follows
                    <# of messages in the folder, 0 for NONE>
+
+   This application has been deprecated in favor of the VMCOUNT()

function
```

Diff of the internal help from Asterisk 1.4 to 1.6:

Not available in Asterisk 1.6

See also `hasnewvoicemail`, and in Appendix C, see `vmcount`.

B.72 IAX2Provision()

Provision a calling IAXy device, optionally using the specified template.

```
IAX2Provision([template])
```

Provisions a calling IAXy device with template. If no template is used, the default is used. IAXy templates are defined in `iaxprov.conf`.

Returns 0 on success or −1 on failure.

```
exten => 123,1,IAX2Provision(default)
```

Asterisk versions:

```
--------| 1.2 |--------| 1.4 |--------| 1.6 |--------
```

Internal help for this application in Asterisk 1.4:

```
    -= Info about application 'IAX2Provision' =-

[Synopsis]
Provision a calling IAXy with a given template

[Description]
  IAX2Provision([template]): Provisions the calling IAXy (assuming
the calling entity is in fact an IAXy) with the given template or
default if one is not specified.  Returns -1 on error or 0 on success.
```

Diff of the internal help from Asterisk 1.2 to 1.4:

None

Diff of the internal help from Asterisk 1.4 to 1.6:

None

B.73 ICES()

Streams the active channel to IceCast server.

```
ICES(config_file.xml)
```

Sends an audio stream of the active channel to an IceCast server.[9]
For a sample *config_file.xml*, see `contrib/asterisk-ices.xml`.

Asterisk versions:

```
--------| 1.2 |--------| 1.4 |--------| 1.6 |--------
```

Internal help for this application in Asterisk 1.4:

```
    -= Info about application 'ICES' =-

[Synopsis]
Encode and stream using 'ices'

[Description]
   ICES(config.xml) Streams to an icecast server using ices
(available separately).  A configuration file must be supplied
for ices (see examples/asterisk-ices.conf).
```

Diff of the internal help from Asterisk 1.2 to 1.4:

None

Diff of the internal help from Asterisk 1.4 to 1.6:

```
--- in Asterisk 1.4
+++ in Asterisk 1.6
@@ -6,4 +6,6 @@
   [Description]
     ICES(config.xml) Streams to an icecast server using ices
   (available separately).  A configuration file must be supplied
-  for ices (see examples/asterisk-ices.conf).
+  for ices (see contrib/asterisk-ices.xml).
+
+  - ICES version 2 cient and server required.
```

B.74 ImportVar()

Sets a variable with the contents of a channel variable from another channel.

```
ImportVar(newVariable=channel,variable)
```

---

9. http://en.wikipedia.org/wiki/Icecast

Sets the variable `newVariable` to the value contained in `variable` in the specified channel. If `newVariable` begins with _, single inheritance is used; if it begins with __, unlimited inheritance is used.

```
; import Caller-ID from channel Zap/1:
exten => 123,1,Answer()
exten => 123,n,ImportVar(cid=Zap/1,CALLERID)
```

Asterisk versions:

```
--------| 1.2 |--------| 1.4 |--------| 1.6 |--------
```

Internal help for this application in Asterisk 1.4:

```
    -= Info about application 'ImportVar' =-

[Synopsis]
Import a variable from a channel into a new variable

[Description]
  ImportVar(newvar=channelname|variable): This application imports a
variable
from the specified channel (as opposed to the current one) and stores
it as
a variable in the current channel (the channel that is calling this
application). Variables created by this application have the same
inheritance
properties as those created with the Set application. See the
documentation for
Set for more information.
```

Diff of the internal help from Asterisk 1.2 to 1.4:

None

Diff of the internal help from Asterisk 1.4 to 1.6:

```
--- in Asterisk 1.4
+++ in Asterisk 1.6
@@ -4,7 +4,7 @@
   Import a variable from a channel into a new variable

   [Description]
-    ImportVar(newvar=channelname|variable): This application imports a
variable
+    ImportVar(newvar=channelname,variable): This application imports a
variable
   from the specified channel (as opposed to the current one) and
stores it as
   a variable in the current channel (the channel that is calling this
   application). Variables created by this application have the same
inheritance
```

See also set.

B.75 Incomplete()

Indicates that the extension as dialed thus far is incomplete.

```
Incomplete(options)
```

Indicates to the calling channel that the dialed extension is not complete; that is, Asterisk needs more digits to make a match. If the option n is set, it will not answer the channel.

```
exten => 123,1,Incomplete(n)

exten => 1234,1,Verbose(1,### Hello world!);
```

Asterisk versions:

```
 |       |         |       |           |  1.6  |--------
```

Internal help for this application in Asterisk 1.6:

```
    -= Info about application 'Incomplete' =-

[Synopsis]
returns AST_PBX_INCOMPLETE value

[Description]
  Incomplete([n]): Signals the PBX routines that the previous matched
extension
is incomplete and that further input should be allowed before matching
can
be considered to be complete.  Can be used within a pattern match when
certain criteria warrants a longer match.
  If the 'n' option is specified, then Incomplete will not attempt to
answer
the channel first.  Note that most channel types need to be in Answer
state
in order to receive DTMF.
```

Diff of the internal help from Asterisk 1.4 to 1.6:

Not available in Asterisk 1.4

See also progress, proceeding.

B.76 JabberSend()

Sends a message via Jabber.

```
JabberSend(jabber_id,recipient,message)
```

Sends an instant message message via Jabber (XMPP, Google Talk) using a Jabber identity *jabber_id* defined in `jabber.conf` to the *recipient* Jabber account.

```
exten => 123,1,JabberSend(asterisk,jbloggs@example.com,Incoming call)
exten => 123,n,Dial(SIP/jbloggs)
exten => 123,n,Hangup()
```

Asterisk versions:

```
  |        |            |  1.4  |--------|  1.6  |--------
```

Internal help for this application in Asterisk 1.4:

```
    -= Info about application 'JabberSend' =-

[Synopsis]
JabberSend(jabber,screenname,message)

[Description]
JabberSend(Jabber,ScreenName,Message)
  Jabber - Client or transport Asterisk uses to connect to Jabber
  ScreenName - User Name to message.
  Message - Message to be sent to the buddy
```

Diff of the internal help from Asterisk 1.2 to 1.4:

Not available in Asterisk 1.2

Diff of the internal help from Asterisk 1.4 to 1.6:

```
--- in Asterisk 1.4
+++ in Asterisk 1.6
@@ -6,5 +6,5 @@
   [Description]
   JabberSend(Jabber,ScreenName,Message)
     Jabber - Client or transport Asterisk uses to connect to Jabber
-    ScreenName - User Name to message.
-    Message - Message to be sent to the buddy
+    ScreenName - XMPP/Jabber JID (Name) of recipient
+    Message - Message to be sent to the budd (UTF8)y
```

See also **jabberstatus**, and in Appendix C, see **jabber_status**, **jabber.conf**.

B.77 JabberStatus()

Returns the status of a Jabber user.

```
JabberStatus(jabber_id,jabber_account,variable)
```

Checks the status of the Jabber user *jabber_account* and sets *variable* accordingly. The Jabber identity *jabber_id* as configured in `jabber.conf` is used for the connection.

For more information, see Appendix C..

```
exten => 123,1,JabberStatus(asterisk,jbloggs@example.com,jabberstatus)
exten => 123,n,Verbose(1,### The Jabber status of Joe Bloggs is
${jabberstatus})
```

Asterisk versions:

```
|       |               | 1.4 |--------| 1.6 |--------
```

Internal help for this application in Asterisk 1.4:

```
    -= Info about application 'JabberStatus' =-

[Synopsis]
JabberStatus(Jabber,ScreenName,Variable)

[Description]
JabberStatus(Jabber,ScreenName,Variable)
  Jabber - Client or transport Asterisk uses to connect to Jabber
  ScreenName - User Name to retrieve status from.
  Variable - Variable to store presence in will be 1-6.
            In order, Online, Chatty, Away, XAway, DND, Offline
            If not in roster variable will = 7
```

Diff of the internal help from Asterisk 1.2 to 1.4:

Not available in Asterisk 1.2

Diff of the internal help from Asterisk 1.4 to 1.6:

```
--- in Asterisk 1.4
+++ in Asterisk 1.6
@@ -8,5 +8,7 @@
    Jabber - Client or transport Asterisk uses to connect to Jabber
    ScreenName - User Name to retrieve status from.
    Variable - Variable to store presence in will be 1-6.
-            In order, Online, Chatty, Away, XAway, DND, Offline
-            If not in roster variable will = 7
+            In order, 1=Online, 2=Chatty, 3=Away, 4=XAway, 5=DND,
6=Offline
+            If not in roster variable will be set to 7
+
+  Note: This application is deprecated. Please use the JABBER_STATUS()

function instead.
```

See also **jabbersend**, and in Appendix C, see **jabber_status**, **jabber.conf**.

B.78 KeepAlive()

Internal helper function.

```
KeepAlive()
```

Warning: This application is used internally by Asterisk. Do not run it in your dialplans!

Asterisk versions:

```
    |     |         |       |           | 1.6 |--------
```

Internal help for this application in Asterisk 1.6:

```
    -= Info about application 'KeepAlive' =-

[Synopsis]
returns AST_PBX_KEEPALIVE value

[Description]
  KeepAlive(): This application is chiefly meant for internal use with
Gosubs.
Please do not run it alone from the dialplan!
```

Diff of the internal help from Asterisk 1.4 to 1.6:

Not available in Asterisk 1.4

B.79 Log()

Sends a specified message to the specified log level.

```
Log(level,message)
```

Delivers a specified message to a defined log level.

- **level**

  One of the following target levels: ERROR, WARNING, NOTICE, DEBUG, VERBOSE, DTMF

- **message**

  The message to be written to the log

Asterisk versions:

```
  |     |            | 1.4 |--------| 1.6 |--------
```

Internal help for this application in Asterisk 1.4:

```
    -= Info about application 'Log' =-

[Synopsis]
Send arbitrary text to a selected log level

[Description]
Log(<level>|<message>)
   level must be one of ERROR, WARNING, NOTICE, DEBUG, VERBOSE, DTMF
```

Diff of the internal help from Asterisk 1.2 to 1.4:

Not available in Asterisk 1.2

Diff of the internal help from Asterisk 1.4 to 1.6:

```
--- in Asterisk 1.4
+++ in Asterisk 1.6
@@ -4,5 +4,5 @@
   Send arbitrary text to a selected log level

   [Description]
-  Log(<level>|<message>)
+  Log(<level>,<message>)
      level must be one of ERROR, WARNING, NOTICE, DEBUG, VERBOSE, DTMF
```

See also noop, dumpchan, verbose.

B.80 LookupBlacklist()

Checks the caller ID of a call against the local number blacklist in the Asterisk database (AstDB).

```
    LookupBlacklist([options])
```

Searches for the caller ID number (or name) of the active channel in the blacklist family of the AstDB. If the option j is given, the number exists in the AstDB and the priority n+101 exists, the channel is handed to that priority. If no caller ID is available, the application does nothing.

Sets the channel variable LOOKUPBLSTATUS to FOUND or NOTFOUND.

To add numbers to the blacklist from the CLI, enter `database put blacklist` "*number*" "1"; similarly `database del blacklist` "*number*" to delete the entry and `database show blacklist` for a listing of all the entries in the database.

```
; Block calls from numbers in the blacklist,
; otherwise dial the number in the variable ${PETER}:
exten => 123,1,Answer()
exten => 123,n,LookupBlacklist()
exten => 123,n,GotoIf($["${LOOKUPBLSTATUS}" = "FOUND"]?black,1)
exten => 123,n,Dial(${PETER},30)

exten => black,1,Playback(tt-allbusy)
exten => black,n,Hangup()
```

You can accomplish the same effect as **LookupBlacklist()** with the following dialplan entries:

```
exten => 123,1,Macro(blacklist,${CALLERID(num)})
exten => 123,n,Dial(IAX2/user:secret@widgets.biz/500)

[macro-blacklist]
; Call: Macro(blacklist,${CALLERID(num)})
exten => s,1,GotoIf(${DB_EXISTS(blacklist/${ARG1})}?black)
exten => s,10(black),NoOp(${ARG1} is in the blacklist)
exten => s,n,Busy(5)
exten => s,n,Hangup()
```

Asterisk versions:

```
--------| 1.2 |--------| 1.4 |         |      |
```

Internal help for this application in Asterisk 1.4:

```
    -= Info about application 'LookupBlacklist' =-

[Synopsis]
Look up Caller*ID name/number from blacklist database

[Description]
  LookupBlacklist(options): Looks up the Caller*ID number on the active
channel in the Asterisk database (family 'blacklist').
The option string may contain the following character:
        'j' -- jump to n+101 priority if the number/name is found in
the blacklist
This application sets the following channel variable upon completion:
        LOOKUPBLSTATUS          The status of the Blacklist lookup as a
text string, one of
                FOUND | NOTFOUND
Example: exten => 1234,1,LookupBlacklist()

This application is deprecated and may be removed from a future
release.
Please use the dialplan function BLACKLIST() instead.
```

Diff of the internal help from Asterisk 1.2 to 1.4:

```
--- in Asterisk 1.2
+++ in Asterisk 1.4
@@ -12,3 +12,6 @@
            LOOKUPBLSTATUS          The status of the Blacklist lookup
as a text string, one of
                    FOUND | NOTFOUND
   Example: exten => 1234,1,LookupBlacklist()
+
+   This application is deprecated and may be removed from a future
release.
+   Please use the dialplan function BLACKLIST() instead.
```

Diff of the internal help from Asterisk 1.4 to 1.6:

Not available in Asterisk 1.6

B.81 LookupCIDName()

Looks up a caller ID name in the AstDB.

```
LookupCIDName()
```

Looks up the caller ID number in the AstDB (family `cidname`), and, if it exists, sets the corresponding caller ID name. This application does nothing if no caller ID is present. **LookupCIDName()** can be useful if you receive number information but no names, or if you want to change the caller ID information certain incoming calls.

Returns 0.

To add entries to the list from the CLI enter `database put cidname "number"` "*name*"; similarly enter `database del cidname "number"` to delete entries and `database show cidname` to print a list of all the entries in the database.

```
exten => 123,1,Answer()
exten => 123,n,LookupCIDName()
```

Asterisk versions:

```
--------| 1.2 |--------| 1.4 |           |        |
```

Internal help for this application in Asterisk 1.4:

```
   -= Info about application 'LookupCIDName' =-

[Synopsis]
Look up CallerID Name from local database

[Description]
  LookupCIDName: Looks up the Caller*ID number on the active
```

```
channel in the Asterisk database (family 'cidname') and sets the
Caller*ID name.  Does nothing if no Caller*ID was received on the
channel.  This is useful if you do not subscribe to Caller*ID
name delivery, or if you want to change the names on some incoming
calls.

LookupCIDName is deprecated.  Please use ${DB(cidname/${CALLERID(num)})}
instead.
```

Diff of the internal help from Asterisk 1.2 to 1.4:

```
--- in Asterisk 1.2
+++ in Asterisk 1.4
@@ -10,3 +10,6 @@
   channel.  This is useful if you do not subscribe to Caller*ID
   name delivery, or if you want to change the names on some incoming
   calls.
+
+  LookupCIDName is deprecated.  Please use
${DB(cidname/${CALLERID(num)})}
+  instead.
```

Diff of the internal help from Asterisk 1.4 to 1.6:

Not available in Asterisk 1.6

B.82 Macro()

Executes a previously defined macro.

```
Macro(macroname[,arg1[,arg2[,...]]])
```

Executes a macro defined in the context `macro-macroname` by handing the channel over to the s extension in the macro and returning after the macro has finished running.

The called extension, context, and priority are passed to the macro in the variables ${MACRO_EXTEN}, ${MACRO_CONTEXT} and ${MACRO_PRIORITY}. The arguments are passed to the macro in ${ARG1}, ${ARG2}, and so on.

Macro() returns –1 if any step in the macro returns –1, otherwise it returns 0. If the variable ${MACRO_OFFSET} is set when the macro finishes, the application will continue executing at priority n+1+MACRO_OFFSET if it exists, otherwise it will continue at n+1.

If **Goto()** is called from within the macro, macro execution ends and the call continues in the priority specified in **Goto()**.

```
; define a macro that counts down from the provided value:
[macro-countdown]
exten => s,1,Set(COUNT=${ARG1})
```

```
exten => s,n,While($[ ${COUNT} > 0])
exten => s,n,SayNumber(${COUNT})
exten => s,n,Set(COUNT=$[ ${COUNT} - 1 ])
exten => s,n,EndWhile()

[default]
exten => 123,1,Macro(countdown,3)    ; call the macro "countdown" with
↳ARG1=3
exten => 124,1,Macro(countdown,5)    ; call the macro "countdown" with
↳ARG1=5
```

Asterisk versions:

--------| 1.2 |--------| 1.4 |--------| 1.6 |--------

Internal help for this application in Asterisk 1.4:

```
    -= Info about application 'Macro' =-

[Synopsis]
Macro Implementation

[Description]
   Macro(macroname|arg1|arg2...): Executes a macro using the context
'macro-<macroname>', jumping to the 's' extension of that context and
executing each step, then returning when the steps end.
The calling extension, context, and priority are stored in
${MACRO_EXTEN},
${MACRO_CONTEXT} and ${MACRO_PRIORITY} respectively.  Arguments become
${ARG1}, ${ARG2}, etc in the macro context.
If you Goto out of the Macro context, the Macro will terminate and
control
will be returned at the location of the Goto.
If ${MACRO_OFFSET} is set at termination, Macro will attempt to
continue
at priority MACRO_OFFSET + N + 1 if such a step exists, and N + 1
otherwise.
Extensions: While a macro is being executed, it becomes the current
context.
            This means that if a hangup occurs, for instance, that the
macro
            will be searched for an 'h' extension, NOT the context from
which
            the macro was called. So, make sure to define all
appropriate
            extensions in your macro! (you can use 'catch' in AEL)
WARNING: Because of the way Macro is implemented (it executes the
priorities
            contained within it via sub-engine), and a fixed per-thread
            memory stack allowance, macros are limited to 7 levels
            of nesting (macro calling macro calling macro, etc.); It
```

```
                  may be possible that stack-intensive applications in deeply
nested macros
                  could cause asterisk to crash earlier than this limit.
```

Diff of the internal help from Asterisk 1.2 to 1.4:

```
--- in Asterisk 1.2
+++ in Asterisk 1.4
@@ -23,5 +23,5 @@
                  contained within it via sub-engine), and a fixed per-thread
                  memory stack allowance, macros are limited to 7 levels
                  of nesting (macro calling macro calling macro, etc.); It
-                 may be possible that stack-intensive applications in deeply
nested
-                 macros could cause asterisk to crash earlier than this
limit.
+                 may be possible that stack-intensive applications in deeply
nested macros
+                 could cause asterisk to crash earlier than this limit.
```

Diff of the internal help from Asterisk 1.4 to 1.6:

```
--- in Asterisk 1.4
+++ in Asterisk 1.6
@@ -4,7 +4,7 @@
   Macro Implementation

   [Description]
-    Macro(macroname|arg1|arg2...): Executes a macro using the context
+    Macro(macroname,arg1,arg2...): Executes a macro using the context
   'macro-<macroname>', jumping to the 's' extension of that context
and
   executing each step, then returning when the steps end.
   The calling extension, context, and priority are stored in
${MACRO_EXTEN},
@@ -18,10 +18,12 @@
                  This means that if a hangup occurs, for instance, that
the macro
                  will be searched for an 'h' extension, NOT the context
from which
                  the macro was called. So, make sure to define all
appropriate
-                 extensions in your macro! (you can use 'catch' in AEL)
+                 extensions in your macro! (Note: AEL does not use
macros)
   WARNING: Because of the way Macro is implemented (it executes the
priorities
                  contained within it via sub-engine), and a fixed per-thread
                  memory stack allowance, macros are limited to 7 levels
                  of nesting (macro calling macro calling macro, etc.); It
                  may be possible that stack-intensive applications in deeply
nested macros
```

```
-            could cause asterisk to crash earlier than this limit.
+            could cause asterisk to crash earlier than this limit. It
is advised that
+            if you need to deeply nest macro calls, that you use the
Gosub application
+                (now allows arguments like a Macro) with explicit Return()
calls instead.
```

See also goto, gosub.

B.83 MacroExclusive()

Executes a previously defined macro but allows only a single instance of the macro to execute at any given point in time.

```
MacroExclusive(macroname[,arg1[,arg2[,...]]])
```

Executes, same as **Macro()**, a macro defined in macro-macroname, by handing the channel over to the s extension in the macro and returning after the macro has finished running, but allows only a single instance to run at any given time! If the same macro is called at the same time from elsewhere in the dialplan, this second instance must wait until the first instance has completed.

The called extension, context and priority are passed to the macro in the variables ${MACRO_EXTEN}, ${MACRO_CONTEXT}, and ${MACRO_PRIORITY}. The arguments are passed to the macro in ${ARG1}, ${ARG2}, and so on.

Macro() returns −1 if any step in the macro returns −1, otherwise it returns 0. If the variable ${MACRO_OFFSET} is set when the macro finishes, the application will continue executing at priority n+1+MACRO_OFFSET if it exists, otherwise it will continue at n+1.

If **Goto()** is called from within the macro, macro execution ends and the call continues in the priority specified in **Goto()**.

```
; define a macro that counts down from the provided value;
[macro-countdown]
exten => s,1,Set(COUNT=${ARG1})
exten => s,n,While($[ ${COUNT} > 0])
exten => s,n,SayNumber(${COUNT})
exten => s,n,Set(COUNT=$[ ${COUNT} - 1 ])
exten => s,n,EndWhile()

[default]
exten => 123,1,MacroExclusive(countdown,3)
; call the macro "countdown" with ARG1=3
exten => 124,1,MacroExclusive(countdown,5)
; call the macro "countdown" with ARG1=5
```

Asterisk versions:

```
|        |              | 1.4 |--------| 1.6 |--------
```

Internal help for this application in Asterisk 1.4:

```
    -= Info about application 'MacroExclusive' =-

[Synopsis]
Exclusive Macro Implementation

[Description]
  MacroExclusive(macroname|arg1|arg2...):
Executes macro defined in the context 'macro-macroname'
Only one call at a time may run the macro.
(we'll wait if another call is busy executing in the Macro)
Arguments and return values as in application Macro()
```

Diff of the internal help from Asterisk 1.2 to 1.4:

Not available in Asterisk 1.2

Diff of the internal help from Asterisk 1.4 to 1.6:

```
--- in Asterisk 1.4
+++ in Asterisk 1.6
@@ -4,7 +4,7 @@
   Exclusive Macro Implementation

   [Description]
-    MacroExclusive(macroname|arg1|arg2...):
+    MacroExclusive(macroname,arg1,arg2...):
   Executes macro defined in the context 'macro-macroname'
   Only one call at a time may run the macro.
   (we'll wait if another call is busy executing in the Macro)
```

See also macro, goto, gosub, doc/macroexclusive.txt.

B.84 MacroExit()

Interrupts execution of a macro.

```
MacroExit()
```

May be used within a macro to end execution of the macro as though there were no further priorities remaining.

Asterisk versions:

```
--------| 1.2 |--------| 1.4 |--------| 1.6 |--------
```

Internal help for this application in Asterisk 1.4:

```
    -= Info about application 'MacroExit' =-

[Synopsis]
Exit From Macro

[Description]
  MacroExit():
Causes the currently running macro to exit as if it had
ended normally by running out of priorities to execute.
If used outside a macro, will likely cause unexpected
behavior.
```

Diff of the internal help from Asterisk 1.2 to 1.4:

None

Diff of the internal help from Asterisk 1.4 to 1.6:

None

See also macro.

B.85 MacroIf()

Conditionally starts different macros.

```
    MacroIf(expression?macronameA[,argA1][:macronameB[,argB1]])
```

Calls a macro depending on a condition (defined in the same way as in **GotoIf()**).

Asterisk versions:

```
--------| 1.2 |--------| 1.4 |--------| 1.6 |--------
```

Internal help for this application in Asterisk 1.4:

```
    -= Info about application 'MacroIf' =-

[Synopsis]
Conditional Macro Implementation

[Description]
  MacroIf(<expr>?macroname_a[|arg1][:macroname_b[|arg1]])
Executes macro defined in <macroname_a> if <expr> is true
(otherwise <macroname_b> if provided)
Arguments and return values as in application macro()
```

Diff of the internal help from Asterisk 1.2 to 1.4:

None

Diff of the internal help from Asterisk 1.4 to 1.6:

```
--- in Asterisk 1.4
+++ in Asterisk 1.6
@@ -4,7 +4,7 @@
   Conditional Macro Implementation

   [Description]
-    MacroIf(<expr>?macroname_a[|arg1][:macroname_b[|arg1]])
+    MacroIf(<expr>?macroname_a[,arg1][:macroname_b[,arg1]])
   Executes macro defined in <macroname_a> if <expr> is true
   (otherwise <macroname_b> if provided)
-  Arguments and return values as in application macro()
+  Arguments and return values as in application Macro()
```

See also macro, gotoif, gosubif.

B.86 mailboxExists()

Checks to see whether the specified voicemail box exists.

```
mailboxExists(mailbox[@context][,options])
```

Checks to see whether the voicemail box defined in *mailbox* exists. A voicemail context may be specified if the mailbox being checked is not in the default context.

Sets the channel variable VMBOXEXISTSSTATUS to SUCCESS (mailbox found) or FAILED (mailbox not found).

Option j enables jumping to priority n+101 on success.

```
exten => 123,1,Answer()
exten => 123,n,mailboxExists(123@default)
exten => 123,n,Goto(box-${VMBOXEXISTSSTATUS})
exten => 123,10(box-SUCCESS),Voicemail(123,u)
exten => 123,20(box-FAILED),Playback(sorry)
```

Asterisk versions:

```
--------| 1.2 |--------| 1.4 |--------| 1.6 |--------
```

Internal help for this application in Asterisk 1.4:

```
   -= Info about application 'MailboxExists' =-

[Synopsis]
Check to see if Voicemail mailbox exists
```

```
[Description]
  MailboxExists(mailbox[@context][|options]): Check to see if the
specified
mailbox exists. If no voicemail context is specified, the 'default'
context
will be used.
  This application will set the following channel variable upon
completion:
    VMBOXEXISTSSTATUS - This will contain the status of the execution
of the
                    MailboxExists application. Possible values
include:
                    SUCCESS | FAILED

  Options:
    j - Jump to priority n+101 if the mailbox is found.
```

Diff of the internal help from Asterisk 1.2 to 1.4:

None

Diff of the internal help from Asterisk 1.4 to 1.6:

```
--- in Asterisk 1.4
+++ in Asterisk 1.6
@@ -4,7 +4,7 @@
   Check to see if Voicemail mailbox exists

  [Description]
-    MailboxExists(mailbox[@context][|options]): Check to see if the
specified
+    MailboxExists(mailbox[@context][,options]): Check to see if the
specified
   mailbox exists. If no voicemail context is specified, the 'default'
context
   will be used.
    This application will set the following channel variable upon
completion:
@@ -12,5 +12,4 @@
                    MailboxExists application. Possible values
include:
                    SUCCESS | FAILED

-    Options:
-      j - Jump to priority n+101 if the mailbox is found.
+    Options: (none)
```

See also Appendix C.

B.87 Math()

Performs simple mathematical operations.

```
Math(output_variable,number1<operator>number2)
```

Warning: Math() is removed as of Asterisk 1.4 and has been replaced by the **MATH()** function.

Asterisk versions:

```
--------| 1.2 |          |     |         |     |
```

Internal help for this application in Asterisk 1.2:

```
     -= Info about application 'Math' =-

[Synopsis]
Performs Mathematical Functions

[Description]
Math(returnvar,<number1><op><number 2>

Perform floating point calculation on number 1 to number 2 and
store the result in returnvar.   Valid ops are:
    +,-,/,*,%%%%,<,>,>=,<=,==
and behave as their C equivalents.
   This application has been deprecated in favor of the MATH function.
```

Diff of the internal help from Asterisk 1.2 to 1.4:

Not available in Asterisk 1.4

See also Appendix C.

B.88 MD5()

Computes the MD5 checksum of a string.

```
MD5(output_variable=string)
```

Warning: The application **MD5()** is removed as of Asterisk 1.4 and has been replaced by the **MD5()** function.

Asterisk versions:

--------| **1.2** | | | | |

Internal help for this application in Asterisk 1.2:

```
    -= Info about application 'MD5' =-

[Synopsis]
Calculate MD5 checksum

[Description]
  MD5(<var>=<string>): Calculates a MD5 checksum on <string>.
Returns hash value in a channel variable.
```

Diff of the internal help from Asterisk 1.2 to 1.4:

Not available in Asterisk 1.4

See also md5check, and in Appendix C, see md5.

B.89 MD5Check()

Verifies the MD5 checksum against a supplied string.

```
MD5Check(MD5checksum|string|options)
```

Warning: MD5Check() is removed as of Asterisk 1.4 and has been replaced by the **MD5()** function.

Asterisk versions:

--------| **1.2** | | | | |

Internal help for this application in Asterisk 1.2:

```
    -= Info about application 'MD5Check' =-

[Synopsis]
Check MD5 checksum

[Description]
  MD5Check(<md5hash>|<string>[|options]): Calculates a MD5 checksum on
  <string>
and compares it with the hash. Returns 0 if <md5hash> is correct for
<string>.
The option string may contain zero or more of the following characters:
```

```
            'j' -- jump to priority n+101 if the hash and string do not
     match
     This application sets the following channel variable upon completion:
            CHECKMD5STATUS  The status of the MD5 check, one of the
     following
                     MATCH | NOMATCH
```

Diff of the internal help from Asterisk 1.2 to 1.4:

Not available in Asterisk 1.4

See also md5, and in Appendix C, see md5.

B.90 MeetMe()

Places the caller in a MeetMe conference.

```
MeetMe([conference][,options[,PIN]])
```

Connects the caller in the current channel to a MeetMe conference defined by *conference*. If this is not specified, the application asks the caller to enter a conference number.

If PIN is correctly set to the PIN (personal identification number) of the conference (set statically in meetme.conf or dynamically by the conference operator) the caller is placed directly into the conference; otherwise, the caller must enter the PIN first.

Warning: MeetMe conferences require a Zaptel interface to be installed in the Asterisk server; these provide a time source for synchronization of the participating channels. If no Zaptel interface is available, the ztdummy driver may be used.

MeetMe conferences always use the ul-aw codec internally. The more conference participants use other codecs such as GSM or a-law, the higher the processor load due to transcoding.

Valid options include the following:

- **a**

  Sets admin mode.

- **A**

  Marks the joining user as a special participant (see w and x).

- **b**

  Starts the AGI script defined in ${MEETME_AGI_BACKGROUND}, conf-background by default. (Works only if all the channels in the conference are Zap channels.) The script is passed all DTMF key-presses; will not work in combination with options that also capture DTMF (such as p).

- **c**

  Announces the number of participants to a joining user.

- **d**

  Dynamically allocates a new conference.

- **D**

  Dynamically allocates a new conference but asks the user to set a PIN. (If no PIN is desired, the user must press #.)

- **e**

  Selects an empty conference.

- **E**

  Selects an empty conference that does not require a PIN.

- **i**

  Announces join and exit of new participants with review (works only with Zap channels).

- **I**

  Announces join and exit of new participants without review (works only with Zap channels).

- **m**

  Listen-only mode.

- **M**

  Music-on-hold mode. Plays music-on-hold if there is only one participant in the conference.

- **o**

  Sets talker optimization. Improves conference quality and reduces transcoding overhead by muting participants who are not currently speaking.

- **p**

  Participants may leave by pressing #.

- **P**

  Requests a PIN even if it is provided in the command.

- **q**

  Quiet mode. Does not play entry/exit notification tones.

- **r**

  Records a conference. File: ${MEETME_RECORDINGFILE}, format: ${MEETME_RECORDINGFORMAT}. Default filename is meetme-conf-rec-

${conference}-${uniqueID}. The default format is WAV. (Works only with Zap channels.)

- **s**

  Switches to menu (user or admin) * is pressed.

- **t**

  Talk-only mode.

- **T**

  Talker detection. (Information is sent to the Manager interface and displayed in the MeetMe list in the CLI.)

- **v**

  Video mode (not yet implemented).

- **w**

  Wait until the marked participant joins the conference. Until this point, the other participants will hear music-on-hold.

- **x**

  Ends the conference when the last marked participant exits (see A).

- **X**

  Participants may exit the conference by dialing a single-digit extension in the ${MEETME_EXIT_CONTEXT} context, or the current context if this variable is not defined. Option X does not work with p or s.

- **1**

  Does not play the "You are currently the only person in this conference" message when the first conference participant enters.

You can use e (or E) together with d (or D) to dynamically open a new conference. This means you will have to find a way of distributing the conference number to the other users, or employ some dialplan logic to accomplish the same objective.

Note: The options d or D *dynamically* open conferences; conferences are defined *statically* in meetme.conf.

```
exten => 123,1,Answer()
; Place the caller in conference 333 (with PIN 1234):
exten => 123,n,MeetMe(333,DpM,1234)
```

See also meetmeadmin, meetmecount.

B.90.1 Commands in the CLI

The following commands are for administering conferences from the CLI. (The value *participant* is the number of participant as displayed in the participant list):

- **MeetMe**

  List all conferences

- **MeetMe list** *conference*

  List the participants in the specified conference

- **MeetMe kick** *conference participant*

  Kicks a participant out of the conference

- **MeetMe kick** *conference*

  Kicks all participants out of the conference

- **MeetMe lock** *conference*

  Locks a conference to new participants

- **MeetMe unlock** *conference*

  Unlocks a previous lock (see above)

- **MeetMe mute** *conference participant*

  Mute a conference participant

- **MeetMe unmute** *conference participant*

  Unmute a conference participant (see above)

Asterisk versions:

```
--------| 1.2 |--------| 1.4 |--------| 1.6 |--------
```

Internal help for this application in Asterisk 1.4:

```
    -= Info about application 'MeetMe' =-

[Synopsis]
MeetMe conference bridge

[Description]
  MeetMe([confno][,[options][,pin]]): Enters the user into a specified
MeetMe
conference.  If the conference number is omitted, the user will be
prompted
to enter one.  User can exit the conference by hangup, or if the 'p'
option
is specified, by pressing '#'.
Please note: The Zaptel kernel modules and at least one hardware driver
(or ztdummy)
```

must be present for conferencing to operate properly. In addition, the chan_zap
          channel driver must be loaded for the 'i' and 'r' options to operate at all.

The option string may contain zero or more of the following characters:
      'a' -- set admin mode
      'A' -- set marked mode
      'b' -- run AGI script specified in ${MEETME_AGI_BACKGROUND}
          Default: conf-background.agi (Note: This does not work with
          non-Zap channels in the same conference)
      'c' -- announce user(s) count on joining a conference
      'd' -- dynamically add conference
      'D' -- dynamically add conference, prompting for a PIN
      'e' -- select an empty conference
      'E' -- select an empty pinless conference
      'F' -- Pass DTMF through the conference.
      'i' -- announce user join/leave with review
      'I' -- announce user join/leave without review
      'l' -- set listen only mode (Listen only, no talking)
      'm' -- set initially muted
      'M' -- enable music on hold when the conference has a single caller
      'o' -- set talker optimization - treats talkers who aren't speaking as
          being muted, meaning (a) No encode is done on transmission and
          (b) Received audio that is not registered as talking is omitted
          causing no buildup in background noise. Note that this option
          will be removed in 1.6 and enabled by default.
      'p' -- allow user to exit the conference by pressing '#'
      'P' -- always prompt for the pin even if it is specified
      'q' -- quiet mode (don't play enter/leave sounds)
      'r' -- Record conference (records as ${MEETME_RECORDINGFILE}
          using format ${MEETME_RECORDINGFORMAT}). Default filename is
          meetme-conf-rec-${CONFNO}-${UNIQUEID} and the default format is
          wav.
      's' -- Present menu (user or admin) when '*' is received ('send' to menu)
      't' -- set talk only mode. (Talk only, no listening)
      'T' -- set talker detection (sent to manager interface and meetme list)
      'w[(<secs>)]'
          -- wait until the marked user enters the conference
      'x' -- close the conference when last marked user exits
      'X' -- allow user to exit the conference by entering a valid single

```
                digit extension ${MEETME_EXIT_CONTEXT} or the current
context
                if that variable is not defined.
           '1' -- do not play message when first person enters
```

Diff of the internal help from Asterisk 1.2 to 1.4:

```
--- in Asterisk 1.2
+++ in Asterisk 1.4
@@ -4,10 +4,10 @@
   MeetMe conference bridge

   [Description]
-     MeetMe([confno][,[options][,pin]]): Enters the user into a
specified MeetMe conference.
-  If the conference number is omitted, the user will be prompted to
enter
-  one.
-  User can exit the conference by hangup, or if the 'p' option is
specified, by pressing '#'.
+     MeetMe([confno][,[options][,pin]]): Enters the user into a
specified MeetMe
+  conference.  If the conference number is omitted, the user will be
prompted
+  to enter one.  User can exit the conference by hangup, or if the 'p'
option
+  is specified, by pressing '#'.
   Please note: The Zaptel kernel modules and at least one hardware
driver (or ztdummy)
                must be present for conferencing to operate properly.
In addition, the chan_zap
                channel driver must be loaded for the 'i' and 'r'
options to operate at all.
@@ -16,22 +16,31 @@
           'a' -- set admin mode
           'A' -- set marked mode
           'b' -- run AGI script specified in ${MEETME_AGI_BACKGROUND}
-                 Default: conf-background.agi
-                 (Note: This does not work with non-Zap channels in the
same conference)
+                 Default: conf-background.agi  (Note: This does not work
with
+                 non-Zap channels in the same conference)
           'c' -- announce user(s) count on joining a conference
           'd' -- dynamically add conference
           'D' -- dynamically add conference, prompting for a PIN
           'e' -- select an empty conference
           'E' -- select an empty pinless conference
-          'i' -- announce user join/leave
-          'm' -- set monitor only mode (Listen only, no talking)
+          'F' -- Pass DTMF through the conference.
+          'i' -- announce user join/leave with review
+          'I' -- announce user join/leave without review
```

```
+        'l' -- set listen only mode (Listen only, no talking)
+        'm' -- set initially muted
         'M' -- enable music on hold when the conference has a single
caller
+        'o' -- set talker optimization - treats talkers who aren't
speaking as
+                being muted, meaning (a) No encode is done on
transmission and
+                (b) Received audio that is not registered as talking is
omitted
+                causing no buildup in background noise.  Note that this
option
+                will be removed in 1.6 and enabled by default.
         'p' -- allow user to exit the conference by pressing '#'
         'P' -- always prompt for the pin even if it is specified
         'q' -- quiet mode (don't play enter/leave sounds)
         'r' -- Record conference (records as ${MEETME_RECORDINGFILE}
                using format ${MEETME_RECORDINGFORMAT}). Default
filename is
-                meetme-conf-rec-${CONFNO}-${UNIQUEID} and the default
format is wav.
+                meetme-conf-rec-${CONFNO}-${UNIQUEID} and the default
format is
+                wav.
         's' -- Present menu (user or admin) when '*' is received
('send' to menu)
         't' -- set talk only mode. (Talk only, no listening)
         'T' -- set talker detection (sent to manager interface and
meetme list)
@@ -41,3 +50,4 @@
         'X' -- allow user to exit the conference by entering a valid
single
                digit extension ${MEETME_EXIT_CONTEXT} or the current
context
                if that variable is not defined.
+        '1' -- do not play message when first person enters
```

Diff of the internal help from Asterisk 1.4 to 1.6:

```
--- in Asterisk 1.4
+++ in Asterisk 1.6
@@ -8,8 +8,8 @@
   conference.  If the conference number is omitted, the user will be
prompted
   to enter one.  User can exit the conference by hangup, or if the 'p'
option
   is specified, by pressing '#'.
-  Please note: The Zaptel kernel modules and at least one hardware
driver (or ztdummy)
-                must be present for conferencing to operate properly.
In addition, the chan_zap
+  Please note: The DAHDI kernel modules and at least one hardware
driver (or dahdi_dummy)
```

```
+                  must be present for conferencing to operate properly.
In addition, the chan_dahdi
                   channel driver must be loaded for the 'i' and 'r'
options to operate at all.

   The option string may contain zero or more of the following
characters:
@@ -17,8 +17,9 @@
          'A' -- set marked mode
          'b' -- run AGI script specified in ${MEETME_AGI_BACKGROUND}
                   Default: conf-background.agi  (Note: This does not work with
-                 non-Zap channels in the same conference)
+                 non-DAHDI channels in the same conference)
          'c' -- announce user(s) count on joining a conference
+         'C' -- continue in dialplan when kicked out of conference
          'd' -- dynamically add conference
          'D' -- dynamically add conference, prompting for a PIN
          'e' -- select an empty conference
@@ -28,13 +29,19 @@
          'I' -- announce user join/leave without review
          'l' -- set listen only mode (Listen only, no talking)
          'm' -- set initially muted
-         'M' -- enable music on hold when the conference has a single
caller
+         'M[(<class>)]'
+             -- enable music on hold when the conference has a single
caller.
+             Optionally, specify a musiconhold class to use.  If one
is not
+             provided, it will use the channel's currently set music
class,
+             or "default".
          'o' -- set talker optimization - treats talkers who aren't
speaking as
                   being muted, meaning (a) No encode is done on
transmission and
                   (b) Received audio that is not registered as talking is
omitted
-                 causing no buildup in background noise.  Note that this
option
-                 will be removed in 1.6 and enabled by default.
-         'p' -- allow user to exit the conference by pressing '#'
+                 causing no buildup in background noise
+         'p[(<keys>)]'
+             -- allow user to exit the conference by pressing '#' (default)
+             or any of the defined keys. If keys contain '*' this
will override
+             option 's'. The key used is set to channel variable
MEETME_EXIT_KEY.
          'P' -- always prompt for the pin even if it is specified
          'q' -- quiet mode (don't play enter/leave sounds)
          'r' -- Record conference (records as ${MEETME_RECORDINGFILE}
```

```
@@ -51,3 +58,10 @@
                digit extension ${MEETME_EXIT_CONTEXT} or the current
context
                if that variable is not defined.
        '1' -- do not play message when first person enters
+       'S(x)' -- Kick the user 'x' seconds *after* he entered into
the conference.
+           'L(x[:y][:z])' - Limit the conference to 'x' ms. Play a
warning when 'y' ms are
+               left. Repeat the warning every 'z' ms. The following
special
+               variables can be used with this option:
+                   * CONF_LIMIT_TIMEOUT_FILE      File to play when time
is up.
+                   * CONF_LIMIT_WARNING_FILE      File to play as
warning if 'y' is defined.
+                                                  The default is to say
the time remaining.
```

See also **meetmecount**, **meetmeadmin**, **meetmechanneladmin**, and in
Appendix C, see **meetme_info**.

B.91 MeetMeAdmin()

Administers a MeetMe conference.

```
MeetMeAdmin(conference,command[,participant])
```

Executes a command in the specified conference. The command may be one of the
following (the *participant* is required only for the kick command (*k*)):

- **K**
  Kicks all participants out of the conference
- **k**
  Kicks *participant* out of the conference
- **e**
  Kicks the last participant who joined out of the conference
- **L**
  Locks the conference to new participants
- **l**
  Unlocks the conference
- **M**
  Mutes a conference participant

- **m**

  Unmutes a muted participant

- **N**

  Mutes everyone except the conference administrator

- **n**

  Unmutes everyone mute by N

```
; Mute participant 3 in conference 333
exten => 123,1,MeetMeAdmin(333,M,3)

; Kick participant 3 out of conference 333:
exten => 123,1,MeetMeAdmin(333,k,3)
```

Asterisk versions:

```
--------| 1.2 |--------| 1.4 |--------| 1.6 |--------
```

Internal help for this application in Asterisk 1.4:

```
    -= Info about application 'MeetMeAdmin' =-

[Synopsis]
MeetMe conference Administration

[Description]
  MeetMeAdmin(confno,command[,user]): Run admin command for conference
        'e' -- Eject last user that joined
        'k' -- Kick one user out of conference
        'K' -- Kick all users out of conference
        'l' -- Unlock conference
        'L' -- Lock conference
        'm' -- Unmute one user
        'M' -- Mute one user
        'n' -- Unmute all users in the conference
        'N' -- Mute all non-admin users in the conference
        'r' -- Reset one user's volume settings
        'R' -- Reset all users volume settings
        's' -- Lower entire conference speaking volume
        'S' -- Raise entire conference speaking volume
        't' -- Lower one user's talk volume
        'T' -- Raise one user's talk volume
        'u' -- Lower one user's listen volume
        'U' -- Raise one user's listen volume
        'v' -- Lower entire conference listening volume
        'V' -- Raise entire conference listening volume
```

Diff of the internal help from Asterisk 1.2 to 1.4:

```
--- in Asterisk 1.2
+++ in Asterisk 1.4
```

```
@@ -10,7 +10,17 @@
             'K' -- Kick all users out of conference
             'l' -- Unlock conference
             'L' -- Lock conference
-            'm' -- Unmute conference
-            'M' -- Mute conference
-            'n' -- Unmute entire conference (except admin)
-            'N' -- Mute entire conference (except admin)
+            'm' -- Unmute one user
+            'M' -- Mute one user
+            'n' -- Unmute all users in the conference
+            'N' -- Mute all non-admin users in the conference
+            'r' -- Reset one user's volume settings
+            'R' -- Reset all users volume settings
+            's' -- Lower entire conference speaking volume
+            'S' -- Raise entire conference speaking volume
+            't' -- Lower one user's talk volume
+            'T' -- Raise one user's talk volume
+            'u' -- Lower one user's listen volume
+            'U' -- Raise one user's listen volume
+            'v' -- Lower entire conference listening volume
+            'V' -- Raise entire conference listening volume
```

Diff of the internal help from Asterisk 1.4 to 1.6:

None

See also meetme, meetmecount.

B.92 MeetMeChannelAdmin()

Channel-specific MeetMe conference administration.

```
MeetMeChannelAdmin(channel, command)
```

Executes the supplied MeetMe command for the specified channel. Works similar to **MeetMeAdmin()** but on a specific channel instead of a conference.
The command *command* can be one of the following:

- **k**

  Kicks the participant on the channel from the conference

- **M**

  Mutes the participant on the channel

- **m**

  Unmutes the participant on the channel

Asterisk versions:

| | | | | | 1.6 |--------|

Internal help for this application in Asterisk 1.6:

```
    -= Info about application 'MeetMeChannelAdmin' =-

[Synopsis]
MeetMe conference Administration (channel specific)

[Description]
   MeetMeChannelAdmin(channel,command): Run admin command for a specific
channel in any coference.
        'k' -- Kick the specified user out of the conference he is in
        'm' -- Unmute the specified user
        'M' -- Mute the specified user
```

Diff of the internal help from Asterisk 1.4 to 1.6:

Not available in Asterisk 1.4

See also meetme, meetmeadmin.

B.93 MeetMeCount()

Counts the number of participants in a MeetMe conference.

```
MeetMeCount(conference[,variablename])
```

Announces the number of participants in the conference. If the variable name is provided, the announcement is skipped and the count written to this variable.

Returns 0 on success, −1 on error.

This example counts the participants in conference 501 and stores the number in ${COUNT}.

```
; Count the participants in conference 501 and store in ${COUNT};
exten => 333,1,MeetMeCount(501,COUNT)
```

Asterisk versions:

--------| 1.2 |--------| 1.4 |--------| 1.6 |--------

Internal help for this application in Asterisk 1.4:

```
    -= Info about application 'MeetMeCount' =-

[Synopsis]
MeetMe participant count

[Description]
```

```
   MeetMeCount(confno[|var]): Plays back the number of users in the
specified
MeetMe conference. If var is specified, playback will be skipped and
the valu
will be returned in the variable. Upon app completion, MeetMeCount will
hangup
the channel, unless priority n+1 exists, in which case priority
progress will
continue.
A ZAPTEL INTERFACE MUST BE INSTALLED FOR CONFERENCING FUNCTIONALITY.
```

Diff of the internal help from Asterisk 1.2 to 1.4:

```
--- in Asterisk 1.2
+++ in Asterisk 1.4
@@ -6,6 +6,7 @@
   [Description]
     MeetMeCount(confno[|var]): Plays back the number of users in the
specified
   MeetMe conference. If var is specified, playback will be skipped and
the value
-  will be returned in the variable. Upon app completion, MeetMeCount
will hangup the
-  channel, unless priority n+1 exists, in which case priority progress
will continue.
+  will be returned in the variable. Upon app completion, MeetMeCount
will hangup
+  the channel, unless priority n+1 exists, in which case priority
progress will
+  continue.
   A ZAPTEL INTERFACE MUST BE INSTALLED FOR CONFERENCING FUNCTIONALITY.
```

Diff of the internal help from Asterisk 1.4 to 1.6:

```
--- in Asterisk 1.4
+++ in Asterisk 1.6
@@ -4,9 +4,8 @@
   MeetMe participant count

   [Description]
-     MeetMeCount(confno[|var]): Plays back the number of users in the
specified
+     MeetMeCount(confno[,var]): Plays back the number of users in the
specified
   MeetMe conference. If var is specified, playback will be skipped and
the value
   will be returned in the variable. Upon app completion, MeetMeCount
will hangup
   the channel, unless priority n+1 exists, in which case priority
progress will
   continue.
-  A ZAPTEL INTERFACE MUST BE INSTALLED FOR CONFERENCING FUNCTIONALITY.
```

See also meetme, meetmeadmin.

B.94 Milliwatt()

Generates a 1,000Hz test tone on a channel.

```
Milliwatt()
```

Milliwatt tone lines are used by telecommunications carriers for testing and measuring line characteristics. They can be used to check for echo, excessive or inadequate volume, or some kinds of line noise. Standard milliwatt test tones are 1,004Hz at 0 dbm (u-law).

```
; generate a 1000 Hz Milliwatt test tone:
exten => 123,1,Milliwatt()
```

Asterisk versions:

```
--------| 1.2 |--------| 1.4 |--------| 1.6 |--------
```

Internal help for this application in Asterisk 1.4:

```
    -= Info about application 'Milliwatt' =-

[Synopsis]
Generate a Constant 1004Hz tone at 0dbm (mu-law)

[Description]
    Milliwatt([options]): Generate a Constant 1004Hz tone at 0dbm.
Previous versions of this application generated the tone at 1000Hz.
If for
some reason you would prefer that behavior, supply the 'o' option to
get the
old behavior.
```

Diff of the internal help from Asterisk 1.2 to 1.4:

```
--- in Asterisk 1.2
+++ in Asterisk 1.4
@@ 1,7 11,10 @@
    -= Info about application 'Milliwatt' =-

  [Synopsis]
- Generate a Constant 1000Hz tone at 0dbm (mu-law)
+ Generate a Constant 1004Hz tone at 0dbm (mu-law)

  [Description]
- Milliwatt(): Generate a Constant 1000Hz tone at 0dbm (mu-law)
+     Milliwatt([options]): Generate a Constant 1004Hz tone at 0dbm.
+ Previous versions of this application generated the tone at 1000Hz.   If
for
+ some reason you would prefer that behavior, supply the 'o' option
to get the
+ old behavior.
```

Diff of the internal help from Asterisk 1.4 to 1.6:

None

See also echo.

B.95 MinivmAccMess()

Records messages for the Mini-Voicemail system.

```
MinivmAccMess(user@domain,option)
```

Records messages for the given user account in the Mini-Voicemail system
(minivm). If the account-specific directories do not exist, they are created.
   Possible values for *option*:

- **u**

  Record unavailable message

- **b**

  Record busy message

- **t**

  Temporary away message to be played prior to regular unavailable or busy
  messages; see minivmgreet

- **n**

  Account name

```
exten => *90,1,Answer()
exten => *90,n,MinivmAccMess(${CALLERID(num)}@localdomain,n)
exten => *90,n,Hangup()

exten => *91,1,Answer()
exten => *91,n,MinivmAccMess(${CALLERID(num)}@localdomain,u)
exten => *91,n,Hangup()

exten => *92,1,Answer()
exten => *92,n,MinivmAccMess(${CALLERID(num)}@localdomain,b)
exten => *92,n,Hangup()

exten => *93,1,Answer()
exten => *93,n,MinivmAccMess(${CALLERID(num)}@localdomain,t)
exten => *93,n,Hangup()
```

Asterisk versions:

```
        |       |         |       |       | 1.6 |--------
```

Internal help for this application in Asterisk 1.6:

```
    -= Info about application 'MinivmAccMess' =-

[Synopsis]
Record account specific messages

[Description]
  MinivmAccmess(username@domain,option):
This application is part of the Mini-Voicemail system, configured
in minivm.conf.
Use this application to record account specific audio/video messages
for
busy, unavailable and temporary messages.
Account specific directories will be created if they do not exist.

The option selects message to be recorded:
    u       Unavailable
    b       Busy
    t       Temporary (overrides busy and unavailable)
    n       Account name

Result is given in channel variable MINIVM_ACCMESS_STATUS
        The possible values are:    SUCCESS  |  FAILED
            FAILED is set if the file can't be created.
```

Diff of the internal help from Asterisk 1.4 to 1.6:

Not available in Asterisk 1.4

See also minivmdelete, minivmgreet, minivmnotify, minivmrecord, and in Appendix C, see minivmaccount, minivmcounter.

B.96 MinivmDelete()

Delete message in Mini Voicemail

```
    MinivmDelete(filename)
```

Deletes the file *filename* in Mini-VoiceMail (minivm).

Asterisk versions:

```
    |      |        |     |        | 1.6 |--------
```

Internal help for this application in Asterisk 1.6:

```
    -= Info about application 'MinivmDelete' =-

[Synopsis]
Delete Mini-Voicemail voicemail messages
```

```
[Description]
  MinivmDelete(filename):
This application is part of the Mini-Voicemail system, configured in
minivm.conf.
It deletes voicemail file set in MVM_FILENAME or given filename.

Result is given in channel variable MINIVM_DELETE_STATUS
        The possible values are:      SUCCESS | FAILED
           FAILED is set if the file does not exist or can't be deleted.
```

Diff of the internal help from Asterisk 1.4 to 1.6:

Not available in Asterisk 1.4

See also minivmaccmess, minivmgreet, minivmnotify, minivmrecord, and in Appendix C, see minivmaccount, minivmcounter.

B.97 MinivmGreet()

Play back Mini-VoiceMail greeting.

```
MinivmGreet(user@domain[,option])
```

Plays the default or user-specific greeting (see minivmaccmess) for Mini-VoiceMail (minivm).

Possible *option* settings:

- **b**

  Play busy message

- **u**

  Play unavailable message

- **s**

  Skip the playback of instructions for leaving a message

Combinations are allowed but b and u together are nonsensical. If a temporary message has been set, it is played and b or u is ignored.

Asterisk versions:

| | | | | **1.6** |--------

Internal help for this application in Asterisk 1.6:

```
   -= Info about application 'MinivmGreet' =-

[Synopsis]
Play Mini-Voicemail prompts
```

```
[Description]
  MinivmGreet(username@domain[,options]):
This application is part of the Mini-Voicemail system, configured
in minivm.conf.
MinivmGreet() plays default prompts or user specific prompts for
an account.
Busy and unavailable messages can be choosen, but will be overridden
if a temporary
message exists for the account.

Result is given in channel variable MINIVM_GREET_STATUS
        The possible values are:     SUCCESS | USEREXIT | FAILED

  Options:
    b    - Play the 'busy' greeting to the calling party.
    s    - Skip the playback of instructions for leaving a message
to the
         calling party.
    u    - Play the 'unavailable greeting.
```

Diff of the internal help from Asterisk 1.4 to 1.6:

Not available in Asterisk 1.4

See also minivmaccmess, minivmdelete, minivmnotify, minivmrecord, Appendix C: minivmaccount, and in Appendix C, see minivmcounter.

B.98 MinivmNotify()

Sends e-mail notification of new Mini-Voicemail messages.

```
MinivmNotify(user@domain,template)
```

Provides e-mail notification of new Mini-VoiceMail (minivm) messages to the specified e-mail address.

Asterisk versions:

```
  |       |         |       |        | 1.6 |--------
```

Internal help for this application in Asterisk 1.6:

```
    -= Info about application 'MinivmNotify' =-

[Synopsis]
Notify voicemail owner about new messages.

[Description]
  MinivmNotify(username@domain[,template]):
This application is part of the Mini-Voicemail system, configured
in minivm.conf.
```

```
MiniVMnotify forwards messages about new voicemail to e-mail and pager.
If there's no user account for that address, a temporary account will
be used with default options (set in minivm.conf).
The recorded file name and path will be read from MVM_FILENAME and the
duration of the message will be accessed from MVM_DURATION (set by
MinivmRecord())
If the channel variable MVM_COUNTER is set, this will be used in the
message file name and available in the template for the message.
If not template is given, the default email template will be used to
send email and
default pager template to send paging message (if the user account is
configured with
a paging address.

Result is given in channel variable MINIVM_NOTIFY_STATUS
        The possible values are:     SUCCESS | FAILED
```

Diff of the internal help from Asterisk 1.4 to 1.6:

Not available in Asterisk 1.4

See also minivmaccmess, minivmdelete, minivmgreet, minivmrecord, and in Appendix C, see minivmaccount, minivmcounter.

B.99 MinivmRecord()

Record a voicemail in Mini-VoiceMail.

```
MinivmRecord(user@domain[,option])
```

Records the caller's voicemail message in the Mini-Voicemail (minivm) system. This is more or less the equivalent to VoiceMail() (voicemail).

Asterisk versions:

```
      |      |       |      |      | 1.6 |--------
```

Internal help for this application in Asterisk 1.6:

```
    -= Info about application 'MinivmRecord' =-

[Synopsis]
Receive Mini-Voicemail and forward via e-mail

[Description]
  MinivmRecord(username@domain[,options]):
This application is part of the Mini-Voicemail system, configured in
minivm.conf.
MiniVM records audio file in configured format and forwards message to
e-mail and pager.
```

If there's no user account for that address, a temporary account will
be used with default options.
The recorded file name and path will be stored in MINIVM_FILENAME and
the
duration of the message will be stored in MINIVM_DURATION

Note: If the caller hangs up after the recording, the only way to send
the message and clean up is to execute in the "h" extension.

The application will exit if any of the following DTMF digits are
received and the requested extension exist in the current context.
     0 - Jump to the 'o' extension in the current dialplan context.
     * - Jump to the 'a' extension in the current dialplan context.

Result is given in channel variable MINIVM_RECORD_STATUS
          The possible values are: SUCCESS | USEREXIT | FAILED

   Options:
     g(#) - Use the specified amount of gain when recording the
voicemail
           message. The units are whole-number decibels (dB).

Diff of the internal help from Asterisk 1.4 to 1.6:

Not available in Asterisk 1.4

See also `minivmaccmess`, `minivmdelete`, `minivmgreet`, `minivmnotify`, and
in Appendix C, see `minivmaccount`, `minivmcounter`.

B.100 MixMonitor()

Records the audio on the current channel but mixes it before writing it to a file.

```
MixMonitor(fileprefix.format[,options[,command]])
```

Starts recording the audio on the current channel. Instead of recording each direction in a separate file the way **Monitor()** would, mixes the two audio streams on-the-fly and writes the result to the specified file.
   Options:

- **a**

   Appends the audio stream to an existing file.

- **b**

   Saves audio to the file only while the channel is bridged; i.e., once a conversation has actually begun, and only until it is hung up.

- **v(x)**

  Adjusts the heard volume by an increment of x (range −4 to 4).

- **V(x)**

  Adjusts the spoken volume by an increment of x (range −4 to 4).

- **W(x)**

  Adjusts both heard and spoken volume by an increment of x (range −4 to 4).

The *command* (if provided) is executed after recording. The variable ${MIXMONITOR_FILENAME} is set to the filename used for the recording.

See also additional information in the description of Monitor().

Asterisk versions:

```
--------| 1.2 |--------| 1.4 |--------| 1.6 |--------
```

Internal help for this application in Asterisk 1.4:

```
   -= Info about application 'MixMonitor' =-

[Synopsis]
Record a call and mix the audio during the recording

[Description]
  MixMonitor(<file>.<ext>[|<options>[|<command>]])

Records the audio on the current channel to the specified file.
If the filename is an absolute path, uses that path, otherwise
creates the file in the configured monitoring directory from
asterisk.conf.

Valid options:
  a       - Append to the file instead of overwriting it.
  b       - Only save audio to the file while the channel is bridged.
            Note: Does not include conferences or sounds played to each
bridged
                 party.
  v(<x>) - Adjust the heard volume by a factor of <x> (range -4 to 4)
  V(<x>) - Adjust the spoken volume by a factor of <x> (range -4 to 4)
  W(<x>) - Adjust the both heard and spoken volumes by a factor of <x>
            (range -4 to 4)

<command> will be executed when the recording is over
Any strings matching ^{X} will be unescaped to ${X}.
All variables will be evaluated at the time MixMonitor is called.
The variable MIXMONITOR_FILENAME will contain the filename used to
record.
```

Diff of the internal help from Asterisk 1.2 to 1.4:

None

Diff of the internal help from Asterisk 1.4 to 1.6:

```
--- in Asterisk 1.4
+++ in Asterisk 1.6
@@ -4,8 +4,7 @@
    Record a call and mix the audio during the recording

    [Description]
-      MixMonitor(<file>.<ext>[|<options>[|<command>]])
-
+      MixMonitor(<file>.<ext>[,<options>[,<command>]]):
    Records the audio on the current channel to the specified file.
    If the filename is an absolute path, uses that path, otherwise
    creates the file in the configured monitoring directory from
```

See also monitor.

B.101 Monitor()

Records the current channel in two separate files.

```
Monitor([format[,fileprefix[,options]]])
```

Starts audio recording on the current channel. Incoming and outgoing audio packets are written to separate files until the channel is hung up or monitoring is
stopped with **StopMonitor()**.

The parameter *format* sets the file format. If this is not specified, wav is used.

The parameter *fileprefix* specifies the filename without extension. If this is not
specified, the filename is assembled out of the channel name and a number, for example, IAX2[foo@bar]-3. Incoming audio is written to *fileprefix-in.format*,
outgoing audio in *fileprefix-out.format*, both in /var/spool/asterisk/
monitor.

Two options may be specified:

▪ **m**

   After recording is complete, mixes the incoming and outgoing audio files into a
   single file and deletes the originals. Requires that **soxmix** from the **sox** package be installed on the server.[10]

---

10. http://sox.sourceforge.net/; see also an explanation in musiconhold.conf, version 12.17.7 or
newer. You may check your installed version with **soxmix -help**.

If the variable ${MONITOR_EXEC} is defined, this application is executed instead of **soxmix** and the original incoming and outgoing audio files are not deleted.[11]

soxmix (or ${MONITOR_EXEC} if specified) is passed three values: the names of the incoming and outgoing audio files and the name of the mixed file, which is the fileprefix without -in/-out. If ${MONITOR_EXEC_ARGS} is set, the contents are used as arguments to ${MONITOR_EXEC}.

Warning: Note that **soxmix** attempts to determine the file type based on the file extension. Formats such as GSM and WAV are normally not a problem, but for other formats such as a-law and u-law, it expects the file extensions .al and .ul, respectively. To resolve this, read the manual pages for **sox** (**/soxmix**) and use ${MONITOR_EXEC_ARGS} or write a small wrapper script that reads the format parameter and call it in ${MONITOR_EXEC}.

Note: If you wanted a single mixed sound file, **MixMonitor()** is usually the better option, as it mixes on-the-fly and thereby avoids a spike in CPU load at the end of the recording.

- **b**

  Saves audio to the file only while the channel is bridged; i.e., once a conversation has actually begun, and only until it is hung up.

Returns 0 on success, or –1 on failure (such as a failure to open the audio files for writing, or the channel is already being monitored, and so on).

```
; record the conversation and mix the audio afterwards:
exten => 123,1,Answer()
exten => 123,n,Monitor(gsm,,mb)
exten => 123,n,SayDigits(123456789)
exten => 123,n,Hangup()

; as above, only with our own wrapper script that calls soxmix:
exten => 123,1,Answer()
exten => 123,n,Set(MONITOR_EXEC=/path/to/my-soxmix-wrapper.sh)
exten => 123,n,Monitor(gsm,,mb)
exten => 123,n,SayDigits(123456789)
exten => 123,n,Hangup()
```

---

11. Depends on Asterisk version; older versions do not delete automatically. It's best to check with a proper test.

Warning: Before recording conversations, make sure you are complying with the relevant legislation in your jurisdiction. In most cases, both parties must be aware they are being recorded.[12]

Note: Some Asterisk users who routinely need to record many conversations (50 to 500) report much better performance if call recordings are written to a RAM disk (fewer and shorter seek operations) before being copied to a hard disk when the conversation is over.

Asterisk versions:

```
--------| 1.2 |--------| 1.4 |--------| 1.6 |--------
```

Internal help for this application in Asterisk 1.4:

```
    -= Info about application 'Monitor' =-

[Synopsis]
Monitor a channel

[Description]
Monitor([file_format[:urlbase]|[fname_base]|[options]]):
Used to start monitoring a channel. The channel's input and output
voice packets are logged to files until the channel hangs up or
monitoring is stopped by the StopMonitor application.
   file_format         optional, if not set, defaults to "wav"
   fname_base          if set, changes the filename used to the one
specified.
   options:
     m     - when the recording ends mix the two leg files into one and
             delete the two leg files.  If the variable MONITOR_EXEC is
set, the
             application referenced in it will be executed instead of
             sox and the raw leg files will NOT be deleted automatically.
             sox or MONITOR_EXEC is handed 3 arguments, the two leg files
             and a target mixed file name which is the same as the leg
file names
             only without the in/out designator.
             If MONITOR_EXEC_ARGS is set, the contents will be passed on
as
             additional arguements to MONITOR_EXEC
             Both MONITOR_EXEC and the Mix flag can be set from the
             administrator interface
```

---

12. See also www.voip-info.org/wiki/view/Monitor+Recording+Legal+Issues.

```
      b    - Don't begin recording unless a call is bridged to another
channel

Returns -1 if monitor files can't be opened or if the channel is
already
monitored, otherwise 0.
```

Diff of the internal help from Asterisk 1.2 to 1.4:

```
--- in Asterisk 1.2
+++ in Asterisk 1.4
@@ -14,8 +14,8 @@
       m    - when the recording ends mix the two leg files into one and
             delete the two leg files.  If the variable MONITOR_EXEC is
set, the
             application referenced in it will be executed instead of
-            soxmix and the raw leg files will NOT be deleted
automatically.
-            soxmix or MONITOR_EXEC is handed 3 arguments, the two leg
files
+            sox and the raw leg files will NOT be deleted
automatically.
+            sox or MONITOR_EXEC is handed 3 arguments, the two leg
files
             and a target mixed file name which is the same as the leg
file names
             only without the in/out designator.
             If MONITOR_EXEC_ARGS is set, the contents will be passed
on as
```

Diff of the internal help from Asterisk 1.4 to 1.6:

```
--- in Asterisk 1.4
+++ in Asterisk 1.6
@@ -4,7 +4,7 @@
   Monitor a channel

   [Description]
-  Monitor([file_format[:urlbase]|[fname_base]|[options]]):
+   Monitor([file_format[:urlbase],[fname_base],[options]]):
   Used to start monitoring a channel. The channel's input and output
   voice packets are logged to files until the channel hangs up or
   monitoring is stopped by the StopMonitor application.
@@ -19,11 +19,13 @@
             and a target mixed file name which is the same as the leg
file names
             only without the in/out designator.
             If MONITOR_EXEC_ARGS is set, the contents will be passed
on as
-            additional arguements to MONITOR_EXEC
+            additional arguments to MONITOR_EXEC
             Both MONITOR_EXEC and the Mix flag can be set from the
             administrator interface
```

```
        b    - Don't begin recording unless a call is bridged to another
channel
+       i    - Skip recording of input stream (disables m option)
+       o    - Skip recording of output stream (disables m option)

   Returns -1 if monitor files can't be opened or if the channel is
already
   monitored, otherwise 0.
```

See also changemonitor, stopmonitor, pausemonitor, unpausemonitor, mixmonitor, record.

B.102 Morsecode()

Transmits the provided string as Morse code.

```
Morsecode(string)
```

Plays back the provided string in Morse code.

```
exten => 123,1,Answer()
exten => 123,n,Morsecode("The dog barks at midnight.")
exten => 123,n,Hangup()
```

Asterisk versions:

```
    |       |       | 1.4 |--------| 1.6 |--------
```

Internal help for this application in Asterisk 1.4:

```
   -= Info about application 'Morsecode' =-

[Synopsis]
Plays morse code

[Description]
Usage: Morsecode(<string>)
Plays the Morse code equivalent of the passed string.  If the variable
MORSEDITLEN is set, it will use that value for the length (in ms) of
the dit
(defaults to 80).  Additionally, if MORSETONE is set, it will use that
tone
(in Hz).  The tone default is 800.
```

Diff of the internal help from Asterisk 1.2 to 1.4:

Not available in Asterisk 1.2

Diff of the internal help from Asterisk 1.4 to 1.6:

```
--- in Asterisk 1.4
+++ in Asterisk 1.6
@@ -4,7 +4,7 @@
   Plays morse code

   [Description]
-  Usage: Morsecode(<string>)
+    Morsecode(<string>):
   Plays the Morse code equivalent of the passed string.  If the
variable
   MORSEDITLEN is set, it will use that value for the length (in ms) of
the dit
   (defaults to 80).  Additionally, if MORSETONE is set, it will use
that tone
```

B.103 MP3Player()

Plays an MP3 file or stream.

```
MP3Player(filename)
```

Uses mpg123[13] to play back an MP3 to the caller. The filename may also be a URL. The caller may interrupt playback by pressing any key.

Asterisk is sensitive to the version of mpg123 used. At the time of this writing the most stable version was 0.59r; using other versions may produce unexpected results.

Note: The popular alternative to mpg123, mpg321, does *not* work reliably with Asterisk.

Returns –1 if the channel is hung up, otherwise returns 0.

```
; play a local MP3 file:
exten => 123,1,Answer()
exten => 123,n,MP3Player(test.mp3)

; play an MP3 stream:
exten => 123,1,Answer()
exten => 123,n,MP3Player(http://server.tld/test.mp3)
```

---

13. http://mpg123.org/, http://sourceforge.net/projects/mpg123/; for Mac OS X, see also http://sourceforge.net/projects/mosx-mpg123/.

Asterisk versions:

```
--------| 1.2 |--------| 1.4 |--------| 1.6 |--------
```

Internal help for this application in Asterisk 1.4:

```
    -= Info about application 'MP3Player' =-

[Synopsis]
Play an MP3 file or stream

[Description]
  MP3Player(location) Executes mpg123 to play the given location,
which typically would be a filename or a URL. User can exit by pressing
any key on the dialpad, or by hanging up.
```

Diff of the internal help from Asterisk 1.2 to 1.4:

None

Diff of the internal help from Asterisk 1.4 to 1.6:

```
--- in Asterisk 1.4
+++ in Asterisk 1.6
@@ -4,6 +4,6 @@
   Play an MP3 file or stream

   [Description]
-    MP3Player(location) Executes mpg123 to play the given location,
+    MP3Player(location): Executes mpg123 to play the given location,
   which typically would be a filename or a URL. User can exit by
pressing
   any key on the dialpad, or by hanging up.
```

B.104 MSet()

Sets one or more variables simultaneously.

```
MSet(variable = value)
```

Asterisk 1.2 and 1.4 used **Set()** to set multiple variables. This functionality was removed from **Set()** in 1.6 and **MSet()** was added to provide it.

```
exten => 123,1,MSet(var1=value1,var2=value2)
```

Warning: MSet() isn't really necessary. If you have to convert your dialplan any-way, we recommend you use multiple **Set()** statements:

```
Set(var1=value1);
Set(var2=value2);
```

Asterisk versions:

```
    |      |           |      |            | 1.6 |--------
```

Internal help for this application in Asterisk 1.6:

```
    -= Info about application 'MSet' =-

[Synopsis]
Set channel variable(s) or function value(s)

[Description]
  MSet(name1=value1,name2=value2,...)
This function can be used to set the value of channel variables or
dialplan
functions. When setting variables, if the variable name is prefixed
with _,
the variable will be inherited into channels created from the current
channel. If the variable name is prefixed with __, the variable will be
inherited into channels created from the current channel and all
children
channels.

MSet behaves in a similar fashion to the way Set worked in 1.2/1.4 and
is thus
prone to doing things that you may not expect. For example, it strips
surrounding
double-quotes from the right-hand side (value). If you need to put a
separator
character (comma or vert-bar), you will need to escape them by
inserting a backslash
before them. Avoid its use if possible.
```

Diff of the internal help from Asterisk 1.4 to 1.6:

Not available in Asterisk 1.4

See also **set**, Appendix C.

B.105 MusicOnHold()

Plays music on the channel.

```
MusicOnHold(class[,duration])
```

Plays music belonging to the specified *class*, as configured in musiconhold
.conf. If *class* is not specified, the default class for the channel is used. To set the
default class for the channel, use the function **MUSICCLASS()**.

The *duration* parameter was added in Asterisk 1.6. If it is omitted, music plays until the channel is hung up. If you want to fix the duration in Asterisk 1.4, use **WaitMusicOnHold()**.

Returns only –1 when the channel is hung up.

```
; send telemarketers to this extension and hope they are very patient:
exten => 123,1,Answer()
exten => 123,n,Playback(tt-allbusy)
exten => 123,n,MusicOnHold(default)
```

Asterisk versions:

```
--------| 1.2 |--------| 1.4 |--------| 1.6 |--------
```

Internal help for this application in Asterisk 1.4:

```
    -= Info about application 'MusicOnHold' =-

[Synopsis]
Play Music On Hold indefinitely

[Description]
MusicOnHold(class): Plays hold music specified by class.
If omitted, the default
music source for the channel will be used. Set the default
class with the SetMusicOnHold() application.
Returns -1 on hangup.
Never returns otherwise.
```

Diff of the internal help from Asterisk 1.2 to 1.4:

None

Diff of the internal help from Asterisk 1.4 to 1.6:

```
--- in Asterisk 1.4
+++ in Asterisk 1.6
@@ -4,8 +4,10 @@
    Play Music On Hold indefinitely

    [Description]
-   MusicOnHold(class): Plays hold music specified by class.
If omitted, the default
-   music source for the channel will be used. Set the default
-   class with the SetMusicOnHold() application.
-   Returns -1 on hangup.
-   Never returns otherwise.
+     MusicOnHold(class[,duration]):
+   Plays hold music specified by class.  If omitted, the default
+   music source for the channel will be used. Change the default
+   class with Set(CHANNEL(musicclass)=...).
+   If duration is given, hold music will be played specified number
```

```
+   of seconds. If duration is ommited, music plays indefinitely.
+   Returns 0 when done, -1 on hangup.
```

See also waitmusiconhold, and in Appendix C, see musicclass.

B.106 NBScat()

Plays back a local NBS stream.

```
NBScat()
```

Uses nbscat8k to retrieve a local NBS (Network Broadcast Sound) stream. (Take a look at the nbs module in Digium's Asterisk CVS for more information.) The caller can exit by pressing any key.

Returns only –1 when the channel is hung up.

```
exten => 123,1,Answer()
exten => 123,n,NBScat()
```

Asterisk versions:

```
--------| 1.2 |--------| 1.4 |--------| 1.6 |--------
```

Internal help for this application in Asterisk 1.4:

```
    -= Info about application 'NBScat' =-

[Synopsis]
Play an NBS local stream

[Description]
  NBScat: Executes nbscat to listen to the local NBS stream.
User can exit by pressing any key
 .
```

Diff of the internal help from Asterisk 1.2 to 1.4:

None

Diff of the internal help from Asterisk 1.4 to 1.6:

```
--- in Asterisk 1.4
+++ in Asterisk 1.6
@@ -4,6 +4,5 @@
   Play an NBS local stream

   [Description]
-   NBScat: Executes nbscat to listen to the local NBS stream.
-  User can exit by pressing any key
-  .
```

```
+    NBScat(): Executes nbscat to listen to the local NBS stream.
+    User can exit by pressing any key.
```

B.107 NoCDR()

Suppresses generation of a Call Detail Record for the call on the current channel.

```
NoCDR()
```

Suppresses generation of a CDR for the current call.

```
; no CDR for calls to INWATS directory assistance:
exten => 18005551212,1,Answer()
exten => 18005551212,n,NoCDR()
exten => 18005551212,n,Dial(Zap/4/18005551212)
```

Asterisk versions:

```
--------| 1.2 |--------| 1.4 |--------| 1.6 |--------
```

Internal help for this application in Asterisk 1.4:

```
    -= Info about application 'NoCDR' =-

[Synopsis]
Tell Asterisk to not maintain a CDR for the current call

[Description]
  NoCDR(): This application will tell Asterisk not to maintain a
CDR for the
current call.
```

Diff of the internal help from Asterisk 1.2 to 1.4:

None

Diff of the internal help from Asterisk 1.4 to 1.6:

None

See also **forkcdr**.

B.108 NoOp()

Does nothing.

```
NoOp(text)
```

This application does absolutely nothing (well, not *exactly*). You can use **NoOp()** to print text to the Asterisk CLI, which can be very useful.

The text need not be between quotation marks. If they are entered, they will be printed on the CLI along with the rest of the text.

Warning: Text from **NoOp()** appears on the CLI at verbose level 3 or higher. You can set this in the CLI with **set verbose 3** or by invoking the Asterisk CLI with **asterisk -vvvr.**

```
exten => 123,1,NoOp(Caller-ID: ${CALLERID})
```

Asterisk versions:

```
--------| 1.2 |--------| 1.4 |--------| 1.6 |--------
```

Internal help for this application in Asterisk 1.4:

```
    -= Info about application 'NoOp' =-

[Synopsis]
Do Nothing

[Description]
  NoOp(): This applicatiion does nothing. However, it is useful
for debugging
purposes. Any text that is provided as arguments to this application
can be
viewed at the Asterisk CLI. This method can be used to see the
evaluations of
variables or functions without having any effect.
```

Diff of the internal help from Asterisk 1.2 to 1.4:

None

Diff of the internal help from Asterisk 1.4 to 1.6:

```
--- in Asterisk 1.4
+++ in Asterisk 1.6
@@ -1,10 +1,11 @@
     -= Info about application 'NoOp' =-

   [Synopsis]
-  Do Nothing
+  Do Nothing (No Operation)

   [Description]
-    NoOp(): This applicatiion does nothing. However, it is useful
for debugging
+    NoOp(): This application does nothing. However, it is useful
for debugging
   purposes. Any text that is provided as arguments to this
```

```
application can be
   viewed at the Asterisk CLI. This method can be used to see the
evaluations of
-  variables or functions without having any effect.
+  variables or functions without having any effect. Alternatively,
see the
+  Verbose() application for finer grain control of output at custom
verbose levels.
```

See also dumpchan, log, verbose.

B.109 ODBCFinish()

Clears the resultset of a multirow ODBC (Open Database Connectivity) query.

```
ODBCFinish(resultset_id)
```

Clears the resultset of an ODBC query (see Appendix C) that returned multiple rows.

Asterisk versions:

```
  |       |          |        |        | 1.6 |--------
```

Internal help for this application in Asterisk 1.6:

```
   -= Info about application 'ODBCFinish' =-

[Synopsis]
Clear the resultset of a successful multirow query

[Description]
ODBCFinish(<result-id>)
   Clears any remaining rows of the specified resultset
```

Diff of the internal help from Asterisk 1.4 to 1.6:

Not available in Asterisk 1.4

See also Appendix C.

B.110 Page()

Pages an extension.

```
Page(technology/resource[&technology2/resource2[&...]][,options])
```

The designated devices are dropped into a dynamically generated conference room with their microphones muted. The only device that can send audio is that of the person paging. When the page is finished the room is struck. Options:

- **d**

  Activates full-duplex audio.

- **q**

  Suppresses the notification tone at the beginning of the page.

- **r**

  The page is recorded to a file.

  ```
  exten => 123,1,Page(SIP/2000&SIP/2001&SIP/2002)
  ```

Tip: Most devices treat a page like any other call. If the device is not explicitly instructed to auto-answer, it will simply ring, and the intended recipient will not hear the message. SIP devices can be told to auto-answer, often by setting a SIP header. Every device is different; review the administration manuals for your SIP phone.

Asterisk versions:

```
--------| 1.2 |--------| 1.4 |--------| 1.6 |--------
```

Internal help for this application in Asterisk 1.4:

```
    -= Info about application 'Page' =-

[Synopsis]
Pages phones

[Description]
Page(Technology/Resource&Technology2/Resource2[|options])
   Places outbound calls to the given technology / resource and dumps
them into a conference bridge as muted participants.  The original
caller is dumped into the conference as a speaker and the room is
destroyed when the original caller leaves.  Valid options are:
        d - full duplex audio
        q - quiet, do not play beep to caller
        r - record the page into a file (see 'r' for app_meetme)
```

Diff of the internal help from Asterisk 1.2 to 1.4:

```
--- in Asterisk 1.2
+++ in Asterisk 1.4
@@ -11,3 +11,4 @@
    destroyed when the original caller leaves.  Valid options are:
            d - full duplex audio
            q - quiet, do not play beep to caller
+           r - record the page into a file (see 'r' for app_meetme)
```

Diff of the internal help from Asterisk 1.4 to 1.6:

```
--- in Asterisk 1.4
+++ in Asterisk 1.6
@@ -4,7 +4,7 @@
    Pages phones

    [Description]
-   Page(Technology/Resource&Technology2/Resource2[|options])
+   Page(Technology/Resource&Technology2/Resource2[,options])
       Places outbound calls to the given technology / resource and dumps
    them into a conference bridge as muted participants.  The original
    caller is dumped into the conference as a speaker and the room is
@@ -12,3 +12,4 @@
            d - full duplex audio
            q - quiet, do not play beep to caller
            r - record the page into a file (see 'r' for app_meetme)
+           s - only dial channel if devicestate says it is not in use
```

B.111 Park()

Parks the current call.

```
Park(extension)
```

Parks the active call, typically in combination with an attended transfer so that you can know where the call has been parked. This application is registered internally, so it shouldn't be necessary to add it into the dialplan, though you should make sure it is enabled in features.conf. You may include the context with the include => parkedcalls statement.

```
; park the call in parking space 701:
include => parkedcalls
exten => 123,1,Answer()
exten => 123,n,Park(701)
```

Asterisk versions:

```
--------| 1.2 |--------| 1.4 |--------| 1.6 |--------
```

Internal help for this application in Asterisk 1.4:

```
    -= Info about application 'Park' =-

[Synopsis]
Park yourself

[Description]
Park():Used to park yourself (typically in combination with a supervised
transfer to know the parking space). This application is always
registered internally and does not need to be explicitly added
into the dialplan, although you should include the 'parkedcalls'
context (or the context specified in features.conf).

If you set the PARKINGEXTEN variable to an extension in your
parking context, park() will park the call on that extension, unless
it already exists. In that case, execution will continue at next
priority.
```

Diff of the internal help from Asterisk 1.2 to 1.4:

```
--- in Asterisk 1.2
+++ in Asterisk 1.4
@@ -8,4 +8,9 @@
   transfer to know the parking space). This application is always
   registered internally and does not need to be explicitly added
   into the dialplan, although you should include the 'parkedcalls'
-  context.
+  context (or the context specified in features.conf).
+
+  If you set the PARKINGEXTEN variable to an extension in your
+  parking context, park() will park the call on that extension, unless
+  it already exists. In that case, execution will continue at next
+  priority.
```

Diff of the internal help from Asterisk 1.4 to 1.6:

```
--- in Asterisk 1.4
+++ in Asterisk 1.6
@@ -4,13 +4,22 @@
   Park yourself

   [Description]
-  Park():Used to park yourself (typically in combination with a supervised
+     Park([timeout,[return_context,[return_exten,[return_priority,[opt
ions]]]]]):Used to park yourself (typically in combination with a
supervised
   transfer to know the parking space). This application is always
   registered internally and does not need to be explicitly added
   into the dialplan, although you should include the 'parkedcalls'
   context (or the context specified in features.conf).
```

```
     If you set the PARKINGEXTEN variable to an extension in your
 -   parking context, park() will park the call on that extension, unless
 +   parking context, Park() will park the call on that extension, unless
     it already exists. In that case, execution will continue at next
     priority.
 +      This application can accept arguments as well.
 +   timeout - A custom parking timeout for this parked call.
 +   return_context - The context to return the call to after it times
 out.
 +   return_exten - The extension to return the call to after it times
 out.
 +   return_priority - The priority to return the call to after it times
 out.
 +   options - A list of options for this parked call.  Valid options
 are:
 +       'r' - Send ringing instead of MOH to the parked call.
 +       'R' - Randomize the selection of a parking space.
 +       's' - Silence announcement of the parking space number.
```

See also **parkandannounce, parkedcall.**

B.112 ParkAndAnnounce()

Parks the current call and announces the parking space on the specified channel.

> ParkAndAnnounce(*template, timeout, channel*[, *return-context*])

Parks the current call in the parking lot, but announces the parking space on another, specified channel. The *template* specifies a colon-delimited list of sound files to be played; the word PARKED in the example below is replaced with the parking space number assigned to the parked call. The *timeout* sets the maximum time the call may be parked before it is placed back in the *return-context*. The *channel* denotes the channel on which the announcement is to be made (in our example, Console/dsp prints the announcements to the Asterisk console). The *return-context* is a label in GOTO() format that determines the context to return the call to if it is not retrieved from the parking lot by a user within the defined *timeout* window. The default return priority is *n*+1 in *return-context*.

```
include => parkedcalls
exten => 123,1,Answer()
exten => 123,n,ParkAndAnnounce(vm-youhave:a:pbx-transfer:at:vm-
extension:PARKED,120,Console/dsp)
exten => 123,n,Playback(vm-nobodyavail)
exten => 123,n,Playback(vm-goodbye)
exten => 123,n,Hangup()
```

Asterisk versions:

```
--------| 1.2 |--------| 1.4 |--------| 1.6 |--------
```

Internal help for this application in Asterisk 1.4:

```
    -= Info about application 'ParkAndAnnounce' =-

[Synopsis]
Park and Announce

[Description]
  ParkAndAnnounce(announce:template|timeout|dial|[return_context]):
Park a call into the parkinglot and announce the call to another
channel.

announce template: Colon-separated list of files to announce.
The word PARKED
                will be replaced by a say_digits of the extension
in which
                the call is parked.
timeout:        Time in seconds before the call returns into the
return
                context.
dial:           The app_dial style resource to call to make the
                announcement.  Console/dsp calls the console.
return_context: The goto-style label to jump the call back into
after
                timeout.  Default <priority+1>.

The variable ${PARKEDAT} will contain the parking extension into
which the
call was placed.  Use with the Local channel to allow the dialplan
to make
use of this information.
```

Diff of the internal help from Asterisk 1.2 to 1.4:

```
--- in Asterisk 1.2
+++ in Asterisk 1.4
@@ -5,9 +5,18 @@

   [Description]
     ParkAndAnnounce(announce:template|timeout|dial|[return_context]):
-  Park a call into the parkinglot and announce the call over the
console.
-  announce template: colon separated list of files to announce, the
word PARKED
-                        will be replaced by a say_digits of the ext the
call is parked in
-  timeout: time in seconds before the call returns into the return
context.
```

```
 -  dial: The app_dial style resource to call to make the announcement.
Console/dsp calls the console.
 -  return_context: the goto style label to jump the call back into
after timeout. default=prio+1
 +  Park a call into the parkinglot and announce the call to another
channel.
 +
 +  announce template: Colon-separated list of files to announce.
The word PARKED
 +                      will be replaced by a say_digits of the extension
in which
 +                      the call is parked.
 +  timeout:            Time in seconds before the call returns into the
return
 +                      context.
 +  dial:               The app_dial style resource to call to make the
 +                      announcement.  Console/dsp calls the console.
 +  return_context:     The goto-style label to jump the call back into
after
 +                      timeout.  Default <priority+1>.
 +
 +  The variable ${PARKEDAT} will contain the parking extension into
which the
 +  call was placed.  Use with the Local channel to allow the dialplan
to make
 +  use of this information.
```

Diff of the internal help from Asterisk 1.4 to 1.6:

```
--- in Asterisk 1.4
+++ in Asterisk 1.6
@@ -4,7 +4,7 @@
   Park and Announce

   [Description]
-    ParkAndAnnounce(announce:template|timeout|dial|[return_context]):
+    ParkAndAnnounce(announce:template,timeout,dial[,return_context]):
   Park a call into the parkinglot and announce the call to another
channel

   announce template: Colon-separated list of files to announce. The
word PARKED
```

See also **park**, **parkedcall**.

B.113 ParkedCall()

Retrieves a parked call.

```
ParkedCall(extension)
```

Connects the channel to a parked call identified by *extension*. This application is
registered internally so it shouldn't be necessary to add it into the dialplan, though
you should make sure it is enabled in `features.conf`. You may include the context with the `include => parkedcalls` statement.

```
; retrieve the call parked at 701:
exten => 123,1,Answer()
exten => 123,n,ParkedCall(701)
```

Asterisk versions:

```
--------| 1.2 |--------| 1.4 |--------| 1.6 |--------
```

Internal help for this application in Asterisk 1.4:

```
    -= Info about application 'ParkedCall' =-

[Synopsis]
Answer a parked call

[Description]
ParkedCall(exten):Used to connect to a parked call.
This application is always
registered internally and does not need to be explicitly added
into the dialplan, although you should include the 'parkedcalls'
context.
```

Diff of the internal help from Asterisk 1.2 to 1.4:

None

Diff of the internal help from Asterisk 1.4 to 1.6:

```
--- in Asterisk 1.4
+++ in Asterisk 1.6
@@ -4,7 +4,8 @@
   Answer a parked call

   [Description]
-  ParkedCall(exten):Used to connect to a parked call.
This application is always
+  ParkedCall(exten): Used to connect to a parked call.
This application is always
   registered internally and does not need to be explicitly added
   into the dialplan, although you should include the 'parkedcalls'
-  context.
+  context.  If no extension is provided, then the first available
+  parked call will be acquired.
```

See also **park, parkandannounce.**

B.114 PauseMonitor()

Stops monitoring of a channel.

```
PauseMonitor()
```

Stops recording of a channel until it is resumed with **UnpauseMonitor()**.

Asterisk versions:

```
|        |              | 1.4 |--------| 1.6 |--------
```

Internal help for this application in Asterisk 1.4:

```
    -= Info about application 'PauseMonitor' =-

[Synopsis]
Pause monitoring of a channel

[Description]
PauseMonitor
Pauses monitoring of a channel until it is re-enabled by a call
to UnpauseMonitor.
```

Diff of the internal help from Asterisk 1.2 to 1.4:

Not available in Asterisk 1.2

Diff of the internal help from Asterisk 1.4 to 1.6:

```
--- in Asterisk 1.4
+++ in Asterisk 1.6
@@ -4,5 +4,5 @@
   Pause monitoring of a channel

   [Description]
-  PauseMonitor
+    PauseMonitor():
   Pauses monitoring of a channel until it is re-enabled by a call
to UnpauseMonitor.
```

See also **monitor, unpausemonitor, stopmonitor, changemonitor.**

B.115 PauseQueueMember()

Pauses a queue device so that it cannot take calls from the queue.

```
PauseQueueMember([queue],interface[,options])
```

Blocks calls for a queue member until expressly reenabled, either with **Unpause-QueueMember()** or through the Manager interface. If no queue is specified, the

device is paused in every queue in which it is a member. If a queue is specified but
the device is not a member in it, or if no queue is specified and the device does not
belong to any queues *and* option j is set, execution proceeds at priority n+101 if it
exists.

Returns 0 on success, otherwise returns −1 if the interface cannot be found or
the jump priority does not exist. Sets the channel variable PQMSTATUS to PAUSED or
NOTFOUND.

```
; If we dial *111002, Agent/1002 is paused on all queues where she
is a member:
exten => *11ZXXX,1,PauseQueueMember(,Agent/${EXTEN:3})

; Dialing *121002 unpauses Agent/1002 again:
exten => *12ZXXX,1,UnpauseQueueMember(,Agent/${EXTEN:3})
```

Asterisk versions:

```
--------| 1.2 |--------| 1.4 |--------| 1.6 |--------
```

Internal help for this application in Asterisk 1.4:

```
    -= Info about application 'PauseQueueMember' =-

[Synopsis]
Pauses a queue member

[Description]
   PauseQueueMember([queuename]|interface[|options]):
Pauses (blocks calls for) a queue member.
The given interface will be paused in the given queue. This prevents
any calls from being sent from the queue to the interface until it is
unpaused with UnpauseQueueMember or the manager interface.  If no
queuename is given, the interface is paused in every queue it is a
member of.  If the interface is not in the named queue, or if no queue
is given and the interface is not in any queue, it will jump to
priority n+101, if it exists and the appropriate options are set.
The application will fail if the interface is not found and no
extension
to jump to exists.
The option string may contain zero or more of the following characters:
       'j' -- jump to +101 priority when appropriate.
  This application sets the following channel variable upon completion:
      PQMSTATUS      The status of the attempt to pause a queue member
as a
                      text string, one of
           PAUSED | NOTFOUND
Example: PauseQueueMember(|SIP/3000)
```

Diff of the internal help from Asterisk 1.2 to 1.4:

None

Diff of the internal help from Asterisk 1.4 to 1.6:

```
--- in Asterisk 1.4
+++ in Asterisk 1.6
@@ -4,21 +4,17 @@
   Pauses a queue member

   [Description]
-      PauseQueueMember([queuename]|interface[|options]):
+      PauseQueueMember([queuename],interface[,options[,reason]]):
   Pauses (blocks calls for) a queue member.
   The given interface will be paused in the given queue.  This
prevents
   any calls from being sent from the queue to the interface until it
is
   unpaused with UnpauseQueueMember or the manager interface.  If no
   queuename is given, the interface is paused in every queue it is a
-  member of.  If the interface is not in the named queue, or if no
queue
-  is given and the interface is not in any queue, it will jump to
-  priority n+101, if it exists and the appropriate options are set.
-  The application will fail if the interface is not found and no
extension
-  to jump to exists.
-  The option string may contain zero or more of the following
characters:
-          'j' -- jump to +101 priority when appropriate.
+  member of. The application will fail if the interface is not found.
+  The reason string is entirely optional and is used to add extra
information
+  to the appropriate queue_log entries and manager events.
   This application sets the following channel variable upon
completion:
          PQMSTATUS      The status of the attempt to pause a queue
member as a
                          text string, one of
              PAUSED | NOTFOUND
-  Example: PauseQueueMember(|SIP/3000)
+  Example: PauseQueueMember(,SIP/3000)
```

See also unpausequeuemember.

B.116 Pickup()

Answer a call directed at another extension.

```
Pickup(extension[@context][&extension2@context2[&...]])
```

Pickup() allows a user to answer a channel that is ringing another extension.

```
exten => 1234,1,Pickup(2000@sales)
```

Asterisk versions:

```
--------| 1.2 |--------| 1.4 |--------| 1.6 |--------
```

Internal help for this application in Asterisk 1.4:

```
    -= Info about application 'Pickup' =-

[Synopsis]
Directed Call Pickup

[Description]
  Pickup(extension[@context][&extension2@context...]): This application
can pickup any ringing channel
that is calling the specified extension. If no context is specified,
the current
context will be used. If you use the special string "PICKUPMARK" for
the context parameter, for example
10@PICKUPMARK, this application tries to find a channel which has
defined a channel variable with the same content
as "extension".
```

Diff of the internal help from Asterisk 1.2 to 1.4:

```
--- in Asterisk 1.2
+++ in Asterisk 1.4
@@ -4,6 +4,8 @@
   Directed Call Pickup

   [Description]
-    Pickup(extension[@context]): This application can pickup any
ringing channel
+    Pickup(extension[@context][&extension2@context...]): This
application can pickup any ringing channel
   that is calling the specified extension. If no context is specified,
the current
-  context will be used.
+  context will be used. If you use the special string "PICKUPMARK" for
the context parameter, for example
+  10@PICKUPMARK, this application tries to find a channel which has
defined a channel variable with the same content
+  as "extension".
```

Diff of the internal help from Asterisk 1.4 to 1.6:

```
--- in Asterisk 1.4
+++ in Asterisk 1.6
@@ -4,8 +4,11 @@
   Directed Call Pickup

   [Description]
-    Pickup(extension[@context][&extension2@context...]): This
application can pickup any ringing channel
```

```
-   that is calling the specified extension. If no context is specified,
the current
-   context will be used. If you use the special string "PICKUPMARK" for
the context parameter, for example
-    10@PICKUPMARK, this application tries to find a channel which has
defined a channel variable with the same content
-   as "extension".
+     Pickup([extension[@context][&extension2@[context]...]]):   This
application can
+   pickup any ringing channel that is calling the specified
extension.   If no
+   context is specified, the current context will be used. If you use
the special
+   string "PICKUPMARK" for the context parameter, for example
10@PICKUPMARK,
+   this application tries to find a channel which has defined a
${PICKUPMARK}
+   channel variable with the same value as "extension" (in this
example, "10").
+   When no parameter is specified, the application will pickup a
channel matching
+   the pickup group of the active channel.
```

See also **pickupchan.**

B.117 PickupChan()

Answer a call to a ringing channel.

```
PickupChan(channel[&channel[...]])
```

Pickup() allows a call directed to another extension to be answered. **PickupChan()** works similarly but it can answer any ringing channel irrespective of extension.

Asterisk versions:

```
    |     |        |     |       | 1.6 |--------
```

Internal help for this application in Asterisk 1.6:

```
    -= Info about application 'PickupChan' =-

[Synopsis]
Pickup a ringing channel

[Description]
   PickupChan(channel[&channel...]): This application can pickup
any ringing channel
```

Diff of the internal help from Asterisk 1.4 to 1.6:

Not available in Asterisk 1.4

See also pickup.

B.118 Playback()

Plays a sound file to the caller.

```
Playback(filename[,options])
```

Plays *filename* (in the directory `/var/lib/asterisk/sounds`) to the caller. The filename does not include an extension; Asterisk automatically selects the format with the lowest transcoding cost. Options may be specified.

The option `skip` causes the message to be skipped if the channel is not in the up state.

Answers the channel before the audio file is played, unless `noanswer` is specified. Note that not all channels support playback when "on hook."

Returns −1 when the channel is hung up. If the file cannot be found, jumps to priority n+101 if it exists.

```
exten => 123,1,Answer()
exten => 123,n,Playback(tt-weasels)
```

Asterisk versions:

```
--------| 1.2 |--------| 1.4 |--------| 1.6 |--------
```

Internal help for this application in Asterisk 1.4:

```
   -= Info about application 'Playback' =-

[Synopsis]
Play a file

[Description]
  Playback(filename[&filename2...][|option]):  Plays back given
filenames (do not put
extension). Options may also be included following a pipe symbol.
The 'skip'
option causes the playback of the message to be skipped if the channel
is not in the 'up' state (i.e. it hasn't been  answered  yet).  If
'skip' is
specified, the application will return immediately should the channel
not be
off hook.  Otherwise, unless 'noanswer' is specified, the channel will
be answered before the sound is played. Not all channels support
playing
```

messages while still on hook. If 'j' is specified, the application
will jump to priority n+101 if present when a file specified to be
played
does not exist.
This application sets the following channel variable upon completion:
  PLAYBACKSTATUS   The status of the playback attempt as a text string,
one of
                 SUCCESS | FAILED
See Also: Background (application) -- for playing soundfiles that are
interruptible
         WaitExten (application) -- wait for digits from caller,
optionally play music on hold

Diff of the internal help from Asterisk 1.2 to 1.4:

--- in Asterisk 1.2
+++ in Asterisk 1.4
@@ -17,3 +17,5 @@
   This application sets the following channel variable upon
completion:
     PLAYBACKSTATUS   The status of the playback attempt as a text
string, one of
                 SUCCESS | FAILED
+ See Also: Background (application) -- for playing soundfiles that
are interruptible
+   WaitExten (application) -- wait for digits from caller,
optionally play music on hold

Diff of the internal help from Asterisk 1.4 to 1.6:

--- in Asterisk 1.4
+++ in Asterisk 1.6
@@ -4,16 +4,14 @@
   Play a file

   [Description]
-    Playback(filename[&filename2...][|option]): Plays back given
filenames (do not put
- extension). Options may also be included following a pipe symbol.
The 'skip'
- option causes the playback of the message to be skipped if the channel
+   Playback(filename[&filename2...][,option]): Plays back given
filenames (do not put
+ extension). Options may also be included following a comma.
+ The 'skip' option causes the playback of the message to be skipped
if the channel
   is not in the 'up' state (i.e. it hasn't been answered yet). If
'skip' is
   specified, the application will return immediately should the
channel not be
   off hook. Otherwise, unless 'noanswer' is specified, the channel
will

```
     be answered before the sound is played. Not all channels support
playing
-    messages while still on hook. If 'j' is specified, the application
-    will jump to priority n+101 if present when a file specified to be
played
-    does not exist.
+    messages while still on hook.
     This application sets the following channel variable upon
completion:
     PLAYBACKSTATUS    The status of the playback attempt as a text
string, one of
                SUCCESS | FAILED
```

See also background.

B.119 Playtones()

Plays back one or more tones.

```
Playtones(tones)
```

Plays a list of one or more tones. **Playtones()** runs in the background; i.e., it will continue to play tones while execution of the dialplan continues. The argument *tones* is either a tone name as defined in indications.conf or a list of tone frequencies and durations. For an explanation on how to define your own tones, see indications.conf.

Use **StopPlaytones()** to stop playing tones.

```
; Two seconds "busy", then two seconds "congestion" tones:
exten => 123,1,Playtones(busy)
exten => 123,n,Wait(2)
exten => 123,n,StopPlaytones()
exten => 123,n,Playtones(congestion)
exten => 123,n,Wait(2)
exten => 123,n,StopPlaytones()
exten => 123,n,Goto(1)
```

Asterisk versions:

```
--------| 1.2 |--------| 1.4 |--------| 1.6 |--------
```

Internal help for this application in Asterisk 1.4:

```
   -= Info about application 'PlayTones' =-

[Synopsis]
Play a tone list

[Description]
PlayTones(arg): Plays a tone list. Execution will continue with the
```

```
next step immediately,
while the tones continue to play.
Arg is either the tone name defined in the indications.conf
configuration file, or a directly
specified list of frequencies and durations.
See the sample indications.conf for a description of the specification
of a tonelist.

Use the StopPlayTones application to stop the tones playing.
```

Diff of the internal help from Asterisk 1.2 to 1.4:

None

Diff of the internal help from Asterisk 1.4 to 1.6:

```
--- in Asterisk 1.4
+++ in Asterisk 1.6
@@ -4,7 +4,7 @@
   Play a tone list

   [Description]
-  PlayTones(arg): Plays a tone list. Execution will continue with
the next step immediately,
+    PlayTones(arg): Plays a tone list. Execution will continue with
the next step immediately,
   while the tones continue to play.
   Arg is either the tone name defined in the indications.conf
configuration file, or a directly
   specified list of frequencies and durations.
```

See also stopplaytones, indications.conf, busy, congestion, progress, ringing.

B.120 PrivacyManager()

Requests the input of the caller's telephone number, if caller ID cannot be obtained.

```
PrivacyManager([maxRetries[,minLength[,options]]])
```

If no caller ID is received, the channel is answered and the caller is asked to enter his own telephone number. If caller ID is present on the line, **PrivacyManager()** has no effect.

The caller has *maxRetries* (default is 3) number of attempts to provide a valid number of at least *minLength* (default is 10) digits in length. The default values are set in privacy.conf, which may contain the following entries:

- **maxretries**

  The maximum number of times the caller can attempt to enter a number complying with the length limit set by *minLength* (usually 3).

- **minlength**

  The minimum number of digits a number entered must have to be accepted
  (usually 10).

If you want to prevent **PrivacyManager()** from reading from the configuration
file every time it is called, you can set values in the command in the dialplan.

If option j is set, it jumps to n+101 if the caller fails to provide a valid number
within the number of allowed attempts.

The channel variable PRIVACYMGRSTATUS is set to SUCCESS or FAILED and indicates whether the privacy manager was able to get a valid phone number from the
caller.

```
exten => 123,1,Answer()
exten => 123,n,PrivacyManager()
exten => 123,n,GotoIf($["${PRIVACYMGRSTATUS}" = "FAILED"]?pm-failed,1)
exten => 123,n,Dial(Zap/1)

exten => pm-failed,1,Playback(sorry)
exten => pm-failed,n,Playback(vm-goodbye)
```

Asterisk versions:

--------| **1.2** |--------| **1.4** |--------| **1.6** |--------

Internal help for this application in Asterisk 1.4:

```
    -= Info about application 'PrivacyManager' =-

[Synopsis]
Require phone number to be entered, if no CallerID sent

[Description]
  PrivacyManager([maxretries[|minlength[|options]]]): If no Caller*ID
is sent, PrivacyManager answers the channel and asks the caller to
enter their phone number. The caller is given 3 attempts to do so.
The application does nothing if Caller*ID was received on the channel.
  Configuration file privacy.conf contains two variables:
   maxretries  default 3  -maximum number of attempts the caller is
allowed
              to input a callerid.
   minlength   default 10 -minimum allowable digits in the input
callerid number.
If you don't want to use the config file and have an i/o operation with
```

```
every call, you can also specify maxretries and minlength as
application
parameters. Doing so supercedes any values set in privacy.conf.
The option string may contain the following character:
  'j' -- jump to n+101 priority after <maxretries> failed attempts to
collect
        the minlength number of digits.
The application sets the following channel variable upon completion:
PRIVACYMGRSTATUS  The status of the privacy manager's attempt to
collect
                a phone number from the user. A text string that is
either:
            SUCCESS | FAILED
```

Diff of the internal help from Asterisk 1.2 to 1.4:

None

Diff of the internal help from Asterisk 1.4 to 1.6:

```
--- in Asterisk 1.4
+++ in Asterisk 1.6
@@ -4,20 +4,14 @@
   Require phone number to be entered, if no CallerID sent

   [Description]
-    PrivacyManager([maxretries[|minlength[|options]]]): If no
Caller*ID
+    PrivacyManager([maxretries][,minlength][,context]): If no
Caller*ID
   is sent, PrivacyManager answers the channel and asks the caller to
-  enter their phone number. The caller is given 3 attempts to do so.
+  enter their phone number. The caller is given 'maxretries' attempts
to do so.
   The application does nothing if Caller*ID was received on the
channel.
-    Configuration file privacy.conf contains two variables:
      maxretries  default 3  -maximum number of attempts the caller is
allowed
                to input a callerid.
      minlength   default 10 -minimum allowable digits in the input
callerid number.
-  If you don't want to use the config file and have an i/o operation
with
-  every call, you can also specify maxretries and minlength as
application
-  parameters. Doing so supercedes any values set in privacy.conf.
-  The option string may contain the following character:
-    'j' -- jump to n+101 priority after <maxretries> failed attempts
to collect
-          the minlength number of digits.
+    context     context to check the given Caller*ID against
patterns.
```

```
      The application sets the following channel variable upon completion:
      PRIVACYMGRSTATUS   The status of the privacy manager's attempt to
collect
                        a phone number from the user. A text string that
      is either:
```

See also **zapateller**.

B.121 Proceeding()

Indicates call is proceeding.

```
      Proceeding()
```

Sends a message to the calling channel indicating that the call is proceeding.

Asterisk versions:

Internal help for this application in Asterisk 1.6:

```
      -= Info about application 'Proceeding' =-

    [Synopsis]
    Indicate proceeding

    [Description]
      Proceeding(): This application will request that a proceeding
    message be provided to the calling channel.
```

Diff of the internal help from Asterisk 1.4 to 1.6:

Not available in Asterisk 1.4

See also **progress**.

B.122 Progress()

Indicates call progress.

```
      Progress()
```

Provides in-band progress indication, if available, to the caller (… so that she knows that "something is happening").

  Returns 0.

```
    ; Indicate progress:
    exten => 123,1,Progress()
```

Asterisk versions:

```
--------| 1.2 |--------| 1.4 |--------| 1.6 |--------
```

Internal help for this application in Asterisk 1.4:

```
    -= Info about application 'Progress' =-

[Synopsis]
Indicate progress

[Description]
   Progress(): This application will request that in-band progress
information
be provided to the calling channel.
```

Diff of the internal help from Asterisk 1.2 to 1.4:

None

Diff of the internal help from Asterisk 1.4 to 1.6:

None

See also busy, congestion, playtones, ringing, proceeding.

B.123 Queue()

Places a call in the specified queue.

```
Queue(queue[,options[,URL[,announcement[,timeout[,AGI]]]]])
```

Passes an incoming call into the specified queue (predefined in `queues.conf`). The following options are available and may be combined:

- **t**

  Allow the called party to transfer the call.

- **T**

  Allow the calling party to transfer the call.

- **d**

  Data-quality (modem) call (minimizes delay).

- **h**

  Called party may end the call by pressing *.

- **H**

  Calling party may end the call by pressing *.

- **n**

  No retries on timeout; proceeds to the next priority.

- **r**

  Rings instead of playing music-on-hold.

- **i**

  Ignore call forward requests from queue members; instead do nothing.

- **w**

  Allow the called party to record the conversation via **Monitor()**.

- **W**

  Allow the calling party to record the conversation via **Monitor()**.

If a call is transferable, it may also be parked and retrieved by another user.

The argument *announcement* specifies a wait announcement to be played back to the caller at intervals until a queue member answers the call; the default announcement, if there is one, is defined in `queues.conf`.

The optional URL is sent to the caller if the calling channel supports it. This could be used on compatible SIP phones to provide queue information to the phone's display panel.

The *timeout* argument causes the call to fail out of the queue after *timeout* number of seconds. The default is 300 seconds (or 5 minutes).

Returns −1 if the originating channel is hung up or if the call is connected to a queue member and ended by one of the parties. If the queue is full, does not exist, or has no member, returns 0; execution continues in the next priority in the dialplan.

```
; pass the caller to the support queue:
exten => 123,1,Answer()
exten => 123,n,Queue(support,t)
```

Asterisk versions:

```
--------| 1.2 |--------| 1.4 |--------| 1.6 |--------
```

Internal help for this application in Asterisk 1.4:

```
    -= Info about application 'Queue' =-

[Synopsis]
Queue a call for a call queue

[Description]
  Queue(queuename[|options[|URL][|announceoverride][|timeout][|AGI]]):
Queues an incoming call in a particular call queue as defined in
queues.conf.
```

This application will return to the dialplan if the queue does not
exist, or
any of the join options cause the caller to not enter the queue.
The option string may contain zero or more of the following characters:
    'd' -- data-quality (modem) call (minimum delay).
    'h' -- allow callee to hang up by hitting *.
    'H' -- allow caller to hang up by hitting *.
    'n' -- no retries on the timeout; will exit this application and
           go to the next step.
    'i' -- ignore call forward requests from queue members and do
nothing
           when they are requested.
    'r' -- ring instead of playing MOH
    't' -- allow the called user transfer the calling user
    'T' -- to allow the calling user to transfer the call.
    'w' -- allow the called user to write the conversation to disk
via Monitor
    'W' -- allow the calling user to write the conversation to disk
via Monitor
  In addition to transferring the call, a call may be parked and then
picked
up by another user.
  The optional URL will be sent to the called party if the channel
supports
it.
  The optional AGI parameter will setup an AGI script to be executed
on the
calling party's channel once they are connected to a queue member.
  The timeout will cause the queue to fail out after a specified number
of
seconds, checked between each queues.conf 'timeout' and 'retry' cycle.
  This application sets the following channel variable upon completion:
    QUEUESTATUS   The status of the call as a text string, one of
           TIMEOUT | FULL | JOINEMPTY | LEAVEEMPTY | JOINUNAVAIL |
LEAVEUNAVAIL

Diff of the internal help from Asterisk 1.2 to 1.4:

```
--- in Asterisk 1,2
+++ in Asterisk 1.4
@@ -4,7 +4,7 @@
   Queue a call for a call queue

   [Description]
-   Queue(queuename[|options[|URL][|announceoverride][|timeout]]):
+   Queue(queuename[|options[|URL][|announceoverride][|timeout][|AGI])
:

  Queues an incoming call in a particular call queue as defined in
queues.conf.
  This application will return to the dialplan if the queue does not
exist, or
  any of the join options cause the caller to not enter the queue.
@@ -13,7 +13,9 @@
```

```
              'h' -- allow callee to hang up by hitting *.
              'H' -- allow caller to hang up by hitting *.
              'n' -- no retries on the timeout; will exit this application
and
-                 go to the next step.
+                 go to the next step.
+             'i' -- ignore call forward requests from queue members and do
nothing
+                 when they are requested.
              'r' -- ring instead of playing MOH
              't' -- allow the called user transfer the calling user
              'T' -- to allow the calling user to transfer the call.
@@ -23,6 +25,8 @@
   up by another user.
      The optional URL will be sent to the called party if the channel
supports
   it.
+    The optional AGI parameter will setup an AGI script to be executed
on the
+  calling party's channel once they are connected to a queue member.
      The timeout will cause the queue to fail out after a specified
number of
   seconds, checked between each queues.conf 'timeout' and 'retry'
cycle.
      This application sets the following channel variable upon
completion:
```

Diff of the internal help from Asterisk 1.4 to 1.6:

```
--- in Asterisk 1.4
+++ in Asterisk 1.6
@@ -4,31 +4,56 @@
   Queue a call for a call queue

   [Description]
-    Queue(queuename[|options[|URL][|announceoverride][|timeout][|AGI]):
+    Queue(queuename[,options[,URL][,announceoverride][,timeout][,AGI][
,macro][,gosub][,rule]):
   Queues an incoming call in a particular call queue as defined in
queues.conf.
   This application will return to the dialplan if the queue does not
exist, or
   any of the join options cause the caller to not enter the queue.
   The option string may contain zero or more of the following
characters:
+         'c' -- continue in the dialplan if the callee hangs up.
          'd' -- data-quality (modem) call (minimum delay).
-         'h' -- allow callee to hang up by hitting *.
-         'H' -- allow caller to hang up by hitting *.
+         'h' -- allow callee to hang up by hitting '*', or whatver
disconnect sequence
+                that is defined in the featuremap section in
features.conf.
```

```
+          'H' -- allow caller to hang up by hitting '*', or whatever
disconnect sequence
+               that is defined in the featuremap section in
features.conf.
          'n' -- no retries on the timeout; will exit this application
and
               go to the next step.
          'i' -- ignore call forward requests from queue members and do
nothing
               when they are requested.
-         'r' -- ring instead of playing MOH
-         't' -- allow the called user transfer the calling user
-         'T' -- to allow the calling user to transfer the call.
+         'r' -- ring instead of playing MOH. Periodic Announcements are
still made, if applicable.
+         't' -- allow the called user transfer the calling user by
pressing '#' or
+               whatever blindxfer sequence defined in the featuremap
section in
+               features.conf
+         'T' -- to allow the calling user to transfer the call by
pressing '#' or
+               whatever blindxfer sequence defined in the featuremap
section in
+               features.conf
          'w' -- allow the called user to write the conversation to
disk via Monitor
+               by pressing the automon sequence defined in the
featuremap section in
+               features.conf
          'W' -- allow the calling user to write the conversation
to disk via Monitor
-    In addition to transferring the call, a call may be parked
and then picked
-  up by another user.
+               by pressing the automon sequence defined in the
featuremap section in
+               features.conf
!         'U'   Allow the called party to enable parking of the call
by sending
+               the DTMF sequence defined for call parking in
features.conf.
+         'K' -- Allow the calling party to enable parking of the call
by sending
+               the DTMF sequence defined for call parking in
 features.conf.
+         'x' -- allow the called user to write the conversation to disk
 via MixMonitor
+               by pressing the automixmon sequence defined in the
featuremap section in
+               features.conf
+         'X' -- allow the calling user to write the conversation to
disk via MixMonitor
```

```
+                 by pressing the automixmon sequence defined in the
 featuremap section in
+                 features.conf
    The optional URL will be sent to the called party if the channel
 supports
   it.
    The optional AGI parameter will setup an AGI script to be executed
 on the
   calling party's channel once they are connected to a queue member.
+    The optional macro parameter will run a macro on the
+  calling party's channel once they are connected to a queue member.
+    The optional gosub parameter will run a gosub on the
+  calling party's channel once they are connected to a queue member.
+    The optional rule parameter will cause the queue's defaultrule
 to be
+  overridden by the rule specified.
    The timeout will cause the queue to fail out after a specified
 number of
   seconds, checked between each queues.conf 'timeout' and 'retry'
 cycle.
    This application sets the following channel variable upon
 completion:
        QUEUESTATUS   The status of the call as a text string, one of
-              TIMEOUT | FULL | JOINEMPTY | LEAVEEMPTY | JOINUNAVAIL |
LEAVEUNAVAIL
+              TIMEOUT | FULL | JOINEMPTY | LEAVEEMPTY | JOINUNAVAIL |
LEAVEUNAVAIL | CONTINUE
```

See also `queuelog` and Chapter 12, "Queues."

B.124 QueueLog()

Writes an entry to the queue log.

    QueueLog(*queue*, *uniqueID*, *agent*, *event*[, *additionalinfo*])

Writes an entry to the queue log (usually `/var/log/asterisk/queue_log`). Lets you define custom events. See Chapter 12 for a more complete explanation.

    QueueLog(support, ${UNIQUEID}, ${AGENT}, LUNCH, Bon appetit)

Asterisk versions:

```
    |     |         | 1.4 |--------| 1.6 |--------
```

Internal help for this application in Asterisk 1.4:

```
   -= Info about application 'QueueLog' =-

[Synopsis]
Writes to the queue_log
```

```
[Description]
    QueueLog(queuename|uniqueid|agent|event[|additionalinfo]):
Allows you to write your own events into the queue log
Example: QueueLog(101|${UNIQUEID}|${AGENT}|WENTONBREAK|600)
```

Diff of the internal help from Asterisk 1.2 to 1.4:

Not available in Asterisk 1.2

Diff of the internal help from Asterisk 1.4 to 1.6:

```
--- in Asterisk 1.4
+++ in Asterisk 1.6
@@ -4,6 +4,6 @@
   Writes to the queue_log

   [Description]
-      QueueLog(queuename|uniqueid|agent|event[|additionalinfo]):
+      QueueLog(queuename,uniqueid,agent,event[,additionalinfo]):
   Allows you to write your own events into the queue log
-  Example: QueueLog(101|${UNIQUEID}|${AGENT}|WENTONBREAK|600)
+  Example: QueueLog(101,${UNIQUEID},${AGENT},WENTONBREAK,600)
```

See also **queue**, Chapter 12.

B.125 RaiseException()

Raises an exception.

```
RaiseException(reason)
```

RaiseException() raises an exception; this causes execution to jump to extension e in the current context where the exception may be handled by using the **EXCEPTION()** function. If no e extension has been defined, the channel is hung up.

```
123 => {
    /*/ ...
    if ("${input}" = "") {
        RaiseException(NOINPUT);
    }
    // ...
}

e => {
    Verbose(1,### An exception occurred);
    if ("${EXCEPTION(reason)}" = "NOINPUT") {
        Playback(nothing-was-entered);
    }
    else if ("${EXCEPTION(reason)}" = "RESPONSETIMEOUT") {
        Playback(you-waited-too-long);
    }
```

```
        }
        else {
            Playback(an-error-occurred);
        }
        Hangup();
    }
```

Asterisk versions:

| | | | | | **1.6** |--------

Internal help for this application in Asterisk 1.6:

```
    -= Info about application 'RaiseException' =-

[Synopsis]
Handle an exceptional condition

[Description]
   RaiseException(<reason>): This application will jump to the "e"
extension
in the current context, setting the dialplan function EXCEPTION().
If the "e"
extension does not exist, the call will hangup.
```

Diff of the internal help from Asterisk 1.4 to 1.6:

Not available in Asterisk 1.4

See also Appendix C.

B.126 Random()

Jumps to a priority randomly.

```
    Random([probability]:[[context,]extension,]priority)
```

Jumps to the specified priority, extension and context based on the *probability* provided, which must be a whole number between 1 and 100; in *probability* percent of cases, the call will jump to the specified priority.

```
; Game of chance with 20% chance of winning:
exten => 123,1,Random(20:won,1)
exten => 123,n,Goto(lost,1)

exten => won,1,Playback(hooray)
exten => won,n,Goto(123,1)

exten => lost,1,Playback(sorry)
exten => lost,n,Goto(123,1)
```

Warning: This application is deprecated as of 1.4; use the Asterisk function
RAND() instead.

Asterisk versions:

```
--------| 1.2 |--------| 1.4 |           |     |
```

Internal help for this application in Asterisk 1.4:

```
    -= Info about application 'Random' =-

[Synopsis]
Conditionally branches, based upon a probability

[Description]
Random([probability]:[[context|]extension|]priority)
  probability := INTEGER in the range 1 to 100
DEPRECATED: Use GotoIf($[${RAND(1,100)} > <number>]?<label>)
```

Diff of the internal help from Asterisk 1.2 to 1.4:

```
--- in Asterisk 1.2
+++ in Asterisk 1.4
@@ -6,3 +6,4 @@
   [Description]
   Random([probability]:[[context|]extension|]priority)
     probability := INTEGER in the range 1 to 100
+  DEPRECATED: Use GotoIf($[${RAND(1,100)} > <number>]?<label>)
```

Diff of the internal help from Asterisk 1.4 to 1.6:

Not available in Asterisk 1.6

See also Appendix C.

B.127 Read()

Reads DTMF input from a caller and assigns the result to a variable.

```
Read(variable[,filename[,maxDigits[,option[,attempts[,timeout]]]]])
```

Reads a DTMF sequence ending in # from the caller and places it in the specified
variable.

   If *filename* is specified (without file extension!), this audio file is played back to
the caller before input is read.

The argument *maxDigits* defines the maximum number of digits allowed. If provided, the application stops reading when this number of digits has been entered, without the caller having to enter # to end the call. The default setting is 0 and means there is no preset limit and # is expected; the absolute maximum number of digits allowed is 255.

The option skip causes the application to end immediately if the channel is inactive. The option noanswer enables reading even if the channel is inactive.

The caller has up to *attempts* attempts within *timeout* seconds to enter a sequence. A *timeout* of zero means there is no time limit.

Returns −1 if the channel is hung up, otherwise returns 0.

```
; Read a 4 digit number, allowing up to 3 attempts, and say this
number back to the caller:
exten => 123,1,Read(NUMBER,,4,3)
exten => 123,n,SayNumber(${NUMBER})
exten => 123,n,Goto(1)
```

Asterisk versions:

--------| **1.2** |--------| **1.4** |--------| **1.6** |--------

Internal help for this application in Asterisk 1.4:

```
    -= Info about application 'Read' =-

[Synopsis]
Read a variable

[Description]
  Read(variable[|filename][|maxdigits][|option][|attempts][|timeout])

Reads a #-terminated string of digits a certain number of times from
the
user in to the given variable.
  filename    -- file to play before reading digits or tone with option
i
  maxdigits   -- maximum acceptable number of digits. Stops reading
after
              maxdigits have been entered (without requiring the
user to
              press the '#' key).
              Defaults to 0 - no limit - wait for the user press
the '#' key.
              Any value below 0 means the same. Max accepted value
is 255.
  option      -- options are 's' , 'i', 'n'
              's' to return immediately if the line is not up,
              'i' to play  filename as an indication tone from
your indications.conf
```

```
                        'n' to read digits even if the line is not up.
  attempts    -- if greater than 1, that many attempts will be made
in the
                        event no data is entered.
  timeout     -- An integer number of seconds to wait for a digit
response. If greater
                        than 0, that value will override the default timeout.
```

Read should disconnect if the function fails or errors out.

Diff of the internal help from Asterisk 1.2 to 1.4:

```
--- in Asterisk 1.2
+++ in Asterisk 1.4
@@ -8,16 +8,19 @@

   Reads a #-terminated string of digits a certain number of times
from the
   user in to the given variable.
-    filename    -- file to play before reading digits.
+    filename    -- file to play before reading digits or tone with
option i
     maxdigits  -- maximum acceptable number of digits. Stops reading
after
                        maxdigits have been entered (without requiring the
user to
                        press the '#' key).
                        Defaults to 0 - no limit - wait for the user press
the '#' key.
                        Any value below 0 means the same. Max accepted value
is 255.
-    option      -- may be 'skip' to return immediately if the line is
not up,
-                        or 'noanswer' to read digits even if the line is not
up.
+    option      -- options are 's' , 'i', 'n'
+                        's' to return immediately if the line is not up,
+                        'i' to play  filename as an indication tone from
your indications.conf
+                        'n' to read digits even if the line is not up.
     attempts   -- if greater than 1, that many attempts will be made
in the
                        event no data is entered.
-    timeout     -- if greater than 0, that value will override the
default timeout.
+    timeout     -- An integer number of seconds to wait for a digit
response. If greater
+                        than 0, that value will override the default
timeout.

   Read should disconnect if the function fails or errors out.
```

Diff of the internal help from Asterisk 1.4 to 1.6:

```
--- in Asterisk 1.4
+++ in Asterisk 1.6
@@ -4,11 +4,11 @@
   Read a variable

   [Description]
-    Read(variable[|filename][|maxdigits][|option][|attempts][|timeout])
+    Read(variable[,filename[&filename2...]][,maxdigits][,option][,atte
mpts][,timeout])

   Reads a #-terminated string of digits a certain number of times from
the
   user in to the given variable.
-    filename   -- file to play before reading digits or tone with
option i
+    filename   -- file(s) to play before reading digits or tone with
option i
    maxdigits  -- maximum acceptable number of digits. Stops reading
after
               maxdigits have been entered (without requiring the
user to
               press the '#' key).
@@ -20,7 +20,9 @@
               'n' to read digits even if the line is not up.
    attempts   -- if greater than 1, that many attempts will be made
in the
               event no data is entered.
-    timeout    -- An integer number of seconds to wait for a digit
response. If greater
-              than 0, that value will override the default
timeout.
-
- Read should disconnect if the function fails or errors out.
+    timeout    -- The number of seconds to wait for a digit response.
If greater
+              than 0, that value will override the default
timeout. Can be floating point.
+ This application sets the following channel variable upon
completion:
+     READSTATUS - This is the status of the read operation.
+               Possible values are:
+               OK | ERROR | HANGUP | INTERRUPTED | SKIPPED |
TIMEOUT
```

See also **senddtmf**.

B.128 ReadExten()

Read DTMF extension.

```
ReadExten(variable[,prompt_filename[,context[,option[,timeout]]]])
```

Prompts the caller to input an extension, listens for DTMF and stores the entered digits in the specified variable.

Asterisk versions:

| | | | | | 1.6 |--------

Internal help for this application in Asterisk 1.6:

```
    -= Info about application 'ReadExten' =-

[Synopsis]
Read an extension into a variable

[Description]
  ReadExten(<variable>[,[<filename>][,[<context>][,[<option>]
[,<timeout>]]]])

Reads a #-terminated string of digits from the user into the given
variable.
   filename  file to play before reading digits or tone with option i
   context   context in which to match extensions
   option    options are:
                 s - Return immediately if the channel is not answered,
                 i - Play filename as an indication tone from your
                     indications.conf
                 n - Read digits even if the channel is not answered.
   timeout   An integer number of seconds to wait for a digit response.
If
             greater than 0, that value will override the default
timeout.

ReadExten will set READEXTENSTATUS on exit with one of the following
statuses:
   OK        A valid extension exists in ${variable}
   TIMEOUT   No extension was entered in the specified time
   INVALID   An invalid extension, ${INVALID_EXTEN}, was entered
   SKIP      Line was not up and the option 's' was specified
   ERROR     Invalid arguments were passed
```

Diff of the internal help from Asterisk 1.4 to 1.6:

Not available in Asterisk 1.4

See also read.

B.129 ReadFile()

Reads a file.

```
ReadFile(variable=filename,length)
```

Reads the contents of *filename* up to *length* characters into the variable *variable*.

Asterisk versions:

```
--------| 1.2 |--------| 1.4 |--------| 1.6 |--------
```

Internal help for this application in Asterisk 1.4:

```
    -= Info about application 'ReadFile' =-

[Synopsis]
ReadFile(varname=file,length)

[Description]
ReadFile(varname=file,length)
  Varname - Result stored here.
  File - The name of the file to read.
  Length - Maximum number of characters to capture.
```

Diff of the internal help from Asterisk 1.2 to 1.4:

None

Diff of the internal help from Asterisk 1.4 to 1.6:

```
--- in Asterisk 1.4
+++ in Asterisk 1.6
@@ -1,10 +1,10 @@
    -= Info about application 'ReadFile' =-

 [Synopsis]
- ReadFile(varname=file,length)
+ Read the contents of a text file into a channel variable

 [Description]
 ReadFile(varname=file,length)
-   Varname - Result stored here.
-   File - The name of the file to read.
-   Length - Maximum number of characters to capture.
+   varname  - Result stored here.
+   file     - The name of the file to read.
+   length   - Maximum number of characters to capture.
```

B.130 RealTime()

Gets configuration information from the realtime configuration database.

```
RealTime(family,column,value[,prefix])
```

Retrieves configuration settings from the realtime configuration database into channel variables. All unique column names are set as channel variables. Optionally, the argument *prefix* is prepended to the column name to form the variable name. (In the example in the internal help, the prefix `var_` to the column name `test` results in the variable `${var_test}`.).\

Sets the channel variable `REALTIMECOUNT` to the number of values read.

In `extconfig.conf`:

```
; Family => DBMS,database,table
sipusers => mysql,asterisk,sip_users
```

In `extensions.conf`:

```
exten => 123,1,RealTime(sipusers,ext,5678,var_)
```

This executes the following SQL query in the database `asterisk`:

```
SELECT * FROM sip_users WHERE ext = 5678
```

Assuming the table has the columns `firstname` and `lastname`, we can print those values to the CLI this way:

```
exten => 123,n,NoOp(The first name of the user at ext. 5678 is:
${var_firstname})
exten => 123,n,NoOp(The last name of the user at ext. 5678 is:
${var_lastname})
```

Asterisk versions:

```
--------| 1.2 |--------| 1.4 |            |      |
```

Internal help for this application in Asterisk 1.4.

```
    -= Info about application 'RealTime' =-

[Synopsis]
Realtime Data Lookup

[Description]
Use the RealTime config handler system to read data into channel
variables.
RealTime(<family>|<colmatch>|<value>[|<prefix>])

All unique column names will be set as channel variables with optional
prefix
```

```
to the name.  For example, a prefix of 'var_' would make the column
'name'
become the variable ${var_name}.  REALTIMECOUNT will be set with the
number
of values read.
```

Diff of the internal help from Asterisk 1.2 to 1.4:

```
--- in Asterisk 1.2
+++ in Asterisk 1.4
@@ -7,5 +7,7 @@
   Use the RealTime config handler system to read data into channel
variables.
   RealTime(<family>|<colmatch>|<value>[|<prefix>])

-  All unique column names will be set as channel variables with
optional prefix to the name.
-  e.g. prefix of 'var_' would make the column 'name' become the
variable ${var_name}
+  All unique column names will be set as channel variables with
optional prefix
+  to the name.  For example, a prefix of 'var_' would make the column
 'name'
+  become the variable ${var_name}.  REALTIMECOUNT will be set with the
number
+  of values read.
```

Diff of the internal help from Asterisk 1.4 to 1.6:

Not available in Asterisk 1.6

See also **realtimeupdate.**

B.131 RealTimeUpdate()

Updates a value in the realtime configuration database.

```
RealTimeUpdate(family,column,value,columnUpdate,valueUpdate)
```

Updates a value in the realtime configuration database. Updates the field *columnUpdate* with the value *valueUpdate* in the record matching *family*, *column*, and *value*.

```
; As in our RealTime() example, we can update the record like so:
exten => 123,1,RealTimeUpdate(sipusers,ext,5678,firstname,Richard)
; resulting SQL command:
; UPDATE sip_users SET firstname = 'Richard' WHERE ext = '5678'
```

Asterisk versions:

```
--------| 1.2 |--------| 1.4 |            |       |
```

Internal help for this application in Asterisk 1.4:

```
    -= Info about application 'RealTimeUpdate' =-

[Synopsis]
Realtime Data Rewrite

[Description]
Use the RealTime config handler system to update a value
RealTimeUpdate(<family>|<colmatch>|<value>|<newcol>|<newval>)

The column <newcol> in 'family' matching column <colmatch>=<value>
will be
updated to <newval>.  REALTIMECOUNT will be set with the number of
rows
updated or -1 if an error occurs.
```

Diff of the internal help from Asterisk 1.2 to 1.4:

```
--- in Asterisk 1.2
+++ in Asterisk 1.4
@@ -7,4 +7,6 @@
   Use the RealTime config handler system to update a value
   RealTimeUpdate(<family>|<colmatch>|<value>|<newcol>|<newval>)

-  The column <newcol> in 'family' matching column <colmatch>=<value>
will be updated to <newval>
+  The column <newcol> in 'family' matching column <colmatch>=<value>
will be
+  updated to <newval>.  REALTIMECOUNT will be set with the number of
rows
+  updated or -1 if an error occurs.
```

Diff of the internal help from Asterisk 1.4 to 1.6:

Not available in Asterisk 1.6

See also **realtime**.

B.132 ReceiveFAX()

Receives a fax.

```
ReceiveFAX(filename[,options])
```

SendFax() and **ReceiveFax()** replace the well-known fax patch commands **TxFax()** and **RxFax()** and, like them, they use the SpanDSP library.

ReceiveFax() receives an incoming fax and stores it as a TIFF file.

```
exten => 123,1,Verbose(1,### Incoming fax ${CDR(uniqueid)})
exten => 123,n,Set(LOCALSTATIONID=My Company)
exten => 123,n,ReceiveFAX(/tmp/fax-${CDR(uniqueid)}.tif)
exten => 123,n,Verbose(1,###        FAXSTATUS: ${FAXSTATUS})
exten => 123,n,Verbose(1,###         FAXERROR: ${FAXERROR})
exten => 123,n,Verbose(1,###          FAXMODE: ${FAXMODE})
exten => 123,n,Verbose(1,###         FAXPAGES: ${FAXPAGES})
exten => 123,n,Verbose(1,###       FAXBITRATE: ${FAXBITRATE})
exten => 123,n,Verbose(1,###    FAXRESOLUTION: ${FAXRESOLUTION})
exten => 123,n,Verbose(1,### REMOTESTATIONID: ${REMOTESTATIONID})
exten => 123,n,Hangup()
```

Asterisk versions:

```
  |     |        |     |        | 1.6 |--------
```

Internal help for this application in Asterisk 1.6:

```
    -= Info about application 'ReceiveFAX' =-

[Synopsis]
Receive a FAX

[Description]
  ReceiveFAX(filename[|options]):
Receives a fax from the channel into the given filename overwriting
the file if it already exists. File created will have TIFF format.
The option string may contain zero or more of the following characters:
     'c' -- makes the application behave as a calling machine
            The default behaviour is to behave as an answering machine.

This application uses following variables:
     LOCALSTATIONID to identify itself to the remote end.
     LOCALHEADERINFO to generate a header line on each page.

This application sets the following channel variables upon completion:
     FAXSTATUS       - status of operation:
                             SUCCESS | FAILED
     FAXERROR        - Error when FAILED
     FAXMODE         - Mode used:
                             audio | T38
     REMOTESTATIONID - CSID of the remote side.
     FAXPAGES        - number of pages sent.
     FAXBITRATE      - transmition rate.
     FAXRESOLUTION   - resolution.

Returns -1 in case of user hang up or any channel error.
Returns 0 on success.
```

Diff of the internal help from Asterisk 1.4 to 1.6:

Not available in Asterisk 1.4

See also `sendfax`.

B.133 Record()

Records audio from a channel to a file.
   As of Asterisk 1.2:

```
Record(basename[.format[,maxSilence[,maxDuration[,options]]]])
```

Records audio from the channel and saves it in the file basename.`format`. If the file exists, it is overwritten.
   Allowed options are as follows:

- **`format`**
  Specifies the file format of the recording (`g723`, `g729`, `gsm`, `h263`, `ulaw`, `alaw`, `wav`, ...).

- **`maxSilence`**
  Defines the maximum duration of silence allowed before the recording is ended.

- **`maxDuration`**
  Defines the maximum duration of the recording. If not provided or if 0, there is no limit.

- **`options`**
  One or more of the following option flags may be set:

  - **`s`**
    Does not record if the call has not been answered.

  - **`n`**
    Does not answer but records even if the call has not been answered.

  - **`a`**
    Appends the recording to an existing file instead of overwriting it.

  - **`t`**
    The * DTMF key ends the call instead of the default # key.

  - **`q`**
    Does not play a beep tone before recording.

  - **`x`**
    Records all DTMF tones, including # and *. The call ends when `maxDuration` is reached or the caller hangs up.

If `basename` contains %d, it is replaced by a number incremented by 1 for each new recording.

The caller may end recording by pressing the # key; if the caller hangs up before recording is complete, the recording is discarded.

Returns −1 on hangup, otherwise returns 0.

```
; Record the caller's name:
exten => 123,1,Playback(please-say-your-name)
exten => 123,n,Record(/tmp/name.gsm,3,10)
exten => 123,n,Playback(/tmp/name)
```

Note the warnings regarding privacy under **Monitor()**.

Asterisk versions:

```
--------| 1.2 |--------| 1.4 |--------| 1.6 |--------
```

Internal help for this application in Asterisk 1.4:

```
    -= Info about application 'Record' =-

[Synopsis]
Record to a file

[Description]
  Record(filename.format|silence[|maxduration][|options])

Records from the channel into a given filename. If the file exists
it will
be overwritten.
- 'format' is the format of the file type to be recorded (wav, gsm,
etc).
- 'silence' is the number of seconds of silence to allow before
returning.
- 'maxduration' is the maximum recording duration in seconds. If
missing
or 0 there is no maximum.
- 'options' may contain any of the following letters:
      'a' : append to existing recording rather than replacing
      'n' : do not answer, but record anyway if line not yet answered
      'q' : quiet (do not play a beep tone)
      's' : skip recording if the line is not yet answered
      't' : use alternate '*' terminator key (DTMF) instead of default
'#'
      'x' : ignore all terminator keys (DTMF) and keep recording until
hangup

If filename contains '%%%%d', these characters will be replaced with a
number
incremented by one each time the file is recorded. A channel variable
named RECORDED_FILE will also be set, which contains the final
filemname.
```

Use 'core show file formats' to see the available formats on your
system

User can press '#' to terminate the recording and continue to the next
priority.

If the user should hangup during a recording, all data will be lost and
the
application will teminate.

Diff of the internal help from Asterisk 1.2 to 1.4:

```
--- in Asterisk 1.2
+++ in Asterisk 1.4
@@ -17,13 +17,14 @@
         'n' : do not answer, but record anyway if line not yet answered
         'q' : quiet (do not play a beep tone)
         's' : skip recording if the line is not yet answered
-        't' : use alternate '*' terminator key instead of default '#'
+        't' : use alternate '*' terminator key (DTMF) instead of
default '#'
+        'x' : ignore all terminator keys (DTMF) and keep recording
until hangup

    If filename contains '%%%%d', these characters will be replaced with
a number
    incremented by one each time the file is recorded. A channel
variable
    named RECORDED_FILE will also be set, which contains the final
filemname.

-  Use 'show file formats' to see the available formats on your system
+  Use 'core show file formats' to see the available formats on your
system

    User can press '#' to terminate the recording and continue to the
next priority.
```

Diff of the internal help from Asterisk 1.4 to 1.6:

```
--- in Asterisk 1.4
+++ in Asterisk 1.6
@@ -4,7 +4,7 @@
    Record to a file

    [Description]
-    Record(filename.format|silence[|maxduration][|options])
+    Record(filename.format,silence[,maxduration][,options])

    Records from the channel into a given filename. If the file
exists it will be overwritten.
```

See also **dictate, monitor, mixmonitor.**

B.134 RemoveQueueMember()

Removes queue members dynamically.

```
RemoveQueueMember(queue[,interface[,options]])
```

Removes a member from the queue dynamically. If *interface* is not provided, the application removes the current interface (*i.e.*, the interface that is active in the current priority) from the specified queue.

If the interface is not in the specified queue and priority n+101 exists, the application jumps to this priority; otherwise returns an error.

Returns −1 on error, otherwise returns 0.

```
; Remove SIP/3000 from the support queue:
exten => 123,1,RemoveQueueMember(support,SIP/3000)
```

Asterisk versions:

```
--------| 1.2 |--------| 1.4 |--------| 1.6 |--------
```

Internal help for this application in Asterisk 1.4:

```
    -= Info about application 'RemoveQueueMember' =-

[Synopsis]
Dynamically removes queue members

[Description]
   RemoveQueueMember(queuename[|interface[|options]]):
Dynamically removes interface to an existing queue
If the interface is NOT in the queue and there exists an n+101 priority
then it will then jump to this priority.  Otherwise it will return an
error
The option string may contain zero or more of the following characters:
       'j' -- jump to +101 priority when appropriate.
  This application sets the following channel variable upon completion:
      RQMSTATUS      The status of the attempt to remove a queue member
as a
                     text string, one of
           REMOVED | NOTINQUEUE | NOSUCHQUEUE
Example: RemoveQueueMember(techsupport|SIP/3000)
```

Diff of the internal help from Asterisk 1.2 to 1.4:

None

Diff of the internal help from Asterisk 1.4 to 1.6:

```
--- in Asterisk 1.4
+++ in Asterisk 1.6
@@ -4,14 +4,11 @@
   Dynamically removes queue members
```

```
    [Description]
-       RemoveQueueMember(queuename[|interface[|options]]):
+       RemoveQueueMember(queuename[,interface[,options]]):
    Dynamically removes interface to an existing queue
-   If the interface is NOT in the queue and there exists an n+101
priority
-   then it will then jump to this priority.  Otherwise it will return
an error
-   The option string may contain zero or more of the following
characters:
-           'j' -- jump to +101 priority when appropriate.
+   If the interface is NOT in the queue it will return an error.
    This application sets the following channel variable upon
completion:
        RQMSTATUS       The status of the attempt to remove a queue
member as a
                        text string, one of
            REMOVED | NOTINQUEUE | NOSUCHQUEUE
-   Example: RemoveQueueMember(techsupport|SIP/3000)
+   Example: RemoveQueueMember(techsupport,SIP/3000)
```

See also **addqueuemember**, **queue**, **queues.conf**.

B.135 ResetCDR()

Resets the Call Detail Record.

```
ResetCDR([options])
```

Resets the Call Detail Record for the active call. If option w is given, the existing CDR is stored before the reset.

Returns 0.

```
; save the current CDR and then reset it:
exten => 123,1,Answer()
exten => 123,n,Playback(tt-monkeys)
exten => 123,n,ResetCDR(w)
exten => 123,n,Playback(tt-monkeys)
```

Asterisk versions:

```
--------| 1.2 |--------| 1.4 |--------| 1.6 |--------
```

Internal help for this application in Asterisk 1.4:

```
    -= Info about application 'ResetCDR' =-

[Synopsis]
Resets the Call Data Record
```

```
[Description]
  ResetCDR([options]):  This application causes the Call Data Record
to be reset.
  Options:
    w -- Store the current CDR record before resetting it.
    a -- Store any stacked records.
    v -- Save CDR variables.
```

Diff of the internal help from Asterisk 1.2 to 1.4:

None

Diff of the internal help from Asterisk 1.4 to 1.6:

```
--- in Asterisk 1.4
+++ in Asterisk 1.6
@@ -10,3 +10,4 @@
       w -- Store the current CDR record before resetting it.
       a -- Store any stacked records.
       v -- Save CDR variables.
+      e -- Enable CDR only (negate effects of NoCDR).
```

See also `forkcdr`, `nocdr`.

B.136 ResponseTimeout()

Sets the maximum time to wait for a response on a channel.

```
ResponseTimeout(seconds)
```

Warning: `ResponseTimeout()` is removed as of Asterisk 1.4 and has been replaced by the **`TIMEOUT()`** function:

```
Set(TIMEOUT(response)=seconds)
```

Asterisk versions:

```
--------| 1.2 |         |      |        |     |
```

Internal help for this application in Asterisk 1.2:

```
    -= Info about application 'ResponseTimeout' =-

[Synopsis]
Set maximum timeout awaiting response

[Description]
  ResponseTimeout(seconds): This will set the maximum amount of time
```

permitted to wait for an extension to dialed (see the WaitExten
application), before the timeout occurs. If this timeout is reached,
dialplan execution will continue at the 't' extension, if it exists.
   ResponseTimeout has been deprecated in favor of
Set(TIMEOUT(response)=timeout)

Diff of the internal help from Asterisk 1.2 to 1.4:

Not available in Asterisk 1.4

See also absolutetimeout, digittimeout, and in Appendix C, see timeout.

B.137 RetryDial()

Attempts to dial and retries if the attempt fails.

```
RetryDial(announcement,sleep,retries,technology/resource[&technology2/
resource2...][,timeout[,options[,URL]]])
```

Attempts, like **Dial()**, to call a device. Plays the file *announcement* (without file
extension!) if no device can be reached, then waits *sleep* seconds before trying
again. The default sleep interval is 10 seconds. After *retries* retries, the call is
handed to the next priority in the dialplan. If *retries* is set to 0 or –1, the application retries indefinitely.

A single-digit extension may be dialed by the caller during the sleep interval. If
this extension exists in the context defined by ${EXITCONTEXT} or in the current
context, the call is passed to that extension.

All arguments following retries are passed directly to **Dial()**.

```
; Try to dial the number three times, repeat after 5 seconds:
exten => 123,1,RetryDial(trying-to-connect-
you,5,3,IAX2/VOIP/18005554148,30)
; If the caller presses 4 while waiting, try the call on Zap/4:
exten => 0,1,RetryDial(trying-to-connect-you,5,3,Zap/4/18005554148,30)
```

Asterisk versions:

```
--------| 1.2 |--------| 1.4 |--------| 1.6 |--------
```

Internal help for this application in Asterisk 1.4:

```
   -= Info about application 'RetryDial' =-

[Synopsis]
Place a call, retrying on failure allowing optional exit extension.

[Description]
   RetryDial(announce|sleep|retries|dialargs): This application will
attempt to
place a call using the normal Dial application. If no channel can be
```

reached,
the 'announce' file will be played. Then, it will wait 'sleep' number
of
seconds before retrying the call. After 'retries' number of attempts,
the
calling channel will continue at the next priority in the dialplan.
If the
'retries' setting is set to 0, this application will retry endlessly.
  While waiting to retry a call, a 1 digit extension may be dialed.
If that
extension exists in either the context defined in ${EXITCONTEXT} or
the current
one, The call will jump to that extension immediately.
  The 'dialargs' are specified in the same format that arguments are
provided
to the Dial application.

Diff of the internal help from Asterisk 1.2 to 1.4:

```
--- in Asterisk 1.2
+++ in Asterisk 1.4
@@ -7,7 +7,7 @@
     RetryDial(announce|sleep|retries|dialargs): This application
will attempt to
   place a call using the normal Dial application. If no channel can
be reached,
   the 'announce' file will be played. Then, it will wait 'sleep'
number of
-  seconds before retying the call. After 'retires' number of attempts,
the
+  seconds before retrying the call. After 'retries' number of
attempts, the
   calling channel will continue at the next priority in the dialplan.
If the
   'retries' setting is set to 0, this application will retry
endlessly.
     While waiting to retry a call, a 1 digit extension may be dialed.
If that
```

Diff of the internal help from Asterisk 1.4 to 1.6:

```
--- in Asterisk 1.4
+++ in Asterisk 1.6
@@ -4,7 +4,7 @@
   Place a call, retrying on failure allowing optional exit extension.

   [Description]
-    RetryDial(announce|sleep|retries|dialargs): This application will
attempt to
+    RetryDial(announce,sleep,retries,dialargs): This application will
attempt to
   place a call using the normal Dial application. If no channel can be
reached,
```

```
    the 'announce' file will be played. Then, it will wait 'sleep'
number of
    seconds before retrying the call. After 'retries' number of
attempts, the
```

See also **dial**.

B.138 Return()

Returns from a subroutine.

```
Return()
```

Returns from a subroutine called by **Gosub()** or **GosubIf()** to the priority immediately following the **Gosub()** or **GosubIf()** that called the subroutine.

```
exten => 123,1,Playback(tt-monkeys)
exten => 123,n,Gosub(my-subroutine,s,1)
exten => 123,n,Playback(tt-monkeys)
exten => 123,n,Hangup()

[my-subroutine]
exten => s,1,Playback(tt-weasels)
exten => s,n,Return()
```

Asterisk versions:

```
--------| 1.2 |--------| 1.4 |--------| 1.6 |--------
```

Internal help for this application in Asterisk 1.4:

```
    -= Info about application 'Return' =-

[Synopsis]
Return from gosub routine

[Description]
Return()
    Jumps to the last label on the stack, removing it.
```

Diff of the internal help from Asterisk 1.2 to 1.4:

None

Diff of the internal help from Asterisk 1.4 to 1.6:

```
--- in Asterisk 1.4
+++ in Asterisk 1.6
@@ -4,5 +4,6 @@
    Return from gosub routine
```

```
      [Description]
 -    Return()
 -      Jumps to the last label on the stack, removing it.
 +      Return([return-value]):
 +   Jumps to the last label on the stack, removing it. The return
 value, if
 +   any, is saved in the channel variable GOSUB_RETVAL.
```

See also **gosub, gosubif.**

B.139 Ringing()

Indicates ringing to the caller.

```
 Ringing()
```

Indicates ringing to the caller. What form this indication takes depends on the channel driver. Note that this does not necessarily result in ringing tones; if you absolutely need those, use **Playtones()**.

Returns 0.

```
 ; Fake ringing:
 exten => 123,1,Ringing()
 exten => 123,n,Wait(5)
 exten => 123,n,Playback(tt-somethingwrong)
```

Asterisk versions:

```
--------| 1.2 |--------| 1.4 |--------| 1.6 |--------
```

Internal help for this application in Asterisk 1.4:

```
     -= Info about application 'Ringing' =-

 [Synopsis]
 Indicate ringing tone

 [Description]
   Ringing(): This application will request that the channel indicate
 a ringing
 tone to the user.
```

Diff of the internal help from Asterisk 1.2 to 1.4:

None

Diff of the internal help from Asterisk 1.4 to 1.6:

None

See also **busy, congestion, progress, ringing, playtones.**

B.140 SayAlpha()

Spells out a string to the caller.

```
SayAlpha(string)
```

Spells out the specified string to the caller according to the language setting for that channel. The language may be set with the Asterisk function LANGUAGE().

```
exten => 123,1,SayAlpha(ABC123)
```

Asterisk versions:

```
--------| 1.2 |--------| 1.4 |--------| 1.6 |--------
```

Internal help for this application in Asterisk 1.4:

```
    -= Info about application 'SayAlpha' =-

[Synopsis]
Say Alpha

[Description]
   SayAlpha(string): This application will play the sounds that
correspond to
the letters of the given string.
```

Diff of the internal help from Asterisk 1.2 to 1.4:

None

Diff of the internal help from Asterisk 1.4 to 1.6:

None

See also saydigits, saynumber, sayphonetic.

B.141 SayDigits()

Says a sequence of digits to the caller.

```
SayDigits(digits)
```

Says the provided sequence of digits to the caller according to the language settings for the channel. The language may be set with the Asterisk function LANGUAGE().

```
exten => 123,1,SayDigits(1234)
```

Asterisk versions:

```
--------| 1.2 |--------| 1.4 |--------| 1.6 |--------
```

Internal help for this application in Asterisk 1.4:

```
      -= Info about application 'SayDigits' =-

   [Synopsis]
   Say Digits

   [Description]
     SayDigits(digits): This application will play the sounds that
   correspond
   to the digits of the given number. This will use the language that is
   currently
   set for the channel. See the LANGUAGE function for more information on
   setting
   the language for the channel.
```

Diff of the internal help from Asterisk 1.2 to 1.4:

None

Diff of the internal help from Asterisk 1.4 to 1.6:

None

See also sayalpha, saynumber, sayphonetic.

B.142 SayNumber()

Says a number.

```
   SayNumber(number[,gender])
```

Says the provided number according to the language settings for the channel. The language may be set with the Asterisk function **LANGUAGE()**.

Whole numbers from one to 99,999,999 are supported in the following languages:

- **da**

  Danish
- **de**

  German
- **en**

  English

- **es**

  Spanish

- **fr**

  French

- **it**

  Italian

- **nl**

  Dutch

- **no**

  Norwegian

- **pl**

  Polish

- **pt**

  Portuguese

- **se**

  Swedish

- **tw**

  Mandarin (Taiwanese)

The gender is optional and depends on the language.

For continental languages such as German, French, Spanish and Portuguese, use f for feminine, m for masculine, and n for neuter.

For Scandinavian languages such as Danish, Swedish and Norwegian, use c for common (en, enn) and n for neuter (et, ett).

To count in German plural, use p.

For these other languages to work, their respective sound files must be present /var/lib/asterisk/sounds/digits (in subdirectories by language; e.g., de/)

```
; Say in English:
exten => 123,1,Set(LANGUAGE=en)
exten => 123,n,SayNumber(1234)
; "one - thousand - two - hundred - and - thirty - four"

; Say in Norwegian:
exten => 123,1,Set(LANGUAGE=no)
exten => 123,n,SayNumber(1234)
; "tusen - to - hundre - tretti - fire"
```

Asterisk versions:

```
--------| 1.2 |--------| 1.4 |--------| 1.6 |--------
```

Internal help for this application in Asterisk 1.4:

```
    -= Info about application 'SayNumber' =-

[Synopsis]
Say Number

[Description]
  SayNumber(digits[,gender]): This application will play the sounds
  that
correspond to the given number. Optionally, a gender may be specified.
This will use the language that is currently set for the channel. See
the
LANGUAGE function for more information on setting the language for the
channel.
```

Diff of the internal help from Asterisk 1.2 to 1.4:

None

Diff of the internal help from Asterisk 1.4 to 1.6:

None

See also `sayalpha`, `saydigits`, `sayphonetic`.

B.143 SayPhonetic()

Spells a string using the NATO phonetic alphabet.

```
    SayPhonetic(string)
```

Spells out the provided string using the NATO phonetic alphabet or its linguistic variations. The language may be set with the Asterisk function LANGUAGE(). Umlauts and other special characters are not yet supported.

```
exten => 123,1,Set(LANGUAGE=en)
exten => 123,n,SayPhonetic(asterisk)
; Alpha Sierra Tango Echo Romeo India Sierra Kilo

exten => 123,n,Set(LANGUAGE=de)
exten => 123,n,SayPhonetic(asterisk)
; Anton Samuel Theodor Emil Richard Ida Samuel Kaufmann
```

Asterisk versions:

```
--------| 1.2 |--------| 1.4 |--------| 1.6 |--------
```

Internal help for this application in Asterisk 1.4:

```
    -= Info about application 'SayPhonetic' =-

[Synopsis]
Say Phonetic

[Description]
   SayPhonetic(string): This application will play the sounds from the
phonetic
alphabet that correspond to the letters in the given string.
```

Diff of the internal help from Asterisk 1.2 to 1.4:

None

Diff of the internal help from Asterisk 1.4 to 1.6:

None

See also sayalpha, saydigits, saynumber.

B.144 SayUnixTime()

Announce the time in a custom format.

```
SayUnixTime([unixtime][,timezone[,format]])
```

Announces the current time according to the specified time zone and format.
Allowed options are as follows:

- **unixtime**

  The time in UNIX time format; that is, the number of seconds that have
  passed since the start of the epoch, which began at 0:00 UTC, January 1, 1970.
  Accepts negative values. The default value is the current time.

- **timezone**

  The time zone. A list may be found in /usr/share/zoneinfo. The default is
  the system time zone.

- **format**

  The time format of the announcement. A list of allowed formats may be found
  in voicemail.conf from the default installation. The default time format is
  ABdY 'digits/at' IMp.

Returns 0 or −1 if the channel is hung up.

```
exten => 123,1,SayUnixTime(,,IMp)
```

Asterisk versions:

```
--------| 1.2 |--------| 1.4 |--------| 1.6 |--------
```

Internal help for this application in Asterisk 1.4:

```
      -= Info about application 'SayUnixTime' =-

[Synopsis]
Says a specified time in a custom format

[Description]
SayUnixTime([unixtime][|[timezone][|format]])
   unixtime: time, in seconds since Jan 1, 1970.  May be negative.
             defaults to now.
   timezone: timezone, see /usr/share/zoneinfo for a list.
             defaults to machine default.
   format:   a format the time is to be said in.  See voicemail.conf.
             defaults to "ABdY 'digits/at' IMp"
```

Diff of the internal help from Asterisk 1.2 to 1.4:

None

Diff of the internal help from Asterisk 1.4 to 1.6:

```
--- in Asterisk 1.4
+++ in Asterisk 1.6
@@ -4,10 +4,10 @@
   Says a specified time in a custom format

   [Description]
-  SayUnixTime([unixtime][|[timezone][|format]])
-    unixtime: time, in seconds since Jan 1, 1970.  May be negative.
+  SayUnixTime([unixtime][,[timezone][,format]])
+    unixtime  - time, in seconds since Jan 1, 1970.  May be negative.
              defaults to now.
-    timezone: timezone, see /usr/share/zoneinfo for a list.
+    timezone  - timezone, see /usr/share/zoneinfo for a list.
              defaults to machine default.
-    format:   a format the time is to be said in.  See voicemail.conf.
+    format    - a format the time is to be said in.  See
voicemail.conf.
              defaults to "ABdY 'digits/at' IMp"
```

B.145 SendDTMF()

Send DTMF digits on the channel.

```
SendDTMF(digits[,timeout_ms])
```

Sends the specified DTMF sequence on the channel. Permitted symbols are *0-9*, ***, *#*, and *A-D*. Additionally *w* is allowed; it indicates a pause of 500ms. The argument *timeout_ms* sets the pause in milliseconds between tones. If not specified, defaults to 250ms.

Returns 0, or –1 if the channel is hung up.

```
exten => 123,1,SendDTMF(123w456w789,200)
```

Asterisk versions:

```
--------| 1.2 |--------| 1.4 |--------| 1.6 |--------
```

Internal help for this application in Asterisk 1.4:

```
    -= Info about application 'SendDTMF' =-

[Synopsis]
Sends arbitrary DTMF digits

[Description]
 SendDTMF(digits[|timeout_ms]): Sends DTMF digits on a channel.
 Accepted digits: 0-9, *#abcd, w (.5s pause)
 The application will either pass the assigned digits or terminate if
it
 encounters an error.
```

Diff of the internal help from Asterisk 1.2 to 1.4:

None

Diff of the internal help from Asterisk 1.4 to 1.6:

```
--- in Asterisk 1.4
+++ in Asterisk 1.6
@@ -4,7 +4,10 @@
    Sends arbitrary DTMF digits

    [Description]
-    SendDTMF(digits[|timeout_ms]): Sends DTMF digits on a channel.
-    Accepted digits: 0-9, *#abcd, w (.5s pause)
+    SendDTMF(digits[,[timeout_ms][,duration_ms]]): Sends DTMF digits
on a channel.
+    Accepted digits: 0-9, *#abcd, (default .25s pause between digits)
    The application will either pass the assigned digits or terminate
if it
    encounters an error.
```

```
+    Optional Params:
+       timeout_ms: pause between digits.
+       duration_ms: duration of each digit.
```

See also read.

B.146 SendFAX()

Sends a fax.

```
SendFAX(filename[,options])
```

Sends a TIFF file as a fax.

SendFax() and **ReceiveFax()** replace the well-known fax patch commands **TxFax()** and **RxFax()** and, like them, they use the SpanDSP library.

```
exten => 123,1,Set(LOCALSTATIONID=My Company)
exten => 123,n,Set(LOCALHEADERINFO=My Company - There when you need us)
exten => 123,n,SendFAX(/tmp/offer.tif)
exten => 123,n,Verbose(1,###       FAXSTATUS: ${FAXSTATUS})
exten => 123,n,Verbose(1,###        FAXERROR: ${FAXERROR})
exten => 123,n,Verbose(1,###         FAXMODE: ${FAXMODE})
exten => 123,n,Verbose(1,###        FAXPAGES: ${FAXPAGES})
exten => 123,n,Verbose(1,###      FAXBITRATE: ${FAXBITRATE})
exten => 123,n,Verbose(1,###   FAXRESOLUTION: ${FAXRESOLUTION})
exten => 123,n,Verbose(1,### REMOTESTATIONID: ${REMOTESTATIONID})
exten => 123,n,Hangup()
```

Asterisk versions:

```
    |      |        |      |        | 1.6 |--------
```

Internal help for this application in Asterisk 1.6:

```
    -= Info about application 'SendFAX' =-

[Synopsis]
Send a FAX

[Description]
  SendFAX(filename[|options]):
Send a given TIFF file to the channel as a FAX.
The option string may contain zero or more of the following characters:
    'a' - makes the application behave as an answering machine
          The default behaviour is to behave as a calling machine.

This application uses following variables:
    LOCALSTATIONID to identify itself to the remote end.
    LOCALHEADERINFO to generate a header line on each page.
```

```
This application sets the following channel variables upon completion:
     FAXSTATUS         - status of operation:
                              SUCCESS | FAILED
     FAXERROR          - Error when FAILED
     FAXMODE           - Mode used:
                              audio | T38
     REMOTESTATIONID - CSID of the remote side.
     FAXPAGES          - number of pages sent.
     FAXBITRATE        - transmition rate.
     FAXRESOLUTION     - resolution.

Returns -1 in case of user hang up or any channel error.
Returns 0 on success.
```

Diff of the internal help from Asterisk 1.4 to 1.6:

Not available in Asterisk 1.4

See also receivefax.

B.147 SendImage()

Sends an image file to the channel.

```
SendImage(filename[,options])
```

Sends an image file to the channel. If images are supported and the transmission fails, the application hangs up; otherwise it continues on the next priority. If option j is set, the application jumps to priority n+101, if it exists.

Returns 0 on successful transmission of the image or if the channel does not support images, otherwise returns –1. Sets the channel variable SENDIMAGESTATUS to OK or NOSUPPORT.

```
exten => 123,1,SendImage(logo.jpg)
```

Asterisk versions:

```
--------| 1.2 |--------| 1.4 |--------| 1.6 |--------
```

Internal help for this application in Asterisk 1.4:

```
     -= Info about application 'SendImage' =-

[Synopsis]
Send an image file

[Description]
  SendImage(filename): Sends an image on a channel.
If the channel supports image transport but the image send
fails, the channel will be hung up. Otherwise, the dialplan
continues execution.
```

```
    The option string may contain the following character:
            'j' -- jump to priority n+101 if the channel doesn't support
    image transport
    This application sets the following channel variable upon completion:
            SENDIMAGESTATUS            The status is the result of the attempt
    as a text string, one of
                    OK | NOSUPPORT
```

Diff of the internal help from Asterisk 1.2 to 1.4:

None

Diff of the internal help from Asterisk 1.4 to 1.6:

```
--- in Asterisk 1.4
+++ in Asterisk 1.6
@@ -5,11 +5,8 @@

    [Description]
      SendImage(filename): Sends an image on a channel.
-   If the channel supports image transport but the image send
-   fails, the channel will be hung up. Otherwise, the dialplan
-   continues execution.
-   The option string may contain the following character:
-           'j' -- jump to priority n+101 if the channel doesn't
    support image transport
-   This application sets the following channel variable upon
    completion:
-           SENDIMAGESTATUS            The status is the result of
    the attempt as a text string, one of
-                   OK | NOSUPPORT
+   Result of transmission will be stored in SENDIMAGESTATUS
+   channel variable:
+       SUCCESS        Transmission succeeded
+       FAILURE        Transmission failed
+       UNSUPPORTED  Image transmission not supported by channel
```

See also **sendtext, sendurl.**

B.148 SendText()

Sends text to the channel.

```
SendText(text[,option])
```

Sends text to the channel (e.g., to print on the display) if the transmission of text is
supported by the channel and the device. The channel is handed to the next priority
afterwards. If text transmission is not supported, option j makes the channel jump
to priority n+101 if it exists.

The text is 7-bit ASCII for most channels.

Returns 0 if the text is transmitted without errors, otherwise −1. Sets the channel variable SENDTEXTSTATUS to SUCCESS, FAILURE, or UNSUPPORTED.

```
exten => 123,1,SendText(Welcome to Asterisk)
```

Asterisk versions:

```
--------| 1.2 |--------| 1.4 |--------| 1.6 |--------
```

Internal help for this application in Asterisk 1.4:

```
    -= Info about application 'SendText' =-

[Synopsis]
Send a Text Message

[Description]
  SendText(text[|options]): Sends text to current channel (callee).
Result of transmission will be stored in the SENDTEXTSTATUS
channel variable:
      SUCCESS       Transmission succeeded
      FAILURE       Transmission failed
      UNSUPPORTED   Text transmission not supported by channel

At this moment, text is supposed to be 7 bit ASCII in most channels.
The option string many contain the following character:
'j' -- jump to n+101 priority if the channel doesn't support
      text transport
```

Diff of the internal help from Asterisk 1.2 to 1.4:

None

Diff of the internal help from Asterisk 1.4 to 1.6:

```
--- in Asterisk 1.4
+++ in Asterisk 1.6
@@ -4,7 +4,7 @@
   Send a Text Message

  [Description]
-    SendText(text[|options]): Sends text to current channel (callee).
+    SendText(text): Sends text to current channel (callee).
  Result of transmission will be stored in the SENDTEXTSTATUS
  channel variable:
        SUCCESS       Transmission succeeded
@@ -12,6 +12,3 @@
        UNSUPPORTED   Text transmission not supported by channel

  At this moment, text is supposed to be 7 bit ASCII in most channels.
-  The option string many contain the following character:
-  'j' -- jump to n+101 priority if the channel doesn't support
-          text transport
```

See also sendimage, sendurl.

B.149 SendURL()

Sends a URL to the channel.

```
SendURL(URL[,options])
```

Sends the channel a URL for the device to call up, and then hands the channel to the next priority.

If the channel does not support URL transmission, option j makes the channel jump to priority n+101 if it exists. Option wait makes the application wait for confirmation that the URL was loaded before continuing to the next priority.

Returns 0 if the URL is transmitted without errors, otherwise −1. Sets the channel variable SENDURLSTATUS to SUCCESS, FAILURE, UNSUPPORTED, or NOLOAD (requires option wait; the client could not load the URL successfully).

```
exten => 123,1,SendURL(http://www.mobileweather.org/index.xml,wait)
```

Asterisk versions:

```
--------| 1.2 |--------| 1.4 |--------| 1.6 |--------
```

Internal help for this application in Asterisk 1.4:

```
    -= Info about application 'SendURL' =-

[Synopsis]
Send a URL

[Description]
  SendURL(URL[|option]): Requests client go to URL (IAX2) or sends the
URL to the client (other channels).
Result is returned in the SENDURLSTATUS channel variable:
    SUCCESS        URL successfully sent to client
    FAILURE        Failed to send URL
    NOLOAD         Client failed to load URL (wait enabled)
    UNSUPPORTED    Channel does not support URL transport

If the option 'wait' is specified, execution will wait for an
acknowledgement that the URL has been loaded before continuing

If jumping is specified as an option (the 'j' flag), the client
does not
support Asterisk "html" transport, and there exists a step with
priority
n + 101, then execution will continue at that step.

SendURL continues normally if the URL was sent correctly or if the
channel
does not support HTML transport.  Otherwise, the channel is hung up.
```

Diff of the internal help from Asterisk 1.2 to 1.4:

```
--- in Asterisk 1.2
+++ in Asterisk 1.4
@@ -9,17 +9,15 @@
   Result is returned in the SENDURLSTATUS channel variable:
        SUCCESS        URL successfully sent to client
        FAILURE        Failed to send URL
-      NOLOAD         Clien failed to load URL (wait enabled)
+      NOLOAD         Client failed to load URL (wait enabled)
       UNSUPPORTED    Channel does not support URL transport

   If the option 'wait' is specified, execution will wait for an
   acknowledgement that the URL has been loaded before continuing
-  and will return -1 if the peer is unable to load the URL

-  Old behaviour (deprecated):
-   If the client does not support Asterisk "html" transport,
-   and there exists a step with priority n + 101, then execution will
-   continue at that step.
-   Otherwise, execution will continue at the next priority level.
-   SendURL only returns 0 if the URL was sent correctly  or if
-   the channel does not support HTML transport, and -1 otherwise.
+  If jumping is specified as an option (the 'j' flag), the client
does not
+  support Asterisk "html" transport, and there exists a step with
priority
+  n + 101, then execution will continue at that step.
+
+  SendURL continues normally if the URL was sent correctly or if
the channel
+  does not support HTML transport.  Otherwise, the channel is hung up.
```

Diff of the internal help from Asterisk 1.4 to 1.6:

```
--- in Asterisk 1.4
+++ in Asterisk 1.6
@@ -4,7 +4,7 @@
   Send a URL.

   [Description]
-    SendURL(URL[|option]): Requests client go to URL (IAX2) or
sends the
+    SendURL(URL[,option]): Requests client go to URL (IAX2) or
sends the
   URL to the client (other channels).
   Result is returned in the SENDURLSTATUS channel variable:
        SUCCESS        URL successfully sent to client
@@ -12,12 +12,8 @@
        NOLOAD         Client failed to load URL (wait enabled)
        UNSUPPORTED    Channel does not support URL transport

-  If the option 'wait' is specified, execution will wait for an
```

```
+   If the option 'w' is specified, execution will wait for an
    acknowledgement that the URL has been loaded before continuing

-   If jumping is specified as an option (the 'j' flag), the client
does not
-   support Asterisk "html" transport, and there exists a step with
priority
-   n + 101, then execution will continue at that step.
-

    SendURL continues normally if the URL was sent correctly or if
the channel
        does not support HTML transport.  Otherwise, the channel is hung up.
```

See also **sendimage, sendtext.**

B.150 Set()

Sets a variable to the specified value.

```
Set(variable=value)
```

Sets the variable to the specified value. If the variable name starts with _, single inheritance is set (i.e., the variable is inherited by any channels opened from this channel); if it begins with __, unlimited inheritance is set (i.e., all children of this channel, regardless of generation, inherit the variable). Up to 24 variables may be set. Variables are valid only in the channel and are canceled when the channel is hung up. Option g sets a global variable (not a function! see note below) in Asterisk 1.2; in 1.4 this is accomplished with the Asterisk function **GLOBAL()**.

```
; Set a variable TEST to "123":
exten => 123,1,Set(TEST=123)
exten => 123,n,SayDigits(${TEST})

; Set a global variable TEST2 to "456":
exten => 123,n,Set(TEST2=456,g)          ; Asterisk 1.2
exten => 123,n,Set(GLOBAL(TEST2)=456)    ; Asterisk 1.4
```

Note: Whether global variables are cleared upon **reload** depends on the setting of clearglobalvars in extensions.conf.

Set() is also used to write to functions (see Appendix C).

```
exten => 123,1,Set(CALLERID(name)=Widgets)  ; set CALLERID(name)
exten => 123,n,Set(CALLERID(name)=)         ; clear CALLERID(name)

exten => 123,n,Set(DB(my/test)=ok)          ; write a value to the
AstDB
exten => 123,n,Set(var=${DB(my/test)})      ; read a value from the
AstDB
```

Note: The option g in Asterisk 1.2 applies *only* to variables; the behavior of a function depends on the function itself.

Asterisk versions:

```
--------| 1.2 |--------| 1.4 |--------| 1.6 |--------
```

Internal help for this application in Asterisk 1.4:

```
    -= Info about application 'Set' =-

[Synopsis]
Set channel variable(s) or function value(s)

[Description]
  Set(name1=value1|name2=value2|..[|options])
This function can be used to set the value of channel variables
or dialplan
functions. It will accept up to 24 name/value pairs. When setting
variables,
if the variable name is prefixed with _, the variable will be
inherited into
channels created from the current channel. If the variable name
is prefixed
with __, the variable will be inherited into channels created from
the current
channel and all children channels.
  Options:
    g - Set variable globally instead of on the channel
        (applies only to variables, not functions)


The use of Set to set multiple variables at once and the g flag
have both
been deprecated.  Please use multiple Set calls and the GLOBAL()
dialplan
function instead.
```

Diff of the internal help from Asterisk 1.2 to 1.4:

```
--- in Asterisk 1.2
+++ in Asterisk 1.4
@@ -14,3 +14,8 @@
     Options:
       g - Set variable globally instead of on the channel
           (applies only to variables, not functions)
+
+
+  The use of Set to set multiple variables at once and the g flag
have both
```

```
+   been deprecated.   Please use multiple Set calls and the GLOBAL()
dialplan
+   function instead.
```

Diff of the internal help from Asterisk 1.4 to 1.6:

```
--- in Asterisk 1.4
+++ in Asterisk 1.6
@@ -1,21 +1,23 @@
    -= Info about application 'Set' =-

  [Synopsis]
-  Set channel variable(s) or function value(s)
+  Set channel variable or function value

  [Description]
-    Set(name1=value1|name2=value2|..[|options])
+    Set(name=value)
  This function can be used to set the value of channel variables
or dialplan
-  functions. It will accept up to 24 name/value pairs. When setting
variables,
-  if the variable name is prefixed with _, the variable will be
inherited into
-  channels created from the current channel. If the variable name
is prefixed
-  with __, the variable will be inherited into channels created
from the current
-  channel and all children channels.
-    Options:
-      g - Set variable globally instead of on the channel
-          (applies only to variables, not functions)
-
-
-  The use of Set to set multiple variables at once and the g flag
have both
-  been deprecated.   Please use multiple Set calls and the GLOBAL()
dialplan
-  function instead.
+  functions. When setting variables, if the variable name is prefixed
with _,
+  the variable will be inherited into channels created from the
current
+  channel. If the variable name is prefixed with __, the variable
will be
+  inherited into channels created from the current channel and all
children
+  channels.
+  Compatibility note: If (and only if), in
/etc/asterisk/asterisk.conf, you have a [compat]
+  category, and you have app_set = 1.6 under that, then the behavior
of this
```

+   app changes, and does not strip surrounding quotes from the right hand side
+   as it did previously in 1.4. The app_set = 1.6 is only inserted if 'make samples'
+   is executed, or if users insert this by hand into the asterisk.conf file.
+   /nThe advantages of not stripping out quoting, and not caring about the
+   separator characters (comma and vertical bar) were sufficient to make these
+   changes in 1.6. Confusion about how many backslashes would be needed to properly
+   protect separators and quotes in various database access strings has been greatly
+   reduced by these changes.

See also importvar, doc/README.variables (1.2)/doc/ channelvariables.txt (1.4), and in Appendix C, see global.

B.151 SetAccount()

Sets the account code field in the CDR for billing purposes.

```
SetAccount(account)
```

Warning: SetAccount() is removed as of Asterisk 1.4 and has been replaced by the **CDR()** function:

```
Set(CDR(accountcode)=account)
```

Asterisk versions:

--------| 1.2 | | | | |

Internal help for this application in Asterisk 1.2:

```
      -= Info about application 'SetAccount' =-

[Synopsis]
Set the CDR Account Code

[Description]
  SetAccount([account]): This application will set the channel
account code for
billing purposes.
  SetAccount has been deprecated in favor of the
Set(CDR(accountcode)=account).
```

Diff of the internal help from Asterisk 1.2 to 1.4:

Not available in Asterisk 1.4

See also Appendix C.

B.152 SetAMAFlags()

Sets AMA-Flags in the Call Detail Record (CDR).

```
SetAMAFlags(flags)
```

Sets AMA (Automatic Message Accounting)[14] flags in the CDR. Overrides any settings found in channel configuration files. Valid settings are default, omit, billing, and documentation.
   Returns 0.

```
exten => 123,1,SetAMAFlags(billing)
```

Asterisk versions:

--------| 1.2 |--------| 1.4 |--------| 1.6 |--------

Internal help for this application in Asterisk 1.4:

```
    -= Info about application 'SetAMAFlags' =-

[Synopsis]
Set the AMA Flags

[Description]
  SetAMAFlags([flag]): This application will set the channel's
AMA Flags for
  billing purposes.
```

Diff of the internal help from Asterisk 1.2 to 1.4:

```
--- in Asterisk 1.2
+++ in Asterisk 1.4
@@ -4,5 +4,5 @@
   Set the AMA Flags
```

---

14. In North America, Automatic Message Accounting has been used for billing in Direct Distance Dialing since it was introduced in the mid-twentieth century. The system was originally used to provide the toll system with information about the originating circuit so that the subscriber could be billed for a long-distance call.

```
    [Description]
-      SetAMAFlags([flag]): This channel will set the channel's AMA
Flags for billing
-    purposes.
+      SetAMAFlags([flag]): This application will set the channel's
AMA Flags for
+      billing purposes.
```

Diff of the internal help from Asterisk 1.4 to 1.6:

None

B.153 SetCallerID()

Sets the caller ID.

```
SetCallerID(caller_id[,options])
```

Warning: SetCIDName() has been replaced with the **CALLERID()** function:

```
Set(CALLERID(all)=Name <number>)
```

Asterisk versions:

```
--------| 1.2 |--------| 1.4 |              |      |
```

Internal help for this application in Asterisk 1.4:

```
    -= Info about application 'SetCallerID' =-

[Synopsis]
Set CallerID

[Description]
  SetCallerID(clid[|a])! Set Caller-ID on a call to a new
value.  Sets ANI as well if a flag is used.
```

Diff of the internal help from Asterisk 1.2 to 1.4:

None

Diff of the internal help from Asterisk 1.4 to 1.6:

Not available in Asterisk 1.6

See also Appendix C.

B.154 SetCallerPres()

Sets caller ID presentation flags.

```
SetCallerPres(presentation)
```

Sets caller ID presentation flags for Q.931 PRI (Primary Rate Interface) connections. Similar to **CallerPres()**, but using words instead of a bitmask.

Valid presentations are as follows:

- **allowed_not_screened**
  Presentation allowed, not screened
- **allowed_passed_screen**
  Presentation allowed, passed screen
- **allowed_passed_screen**
  Presentation allowed, failed screen
- **allowed**
  Presentation allowed, network number
- **prohib_not_screened**
  Presentation prohibited, not screened
- **prohib_passed_screen**
  Presentation prohibited, passed screen
- **prohib_failed_screen**
  Presentation prohibited, failed screen
- **prohib**
  Presentation prohibited, network number
- **unavailable**
  Number unavailable

Returns 0.

```
exten => 123,1,SetCallerPres(allowed_not_screened)
exten => 123,n,Dial(Zap/4/18775558190)
```

You may need to set usecallingpres=yes in zapata.conf for this to work.

Asterisk versions:

```
--------| 1.2 |--------| 1.4 |--------| 1.6 |--------
```

Internal help for this application in Asterisk 1.4:

```
    -= Info about application 'SetCallerPres' =-

[Synopsis]
Set CallerID Presentation

[Description]
  SetCallerPres(presentation): Set Caller*ID presentation on a call.
  Valid presentations are:

        allowed_not_screened    : Presentation Allowed, Not Screened
        allowed_passed_screen   : Presentation Allowed, Passed Screen
        allowed_failed_screen   : Presentation Allowed, Failed Screen
        allowed                 : Presentation Allowed, Network Number
        prohib_not_screened     : Presentation Prohibited, Not Screened
        prohib_passed_screen    : Presentation Prohibited, Passed Screen
        prohib_failed_screen    : Presentation Prohibited, Failed Screen
        prohib                  : Presentation Prohibited, Network Number
        unavailable             : Number Unavailable
```

Diff of the internal help from Asterisk 1.2 to 1.4:

None

Diff of the internal help from Asterisk 1.4 to 1.6:

None

B.155 SetCDRUserField()

Sets the value of the "user" in the CDR.

```
    SetCDRUserField(string)
```

Sets the value of the "user" in the CDR. This field allows nonstandard information to be recorded in the CDR.

Asterisk versions:

```
--------| 1.2 |--------| 1.4 |             |       |
```

Internal help for this application in Asterisk 1.4:

```
    -= Info about application 'SetCDRUserField' =-

[Synopsis]
Set the CDR user field
```

```
[Description]
[Synopsis]
SetCDRUserField(value)

[Description]
SetCDRUserField(value): Set the CDR 'user field' to value
        The Call Data Record (CDR) user field is an extra field you
        can use for data not stored anywhere else in the record.
        CDR records can be used for billing or storing other arbitrary
data
        (I.E. telephone survey responses)
        Also see AppendCDRUserField().

This application is deprecated in favor of Set(CDR(userfield)=...)
```

Diff of the internal help from Asterisk 1.2 to 1.4:

```
--- in Asterisk 1.2
+++ in Asterisk 1.4
@@ -14,3 +14,5 @@
        CDR records can be used for billing or storing other
arbitrary data
        (I.E. telephone survey responses)
        Also see AppendCDRUserField().
+
+    This application is deprecated in favor of Set(CDR(userfield)=...)
```

Diff of the internal help from Asterisk 1.4 to 1.6:

Not available in Asterisk 1.6

Warning: Although it is not noted in the internal help, this application is deprecated. See the source code:

```
ast_log(LOG_WARNING, "SetCDRUserField is deprecated.
  Please use CDR(userfield) instead.\n");
```

See also Appendix C.

B.156 SetCIDName()

Sets the caller-id NAME.

```
SetCIDName(name)
```

Warning: `SetCIDName()` is no longer available as of Asterisk 1.4 and has been replaced by the **CALLERID()** function:

```
Set(CALLERID(name)=Bob)
```

Asterisk versions:

```
--------| 1.2 |        |      |       |      |
```

Internal help for this application in Asterisk 1.2:

```
    -= Info about application 'SetCIDName' =-

[Synopsis]
Set CallerID Name

[Description]
   SetCIDName(cname[|a]): Set Caller*ID Name on a call to a new
value, while preserving the original Caller*ID number.  This is
useful for providing additional information to the called
party.
SetCIDName has been deprecated in favor of the function
CALLERID(name)
```

Diff of the internal help from Asterisk 1.2 to 1.4:

Not available in Asterisk 1.4

See also `setcidnum`, and in Appendix C, see `callerid`.

B.157 SetCIDNum()

Sets the caller ID number.

```
SetCIDNum(number, options)
```

Warning: `SetCIDNum()` is no longer available as of Asterisk 1.4 and was replaced by the **CALLERID()** function:

```
Set(CALLERID(num)=4035551212)
Set(CALLERID(ani)=8005551212)
```

Asterisk versions:

```
--------| 1.2 |           |     |        |     |
```

Internal help for this application in Asterisk 1.2:

```
    -= Info about application 'SetCIDNum' =-

[Synopsis]
Set CallerID Number

[Description]
  SetCIDNum(cnum[|a]): Set Caller*ID Number on a call to a new
value, while preserving the original Caller*ID name.  This is
useful for providing additional information to the called
party. Sets ANI as well if a flag is used.
SetCIDNum has been deprecated in favor of the function
CALLERID(number)
```

Diff of the internal help from Asterisk 1.2 to 1.4:

Not available in Asterisk 1.4

See also `setcidname`, and in Appendix C, see `callerid`.

B.158 SetGlobalVar()

Sets the value of a global variable.

```
SetGlobalVar(variable=value)
```

Sets the value of a global variable. If the variable does not exist, it is created.

Warning: `SetGlobalVar()` is removed as of Asterisk 1.6 and should not be used in 1.4 either. A global variable is set through the application `Set()` (set) via the **GLOBAL()** function:

```
Set(GLOBAL(variable)=value)
```

Use **Set()** with the option g in Asterisk 1.2 and **the GLOBAL()** function for Asterisk 1.4 and up.

Asterisk versions:

```
--------| 1.2 |--------| 1.4 |          |     |
```

Internal help for this application in Asterisk 1.4:

```
    -= Info about application 'SetGlobalVar' =-

[Synopsis]
Set a global variable to a given value

[Description]
  SetGlobalVar(variable=value): This application sets a given
global variable to
the specified value.


This application is deprecated in favor of Set(GLOBAL(var)=value)
```

Diff of the internal help from Asterisk 1.2 to 1.4:

```
--- in Asterisk 1.2
+++ in Asterisk 1.4
@@ -6,3 +6,6 @@
   [Description]
     SetGlobalVar(variable=value): This application sets a given
global variable to
   the specified value.
+
+
+  This application is deprecated in favor of Set(GLOBAL(var)=value)
```

Diff of the internal help from Asterisk 1.4 to 1.6:

Not available in Asterisk 1.6

Warning: Although it is not noted in the internal help, this application is deprecated. See the source code:

```
ast_log(LOG_WARNING, "SetGlobalVar is deprecated.
  Please use Set(GLOBAL(%s)=%s) instead.\n", name, stringp);
```

See also set, and in Appendix C, see global.

B.159 SetGroup()

Sets the group for the active channel.

```
SetGroup(group_name)
```

Warning: `SetGroup()` is removed as of Asterisk 1.4 and has been replaced by the **GROUP()** function.

Asterisk versions:

```
--------| 1.2 |         |     |        |     |
```

Internal help for this application in Asterisk 1.2:

```
    -= Info about application 'SetGroup' =-

[Synopsis]
Set the channel's group

[Description]
Usage: SetGroup(groupname[@category])
   Sets the channel group to the specified value.  Equivalent to
Set(GROUP=group).  Always returns 0.
```

Diff of the internal help from Asterisk 1.2 to 1.4:

Not available in Asterisk 1.4

See also checkgroup, getgroupcount, getgroupmatchcount, and in Appendix C, see group.

B.160 SetLanguage()

Sets the language of the active channel.

```
SetLanguage(language)
```

Warning: `SetLanguage()` is removed as of Asterisk 1.4 and has been replaced by the **CHANNEL()** function:

```
Set(CHANNEL(language)=de)
```

Asterisk versions:

```
--------| 1.2 |              |       |          |       |
```

Internal help for this application in Asterisk 1.2:

```
    -= Info about application 'SetLanguage' =-

[Synopsis]
Set the channel's preferred language

[Description]
  SetLanguage(language): This will set the channel language to the
given value.
This information is used for the syntax in generation of numbers,
and to choose
a sound file in the given language, when it is available.
  For example, if language is set to 'fr' and the file 'demo-congrats'
is
requested to be played, if the file 'fr/demo-congrats' exists, then
it will play that file. If not, it will play the normal
'demo-congrats'.
For some language codes, SetLanguage also changes the syntax of some
Asterisk functions, like SayNumber.
  SetLanguage has been deprecated in favor of Set(LANGUAGE()=language)
```

Diff of the internal help from Asterisk 1.2 to 1.4:

Not available in Asterisk 1.4

See also Appendix C.

B.161 SetMusicOnHold()

Sets the default class for Music On Hold (MOH).

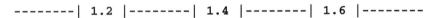

Sets the default class for Music On Hold (MOH).

Asterisk versions:

```
--------| 1.2 |--------| 1.4 |--------| 1.6 |--------
```

Internal help for this application in Asterisk 1.4:

```
    -= Info about application 'SetMusicOnHold' =-

[Synopsis]
Set default Music On Hold class
```

```
[Description]
SetMusicOnHold(class): Sets the default class for music on hold for
a given channel.  When
music on hold is activated, this class will be used to select which
music is played.
```

Diff of the internal help from Asterisk 1.2 to 1.4:

None

Diff of the internal help from Asterisk 1.4 to 1.6:

```
--- in Asterisk 1.4
+++ in Asterisk 1.6
@@ -4,6 +4,12 @@
   Set default Music On Hold class

   [Description]
-  SetMusicOnHold(class): Sets the default class for music on hold
for a given channel.  When
+    SetMusicOnHold(class):
+
+      !!! DEPRECATED. USe Set(CHANNEL(musicclass)=...) instead !!!
+
+  Sets the default class for music on hold for a given channel.  When
    music on hold is activated, this class will be used to select which
    music is played.
+
+      !!! DEPRECATED. USe Set(CHANNEL(musicclass)=...) instead !!!
```

Warning: Superseded by the Asterisk function CHANNEL(musicclass) in 1.4
and above.

See also Appendix C.

B.162 SetRDNIS()

Sets the RDNIS number on the active call.

```
SetRDNIS(number)
```

Warning: SetRDNIS() is removed as of Asterisk 1.4 and has been replaced by
the **CALLERID()** function:

```
Set(CALLERID(rdnis)=number)
```

Asterisk versions:

```
--------| 1.2 |         |       |         |      |
```

Internal help for this application in Asterisk 1.2:

```
    -= Info about application 'SetRDNIS' =-

[Synopsis]
Set RDNIS Number

[Description]
  SetRDNIS(cnum): Set RDNIS Number on a call to a new
value.
SetRDNIS has been deprecated in favor of the function
CALLERID(rdnis)
```

Diff of the internal help from Asterisk 1.2 to 1.4:

Not available in Asterisk 1.4

See also Appendix C.

B.163 SetTransferCapability()

Sets the ISDN transfer capability.

```
SetTransferCapability(transferCapability)
```

Allowed values:

- **SPEECH**

  0x00 - Speech (default, voice calls)

- **DIGITAL**

  0x08 - Unrestricted digital information (data calls)

- **RESTRICTED_DIGITAL**

  0x09 - Restricted digital information

- **3K1AUDIO**

  0x10 - 3.1 kHz Audio (fax calls)

- **DIGITAL_W_TONES**

  0x11 - Unrestricted digital information with tones/announcements

- **VIDEO**

  0x18 - Video

Asterisk versions:

```
--------| 1.2 |--------| 1.4 |           |       |
```

Internal help for this application in Asterisk 1.4:

```
    -= Info about application 'SetTransferCapability' =-

[Synopsis]
Set ISDN Transfer Capability

[Description]
  SetTransferCapability(transfercapability): Set the ISDN Transfer
Capability of a call to a new value.
Valid Transfer Capabilities are:

  SPEECH             : 0x00 - Speech (default, voice calls)
  DIGITAL            : 0x08 - Unrestricted digital information (data
calls)
  RESTRICTED_DIGITAL : 0x09 - Restricted digital information
  3K1AUDIO           : 0x10 - 3.1kHz Audio (fax calls)
  DIGITAL_W_TONES    : 0x11 - Unrestricted digital information with
tones/announcements
  VIDEO              : 0x18 - Video

This application is deprecated in favor of
Set(CHANNEL(transfercapability)=...)
```

Diff of the internal help from Asterisk 1.2 to 1.4:

```
--- in Asterisk 1.2
+++ in Asterisk 1.4
@@ -13,4 +13,6 @@
     RESTRICTED_DIGITAL : 0x09 - Restricted digital information
     3K1AUDIO           : 0x10 - 3.1kHz Audio (fax calls)
     DIGITAL_W_TONES    : 0x11 - Unrestricted digital information
with tones/announcements
-    VIDEO              : 0x18 - Video:
+    VIDEO              : 0x18 - Video
+
+  This application is deprecated in favor of
Set(CHANNEL(transfercapability)=...)
```

Diff of the internal help from Asterisk 1.4 to 1.6:

Not available in Asterisk 1.6

B.164 SetVar()

Sets a variable.

```
SetVar(variable=value)
```

Warning: SetVar() is removed as of Asterisk 1.4 and has been replaced by the **Set()** function.

Asterisk versions:

```
--------| 1.2 |            |       |           |       |
```

Internal help for this application in Asterisk 1.2:

```
    -= Info about application 'SetVar' =-

[Synopsis]
Set channel variable(s)

[Description]
  SetVar(name1=value1|name2=value2|..[|options]): This application
has been
deprecated in favor of using the Set application.
```

Diff of the internal help from Asterisk 1.2 to 1.4:

Not available in Asterisk 1.4

See also set.

B.165 SIPAddHeader()

Adds a SIP header to the SIP dialog.

```
SIPAddHeader(header: value)
```

Adds a SIP header as specified to SIP calls initiated using **Dial()**. Nonstandard SIP headers should be preceded with an X- as in X-Asterisk-Accountcode:.

Should be used with caution as different SIP devices expect different headers and respond differently to them. May produce unexpected behavior.

Returns 0.

```
exten => 123,1,SIPAddHeader(X-Asterisk-Account: ${CDR(accountcode)})
exten => 123,n,Dial(SIP/123)
```

Asterisk versions:

```
--------| 1.2 |--------| 1.4 |--------| 1.6 |--------
```

Internal help for this application in Asterisk 1.4:

```
    -= Info about application 'SIPAddHeader' =-

[Synopsis]
Add a SIP header to the outbound call

[Description]
  SIPAddHeader(Header: Content)
Adds a header to a SIP call placed with DIAL.
Remember to user the X-header if you are adding non-standard SIP
headers, like "X-Asterisk-Accountcode:". Use this with care.
Adding the wrong headers may jeopardize the SIP dialog.
Always returns 0
```

Diff of the internal help from Asterisk 1.2 to 1.4:

None

Diff of the internal help from Asterisk 1.4 to 1.6:

```
--- in Asterisk 1.4
+++ in Asterisk 1.6
@@ -4,7 +4,7 @@
   Add a SIP header to the outbound call

   [Description]
-    SIPAddHeader(Header: Content)
+    SIPAddHeader(Header: Content):
   Adds a header to a SIP call placed with DIAL.
   Remember to user the X-header if you are adding non-standard SIP
   headers, like "X-Asterisk-Accountcode:". Use this with care.
```

See also Appendix C.

B.166 SIPdtmfMode()

Changes the DTMF mode for SIP calls.

```
SIPdtmfMode(mode)
```

Changes the DTMF mode for outgoing SIP calls. DTMF can be signaled out-of-band using rfc2833, in-band using inband (RTP), or out-of-band using info (RFC 2976).

```
exten => 123,n,SIPdtmfMode(rfc2833)
```

Asterisk versions:

```
--------| 1.2 |--------| 1.4 |--------| 1.6 |--------
```

Internal help for this application in Asterisk 1.4:

```
        -= Info about application 'SIPDtmfMode' =-

    [Synopsis]
    Change the dtmfmode for a SIP call

    [Description]
    SIPDtmfMode(inband|info|rfc2833): Changes the dtmfmode for a SIP call
```

Diff of the internal help from Asterisk 1.2 to 1.4:

None

Diff of the internal help from Asterisk 1.4 to 1.6:

```
    --- in Asterisk 1.4
    +++ in Asterisk 1.6
    @@ -4,4 +4,4 @@
       Change the dtmfmode for a SIP call

       [Description]
    -   SIPDtmfMode(inband|info|rfc2833): Changes the dtmfmode for a SIP
    call
    +     SIPDtmfMode(inband|info|rfc2833): Changes the dtmfmode for a SIP
    call
```

See also RFC 2833,[15] RFC 2976.[16]

B.167 SIPGetHeader()

Retrieves a SIP header from an incoming call.

```
    SIPGetHeader(variable=header_name/options)
```

Warning: **SIPGetHeader()** is removed as of Asterisk 1.4 and has been replaced by the **SIP_HEADER()** function.

---

15. www.ietf.org/rfc/rfc2833.txt
16. www.ietf.org/rfc/rfc2976.txt

Asterisk versions:

```
--------| 1.2 |            |       |           |       |
```

Internal help for this application in Asterisk 1.2:

```
    -= Info about application 'SIPGetHeader' =-

[Synopsis]
Get a SIP header from an incoming call

[Description]
  SIPGetHeader(var=headername[|options]):
Sets a channel variable to the content of a SIP header
  Options:
    j - Jump to priority n+101 if the requested header isn't found.
  This application sets the following channel variable upon completion:
      SIPGETSTATUS - This variable will contain the status of the
attempt
                    FOUND | NOTFOUND
  This application has been deprecated in favor of the SIP_HEADER
function.
```

Diff of the internal help from Asterisk 1.2 to 1.4:

Not available in Asterisk 1.4

See also Appendix C.

B.168 SMS()

Sends or receives SMS (Short Message System) messages.

```
SMS(queue[,options])
```

Manages the exchange of SMS messages with an SMS-capable telephone or through an SMS service center supporting ETSI[17] ES 201 912 SMS protocol on analog or ISDN lines. Because the shell client **smsq** uses FSK[18] this is unlikely to work over compressed codecs such as GSM.

Options:

- **a**

  Act as recipient

- **s**

  Act as transmitter (SMS service center) to communicate with a telephone set

---

17. ETSI (European Telecommunications Standards Institute)
18. Frequency Shift Keying

All send and receive queues are stored in `/var/spool/asterisk/sms/`: message coming from the service center to the device in `sc-me.queue/`, the message from the device to the service center in `me-sc.queue/`. A log is written to `/var/log/asterisk/sms`.

When connecting as a recipient (a), messages residing in `me-sc.queue/` are sent and then deleted; received messages are written to `sc-me.queue/` with a timestamp in the filename. When connecting as a transmitter (SMS service center), the reverse is true.

Message files are in the format described below. Absent parameters imply a default:

```
oa=Originating Address (sender's number)
```

This contains the complete national direct-dial number, including country code and preceded by the +. For example, +19255554101 would be a valid number.

```
da=Destination Address (recipient's number)
```

Again, a complete national direct-dial number with preceding +.

```
scts=Service Centre Time Stamp (timestamp)
```

Uses the format YYYY-MM-DD HH:MM:SS

```
pid=Protocol Identifier (decimal octet value)
dcs=Data coding scheme (decimal octet value)
mr=Message reference (decimal octet value)
ud=Message Text
```

If characters other than 10, 13, 32-126, 128-255 (decimal) occur in the message, `ud=` is replaced by `ud#` and the message contents are coded in hexadecimal.

```
srr=Status Report Request (0|1)
rp=Return Path (0|1)
vp=Validity Period (minutes)
```

When sending to an SMSC, only `da` and `ud` are used; `oa` is ignored. When sending to a device, only `oa` and `ud` are necessary; `da` is ignored.

An extension for receiving SMS messages might look like this (where 4165553331 is the number of our SMSC):

```
[incoming]
exten => _X.,1,GotoIf($["${CALLERIDNUM}" = "4165553331"]?sms-me-
in,${EXTEN},1)
; or like so:
;exten => _X./_0193010.,1,Goto(sms-me-in,${EXTEN},1)

[sms-me-in]
exten => _X.,1,Wait(1)
```

```
exten => _X.,n,SMS(me-incoming,a)
exten => _X.,n,System(handleincomingsms)
exten => _X.,n,Hangup()
```

where `handleincomingsms` might be a wrapper or command containing; for example, `smsq--process=queue--queue=me-incoming` which is executed for each incoming message.

Outgoing messages are written to files, but may also be generated with the following (outdated) sequence (4165553331 is the number of the SMSC):

```
[outgoing]
exten = 4165553331,1,Goto(sms-me-out,${CALLERIDNUM},1)

[sms-me-out]
exten => _X.,1,Set(CDR(accountcode)=SMS)
exten => _X.,n,Set(smsFrom=${CALLERIDNUM})
exten => _X.,n,SMS(${smsFrom},s,${EXTEN},${smsText})   ; Generate SMS
exten => _X.,n,SMS(${smsFrom},s)                       ; Send SMS
exten => _X.,n,Hangup()
```

Further information and many additional examples may be found at www .voip-info.org/wiki/view/Asterisk+cmd+Sms. With the extremely wide variation in SMS implementation, it is best not to expect **SMS()** to work right "out of the box."

Asterisk versions:

--------| **1.2** |--------| **1.4** |--------| **1.6** |--------

Internal help for this application in Asterisk 1.4:

```
    -= Info about application 'SMS' =-

[Synopsis]
Communicates with SMS service centres and SMS capable analogue phones

[Description]
  SMS(name|[a][s]):  SMS handles exchange of SMS data with a call
to/from SMS capabale
phone or SMS PSTN service center. Can send and/or receive SMS messages.
Works to ETSI ES 201 912 compatible with BT SMS PSTN service in UK
Typical usage is to use to handle called from the SMS service centre
CLI,
or to set up a call using 'outgoing' or manager interface to connect
service centre to SMS()
name is the name of the queue used in /var/spool/asterisk/sms
Arguments:
 a: answer, i.e. send initial FSK packet.
 s: act as service centre talking to a phone.
Messages are processed as per text file message queues.
smsq (a separate software) is a command to generate message
queues and send messages.
```

Diff of the internal help from Asterisk 1.2 to 1.4:

None

Diff of the internal help from Asterisk 1.4 to 1.6:

```
--- in Asterisk 1.4
+++ in Asterisk 1.6
@@ -4,16 +4,26 @@
   Communicates with SMS service centres and SMS capable analogue
phones

   [Description]
-    SMS(name|[a][s]):  SMS handles exchange of SMS data with a call
to/from SMS capabale
+    SMS(name,[a][s][t][p(d)][r][o],addr,body):
+   SMS handles exchange of SMS data with a call to/from SMS capable
   phone or SMS PSTN service center. Can send and/or receive SMS messages.
-  Works to ETSI ES 201 912 compatible with BT SMS PSTN service in UK
-  Typical usage is to use to handle called from the SMS service centre
CLI,
+  Works to ETSI ES 201 912; compatible with BT SMS PSTN service in UK
+  and Telecom Italia in Italy.
+  Typical usage is to use to handle calls from the SMS service centre
CLI,
   or to set up a call using 'outgoing' or manager interface to connect
   service centre to SMS()
   name is the name of the queue used in /var/spool/asterisk/sms
   Arguments:
-  a: answer, i.e. send initial FSK packet.
-  s: act as service centre talking to a phone.
+  a  - answer, i.e. send initial FSK packet.
+  s  - act as service centre talking to a phone.
+  t  - use protocol 2 (default used is protocol 1).
+  p(N)  - set the initial delay to N ms (default is 300).
+         addr and body are a deprecated format to send messages out.
+  r  - set the Status Report Request (SRR) bit.
+  o  - the body should be coded as octets not 7-bit symbols.
  Messages are processed as per text file message queues.
   smsq (a separate software) is a command to generate message
   queues and send messages.
+  NOTE: the protocol has tight delay bounds. Please use short frames
+  and disable/keep short the jitter buffer on the ATA to make sure
that
+  respones (ACK etc.) are received in time.
```

B.169 SoftHangup()

Hangs up the specified channel.

```
SoftHangup(technology/resource[,options])
```

Hangs up the specified channel. Returns 0. Allowed *options* include a, which hangs up all channels on the specified device, not just a single resource.

```
; hang up all channels using Zap/4:
exten => 123,1,SoftHangup(Zap/4,a)
exten => 123,n,Wait(2)
exten => 123,n,Dial(Zap/4/6045551538)
```

Asterisk versions:

--------| **1.2** |--------| **1.4** |--------| **1.6** |--------

Internal help for this application in Asterisk 1.4:

```
    -= Info about application 'SoftHangup' =-

[Synopsis]
Soft Hangup Application

[Description]
  SoftHangup(Technology/resource|options)
Hangs up the requested channel.  If there are no channels to hangup,
the application will report it.
- 'options' may contain the following letter:
     'a' : hang up all channels on a specified device instead of a
single resource
```

Diff of the internal help from Asterisk 1.2 to 1.4:

None

Diff of the internal help from Asterisk 1.4 to 1.6:

```
--- in Asterisk 1.4
+++ in Asterisk 1.6
@@ -4,8 +4,8 @@
   Soft Hangup Application

   [Description]
-    SoftHangup(Technology/resource|options)
+    SoftHangup(Technology/resource[,options]):
   Hangs up the requested channel.  If there are no channels to hangup,
   the application will report it.
-  - 'options' may contain the following letter:
-       'a' : hang up all channels on a specified device instead of a
single resource
+    Options:
+        'a'  - hang up all channels on a specified device instead of a
single resource
```

See also hangup.

B.170 Sort()

Sorts a list of keys and values.

```
Sort(key1:value1[,key2:value2[,...]])
```

Warning: Sort() is removed as of Asterisk 1.4 and has been replaced by the
SORT() function.

Asterisk versions:

--------| 1.2 | | | | |

Internal help for this application in Asterisk 1.2:

```
      -= Info about application 'Sort' =-

[Synopsis]
Sorts a list of keywords and values

[Description]
  Sort(newvar=key1:val1[,key2:val2[[...],keyN:valN]]): This application
will
sort the list provided in ascending order. The result will be stored in
the
specified variable name.
  This application has been deprecated in favor of the SORT function.
```

Diff of the internal help from Asterisk 1.2 to 1.4:

Not available in Asterisk 1.4

See also Appendix C.

B.171 StackPop()

Removes a return address from the **Gosub()** stack.

```
StackPop()
```

Removes a return address from a stack of nested **Gosub()** calls.

Asterisk versions:

--------| 1.2 |--------| 1.4 |--------| 1.6 |--------

Internal help for this application in Asterisk 1.4:

```
   -= Info about application 'StackPop' =-

[Synopsis]
Remove one address from gosub stack

[Description]
StackPop()
  Removes last label on the stack, discarding it.
```

Diff of the internal help from Asterisk 1.2 to 1.4:

None

Diff of the internal help from Asterisk 1.4 to 1.6:

```
--- in Asterisk 1.4
+++ in Asterisk 1.6
@@ -4,5 +4,5 @@
   Remove one address from gosub stack

   [Description]
-  StackPop()
-    Removes last label on the stack, discarding it.
+    StackPop():
+  Removes last label on the stack, discarding it.
```

See also gosub.

B.172 StartMusicOnHold()

Starts playing music.

```
StartMusicOnHold([class])
```

Plays music from the specified *class* as defined in `musiconhold.conf`. If *class* is not specified, the default class for the channel is used.

Asterisk versions:

```
--------| 1.2 |--------| 1.4 |--------| 1.6 |--------
```

Internal help for this application in Asterisk 1.4:

```
   -= Info about application 'StartMusicOnHold' =-

[Synopsis]
Play Music On Hold
```

```
[Description]
StartMusicOnHold(class): Starts playing music on hold, uses default
music class for channel.
Starts playing music specified by class.  If omitted, the default
music source for the channel will be used.  Always returns 0.
```

Diff of the internal help from Asterisk 1.2 to 1.4:

None

Diff of the internal help from Asterisk 1.4 to 1.6:

```
--- in Asterisk 1.4
+++ in Asterisk 1.6
@@ -4,6 +4,7 @@
   Play Music On Hold

   [Description]
-  StartMusicOnHold(class): Starts playing music on hold, uses default
music class for channel.
+     StartMusicOnHold(class):
+  Starts playing music on hold, uses default music class for channel.
   Starts playing music specified by class.  If omitted, the default
   music source for the channel will be used.  Always returns 0.
```

See also stopmusiconhold, musiconhold.

B.173 StopMixMonitor()

Stops recording of a call.

```
StopMixMonitor()
```

Stops recording initiated by MixMonitor() of a call on the active channel.

Asterisk versions:

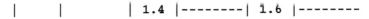

```
        |        |          | 1.4 |--------| 1.6 |--------
```

Internal help for this application in Asterisk 1.4:

```
    -= Info about application 'StopMixMonitor' =-

[Synopsis]
Stop recording a call through MixMonitor

[Description]
  StopMixMonitor()

Stops the audio recording that was started with a call to MixMonitor()
on the current channel.
```

Diff of the internal help from Asterisk 1.2 to 1.4:

Not available in Asterisk 1.2

Diff of the internal help from Asterisk 1.4 to 1.6:

```
--- in Asterisk 1.4
+++ in Asterisk 1.6
@@ -4,7 +4,6 @@
   Stop recording a call through MixMonitor

   [Description]
-     StopMixMonitor()
-
+     StopMixMonitor():
   Stops the audio recording that was started with a call to
MixMonitor()
   on the current channel.
```

See also mixmonitor.

B.174 StopMonitor()

Stops recording of a channel.

```
StopMonitor()
```

Stops recording of a channel. Has no effect if the channel is not currently monitored.

```
exten => 123,1,Answer()
exten => 123,n,Monitor(wav,monitor_test,mb)
exten => 123,n,SayDigits(12345678901234567890)
exten => 123,n,StopMonitor()
```

Asterisk versions:

```
--------| 1.2 |--------| 1.4 |--------| 1.6 |--------
```

Internal help for this application in Asterisk 1.4:

```
   -= Info about application 'StopMonitor' =-

[Synopsis]
Stop monitoring a channel

[Description]
StopMonitor
Stops monitoring a channel. Has no effect if the channel is not
monitored
```

Diff of the internal help from Asterisk 1.2 to 1.4:

None

Diff of the internal help from Asterisk 1.4 to 1.6:

```
--- in Asterisk 1.4
+++ in Asterisk 1.6
@@ -4,5 +4,5 @@
   Stop monitoring a channel

   [Description]
-  StopMonitor
+   StopMonitor():
   Stops monitoring a channel. Has no effect if the channel is not
monitored
```

See also monitor, pausemonitor.

B.175 StopMusicOnHold()

Stops playing music.

```
StopMusicOnHold()
```

Stops playing music that was started by **StartMusicOnHold()**.

Asterisk versions:

`--------| 1.2 |--------| 1.4 |--------| 1.6 |--------`

Internal help for this application in Asterisk 1.4:

```
   -= Info about application 'StopMusicOnHold' =-

[Synopsis]
Stop Playing Music On Hold

[Description]
StopMusicOnHold: Stops playing music on hold.
```

Diff of the internal help from Asterisk 1.2 to 1.4:

None

Diff of the internal help from Asterisk 1.4 to 1.6:

```
--- in Asterisk 1.4
+++ in Asterisk 1.6
@@ -4,4 +4,4 @@
   Stop Playing Music On Hold
```

```
      [Description]
 -    StopMusicOnHold: Stops playing music on hold.
 +       StopMusicOnHold(): Stops playing music on hold.
```

See also **startmusiconhold, musiconhold.**

B.176 StopPlaytones()

Interrupts playback of a tone list.

```
    StopPlaytones()
```

Stops playback of a currently playing tone list.

```
exten => 123,1,Playtones(busy)
exten => 123,n,Wait(2)
exten => 123,n,StopPlaytones()
exten => 123,n,Playtones(congestion)
exten => 123,n,Wait(2)
exten => 123,n,StopPlaytones()
exten => 123,n,Goto(1)
```

Asterisk versions:

```
--------| 1.2 |--------| 1.4 |--------| 1.6 |--------
```

Internal help for this application in Asterisk 1.4:

```
    -= Info about application 'StopPlayTones' =-

[Synopsis]
Stop playing a tone list

[Description]
Stop playing a tone list
```

Diff of the internal help from Asterisk 1.2 to 1.4:

None

Diff of the internal help from Asterisk 1.4 to 1.6:

```
--- in Asterisk 1.4
+++ in Asterisk 1.6
@@ -4,4 +4,4 @@
   Stop playing a tone list

   [Description]
-  Stop playing a tone list
+     StopPlayTones(): Stop playing a tone list
```

See also **playtones, indications.conf.**

B.177 System()

Executes a shell command.

```
System(command)
```

Uses the C **system()** function to execute a command on the system shell (**sh** or its equivalents).

This is very similar to **TrySystem()** except that it returns −1 if it is unable to run the system command where as **TrySystem()** always returns 0.

Sets the channel variable SYSTEMSTATUS to SUCCESS, FAILURE, or APPERROR. (This is undocumented; the command was executed but returned an exit code other than zero.)

```
exten => s,1,System(echo '${DATETIME} - ${CALLERID} - ${CHANNEL}' >> /var/
log/asterisk/calls)
```

See also trysystem.

Note: An alternative is **Backticks()** application or the function **BACKTICKS()** from the app_backticks module.[19]
   This returns the output of the command.

Asterisk versions:

--------| 1.2 |--------| 1.4 |--------| 1.6 |--------

Internal help for this application in Asterisk 1.4:

```
    -= Info about application 'System' =-

[Synopsis]
Execute a system command

[Description]
  System(command): Executes a command  by  using  system(). If the
command
fails, the console should report a fallthrough.
Result of execution is returned in the SYSTEMSTATUS channel variable:
    FAILURE     Could not execute the specified command
    SUCCESS     Specified command successfully executed

Old behaviour:
If the command itself executes but is in error, and if there exists
```

---

19. From www.pbxfreeware.org/archives/2005/06/index.html or www.pbxfreeware.org/

```
a priority n + 101, where 'n' is the priority of the current instance,
then  the  channel  will  be  setup to continue at that priority level.
Note that this jump functionality has been deprecated and will only
occur
if the global priority jumping option is enabled in extensions.conf.
```

Diff of the internal help from Asterisk 1.2 to 1.4:

None

Diff of the internal help from Asterisk 1.4 to 1.6:

```
--- in Asterisk 1.4
+++ in Asterisk 1.6
@@ -9,10 +9,3 @@
   Result of execution is returned in the SYSTEMSTATUS channel
variable:
        FAILURE      Could not execute the specified command
        SUCCESS      Specified command successfully executed
-
- Old behaviour:
- If the command itself executes but is in error, and if there exists
- a priority n + 101, where 'n' is the priority of the current
instance,
- then  the  channel  will  be  setup to continue at that priority
level.
- Note that this jump functionality has been deprecated and will only
occur
- if the global priority jumping option is enabled in extensions.conf.
```

B.178 Transfer()

Transfers the call to another extension.

```
Transfer([technology/]destination[,options])
```

Requests transfer of the caller to the specified extension or device. If the technology is specified (e.g., SIP, IAX2), only calls using the same technology will be transferred. In the case of SIP channels that have not yet been answered, this happens via a 302-REDIRECT message to the caller; if the call has already been answered, through a REFER message. The destination may also be a specific address, such as 8885551212@sip.provider.tld.

If option j is set, jumps to priority n+101 if the transfer fails.

Sets the channel variable TRANSFERSTATUS to SUCCESS, FAILURE, or UNSUPPORTED (meaning the channel driver does not support transfers).

```
; Transfer calls intended for Extension 123 to Extension 130:
exten => 123,1,Transfer(130)
```

Asterisk versions:

```
--------| 1.2 |--------| 1.4 |--------| 1.6 |--------
```

Internal help for this application in Asterisk 1.4:

```
    -= Info about application 'Transfer' =-

[Synopsis]
Transfer caller to remote extension

[Description]
  Transfer([Tech/]dest[|options]):  Requests the remote caller be
transferred
to a given destination. If TECH (SIP, IAX2, LOCAL etc) is used, only
an incoming call with the same channel technology will be transfered.
Note that for SIP, if you transfer before call is setup, a 302 redirect
SIP message will be returned to the caller.

The result of the application will be reported in the TRANSFERSTATUS
channel variable:
        SUCCESS      Transfer succeeded
        FAILURE      Transfer failed
        UNSUPPORTED  Transfer unsupported by channel driver
The option string many contain the following character:
'j' -- jump to n+101 priority if the channel transfer attempt
        fails
```

Diff of the internal help from Asterisk 1.2 to 1.4:

None

Diff of the internal help from Asterisk 1.4 to 1.6:

```
--- in Asterisk 1.4
+++ in Asterisk 1.6
@@ -4,7 +4,7 @@
   Transfer caller to remote extension

   [Description]
-    Transfer([Tech/]dest[|options]):  Requests the remote caller be
transferred
+    Transfer([Tech/]dest):  Requests the remote caller be transferred
   to a given destination. If TECH (SIP, IAX2, LOCAL etc) is used, only
   an incoming call with the same channel technology will be
transfered.
   Note that for SIP, if you transfer before call is setup, a 302
redirect
@@ -15,6 +15,3 @@
         SUCCESS      Transfer succeeded
         FAILURE      Transfer failed
         UNSUPPORTED  Transfer unsupported by channel driver
```

```
    -    The option string many contain the following character:
    -    'j' -- jump to n+101 priority if the channel transfer attempt
    -            fails
```

B.179 TryExec()

Tries to execute a dialplan application.

```
    TryExec(application(arguments))
```

Tries, like **Exec()**, to execute a dialplan application, but does not terminate the call if the attempt fails (either because the application does not exist or because it returned an error code). Instead, the channel variable TRYSTATUS is set to one of the following values:

- **SUCCESS**

  The application returned 0.

- **FAILED**

  The application returned a value other than 0.

- **NOAPP**

  The application could not be found.

For more information, see exec.

Asterisk versions:

```
    |      |          | 1.4 |--------| 1.6 |--------
```

Internal help for this application in Asterisk 1.4:

```
      -= Info about application 'TryExec' =-

[Synopsis]
Executes dialplan application, always returning

[Description]
Usage: TryExec(appname(arguments))
  Allows an arbitrary application to be invoked even when not
hardcoded into the dialplan. To invoke external applications
see the application System.  Always returns to the dialplan.
The channel variable TRYSTATUS will be set to:
    SUCCESS    if the application returned zero
    FAILED     if the application returned non-zero
    NOAPP      if the application was not found or was not specified
```

Diff of the internal help from Asterisk 1.2 to 1.4:

Not available in Asterisk 1.2

Diff of the internal help from Asterisk 1.4 to 1.6:

```
--- in Asterisk 1.4
+++ in Asterisk 1.6
@@ -4,11 +4,11 @@
   Executes dialplan application, always returning

   [Description]
-  Usage: TryExec(appname(arguments))
-    Allows an arbitrary application to be invoked even when not
+    TryExec(appname(arguments)):
+  Allows an arbitrary application to be invoked even when not
   hardcoded into the dialplan. To invoke external applications
   see the application System.  Always returns to the dialplan.
-  The channel variable TRYSTATUS will be set to:
+  The channel variable TRYSTATUS will be set to one of:
        SUCCESS   if the application returned zero
        FAILED    if the application returned non-zero
        NOAPP     if the application was not found or was not specified
```

See also **exec**, **execif**, **trysystem**.

B.180 TrySystem()

Tries to execute a shell command.

```
TrySystem(command)
```

Like **System()**, it executes a command on the shell (**sh** or its equivalents), but always returns 0. In contrast, **System()** returns –1 on error.

Sets the channel variable SYSTEMSTATUS to SUCCESS, FAILURE, or APPERROR (command was run but returned an exit code other than 0).

```
exten => 123,1,TrySystem(echo 'Hey World' > /tmp/hey.txt)
```

Asterisk versions:

```
--------| 1.2 |--------| 1.4 |--------| 1.6 |--------
```

Internal help for this application in Asterisk 1.4:

```
   -= Info about application 'TrySystem' =-

[Synopsis]
Try executing a system command
```

```
[Description]
  TrySystem(command): Executes a command  by  using  system().
on any situation.
Result of execution is returned in the SYSTEMSTATUS channel variable:
    FAILURE      Could not execute the specified command
    SUCCESS      Specified command successfully executed
    APPERROR     Specified command successfully executed, but returned
error code

Old behaviour:
If  the command itself executes but is in error, and if
there exists a priority n + 101, where 'n' is the priority of the
current
instance, then  the  channel  will  be  setup  to continue at that
priority level.  Otherwise, System will terminate.
```

Diff of the internal help from Asterisk 1.2 to 1.4:

None

Diff of the internal help from Asterisk 1.4 to 1.6:

```
--- in Asterisk 1.4
+++ in Asterisk 1.6
@@ -10,9 +10,3 @@
     FAILURE      Could not execute the specified command
     SUCCESS      Specified command successfully executed
     APPERROR     Specified command successfully executed, but
returned error code
-
-  Old behaviour:
-  If  the command itself executes but is in error, and if
-  there exists a priority n + 101, where 'n' is the priority of the
current
-  instance, then  the  channel  will  be  setup  to continue at that
-  priority level.  Otherwise, System will terminate.
```

See also **system**.

B.181 TXTCIDName()

Looks up the caller name via a DNS TXT record.

```
TXTCIDName(caller_id_number|options)
```

If the option j is set, the call jumps to priority n+101 if the lookup fails.

Warning: **TXTCIDName()** is removed as of Asterisk 1.4 and has been replaced by the **TXTCIDNAME()** function.

Asterisk versions:

`--------| 1.2 |          |       |          |       |`

Internal help for this application in Asterisk 1.2:

```
      -= Info about application 'TXTCIDName' =-

[Synopsis]
Lookup caller name from TXT record

[Description]
  TXTCIDName(<CallerIDNumber>[|options]):  Looks up a Caller Name
via DNS and sets
the variable 'TXTCIDNAME'. TXTCIDName will either be blank
or return the value found in the TXT record in DNS.
The option string may contain the following character:
'j' -- jump to n+101 priority if the lookup fails
This application sets the following channel variable upon completion:
  TXTCIDNAMESTATUS The status of the lookup as a text string, one of
       SUCCESS | FAILED
```

Diff of the internal help from Asterisk 1.2 to 1.4:

Not available in Asterisk 1.4

See also Appendix C.

B.182 UnpauseMonitor()

Resumes recording of a channel.

```
UnpauseMonitor()
```

Resumes recording of a channel after it has been paused with **PauseMonitor()**.

Asterisk versions:

`        |       |          | 1.4 |--------| 1.6 |--------`

Internal help for this application in Asterisk 1.4:

```
      -= Info about application 'UnpauseMonitor' =-

[Synopsis]
Unpause monitoring of a channel

[Description]
UnpauseMonitor
Unpauses monitoring of a channel on which monitoring had
previously been paused with PauseMonitor.
```

Diff of the internal help from Asterisk 1.2 to 1.4:

Not available in Asterisk 1.2

Diff of the internal help from Asterisk 1.4 to 1.6:

```
--- in Asterisk 1.4
+++ in Asterisk 1.6
@@ -4,6 +4,6 @@
   Unpause monitoring of a channel

   [Description]
-  UnpauseMonitor
+    UnpauseMonitor():
   Unpauses monitoring of a channel on which monitoring had
   previously been paused with PauseMonitor.
```

See also monitor, pausemonitor.

B.183 UnpauseQueueMember()

Resumes calls to a paused member of a queue.

```
UnpauseQueueMember([queue,]interface[,options])
```

"Unpauses" a queue member previously "paused" with **PauseQueueMember()** (see an example there) so that the member can receive calls again.

Sets the channel variable UPQMSTATUS to UNPAUSED or NOTFOUND.

Asterisk versions:

```
--------| 1.2 |--------| 1.4 |--------| 1.6 |--------
```

Internal help for this application in Asterisk 1.4:

```
    -= Info about application 'UnpauseQueueMember' =-

[Synopsis]
Unpauses a queue member

[Description]
    UnpauseQueueMember([queuename]|interface[|options]):
Unpauses (resumes calls to) a queue member.
This is the counterpart to PauseQueueMember and operates exactly the
same way, except it unpauses instead of pausing the given interface.
The option string may contain zero or more of the following characters:
        'j' -- jump to +101 priority when appropriate.
```

```
         This application sets the following channel variable upon completion:
             UPQMSTATUS        The status of the attempt to unpause a queue
                               member as a text string, one of
                     UNPAUSED | NOTFOUND
       Example: UnpauseQueueMember(|SIP/3000)
```

Diff of the internal help from Asterisk 1.2 to 1.4:

None

Diff of the internal help from Asterisk 1.4 to 1.6:

```
   --- in Asterisk 1.4
   +++ in Asterisk 1.6
   @@ -4,14 +4,14 @@
      Unpauses a queue member

      [Description]
   -    UnpauseQueueMember([queuename]|interface[|options]):
   +    UnpauseQueueMember([queuename],interface[,options[,reason]]):
      Unpauses (resumes calls to) a queue member.
      This is the counterpart to PauseQueueMember and operates exactly the
      same way, except it unpauses instead of pausing the given interface.
   -  The option string may contain zero or more of the following
   characters:
   -         'j' -- jump to +101 priority when appropriate.
   +  The reason string is entirely optional and is used to add extra
   information
   +  to the appropriate queue_log entries and manager events.
         This application sets the following channel variable upon
   completion:
             UPQMSTATUS        The status of the attempt to unpause a queue
                               member as a text string, one of
                     UNPAUSED | NOTFOUND
   -  Example: UnpauseQueueMember(|SIP/3000)
   +  Example: UnpauseQueueMember(,SIP/3000)
```

See also pausequeuemember.

B.184 UserEvent()

Sends an arbitrary event to the manager interface.

```
   UserEvent(eventname[,body])
```

Sends an event of your choosing to the manager interface. The resulting event packet has the following format:

```
   Event: UserEvent eventname
   Channel: channelname
```

```
Uniqueid: call-identifier
[body]
```

Additional lines in the form `fieldname:` value may be specified in the body. Multiple lines are separated with the pipe (|) character (in older versions of Asterisk, with , or ^).

Returns 0.

```
exten => 123,1,UserEvent(Test,Note: I am calling ${XY} now.)
exten => 123,n,Dial(${XY})
```

Asterisk versions:

```
--------| 1.2 |--------| 1.4 |--------| 1.6 |--------
```

Internal help for this application in Asterisk 1.4:

```
    -= Info about application 'UserEvent' =-

[Synopsis]
Send an arbitrary event to the manager interface

[Description]
  UserEvent(eventname[|body]): Sends an arbitrary event to the manager
interface, with an optional body representing additional
arguments.   The
body may be specified as a | delimeted list of headers. Each additional
argument will be placed on a new line in the event. The format of the
event will be:
    Event: UserEvent
    UserEvent: <specified event name>
    [body]
If no body is specified, only Event and UserEvent headers will be
present.
```

Diff of the internal help from Asterisk 1.2 to 1.4:

```
--- in Asterisk 1.2
+++ in Asterisk 1.4
@@ -4,12 +4,12 @@
    Send an arbitrary event to the manager interface

    [Description]
-     UserEvent(eventname[|body]): Sends an arbitrary event to the
-   manager interface, with an optional body representing additional
-   arguments.   The format of the event will be:
-       Event: UserEvent<specified event name>
-       Channel: <channel name>
-       Uniqueid: <call uniqueid>
+     UserEvent(eventname[|body]): Sends an arbitrary event to the
manager
```

```
+   interface, with an optional body representing additional
arguments.  The
+   body may be specified as a | delimeted list of headers. Each
additional
+   argument will be placed on a new line in the event. The format of
the
+   event will be:
+         Event: UserEvent
+         UserEvent: <specified event name>
          [body]
-   If the body is not specified, only Event, Channel, and Uniqueid
fields
-   will be present.  Returns 0.
+   If no body is specified, only Event and UserEvent headers will be
present.
```

Diff of the internal help from Asterisk 1.4 to 1.6:

```
--- in Asterisk 1.4
+++ in Asterisk 1.6
@@ -4,7 +4,7 @@
   Send an arbitrary event to the manager interface

   [Description]
-     UserEvent(eventname[|body]): Sends an arbitrary event to the
manager
+     UserEvent(eventname[,body]): Sends an arbitrary event to the
manager
   interface, with an optional body representing additional
arguments.  The
   body may be specified as a | delimeted list of headers. Each
additional
   argument will be placed on a new line in the event. The format of
the
```

See also **manager.conf**, **Asterisk Manager interface.**

B.185 Verbose()

Sends arbitrary text to the CLI at the verbose level specified.

```
Verbose([level,]message)
```

Sends the specified message to the CLI. The *level* is an integer; the message will only appear at verbose levels equal to or greater than this number.[20]

If level is not specified, 0 is assumed.

---

20. e.g., **asterisk -vvvr** for verbose level 3 - or enter **set verbose 3** in the CLI

Returns 0.

```
exten => 123,1,Verbose(1,Someone is calling extension 123.)
exten => 123,n,Playback(extension)
exten => 123,n,SayDigits(${EXTEN})
```

Asterisk versions:

```
--------| 1.2 |--------| 1.4 |--------| 1.6 |--------
```

Internal help for this application in Asterisk 1.4:

```
    -= Info about application 'Verbose' =-

[Synopsis]
Send arbitrary text to verbose output

[Description]
Verbose([<level>|]<message>)
   level must be an integer value.  If not specified, defaults to 0.
```

Diff of the internal help from Asterisk 1.2 to 1.4:

None

Diff of the internal help from Asterisk 1.4 to 1.6:

```
--- in Asterisk 1.4
+++ in Asterisk 1.6
@@ -4,5 +4,5 @@
   Send arbitrary text to verbose output

   [Description]
-  Verbose([<level>|]<message>)
+  Verbose([<level>,]<message>)
      level must be an integer value.  If not specified, defaults to 0.
```

See also noop, log, dumpchan.

B.186 VMAuthenticate()

Authenticates the caller using the voicemail password of the specified mailbox.

```
VMAuthenticate([mailbox][@context][,options])
```

Behaves just like **Authenticate()** except that the passwords are taken from the configuration (and optional context) in `voicemail.conf`.

If a mailbox is specified, only the password for that mailbox will be accepted. If it is not provided, any voicemail password will be accepted! The channel variable ${AUTH_MAILBOX} is then populated with the name of the authenticated mailbox.

Option s suppresses the prompt.

```
; use the dialed extension as the reference mailbox and authenticate:
exten => 123,1,VMAuthenticate(${EXTEN}@sales)
exten => 123,n,SayDigits(${AUTH_MAILBOX})
```

Asterisk versions:

```
--------| 1.2 |--------| 1.4 |--------| 1.6 |--------
```

Internal help for this application in Asterisk 1.4:

```
    -= Info about application 'VMAuthenticate' =-

[Synopsis]
Authenticate with Voicemail passwords

[Description]
  VMAuthenticate([mailbox][@context][|options]): This application
behaves the
same way as the Authenticate application, but the passwords are taken
from
voicemail.conf.
  If the mailbox is specified, only that mailbox's password will be
considered
valid. If the mailbox is not specified, the channel variable
AUTH_MAILBOX will
be set with the authenticated mailbox.

  Options:
    s - Skip playing the initial prompts.
```

Diff of the internal help from Asterisk 1.2 to 1.4:

None

Diff of the internal help from Asterisk 1.4 to 1.6:

```
--- in Asterisk 1.4
+++ in Asterisk 1.6
@@ -4,7 +4,7 @@
   Authenticate with Voicemail passwords

   [Description]
-    VMAuthenticate([mailbox][@context][|options]): This application
behaves the
+    VMAuthenticate([mailbox][@context][,options]): This application
behaves the
   same way as the Authenticate application, but the passwords are
taken from
     voicemail.conf.
     If the mailbox is specified, only that mailbox's password will be
considered
```

See also authenticate, voicemail.conf.

B.187 VoiceMail()

Allows the caller to leave a voice mail message in the specified mailbox.

```
VoiceMail(mailbox[@context][&mailbox[@context][&...]],options)
```

Old syntax:

```
VoiceMail([s|u|b]mailbox[@context][&mailbox[@context][&...]])
```

Allows the caller to leave a voice mail message in the specified mailbox. The mailbox must already be configured in `voicemail.conf`. If more than mailbox is listed, the greeting from the first mailbox is the one that is played. If the mailbox does not exist, dialplan execution ends.

The option s (silent) suppresses the prompt. The option u plays the "unavailable" message, if it exists. The option b plays (busy) message, if it exists (file `busy` instead of `unavail`).

Warning: You cannot mix syntax types. If you do, the application will fail as though the mailbox does not exist. We recommend always using the new syntax.

If option j is set, jumps to extension n+101, if it exists, on failure.

If the caller presses 0 during the prompt, the call goes to extension o (small letter o, for operator) in the current context.

If the caller presses * during the prompt, the call goes to extension a (small letter a, for assistant) in the current context.

Returns –1 in case of error (the mailbox could not be found or the caller hung up) otherwise returns 0. Sets the channel variable VMSTATUS to SUCCESS, USEREXIT (the caller canceled the message) or FAILED.

```
; send the caller to mailbox 123, play the unavailable message:
exten => 123,1,VoiceMail(123,u)
```

Asterisk versions:

```
--------| 1.2 |--------| 1.4 |--------| 1.6 |--------
```

Internal help for this application in Asterisk 1.4:

```
    -= Info about application 'VoiceMail' =-

[Synopsis]
Leave a Voicemail message

[Description]
  VoiceMail(mailbox[@context][&mailbox[@context]][...][|options]): This
application allows the calling party to leave a message for the
specified
```

list of mailboxes. When multiple mailboxes are specified, the greeting
will
be taken from the first mailbox specified. Dialplan execution will stop
if the
specified mailbox does not exist.
   The Voicemail application will exit if any of the following DTMF
digits are
received:
      0 - Jump to the 'o' extension in the current dialplan context.
      * - Jump to the 'a' extension in the current dialplan context.
   This application will set the following channel variable upon
completion:
      VMSTATUS - This indicates the status of the execution of the
VoiceMail
                 application. The possible values are:
                 SUCCESS | USEREXIT | FAILED

   Options:
      b   - Play the 'busy' greeting to the calling party.
      g(#) - Use the specified amount of gain when recording the
voicemail
             message. The units are whole-number decibels (dB).
             Only works on supported technologies, which is Zap only.
      s   - Skip the playback of instructions for leaving a message to
the
             calling party.
      u   - Play the 'unavailable' greeting.
      j   - Jump to priority n+101 if the mailbox is not found or some
other
             error occurs.

Diff of the internal help from Asterisk 1.2 to 1.4:

```
--- in Asterisk 1.2
+++ in Asterisk 1.4
@@ -22,6 +22,7 @@
        b    - Play the 'busy' greeting to the calling party.
        g(#) - Use the specified amount of gain when recording the
voicemail
               message. The units are whole-number decibels (dB).
+              Only works on supported technologies, which is Zap only.
        s    - Skip the playback of instructions for leaving a message
to the
               calling party.
        u    - Play the 'unavailable' greeting.
```

Diff of the internal help from Asterisk 1.4 to 1.6:

```
--- in Asterisk 1.4
+++ in Asterisk 1.6
@@ -4,7 +4,7 @@
   Leave a Voicemail message
```

```
    [Description]
-     VoiceMail(mailbox[@context][&mailbox[@context]][...][|options]):
This
+     VoiceMail(mailbox[@context][&mailbox[@context]][...][,options]):
This
    application allows the calling party to leave a message for the
 specified
    list of mailboxes. When multiple mailboxes are specified, the
greeting will
    be taken from the first mailbox specified. Dialplan execution will
stop if the
@@ -20,11 +20,13 @@

    Options:
        b    - Play the 'busy' greeting to the calling party.
+       d([c]) - Accept digits for a new extension in context c, if
played during
+              the greeting.  Context defaults to the current context.
        g(#) - Use the specified amount of gain when recording the
voicemail
               message. The units are whole-number decibels (dB).
-              Only works on supported technologies, which is Zap only.
+              Only works on supported technologies, which is DAHDI
only.
        s    - Skip the playback of instructions for leaving a message
to the
               calling party.
        u    - Play the 'unavailable' greeting.
-       j    - Jump to priority n+101 if the mailbox is not found or
some other
-              error occurs.
+       U    - Mark message as Urgent.
+       P    - Mark message as PRIORITY.
```

See also voicemailmain, voicemail.conf.

B.188 VoiceMailMain()

Allows the caller to check voice mail messages.

```
VoiceMailMain([mailbox][@context][,options])
```

Old syntax:

```
VoiceMailMain([[s|p]mailbox][@context])
```

Allows access to the mailbox for listening to messages. If the mailbox number is not specified, the system prompts the caller for the mailbox number.

Option s skips the password prompt. Option p (prefix) prompts the caller to enter a mailbox number; the number specified in the command is then used as a prefix to the number provided by the caller and the resulting string is used as the mailbox number. This can be useful with virtual mailbox hosting. Option a (*folder*) sends the caller directly to the specified folder (default: INBOX).

If a context is specified, only mailboxes in the specified context are accessible. Returns −1 if the caller hangs up, otherwise 0.

```
; go to the voicemail menu for mailbox 123 in the default context:
exten => 123,1,VoiceMailMain(123@default)
```

Asterisk versions:

```
--------| 1.2 |--------| 1.4 |--------| 1.6 |--------
```

Internal help for this application in Asterisk 1.4:

```
    -= Info about application 'VoiceMailMain' =-

[Synopsis]
Check Voicemail messages

[Description]
  VoiceMailMain([mailbox][@context][|options]): This application allows
the
calling party to check voicemail messages. A specific mailbox, and
optional
corresponding context, may be specified. If a mailbox is not provided,
the
calling party will be prompted to enter one. If a context is not
specified,
the 'default' context will be used.

  Options:
      p   - Consider the mailbox parameter as a prefix to the mailbox
that
            is entered by the caller.
    g(#) - Use the specified amount of gain when recording a voicemail
            message. The units are whole-number decibels (dB).
      s   - Skip checking the passcode for the mailbox.
    a(#) - Skip folder prompt and go directly to folder specified.
            Defaults to INBOX
```

Diff of the internal help from Asterisk 1.2 to 1.4:

```
--- in Asterisk 1.2
+++ in Asterisk 1.4
@@ -16,3 +16,5 @@
        g(#) - Use the specified amount of gain when recording a
voicemail
            message. The units are whole-number decibels (dB).
        s   - Skip checking the passcode for the mailbox.
+      a(#) - Skip folder prompt and go directly to folder specified.
+            Defaults to INBOX
```

Diff of the internal help from Asterisk 1.4 to 1.6:

```
--- in Asterisk 1.4
+++ in Asterisk 1.6
@@ -4,7 +4,7 @@
   Check Voicemail messages

   [Description]
-    VoiceMailMain([mailbox][@context][|options]): This application
allows the
+    VoiceMailMain([mailbox][@context][,options]): This application allows
the
   calling party to check voicemail messages. A specific mailbox, and
optional
   corresponding context, may be specified. If a mailbox is not
provided, the
   calling party will be prompted to enter one. If a context is not
specified,
```

See also voicemail, voicemail.conf.

B.189 Wait()

Waits for the specified number of seconds.

```
Wait(seconds)
```

Waits the specified number of seconds, then returns 0. Fractions are allowed (for example, 1.5).

```
exten => s,1,Answer()
exten => s,n,Wait(1.5)
exten => s,n,Background(enter-ext-of-person)
```

Asterisk versions:

```
--------| 1.2 |--------| 1.4 |--------| 1.6 |--------
```

Internal help for this application in Asterisk 1.4:

```
   -= Info about application 'Wait' =-

[Synopsis]
Waits for some time

[Description]
  Wait(seconds): This application waits for a specified number of
seconds.
Then, dialplan execution will continue at the next priority.
  Note that the seconds can be passed with fractions of a second. For
example,
'1.5' will ask the application to wait for 1.5 seconds.
```

Diff of the internal help from Asterisk 1.2 to 1.4:

None

Diff of the internal help from Asterisk 1.4 to 1.6:

None

See also waitexten.

B.190 WaitExten()

Waits for an extension to be dialed.

```
WaitExten([seconds][,options])
```

Waits the specified number of seconds for the caller to dial a new extension, then returns 0. Fractions are allowed (for example, 1.5). If no time is specified, the default extension timeout is used.

Option m plays music-on-hold to the caller while waiting for input. The music-on-hold class may be specified in parentheses (e.g., m (rock)).

```
; Wait 10 seconds for an extension:
exten => s,1,Answer()
exten => s,n,Playback(enter-ext-of-person)
exten => s,n,WaitExten(10)
```

Asterisk versions:

```
--------| 1.2 |--------| 1.4 |--------| 1.6 |--------
```

Internal help for this application in Asterisk 1.4:

```
    -= Info about application 'WaitExten' =-

[Synopsis]
Waits for an extension to be entered

[Description]
  WaitExten([seconds][|options]): This application waits for the
user to enter
a new extension for a specified number of seconds.
  Note that the seconds can be passed with fractions of a second.
For example,
'1.5' will ask the application to wait for 1.5 seconds.
  Options:
    m[(x)] - Provide music on hold to the caller while waiting for
an extension.
             Optionally, specify the class for music on hold
within parentheses.
See also Playback(application), Background(application).
```

Diff of the internal help from Asterisk 1.2 to 1.4:

```
--- in Asterisk 1.2
+++ in Asterisk 1.4
@@ -11,3 +11,4 @@
     Options:
        m[(x)] - Provide music on hold to the caller while waiting
for an extension.
                    Optionally, specify the class for music on hold
within parenthesis.
+   See Also: Playback(application), Background(application).
```

Diff of the internal help from Asterisk 1.4 to 1.6:

```
--- in Asterisk 1.4
+++ in Asterisk 1.6
@@ -4,7 +4,7 @@
   Waits for an extension to be entered

   [Description]
-    WaitExten([seconds][|options]): This application waits for the
user to enter
+    WaitExten([seconds][,options]): This application waits for the
user to enter
    a new extension for a specified number of seconds.
       Note that the seconds can be passed with fractions of a second.
For example,
       '1.5' will ask the application to wait for 1.5 seconds.
```

See also wait.

B.191 WaitForNoise()

Waits for noise on the active channel.

```
WaitForNoise(duration[,iterations[,timeout]])
```

Waits for noise of *duration* milliseconds. If *iterations* are specified, it will wait for that number of noise events of *duration* milliseconds, otherwise waits for one instance. If *timeout* is specified, call is hung up after that number of seconds.

```
; listen for 3 noises of 500 ms duration:
exten => 123,1,Answer()
exten => 123,n,WaitForNoise(500,3)
exten => 123,n,Playback(say-something-already)
```

Asterisk versions:

```
    |       |           |       |           | 1.6 |--------
```

Internal help for this application in Asterisk 1.6:

```
   -= Info about application 'WaitForNoise' =-

[Synopsis]
Waits for a specified amount of noise

[Description]
WaitForNoise(noiserequired[,iterations][,timeout])
Wait for Noise: The same as Wait for Silence but waits for noise
that is above the threshold specified
```

Diff of the internal help from Asterisk 1.4 to 1.6:

Not available in Asterisk 1.4

See also waitforsilence.

B.192 WaitForRing()

Waits the specified number of seconds for a ring signal.

```
WaitForRing(timeout)
```

Waits a maximum of *timeout* seconds for a ring signal, which is only treated as valid until the second ring has completed.

Returns 0 on success, or –1 if the channel is hung up.

```
; Wait 5 seconds for ring, then send DTMF:
exten => 123,1,Answer()
exten => 123,n,WaitForRing(5)
exten => 123,n,SendDTMF(1234)
```

Asterisk versions:

```
--------| 1.2 |--------| 1.4 |--------| 1.6 |--------
```

Internal help for this application in Asterisk 1.4:

```
   -= Info about application 'WaitForRing' =-

[Synopsis]
Wait for Ring Application

[Description]
  WaitForRing(timeout)
Returns 0 after waiting at least timeout seconds. and
only after the next ring has completed.  Returns 0 on
success or -1 on hangup
```

Diff of the internal help from Asterisk 1.2 to 1.4:

None

Diff of the internal help from Asterisk 1.4 to 1.6:

```
--- in Asterisk 1.4
+++ in Asterisk 1.6
@@ -4,7 +4,7 @@
   Wait for Ring Application

   [Description]
-    WaitForRing(timeout)
+    WaitForRing(timeout):
   Returns 0 after waiting at least timeout seconds. and
   only after the next ring has completed.  Returns 0 on
   success or -1 on hangup
```

B.193 WaitForSilence()

Waits for silence of a specified duration.

```
WaitForSilence(duration[,repeats[,timeout]])
```

Waits for *duration* milliseconds of silence. If *repeats* are specified, the application waits until it hears at least that many instances of silence of the specified duration. If *timeout* is specified, the call terminates after that many seconds.

```
; Wait for 2 silences of at least 500 ms:
exten => 123,1,Answer()
exten => 123,n,WaitForSilence(500,2)
exten => 123,n,Playback(the-sound-of-silence)
```

Asterisk versions:

```
--------| 1.2 |--------| 1.4 |--------| 1.6 |--------
```

Internal help for this application in Asterisk 1.4:

```
   -= Info about application 'WaitForSilence' =-

[Synopsis]
Waits for a specified amount of silence

[Description]
  WaitForSilence(silencerequired[|iterations][|timeout])
Wait for Silence: Waits for up to 'silencerequired'
milliseconds of silence, 'iterations' times or once if omitted.
An optional timeout specified the number of seconds to return
after, even if we do not receive the specified amount of silence.
```

Use 'timeout' with caution, as it may defeat the purpose of this
application, which is to wait indefinitely until silence is detected
on the line. This is particularly useful for reverse-911-type
call broadcast applications where you need to wait for an answering
machine to complete its spiel before playing a message.
The timeout parameter is specified only to avoid an infinite loop in
cases where silence is never achieved. Typically you will want to
include two or more calls to WaitForSilence when dealing with an
answering
machine; first waiting for the spiel to finish, then waiting for the
beep, etc.

Examples:
  - WaitForSilence(500|2) will wait for 1/2 second of silence, twice
  - WaitForSilence(1000) will wait for 1 second of silence, once
  - WaitForSilence(300|3|10) will wait for 300ms silence, 3 times,
      and returns after 10 sec, even if silence is not detected

Sets the channel variable WAITSTATUS with to one of these values:
SILENCE - if exited with silence detected
TIMEOUT - if exited without silence detected after timeout

Diff of the internal help from Asterisk 1.2 to 1.4:

```
--- in Asterisk 1.2
+++ in Asterisk 1.4
@@ -4,8 +4,27 @@
   Waits for a specified amount of silence

   [Description]
-   WaitForSilence(x[|y]) Wait for Silence: Waits for up to 'x'
-   milliseconds of silence, 'y' times or 1 if omitted
-   Set the channel variable WAITSTATUS with to one of these
values:SILENCE - if silence of x ms was detectedTIMEOUT - if silence of
x ms was not detected.Examples:
+   WaitForSilence(silencerequired[|iterations][|timeout])
+   Wait for Silence: Waits for up to 'silencerequired'
+   milliseconds of silence, 'iterations' times or once if omitted.
+   An optional timeout specified the number of seconds to return
+   after, even if we do not receive the specified amount of silence.
+   Use 'timeout' with caution, as it may defeat the purpose of this
+   application, which is to wait indefinitely until silence is detected
+   on the line.  This is particularly useful for reverse-911-type
+   call broadcast applications where you need to wait for an answering
+   machine to complete its spiel before playing a message.
+   The timeout parameter is specified only to avoid an infinite loop in
+   cases where silence is never achieved.  Typically you will want to
+   include two or more calls to WaitForSilence when dealing with an
answering
+   machine; first waiting for the spiel to finish, then waiting for
the beep, etc.
+
```

```
+   Examples:
      - WaitForSilence(500|2) will wait for 1/2 second of silence, twice
      - WaitForSilence(1000) will wait for 1 second of silence, once
+     - WaitForSilence(300|3|10) will wait for 300ms silence, 3 times,
+         and returns after 10 sec, even if silence is not detected
+
+   Sets the channel variable WAITSTATUS with to one of these values:
+   SILENCE - if exited with silence detected
+   TIMEOUT - if exited without silence detected after timeout
```

Diff of the internal help from Asterisk 1.4 to 1.6:

```
--- in Asterisk 1.4
+++ in Asterisk 1.6
@@ -4,7 +4,7 @@
   Waits for a specified amount of silence

   [Description]
-    WaitForSilence(silencerequired[|iterations][|timeout])
+    WaitForSilence(silencerequired[,iterations][,timeout]):
   Wait for Silence: Waits for up to 'silencerequired'
   milliseconds of silence, 'iterations' times or once if omitted.
   An optional timeout specified the number of seconds to return
@@ -20,9 +20,9 @@
   machine; first waiting for the spiel to finish, then waiting for
the beep, etc.

   Examples:
-     - WaitForSilence(500|2) will wait for 1/2 second of silence, twice
+     - WaitForSilence(500,2) will wait for 1/2 second of silence, twice
      - WaitForSilence(1000) will wait for 1 second of silence, once
-     - WaitForSilence(300|3|10) will wait for 300ms silence, 3 times,
+     - WaitForSilence(300,3,10) will wait for 300ms silence, 3 times,
          and returns after 10 sec, even if silence is not detected

   Sets the channel variable WAITSTATUS with to one of these values:
```

B.194 WaitMusicOnHold()

Play music-on-hold while waiting for the specified number of seconds.

```
WaitMusicOnHold(duration)
```

Plays music-on-hold while waiting for the specified number of seconds. If no hold music is available, it waits anyway, but without playing music.

Returns 0 on completion, or −1 if the channel is hung up.

```
; 5 minutes of hold music:
exten => 123,1,Answer()
exten => 123,n,WaitMusicOnHold(300)
exten => 123,n,Hangup()
```

Asterisk versions:

```
--------| 1.2 |--------| 1.4 |--------| 1.6 |--------
```

Internal help for this application in Asterisk 1.4:

```
    -= Info about application 'WaitMusicOnHold' =-

[Synopsis]
Wait, playing Music On Hold

[Description]
WaitMusicOnHold(delay): Plays hold music specified number of
seconds.  Returns 0 when
done, or -1 on hangup.  If no hold music is available, the delay will
still occur with no sound.
```

Diff of the internal help from Asterisk 1.2 to 1.4:

None

Diff of the internal help from Asterisk 1.4 to 1.6:

```
--- in Asterisk 1.4
+++ in Asterisk 1.6
@@ -4,6 +4,12 @@
    Wait, playing Music On Hold

    [Description]
-  WaitMusicOnHold(delay): Plays hold music specified number of
seconds.  Returns 0 when
+    WaitMusicOnHold(delay):
+
+    !!! DEPRECATED. Use MusicOnHold instead !!!
+
+  Plays hold music specified number of seconds.  Returns 0 when
    done, or -1 on hangup.  If no hold music is available, the delay
will
    still occur with no sound.

+    !!! DEPRECATED. Use MusicOnHold instead !!!
```

See also musiconhold.conf.

B.195 WaitUntil()

Waits until the specified time.

```
WaitUntil(unix_time)
```

Waits until the specified number of seconds from the start of the epoch (also called "UNIX time").

Sets the channel variable `WAITUNTILSTATUS` to one of the following values:

- **OK**

  Time elapsed.

- **FAILURE**

  Invalid time specification.

- **HANGUP**

  Time elapsed.

- **PAST**

  Time elapsed.

```
exten => s,1,Answer()
exten => s,n,WaitUntil(${STRPTIME(2008-12-06 14:30:00|Europe/
Berlin|%Y-%m-%d %H:%M:%S)})
exten => s,n,SayUnixTime()
```

Asterisk versions:

```
 |     |       |     |       | 1.6 |--------
```

Internal help for this application in Asterisk 1.6:

```
    -= Info about application 'WaitUntil' =-

[Synopsis]
Wait (sleep) until the current time is the given epoch

[Description]
  WaitUntil(<epoch>): Waits until the given time.  Sets
WAITUNTILSTATUS to
one of the following values:
   OK       Wait succeeded
   FAILURE  Invalid argument
   HANGUP   Channel hung up before time elapsed
   PAST     The time specified was already past
```

Diff of the internal help from Asterisk 1.4 to 1.6:

Not available in Asterisk 1.4

See also waituntil.

B.196 While()

Starts a while loop.

```
While(expression)
```

Starts a while loop. The application returns to this point if **EndWhile()** is encountered as long as expression is true; if it is false, execution continues after **EndWhile()**.

```
exten => 123,1,Answer()
exten => 123,n,Set(i=1)
exten => 123,n,While($[${i} < 5])
exten => 123,n,SayNumber(${i})
exten => 123,n,Set(i=$[${i} + 1])
exten => 123,n,EndWhile()
exten => 123,n,Hangup()
```

Asterisk versions:

```
--------| 1.2 |--------| 1.4 |--------| 1.6 |--------
```

Internal help for this application in Asterisk 1.4:

```
    -= Info about application 'While' =-

[Synopsis]
Start a while loop

[Description]
Usage:  While(<expr>)
Start a While Loop.  Execution will return to this point when
EndWhile is called until expr is no longer true.
```

Diff of the internal help from Asterisk 1.2 to 1.4:

```
--- in Asterisk 1.2
+++ in Asterisk 1.4
@@ -1,7 +1,7 @@
    -= Info about application 'While' =-

  [Synopsis]
- Start A While Loop
+ Start a while loop

  [Description]
  Usage:  While(<expr>)
```

Diff of the internal help from Asterisk 1.4 to 1.6:

```
--- in Asterisk 1.4
+++ in Asterisk 1.6
@@ -4,6 +4,5 @@
  Start a while loop
```

```
      [Description]
  -   Usage:   While(<expr>)
  -   Start a While Loop.  Execution will return to this point when
  -   EndWhile is called until expr is no longer true.
  +     While(<expr>): Start a While Loop.  Execution will return to this
  +   point when EndWhile() is called until expr is no longer true.
```

See also **endwhile, exitwhile, continuewhile, gotoif.**

B.197 Zapateller()

Generates the "Special Information Tone" to block advance dialing telemarketing systems.

```
      Zapateller(options)
```

Generates the "Special Information Tone" to indicate the number is not valid or reachable. Some predictive dialing systems will automatically delete a number from the calling list if they receive this tone.

The following pipe-delimited options are accepted:

- **answer**

  Answers the line before playing the tone sequence.

- **nocallerid**

  Plays the tone only if no caller ID information is received.

```
[incoming]
; Play the SIT sequence if no caller ID is present:
exten => s,1,Zapateller(nocallerid)
exten => s,n,Wait(3)
exten => s,n,Answer()
```

Asterisk versions:

```
--------| 1.2 |--------| 1.4 |--------| 1.6 |--------
```

Internal help for this application in Asterisk 1.4:

```
   -= Info about application 'Zapateller' =-

[Synopsis]
Block telemarketers with SIT

[Description]
  Zapateller(options):  Generates special information tone to block
telemarketers from calling you.  Options is a pipe-delimited list of
options.  The following options are available:
'answer' causes the line to be answered before playing the tone,
```

```
'nocallerid' causes Zapateller to only play the tone if there
is no callerid information available.  Options should be separated by |
characters
```

Diff of the internal help from Asterisk 1.2 to 1.4:

None

Diff of the internal help from Asterisk 1.4 to 1.6:

```
--- in Asterisk 1.4
+++ in Asterisk 1.6
@@ -7,7 +7,12 @@
    Zapateller(options):  Generates special information tone to block
  telemarketers from calling you.  Options is a pipe-delimited list of
  options.  The following options are available:
-  'answer' causes the line to be answered before playing the tone,
-  'nocallerid' causes Zapateller to only play the tone if there
-  is no callerid information available.  Options should be separated
by |
-  characters
+      'answer'      - causes the line to be answered before playing the
tone,
+      'nocallerid' - causes Zapateller to only play the tone if there
is no
+                     callerid information available.  Options should
be
+                     separated by , characters
+
+    This application will set the following channel variable upon
completion:
+      ZAPATELLERSTATUS - This will contain the last action
accomplished by the
+                         Zapateller application. Possible values
include:
+                         NOTHING | ANSWERED | ZAPPED
```

See also **privacymanager**.

B.198 ZapBarge()

Allows eavesdropping on a Zap channel.

```
ZapBarge([channel])
```

Allows eavesdropping on a Zap channel. Other participants in the call do not hear the eavesdropper and do not receive any indication that the channel is being monitored.

If the channel is not provided, the user is prompted to enter it, following by the # key. In this case, to barge on Zap/4, press 4 and #.

Returns –1, if the caller hangs up, whether or not a channel is being monitored.

```
exten => 123,1,ZapBarge(Zap/2)
exten => 123,n,Hangup()
```

Asterisk versions:

```
--------| 1.2 |--------| 1.4 |--------| 1.6*|--------
```

(* different name)

Internal help for this application in Asterisk 1.4:

```
   -= Info about application 'ZapBarge' =-

[Synopsis]
Barge in (monitor) Zap channel

[Description]
  ZapBarge([channel]): Barges in on a specified zap
channel or prompts if one is not specified.  Returns
-1 when caller user hangs up and is independent of the
state of the channel being monitored.
```

Diff of the internal help from Asterisk 1.2 to 1.4:

None

Diff of the internal help from Asterisk 1.4 to 1.6 (dahdibarge):

```
--- in Asterisk 1.4
+++ in Asterisk 1.6
@@ -1,10 +1,10 @@
-     -= Info about application 'ZapBarge' =-
+     -= Info about application 'DAHDIBarge' =-

   [Synopsis]
- Barge in (monitor) Zap channel
+ Barge in (monitor) DAHDI channel

   [Description]
-   ZapBarge([channel]): Barges in on a specified zap
+   DAHDIBarge([channel]): Barges in on a specified DAHDI
   channel or prompts if one is not specified.  Returns
   -1 when caller user hangs up and is independent of the
   state of the channel being monitored.
```

See also zapscan, chanspy.

B.199 ZapRAS()

Starts the Zaptel ISDN Remote Access Server.

```
ZapRAS(args)
```

Starts an ISDN RAS Server using **pppd** on the current channel. The channel must be available and a Zap channel. There is no modem emulation.

The point-to-point daemon **pppd** must be configured to recognize a Zap interface. The *args* are a pipe-delimited list of parameters.[21]

Returns −1.

This application is only intended for use with ISDN lines, and your kernel must be patched and configured to support **ZapRAS()**, as well as be configured to support PPP.

```
exten => 123,1,Answer()
exten => 123,n,ZapRAS(debug|64000|noauth|netmask|255.255.255.0|
10.0.0.1:10.0.0.2)
```

Asterisk versions:

```
--------| 1.2 |--------| 1.4 |--------| 1.6*|--------
```

(* different name)

Internal help for this application in Asterisk 1.4:

```
    -= Info about application 'ZapRAS' =-

[Synopsis]
Executes Zaptel ISDN RAS application

[Description]
  ZapRAS(args): Executes a RAS server using pppd on the given channel.
The channel must be a clear channel (i.e. PRI source) and a Zaptel
channel to be able to use this function (No modem emulation is
included).
Your pppd must be patched to be zaptel aware. Arguments should be
separated by | characters.
```

Diff of the internal help from Asterisk 1.2 to 1.4:

None

---

21. The list is long and complex and would not be appropriate in this summary. For more details, see www.voip-info.org/wiki/view/Asterisk+cmd+ZapRAS.

Diff of the internal help from Asterisk 1.4 to 1.6 (dahdiras):

```
--- in Asterisk 1.4
+++ in Asterisk 1.6
@@ -1,11 +1,11 @@
-     -= Info about application 'ZapRAS' =-
+     -= Info about application 'DAHDIRAS' =-

   [Synopsis]
-  Executes Zaptel ISDN RAS application
+  Executes DAHDI ISDN RAS application

   [Description]
-     ZapRAS(args): Executes a RAS server using pppd on the given
channel.
-  The channel must be a clear channel (i.e. PRI source) and a Zaptel
+     DAHDIRAS(args): Executes a RAS server using pppd on the given
channel.
+  The channel must be a clear channel (i.e. PRI source) and a DAHDI
   channel to be able to use this function (No modem emulation is
included).
-  Your pppd must be patched to be zaptel aware. Arguments should be
-  separated by | characters.
+  Your pppd must be patched to be DAHDI aware. Arguments should be
+  separated by , characters.
```

B.200 ZapScan()

Scans through Zap channels for eavesdropping.

```
ZapScan([group])
```

Enables a call center manager to monitor Zap channels quickly and conveniently.
Press # to switch to the next channel or * to exit. You may limit the available channels by specifying group.

```
exten => 123,1,ZapScan()
```

Asterisk versions:

```
--------| 1.2 |--------| 1.4 |--------| 1.6*|--------
```

(* different name)

Internal help for this application in Asterisk 1.4:

```
   -= Info about application 'ZapScan' =-

[Synopsis]
Scan Zap channels to monitor calls
```

```
[Description]
  ZapScan([group]) allows a call center manager to monitor Zap
channels in
a convenient way.  Use '#' to select the next channel and use '*'
to exit
Limit scanning to a channel GROUP by setting the option group argument.
```

Diff of the internal help from Asterisk 1.2 to 1.4:

None

Diff of the internal help from Asterisk 1.4 to 1.6 (dahdiscan):

```
--- in Asterisk 1.4
+++ in Asterisk 1.6
@@ -1,9 +1,9 @@
-    -= Info about application 'ZapScan' =-
+    -= Info about application 'DAHDIScan' =-

   [Synopsis]
-  Scan Zap channels to monitor calls
+  Scan DAHDI channels to monitor calls

   [Description]
-    ZapScan([group]) allows a call center manager to monitor Zap
channels in
+    DAHDIScan([group]) allows a call center manager to monitor DAHDI
channels in
   a convenient way.  Use '#' to select the next channel and use '*' to
exit
   Limit scanning to a channel GROUP by setting the option group
argument.
```

See also zapbarge, chanspy, and in Appendix C, see group.

B.201 ZapSendKeypadFacility()

Sends digits out-of-band on an ISDN PRI connection.

```
ZapSendKeypadFacility(digits)
```

Sends the supplied digits (0-9, *, #) out-of-band using the Q.931 "keypad facility" information element on a PRI connection.

```
exten => 123,1,ZapSendKeypadFacility(*456)
```

Asterisk versions:

```
    |        |        | 1.4 |--------| 1.6*|--------
```

(* different name)

Internal help for this application in Asterisk 1.4:

```
   -= Info about application 'ZapSendKeypadFacility' =-

[Synopsis]
Send digits out of band over a PRI

[Description]
  ZapSendKeypadFacility(): This application will send the given
string of digits in a Keypad Facility
  IE over the current channel.
```

Diff of the internal help from Asterisk 1.2 to 1.4:

Not available in Asterisk 1.2

Diff of the internal help from Asterisk 1.4 to 1.6 (dahdisendkeypadfacility):

```
--- in Asterisk 1.4
+++ in Asterisk 1.6
@@ -1,8 +1,8 @@
-     -= Info about application 'ZapSendKeypadFacility' =-
+     -= Info about application 'DAHDISendKeypadFacility' =-

   [Synopsis]
   Send digits out of band over a PRI

   [Description]
-     ZapSendKeypadFacility(): This application will send the given
string of digits in a Keypad Facility
+     DAHDISendKeypadFacility(): This application will send the given
string of digits in a Keypad Facility
      IE over the current channel.
```

See also **dahdisendkeypadfacility**.

Dialplan Functions

In addition to dialplan applications, which have been part of Asterisk almost from the very beginning, Asterisk supports functions as of Asterisk 1.2. This is part of a long-standing effort to make Asterisk behave more like a programming environment. In contrast to applications, functions may not be called directly. Instead, they are called inside applications and return a value, or—in a departure from the classical definition of a function—they may even be written to using the application **Set()** (see Appendix B, "Dialplan Applications"). Function names are always written in uppercase letters. Surprisingly, functions are written in the same way as variables, inside curly braces and preceded by a $ character (${}). This is necessary because strings are not always bounded by quotation marks.

We could be forgiven for criticizing the less-than-intuitive distinction between Asterisk applications, functions, and even variables. Nor is there a consistently applied naming convention: for example, **SIP_HEADER()** is broken by an underscore (_) but **SIPCHANINFO()** is not. This is a problem with many programming languages and environments; these differences in convention add no useful information but make learning more difficult. In addition to this, the use of the delimiters—comma (,), ampersand (&), and pipe (I)—appears arbitrary. The concept of writing to a function in the same way one might write to a variable goes counter to the basic definition of a function in nearly every other programming language and continues to cause confusion, particularly among new Asterisk users with programming backgrounds.

A better way to imagine a function in Asterisk is as an associative array[1] (a hash table is a form of associative array). For example:

```
Set(CHANNEL(language)=de);
NoOp(Language: ${CHANNEL(language)});
```

would look like this in PHP:

```
$channel["language"] = "de";
echo "Language:", $channel["language"];
```

Many functions allow the key in our associative array analogy to be empty. For example:

```
Set(GROUP()=outgoing);
```

Despite considerable improvements in Version 1.4, the dialplan programming remains rather inflexible when compared with "real" programming languages. If this proves bothersome, you might consider exploring Asterisk Extension Language (AEL) in more depth. It uses the same functions and applications but has a more robust structure and is often easier to interpret.

To find out which functions are currently available in your installation, enter **core show functions** and **core show function** *FUNCTIONNAME* (or **show functions** and **show function** *FUNCTIONNAME* for Asterisk 1.2) in the CLI.[2]

Note that these commands are case sensitive. Function names must be written entirely in uppercase letters.

C.1 AGENT()

```
AGENT(agent_number[:field])
```

Returns information about an agent identified by agentNumber. The following fields may be queried:

- **status**

  (default) The status of the agent, either LOGGEDIN or LOGGEDOUT.

- **password**

  The agent's password.

---

1. http://en.wikipedia.org/wiki/Associative_array

2. Command-line interface. This may be invoked with **asterisk -r**.

- **name**

  The agent's name.

- **mohclass**

  The music-on-hold class.

- **exten**

  The callback extension for the agent. This is used by **AgentCallbackLogin()**.

- **channel**

  The name of the agent's active channel (used by **AgentLogin()**).

```
; set the variable foo to the name of Agent 42:
exten => 123,1,Set(foo=${AGENT(42:name)})
```

Asterisk versions:

```
  |     |              |1.4|---------|1.6|--------
```

Internal help for this function in Asterisk 1.4:

```
      -= Info about function 'AGENT' =-

[Syntax]
AGENT(<agentid>[:item])

[Synopsis]
Gets information about an Agent

[Description]
The valid items to retrieve are:
- status (default)     The status of the agent
                          LOGGEDIN | LOGGEDOUT
- password             The password of the agent
- name                 The name of the agent
- mohclass             MusicOnHold class
- exten                The callback extension for the Agent
(AgentCallbackLogin)
- channel              The name of the active channel for the Agent
(AgentLogin)
```

Diff of the internal help from Asterisk 1.2 to 1.4:

Not available in Asterisk 1.2

Diff of the internal help from Asterisk 1.4 to 1.6:

None

C.2 ARRAY()

```
ARRAY(var1[,var2[,...]])
```

Sets multiple variables simultaneously. (The name is misleading; programmers should note that this does not behave the way an array normally would.) **ARRAY()** can only be used for writing, not reading. Remember to separate values with commas *and* to escape the commas with a backslash (\\), or Asterisk will treat the following characters as an additional parameter for **Set()**!

```
; Set var1 to 1 and var2 to 2:
exten => 123,1,Set(ARRAY(var1,var2)=1\,2)
ARRAY(var1[,var2[,...]])
```

Important: **ARRAY()** is a misleading function that you don't actually need. Use multiple **Set()** statements instead:

```
Set(var1=word1);
Set(var2=wword2);
```

Asterisk versions:

```
    |       |            |1.4|--------|1.6|--------
```

Internal help for this function in Asterisk 1.4:

```
        -= Info about function 'ARRAY' =-

[Syntax]
ARRAY(var1[|var2[...][|varN]])

[Synopsis]
Allows setting multiple variables at once

[Description]
The comma-separated list passed as a value to which the function is set
will be interpreted as a set of values to which the comma-separated list
of variable names in the argument should be set.
Hence, Set(ARRAY(var1|var2)=1\,2) will set var1 to 1 and var2 to 2
Note: remember to either backslash your commas in extensions.conf or
quote the entire argument, since Set can take multiple arguments itself.
```

Diff of the internal help from Asterisk 1.2 to 1.4:

Not available in Asterisk 1.2

Diff of the internal help from Asterisk 1.4 to 1.6:

```
--- in Asterisk 1.4
+++ in Asterisk 1.6
@@ -1,7 +1,7 @@
      -= Info about function 'ARRAY' =-

   [Syntax]
-  ARRAY(var1[|var2[...][|varN]])
+  ARRAY(var1[,var2[...][,varN]])

   [Synopsis]
   Allows setting multiple variables at once
@@ -10,6 +10,4 @@
   The comma-separated list passed as a value to which the function is
set will be interpreted as a set of values to which the comma-separated
list of variable names in the argument should be set.
-  Hence, Set(ARRAY(var1|var2)=1\,2) will set var1 to 1 and var2 to 2
-  Note: remember to either backslash your commas in extensions.conf or
quote the entire argument, since Set can take multiple arguments itself.
+  Hence, Set(ARRAY(var1,var2)=1,2) will set var1 to 1 and var2 to 2.
```

C.3 BASE64_DECODE()

```
BASE64_DECODE(base64_string)
```

Decodes a base64-encoded string.

```
exten => 123,1,Set(foo=${BASE64_DECODE("SGFsbG8gV2VsdA==")})
```

Asterisk versions:

```
|        |         |1.4|--------|1.6|--------
```

Internal help for this function in Asterisk 1.4:

```
     -= Info about function 'BASE64_DECODE' =-

[Syntax]
BASE64_DECODE(<base64_string>)

[Synopsis]
Decode a base64 string

[Description]
Returns the plain text string
```

Diff of the internal help from Asterisk 1.2 to 1.4:

Not available in Asterisk 1.2

Diff of the internal help from Asterisk 1.4 to 1.6:

None

C.4 BASE64_ENCODE()

```
BASE64_ENCODE(string)
```

Encodes a string in base64.

```
exten => 123,1,Set(foo=${BASE64_ENCODE("Hey World")})
```

Asterisk versions:

```
   |        |              |1.4|--------|1.6|--------
```

Internal help for this function in Asterisk 1.4:

```
        -= Info about function 'BASE64_ENCODE' =-

[Syntax]
BASE64_ENCODE(<string>)

[Synopsis]
Encode a string in base64

[Description]
Returns the base64 string
```

Diff of the internal help from Asterisk 1.2 to 1.4:

Not available in Asterisk 1.2

Diff of the internal help from Asterisk 1.4 to 1.6:

None

C.5 BLACKLIST

```
BLACKLIST()
```

This function checks to see whether the caller ID on the channel is on the blacklist. The Asterisk database (AstDB) is used. The caller ID (NAME) or caller ID (NUMBER) must be stored in the database in the `blacklist` family. The key name is irrelevant.

BLACKLIST() only tests, it does not perform any operations. If the caller ID is found to match, it returns 1, otherwise it returns 0.

```
; Check to see whether the caller is on the blacklist.
; If yes, play monkeys!
exten => 123,1,GotoIf(${BLACKLIST()}?black,1)
exten => 123,n,Dial(SIP/200)
exten => 123,n,Hangup()
exten => black,1,Playback(tt-monkeys)
exten => black,n,Hangup()
```

Asterisk versions:

```
   |       |              |1.4|--------|1.6|--------
```

Internal help for this function in Asterisk 1.4:

```
    -= Info about function 'BLACKLIST' =-

[Syntax]
BLACKLIST()

[Synopsis]
Check if the callerid is on the blacklist

[Description]
Uses astdb to check if the Caller*ID is in family 'blacklist'. Returns
1 or 0.
```

Diff of the internal help from Asterisk 1.2 to 1.4:

Not available in Asterisk 1.2

Diff of the internal help from Asterisk 1.4 to 1.6:

None

See also **lookupblacklist** in Appendix B.

C.6 CALLERID()

CALLERID(*field*)

Returns or sets information about the caller. The *field* is one of the following:

- **name**

  Name of the caller, as an alphanumeric string. Keeping this string short is recommended (e.g., 15 characters).

- **num**

  Number of the caller, digits only. (Sometimes also found in field `number`, perhaps depending on the Asterisk version.)

- **all**

  Name and number with the number in angle brackets (e.g., `"Robert Cossack <2125558721>"`).

- **ani**

  ANI,[3] for outgoing calls.

- **dnid**

  DNID[4] number. Corresponds to the dialed number. (Sometimes also found in field dnis, perhaps depending on the Asterisk version.)

- **rdnis**

  RDNIS[5] number. The number which was forwarded to the current extension. (This is useful, for example, if the number of the active mailbox does not correspond to that of the dialed extension.)

The old channel variable `${CALLERIDNUM}` is replaced by the function `${CALLERID(num)}` as of Asterisk 1.4 (Similarly, `${RDNIS}` is replaced by `$(CALLERID(rdnis))` etc.). The application **SetCIDName()** is replaced by **Set(CALLERID(name)=*Name*)** (Similarly, **SetCallerID()** is replaced by **Set(CALLERID(all)=*Name* <*Number*>)** and so on).

```
; Set the variable foo to the complete caller ID:
exten => 123,1,Set(foo=${CALLERID(all)})

; Set the caller name to "Robert Cossack":
exten => 123,1,Set(CALLERID(name)="Robert Cossack")
```

Asterisk versions:

```
--------|1.2|--------|1.4|--------|1.6|--------
```

---

3. Automatic Number Identification

4. Dialed/Destination Number Identification Service

5. Redirected Dialed Number Identification Service

Internal help for this function in Asterisk 1.4:

```
    -= Info about function 'CALLERID' =-

[Syntax]
CALLERID(datatype[,<optional-CID>])

[Synopsis]
Gets or sets Caller*ID data on the channel.

[Description]
Gets or sets Caller*ID data on the channel.  The allowable datatypes
are "all", "name", "num", "ANI", "DNID", "RDNIS".
Uses channel callerid by default or optional callerid, if specified.
```

Diff of the internal help from Asterisk 1.2 to 1.4:

```
--- in Asterisk 1.2
+++ in Asterisk 1.4
@@ -1,7 +1,7 @@
     -= Info about function 'CALLERID' =-

   [Syntax]
-  CALLERID(datatype)
+  CALLERID(datatype[,<optional-CID>])

   [Synopsis]
   Gets or sets Caller*ID data on the channel.
@@ -9,3 +9,4 @@
   [Description]
   Gets or sets Caller*ID data on the channel.  The allowable datatypes
   are "all", "name", "num", "ANI", "DNID", "RDNIS".
+  Uses channel callerid by default or optional callerid, if specified.
```

Diff of the internal help from Asterisk 1.4 to 1.6:

```
--- in Asterisk 1.4
+++ in Asterisk 1.6
@@ -8,5 +8,6 @@

   [Description]
   Gets or sets Caller*ID data on the channel.  The allowable datatypes
-  are "all", "name", "num", "ANI", "DNID", "RDNIS".
+  are "all", "name", "num", "ANI", "DNID", "RDNIS", "pres",
+  and "ton".
   Uses channel callerid by default or optional callerid, if specified.
```

C.7 CDR()

CDR(*field*)

Reads or sets CDR[6] fields. The *field* is one of the following (only reading is possible unless otherwise noted):

- **clid**
  Caller ID
- **src**
  The source number (caller ID number)
- **dst**
  The call destination
- **dcontext**
  Destination context
- **channel**
  Channel name
- **dstchannel**
  Destination channel, if applicable
- **lastapp**
  The last executed application
- **lastdata**
  The arguments to the last executed application
- **start**
  Time the call started
- **answer**
  Time the call was answered
- **end**
  Time the call ended
- **duration**
  Duration of the call in seconds
- **billsec**
  Duration of the call since the call was answered (in other words, the billable duration) in seconds

---

6. Call Data Record

- **disposition**

  Status of the call: ANSWERED, NO ANSWER, BUSY, or FAILED

- **amaflags**

  The AMA[7] flags. Possible flags are DEFAULT, BILLING, DOCUMENTATION, and
  OMIT. (Sometimes BILLING and OMIT are replaced by BILL and IGNORE, perhaps depending on the Asterisk version.)

- **accountcode**

  The alphanumeric ID of the billing account, maximum 20 characters. May be
  set as well as read.

- **uniqueid**

  The unique ID of the channel (maximum 32 characters).

- **userfield**

  A user field for storing arbitrary information (maximum 255 characters). May
  be set as well as read.

```
; Set foo to the duration of the call:
exten => 123,1,Set(foo=${CDR(duration)})

; Set the user field to "my information":
exten => 123,1,Set(CDR(userfield)=my information)
```

Asterisk versions:

```
--------|1.2|--------|1.4|--------|1.6|--------
```

Internal help for this function in Asterisk 1.4:

```
        -= Info about function 'CDR' =-

[Syntax]
CDR(<name>[|options])

[Synopsis]
Gets or sets a CDR variable

[Description]
Options:
  'l' uses the most recent CDR on a channel with multiple records
  'r' searches the entire stack of CDRs on the channel
  'u' retrieves the raw, unprocessed value
  For example, 'start', 'answer', and 'end' will be retrieved as epoch
  values, when the 'u' option is passed, but formatted as YYYY-MM-DD HH:MM:SS
  otherwise. Similarly, disposition and amaflags will return their raw
  integral values.
```

---

7. Automated Message Accounting

```
    Here is a list of all the available cdr field names:
+    clid           lastdata        disposition
     src            start           amaflags
     dst            answer          accountcode
     dcontext       end             uniqueid
     dstchannel     duration        userfield
     lastapp        billsec         channel
    All of the above variables are read-only, except for accountcode,
    userfield, and amaflags. You may, however, supply
    a name not on the above list, and create your own
    variable, whose value can be changed with this function,
    and this variable will be stored on the cdr.
    raw values for disposition:
        1 = NO ANSWER
        2 = BUSY
        3 = FAILED
        4 = ANSWERED
    raw values for amaflags:
        1 = OMIT
        2 = BILLING
        3 = DOCUMENTATION
```

Diff of the internal help from Asterisk 1.2 to 1.4:

```
--- in Asterisk 1.2
+++ in Asterisk 1.4
@@ -7,4 +7,32 @@
    Gets or sets a CDR variable

    [Description]
-   Option 'r' searches the entire stack of CDRs on the channel
+   Options:
+   'l' uses the most recent CDR on a channel with multiple records
+   'r' searches the entire stack of CDRs on the channel
+   'u' retrieves the raw, unprocessed value
+   For example, 'start', 'answer', and 'end' will be retrieved as epoch
+   values, when the 'u' option is passed, but formatted as YYYY-MM-DD HH:MM:SS
+   otherwise. Similarly, disposition and amaflags will return their raw
+   integral values.
+   Here is a list of all the available cdr field names:
+    clid           lastdata        disposition
+    src            start           amaflags
+    dst            answer          accountcode
+    dcontext       end             uniqueid
+    dstchannel     duration        userfield
+    lastapp        billsec         channel
+   All of the above variables are read-only, except for accountcode,
+   userfield, and amaflags. You may, however, supply
+   a name not on the above list, and create your own
+   variable, whose value can be changed with this function,
```

```
+   and this variable will be stored on the cdr.
+    raw values for disposition:
+        1 = NO ANSWER
+        2 = BUSY
+        3 = FAILED
+        4 = ANSWERED
+    raw values for amaflags:
+        1 = OMIT
+        2 = BILLING
+        3 = DOCUMENTATION
```

Diff of the internal help from Asterisk 1.4 to 1.6:

```
--- in Asterisk 1.4
+++ in Asterisk 1.6
@@ -1,7 +1,7 @@
     -= Info about function 'CDR' =-

    [Syntax]
-   CDR(<name>[|options])
+   CDR(<name>[,options])

    [Synopsis]
    Gets or sets a CDR variable
@@ -10,6 +10,8 @@
    Options:
       'l' uses the most recent CDR on a channel with multiple records
       'r' searches the entire stack of CDRs on the channel
+      's' skips any CDR's that are marked 'LOCKED' due to forkCDR() calls.
+          (on setting/writing CDR vars only)
       'u' retrieves the raw, unprocessed value
       For example, 'start', 'answer', and 'end' will be retrieved as epoch
       values, when the 'u' option is passed, but formatted as YYYY-MM-DD HH:MM:SS
@@ -27,11 +29,13 @@
       a name not on the above list, and create your own
       variable, whose value can be changed with this function,
       and this variable will be stored on the cdr.
+      For setting CDR values, the 'l' flag does not apply to
+      setting the accountcode, userfield, or amaflags.
        raw values for disposition:
            1 = NO ANSWER
-           2 = BUSY
-           3 = FAILED
-           4 = ANSWERED
+           2 = BUSY
+           3 = FAILED
+           4 = ANSWERED
         raw values for amaflags:
            1 = OMIT
            2 = BILLING
```

C.8 CHANNEL()

CHANNEL(*field*)

Reads/sets specific channel parameters. The *field* is one of the following (only reading is possible unless otherwise noted):

- **audioreadformat**
  The format for incoming audio on the channel.

- **audionativeformat**
  The native audio format of the channel.

- **audiowriteformat**
  The format for outgoing audio on the channel.

- **callgroup**
  Extensions in Asterisk can be sorted into call groups numbered from 0 to 63 (e.g., as a client number).[8]

- **channeltype**
  The channel driver, or "technology," of the current channel (e.g., IAX or SIP).

- **language**
  The language for voice prompts. May be set as well as read.

- **musicclass**
  The music-on-hold class, as defined in musiconhold.conf. May be set as well as read.

- **state**
  State of the channel (Down, Rsrvd, OffHook, Dialing, Ring, Ringing, Up, Busy, Dialing Offhook, Pre-ring, Unknown).

- **tonezone**
  The "tone zone" defines the standard tone indications (dialing, ringing, busy, etc.) for specific regions and countries. This is set in the configuration file for the channel driver (e.g., zaptel.conf) with the parameters loadzone and defaultzone. Possible values are (as defined in indications.conf): at, au, be, br, ch, cl, cn, cz, de, dk, ee, es, fi, fr, gr, hu, it, lt, mx, ml, no, nz, pl, pt, ru, se, sg, uk, us, us-old, tw, ve, za.

- **videonativeformat**
  The native video format of the channel.

---

8. This limit of 64 call groups appears to be completely arbitrary and may not be sufficient for all users

In addition to the field described above, specific channel drivers can make others available. To learn more about these, look in the documentation for the specific channel driver. Fields that are unavailable on the current channel return an empty string.

```
; Query the channel type:
exten => 123,1,Set(foo=${CHANNEL(channeltype)})

; Change language to English:
exten => 123,1,Set(CHANNEL(language)=en)
```

Asterisk versions:

```
    |      |              |1.4|--------|1.6|--------
```

Internal help for this function in Asterisk 1.4:

```
      -= Info about function 'CHANNEL' =-

[Syntax]
CHANNEL(item)

[Synopsis]
Gets/sets various pieces of information about the channel.

[Description]
Gets/set various pieces of information about the channel.
Standard items (provided by all channel technologies) are:
R/O      audioreadformat    format currently being read
R/O      audionativeformat  format used natively for audio
R/O      audiowriteformat   format currently being written
R/W      callgroup          call groups for call pickup
R/O      channeltype        technology used for channel
R/W      language           language for sounds played
R/W      musicclass         class (from musiconhold.conf) for hold music
R/W      rxgain             set rxgain level on channel drivers that
                            support it
R/O      state              state for channel
R/W      tonezone           zone for indications played
R/W      txgain             set txgain level on channel drivers that
                            support it
R/O      videonativeformat  format used natively for video

chan_sip provides the following additional options:
R/O      rtpqos         Get QOS information about the RTP stream
         This option takes two additional arguments:
  Argument 1:
    audio                    Get data about the audio stream
    video                    Get data about the video stream
  Argument 2:
    local_ssrc           Local SSRC (stream ID)
    local_lostpackets    Local lost packets
```

```
local_jitter            Local calculated jitter
local_count             Number of received packets
remote_ssrc             Remote SSRC (stream ID)
remote_lostpackets      Remote lost packets
remote_jitter           Remote reported jitter
remote_count            Number of transmitted packets
rtt                     Round trip time
all                     All statistics (in a form suited to logging,
                        but not for parsing)
```

Additional items may be available from the channel driver providing
the channel; see its documentation for details.

Any item requested that is not available on the current channel will
return an empty string.

Diff of the internal help from Asterisk 1.2 to 1.4:

Not available in Asterisk 1.2

Diff of the internal help from Asterisk 1.4 to 1.6:

```
--- in Asterisk 1.4
+++ in Asterisk 1.6
@@ -16,6 +16,7 @@
   R/O     channeltype     technology used for channel
   R/W     language        language for sounds played
   R/W     musicclass      class (from musiconhold.conf) for hold music
+  R/W     parkinglot      parkinglot for parking
   R/W     rxgain          set rxgain level on channel drivers that
                           support it
   R/O     state           state for channel
   R/W     tonezone        zone for indications played
@@ -23,22 +24,52 @@
   R/O videonativeformat   format used natively for video

   chan_sip provides the following additional options:
+  R/O     peerip          Get the IP address of the peer
|  R/O     recvip          Get the source IP address of the peer
+  R/O     from            Get the URI from the From: header
+  R/O     uri             Get the URI from the Contact: header
+  R/O     useragent       Get the useragent
+  R/O     peername        Get the name of the peer
+  R/O     t38passthrough  1 if T38 is offered or enabled in this
                           channel, otherwise
   R/O     rtpqos          Get QOS information about the RTP stream
           This option takes two additional arguments: Argument 1:
     Argument 1:
       audio               Get data about the audio stream
       video               Get data about the video stream
+     text                Get data about the text stream
     Argument 2:
       local_ssrc          Local SSRC (stream ID)
```

```
          local_lostpackets    Local lost packets
          local_jitter         Local calculated jitter
+         local_maxjitter      Local calculated jitter (maximum)
+         local_minjitter      Local calculated jitter (minimum)
+         local_normdevjitter  Local calculated jitter (normal deviation)
+         local_stdevjitter    Local calculated jitter (standard deviation)
          local_count          Number of received packets
          remote_ssrc          Remote SSRC (stream ID)
          remote_lostpackets Remote lost packets
          remote_jitter        Remote reported jitter
+         remote_maxjitter     Remote calculated jitter (maximum)
+         remote_minjitter     Remote calculated jitter (minimum)
+         remote_normdevjitter Remote calculated jitter (normal deviation)
+         remote_stdevjitter   Remote calculated jitter (standard deviation)
          remote_count         Number of transmitted packets
          rtt                  Round trip time
+         maxrtt               Round trip time (maximum)
+         minrtt               Round trip time (minimum)
+         normdevrtt           Round trip time (normal deviation)
+         stdevrtt             Round trip time (standard deviation)
          all                  All statistics (in a form suited to logging, but
                               not for parsing)
+   R/O    rtpdest             Get remote RTP destination information
+         This option takes one additional argument:
+   Argument 1:
+     audio                Get audio destination
+     video                Get video destination
+
+   chan_iax2 provides the following additional options:
+   R/W      osptoken        Get or set the OSP token information for a
call
+   R/O      peerip          Get the peer's ip address
+   R/O      peername        Get the peer's username

     Additional items may be available from the channel driver providing
     the channel; see its documentation for details.
```

C.9 CHECKSIPDOMAIN()

CHECKSIPDOMAIN(domain)

Checks to see whether the specified SIP domain name (may also be an IP address) is local (see sip.conf). Returns the domain name, IP address, or empty string.

```
    exten => 123,1,Set(foo=${CHECKSIPDOMAIN(123.45.67.89)})
```

Asterisk versions:

```
--------| 1.2 |--------| 1.4 |--------| 1.6 |--------
```

Internal help for this function in Asterisk 1.4:

```
      -= Info about function 'CHECKSIPDOMAIN' =-

[Syntax]
CHECKSIPDOMAIN(<domain|IP>)

[Synopsis]
Checks if domain is a local domain

[Description]
This function checks if the domain in the argument is configured
as a local SIP domain that this Asterisk server is configured to handle.
Returns the domain name if it is locally handled, otherwise an empty
string. Check the domain= configuration in sip.conf
```

Diff of the internal help from Asterisk 1.2 to 1.4:

None

Diff of the internal help from Asterisk 1.4 to 1.6:

None

C.10 CURL()

```
CURL(URL [|POST-data])
```

Loads a web page from the specified *URL* using GET. If POST-data are provided, these are sent with POST. Returns the page as a string.

```
; Retrieve http://example.com/page.php?id=1&action=view :
exten => 123,1,Set(foo=${CURL(http://example.com/
page.php?id=1&action=view)})
```

Asterisk versions:

```
--------| 1.2 |--------| 1.4 |--------| 1.6 |--------
```

Internal help for this function in Asterisk 1.4:

```
      -= Info about function 'CURL' =-

[Syntax]
CURL(url[|post-data])

[Synopsis]
Retrieves the contents of a URL
```

```
[Description]
    url          - URL to retrieve
    post-data    - Optional data to send as a POST (GET is default action)
```

Diff of the internal help from Asterisk 1.2 to 1.4:

None

Diff of the internal help from Asterisk 1.4 to 1.6:

```
--- in Asterisk 1.4
+++ in Asterisk 1.6
@@ -1,7 +1,7 @@
    -= Info about function 'CURL' =-

    [Syntax]
-   CURL(url[|post-data])
+   CURL(url[,post-data])

    [Synopsis]
    Retrieves the contents of a URL
```

C.11 CUT()

CUT(*variablename, delimiter, field*)

(As of Asterisk 1.2.8, use a pipe (|) character rather than commas as a parameter delimiter.)

Processes a string in a variable according to a specified delimiter (default: –) and returns the requested fields. The *field* may also be a range of fields (e.g., 2–4) or multiple fields and ranges, separated with & (e.g., 2–4&6; ranges such as 3– (everything from field 3 on), or –3 (everything up to field 3) is possible.

If a comma is used as a delimiter, it must first be escaped with a backslash; e.g., CUT(var,\,,2).

```
exten => 123,1,Set(var=1-2-3-4-5)           ; var is "1-2-3-4-5"
exten => 123,n,Set(var=${CUT(var,,1-3&5)})  ; var is "1-2-3-5"
```

Important: The parameter variablename must be the name of a variable, and not a string. If foo is the variable name and bar the contents, the following example would be incorrect: CUT(${bar},,3)

See also **fieldqty.**

Asterisk versions:

```
--------| 1.2 |--------| 1.4 |--------| 1.6 |--------
```

Internal help for this function in Asterisk 1.4:

```
    -= Info about function 'CUT' =-

[Syntax]
CUT(<varname >,<char-delim >,<range-spec >)

[Synopsis]
Slices and dices strings, based upon a named delimiter.

[Description]
   varname    - variable you want cut
   char-delim - defaults to '-'
   range-spec - number of the field you want (1-based offset)
               may also be specified as a range (with -)
               or group of ranges and fields (with &)
```

Diff of the internal help from Asterisk 1.2 to 1.4:

None

Diff of the internal help from Asterisk 1.4 to 1.6:

None

C.12 DB()

```
DB(family/key)
```

Reads/sets a value in the Asterisk DB (AstDB). When reading, either a value is returned, or an empty string if the key does not exist. The output can be found in the variable DB_RESULT.

```
; Set open/source in the AstDB and then query it:
exten => 123,1,Set(DB(open/source)=${yes})
exten => 123,n,Set(var=${DB(open/source)})
exten => 123,n,GotoIf($[[${DB(open/source)} = 1]?opensource:closedsource)
```

Asterisk versions:

```
--------| 1.2 |--------| 1.4 |--------| 1.6 |--------
```

Internal help for this function in Asterisk 1.4:

```
    -= Info about function 'DB' =-

[Syntax]
DB(<family>/<key>)

[Synopsis]
Read from or write to the Asterisk database

[Description]
This function will read from or write a value to the Asterisk database.
On a read, this function returns the corresponding value from the database,
or blank if it does not exist. Reading a database value will also set the
variable DB_RESULT. If you wish to find out if an entry exists, use the
DB_EXISTS function.
```

Diff of the internal help from Asterisk 1.2 to 1.4:

None

Diff of the internal help from Asterisk 1.4 to 1.6:

None

See also db_exists, db_delete, and in Appendix B, see dbdeltree.

C.13 DB_DELETE()

DB_DELETE(*family/key*)

Deletes a value from the AstDB. Upon completion, the variable DB_RESULT is set to
this value, if it exists.

```
; delete cidnums/4045559814:
exten => 123,1,Set(ignored=${DB_DELETE(cidnums/4045559814)})
```

Note: For versions prior to Asterisk 1.4, use the application **DBdel()**.

Asterisk versions:

```
    |      |           | 1.4 |--------| 1.6 |--------
```

Internal help for this function in Asterisk 1.4:

```
    -= Info about function 'DB_DELETE' =-

[Syntax]
DB_DELETE(<family > /<key >)

[Synopsis]
Return a value from the database and delete it

[Description]
This function will retrieve a value from the Asterisk database
 and then remove that key from the database. DB_RESULT
will be set to the key's value if it exists.
```

Diff of the internal help from Asterisk 1.2 to 1.4:

Not available in Asterisk 1.2

Diff of the internal help from Asterisk 1.4 to 1.6:

None

See also db, db_exists, and in Appendix B, see dbdel, dbdeltree.

C.14 DB_EXISTS()

DB_EXISTS(*family/key*)

Tests to see whether a key exists in the AstDB. Returns 1 or 0. Sets the variable
DB_RESULT to the value of the key, if it exists.

```
; Query if cidnums/4045559814 exists:
exten => 123,1,Set(foo=${DB_EXISTS(cidnums/4045559814)})
```

This is one way to replace the application **LookupBlacklist()**. This code
example causes Asterisk to jump to the context blacklisted, extension s, priority
1, if the CID can be found in the blacklist:

```
exten => 123,1,GotoIf(${DB_EXISTS(blacklist/
${CALLERID(num)})}?blacklisted,s,1)
exten => 123,n,Dial(IAX2/user:password@example.com/500)

[blacklisted]
exten => s,1,NoOp(${CALLERID(num)} is in the blacklist)
exten => s,n,Hangup()
```

Asterisk versions:

```
--------| 1.2 |--------| 1.4 |--------| 1.6 |--------
```

Internal help for this function in Asterisk 1.4:

```
      -= Info about function 'DB_EXISTS' =-

[Syntax]
DB_EXISTS(<family > /<key >)

[Synopsis]
Check to see whether a key exists in the Asterisk database

[Description]
This function will check to see whether a key exists in the Asterisk
database. If it exists, the function will return "1". If not,
it will return "0". Checking for existence of a database key will
also set the variable DB_RESULT to the key's value if it exists.
```

Diff of the internal help from Asterisk 1.2 to 1.4:

None

Diff of the internal help from Asterisk 1.4 to 1.6:

None

See also db, db_delete, and in Appendix B, see dbdeltree.

C.15 DUNDILOOKUP()

```
DUNDILOOKUP(number[|DUNDi-context[|options]])
```

Looks up a telephone number with DUNDi. If DUNDi-context is specified, e164 is assumed. The option b (bypass) will cause Asterisk to bypass the internal DUNDi cache. Returns the found entry in the form technology/resource, if it exists, otherwise returns an empty string.

```
; look up number 5145550123 in DUNDi:
exten => 123,1,Set(foo=${DUNDILOOKUP(5145550123)})
```

Asterisk versions:

```
--------| 1.2 |--------| 1.4 |--------| 1.6 |--------
```

Internal help for this function in Asterisk 1.4:

```
    -= Info about function 'DUNDILOOKUP' =-

[Syntax]
DUNDILOOKUP(number[|context[|options]])

[Synopsis]
Do a DUNDi lookup of a phone number.

[Description]
This will do a DUNDi lookup of the given phone number.
If no context is given, the default will be e164. The result of
this function will the Technology/Resource found in the DUNDi
lookup. If no results were found, the result will be blank.
If the 'b' option is specified, the internal DUNDi cache will
be bypassed.
```

Diff of the internal help from Asterisk 1.2 to 1.4:

None

Diff of the internal help from Asterisk 1.4 to 1.6:

```
--- in Asterisk 1.4
+++ in Asterisk 1.6
@@ -9,7 +9,7 @@
   [Description]
   This will do a DUNDi lookup of the given phone number.
   If no context is given, the default will be e164. The result of
-  this function will the Technology/Resource found in the DUNDi
-  lookup. If no results were found, the result will be blank.
+  this function will return the Technology/Resource found in the first
result
+  in the DUNDi lookup. If no results were found, the result will be
   blank.
   If the 'b' option is specified, the internal DUNDi cache will
   be bypassed.
```

See also dundi.conf.

C.16 ENUMLOOKUP()

Asterisk 1.2:

```
ENUMLOOKUP(number [, service [, optionsANDentrynumber [, zone-suffix]]])
```

Asterisk 1.4:

```
ENUMLOOKUP(number [, service [, options, entrynumber [, zone-suffix]]])
```

Looks up a number with ENUM. The *service* can be sip (default), iax2, h323, tel, or ALL. The option c returns the number of entries. The *entrynumber* (default 1) selects the entry from the list of results. The *zone-suffix* (default: e164.arpa) is the ENUM zone. Comprehensive descriptions and examples may be found in doc/ README.enum (**1.2**)/doc/enum.txt (**1.4**).

```
; in Asterisk 1.2:
exten => 123,1,Set(foo=${ENUMLOOKUP(+${CALLERID(num)},sip,1,freenum.org)})

; in Asterisk 1.4:
exten => 123,1,Set(foo=${ENUMLOOKUP(+${CALLERID(num)},sip
↳,,1,freenum.org)})
```

Asterisk versions:

```
--------| 1.2 |--------| 1.4 |--------| 1.6 |--------
```

Internal help for this function in Asterisk 1.4:

```
    -= Info about function 'ENUMLOOKUP' =-

[Syntax]
ENUMLOOKUP(number[|Method-type[|options[|record#[|zone-suffix]]]])

[Synopsis]
ENUMLOOKUP allows for general or specific querying of NAPTR records
or counts of NAPTR types for E

[Description]
Option 'c' returns an integer count of the number of NAPTRs of a certain
RR type.
Combination of 'c' and Method-type of 'ALL' will return a count of all
NAPTRs for the rec
Defaults are: Method-type=sip, no options, record=1, zone-suffix=e164.arpa

For more information, see doc/enum.txt
```

Diff of the internal help from Asterisk 1.2 to 1.4:

```
--- in Asterisk 1.2
+++ in Asterisk 1.4
@@ -1,7 +1,7 @@
    -= Info about function 'ENUMLOOKUP' =-

 [Syntax]
- ENUMLOOKUP(number[,Method-type[,options|record#[,zone-suffix]]])
+ ENUMLOOKUP(number[|Method-type[|options[|record#[|zone-suffix]]]])

 [Synopsis]
 ENUMLOOKUP allows for general or specific querying of NAPTR records
or counts of NAPTR types for
@@ -11,4 +11,4 @@
```

```
    Combination of 'c' and Method-type of 'ALL' will return a count of
all NAPTRs for the r
    Defaults are: Method-type=sip, no options, record=1, zone-suffix=e164.arpa

-   For more information, see README.enum
+   For more information, see doc/enum.txt
```

Diff of the internal help from Asterisk 1.4 to 1.6:

```
--- in Asterisk 1.4
+++ in Asterisk 1.6
@@ -1,14 +1,18 @@
    -= Info about function 'ENUMLOOKUP' =-

[Syntax]
-   ENUMLOOKUP(number[|Method-type[|options[|record#[|zone-suffix]]]])
+   ENUMLOOKUP(number[,Method-type[,options[,record#[,zone-suffix]]]])

[Synopsis]
-   ENUMLOOKUP allows for general or specific querying of NAPTR records or
counts of NAPTR types fo
+   General or specific querying of NAPTR records for ENUM or ENUM-like DNS
pointers

[Description]
    Option 'c' returns an integer count of the number of NAPTRs of a
certain
RR type.
    Combination of 'c' and Method-type of 'ALL' will return a count of all
NAPTRs for the r
+   Option 'u' returns the full URI and does not strip off the URI-scheme.
+   Option 's' triggers ISN specific rewriting
+   Option 'i' looks for branches into an Infrastructure ENUM tree
+   Option 'd' for a direct DNS lookup without any flipping of digits
    Defaults are: Method-type=sip, no options, record=1, zone-
suffix=e164.arpa

-   For more information, see doc/enum.txt
+   For more information, see doc/asterisk.pdf
```

See also **enum.conf**.

C.17 ENV()

```
ENV(variable)
```

Reads/sets an environment variable (a variable in the operating system environment; these can be viewed from the shell with **echo $variable**). Environment
variables are case sensitive and are almost always written all uppercase.

```
; read HOME:
exten => 123,1,Set(foo=${ENV(HOME)})

; set HOME:
exten => 123,1,Set(ENV(HOME)=/myAst)
```

Asterisk versions:

--------| **1.2** |--------| **1.4** |--------| **1.6** |--------

Internal help for this function in Asterisk 1.4:

```
-= Info about function 'ENV' =-

[Syntax]
ENV(<envname>)

[Synopsis]
Gets or sets the environment variable specified

[Description]
Not available
```

Diff of the internal help from Asterisk 1.2 to 1.4:

None

Diff of the internal help from Asterisk 1.4 to 1.6:

None

C.18 EVAL()

```
EVAL(variable)
```

Evaluates a variable twice. An example is useful: If the variable ${VAR} contains a string "${VAR2}", that is what is returned when ${VAR} is called. If **Eval()** is used, the nested variable is also evaluated, and the contents of ${VAR2} are also returned.

```
; If VAR contains the string "${VAR2}" and VAR2 contains the string "Hello
World":
exten => 123,1,Set(foo=${EVAL(${VAR})})
; now foo is "Hello World"
```

Asterisk versions:

--------| **1.2** |--------| **1.4** |--------| **1.6** |--------

Internal help for this function in Asterisk 1.4:

```
     -= Info about function 'EVAL' =-

[Syntax]
EVAL(<variable >)

[Synopsis]
Evaluate stored variables.

[Description]
Using EVAL basically causes a string to be evaluated twice.
When a variable or expression is in the dialplan, it will be
evaluated at runtime. However, if the result of the evaluation
is in fact a variable or expression, using EVAL will have it
evaluated a second time. For example, if the variable ${MYVAR}
contains "${OTHERVAR}", then the result of putting ${EVAL(${MYVAR})}
in the dialplan will be the contents of the variable, OTHERVAR.
Normally, by just putting ${MYVAR} in the dialplan, you would be
left with "${OTHERVAR}".
```

Diff of the internal help from Asterisk 1.2 to 1.4:

None

Diff of the internal help from Asterisk 1.4 to 1.6:

None

C.19 EXISTS()

```
EXISTS(variable)
```

Checks to see whether a variable is defined. Returns 1 or 0.

```
exten => 123,1,Set(Var1=test)
exten => 123,n,Set(Var2=)
exten => 123,n,Set(foo=${EXISTS(${Var1})})    ; foo is 1
exten => 123,n,Set(foo=${EXISTS(${Var2})})    ; foo is 0
```

Asterisk versions:

```
--------| 1.2 |--------| 1.4 |--------| 1.6 |--------
```

Internal help for this function in Asterisk 1.4:

```
     -= Info about function 'EXISTS' =-

[Syntax]
EXISTS(<data>)
```

```
[Synopsis]
Existence Test: Returns 1 if exists, 0 otherwise

[Description]
Not available
```

Diff of the internal help from Asterisk 1.2 to 1.4:

None

Diff of the internal help from Asterisk 1.4 to 1.6:

None

C.20 FIELDQTY()

```
FIELDQTY(variablename, delimiter)
```

Returns the number of fields which exist if *variablename* is partitioned using *delimiter.*

```
exten => 123,1,Set(Var=hello#you#there#on#the#telephone)
exten => 123,n,Set(Count=${FIELDQTY(Var,#)}) ; Count is 6
```

Asterisk versions:

```
--------| 1.2 |--------| 1.4 |--------| 1.6 |--------
```

Internal help for this function in Asterisk 1.4:

```
    -= Info about function 'FIELDQTY' =-

[Syntax]
FIELDQTY(<varname > |<delim >)

[Synopsis]
Count the fields, with an arbitrary delimiter

[Description]
Not available
```

Diff of the internal help from Asterisk 1.2 to 1.4:

```
--- in Asterisk 1.2
+++ in Asterisk 1.4
@@ -1,7 +1,7 @@
    -= Info about function 'FIELDQTY' =-

  [Syntax]
- FIELDQTY(<varname >,<delim >)
+ FIELDQTY(<varname > |<delim >)
```

```
        [Synopsis]
        Count the fields, with an arbitrary delimiter
```

Diff of the internal help from Asterisk 1.4 to 1.6:

```
    --- in Asterisk 1.4
    +++ in Asterisk 1.6
    @@ -1,7 +1,7 @@
        -= Info about function 'FIELDQTY' =-

        [Syntax]
    -   FIELDQTY(<varname > |<delim >)
    +   FIELDQTY(<varname >,<delim >)

        [Synopsis]
        Count the fields, with an arbitrary delimiter
```

See also cut.

C.21 FILTER()

```
FILTER(allowed_characters, string)
```

Filters *string* so that only the allowed characters are returned.

```
    ; allow only 0123456789 from ${cdrnum}:
    exten => 123,1,Set(foo=${FILTER(0123456789,${cdrnum})})
```

Asterisk versions:

```
    |      |          | 1.4 |--------| 1.6 |--------
```

Internal help for this function in Asterisk 1.4:

```
        -= Info about function 'FILTER' =-

    [Syntax]
    FILTER(<allowed-chars>|<string>)

    [Synopsis]
    Filter the string to include only the allowed characters

    [Description]
    Not available
```

Diff of the internal help from Asterisk 1.2 to 1.4:

Not available in Asterisk 1.2

Diff of the internal help from Asterisk 1.4 to 1.6:

```
--- in Asterisk 1.4
+++ in Asterisk 1.6
@@ -1,10 +1,15 @@
     -= Info about function 'FILTER' =-

  [Syntax]
- FILTER(<allowed-chars > |<string >)
+ FILTER(<allowed-chars >,<string >)

  [Synopsis]
  Filter the string to include only the allowed characters

  [Description]
- Not available
+ Permits all characters listed in <allowed-chars>, filtering all
others out.
+ In addition to literally listing the characters, you may also use
ranges of
+ characters (delimited by a '-'), as well as hexadecimal characters
started
+ with a \x (i.e. \x20) and octal characters started with \0 (i.e. \040).
+ Also, \t, \n, and \r are recognized. If you want a literal '-' character,
+ simply prefix it with a '\'
```

C.22 GLOBAL()

GLOBAL(*variablename*)

Used to declare a variable global; i.e., valid beyond the active life of the current channel. Asterisk 1.2 users use **Set()** (see Appendix B) with the option g.

```
; define global variable ${myvariable}:
exten => 123,1,Set(GLOBAL(myvariable)=Test)
```

Note: Whether global variables persist through a **reload** on the Asterisk console depends whether clearglobalvars is set in extensions.conf.

Asterisk versions:

```
   |      |        | 1.4 |--------| 1.6 |--------
```

Internal help for this function in Asterisk 1.4:

```
     -= Info about function 'GLOBAL' =-

[Syntax]
GLOBAL(<varname >)

[Synopsis]
Gets or sets the global variable specified

[Description]
Not available
```

Diff of the internal help from Asterisk 1.2 to 1.4:

Not available in Asterisk 1.2

Diff of the internal help from Asterisk 1.4 to 1.6:

None

C.23 GROUP()

```
GROUP([category])
```

Reads/sets the group for the channel. (Channels may be grouped as required.)

```
exten => 123,1,Set(GROUP()=outgoing)                  ; set group
exten => 123,n,GotoIf($[${GROUP_COUNT()} > 10]?200)   ; too many outgoing
                                                      ; calls?
exten => 123,n,Dial(7785553233)                       ; dial
exten => 123,200,SetVar(DIALSTATUS=CHANUNAVAIL)       ; too many outgoing
                                                      ; calls, refuse
```

Asterisk versions:

```
--------| 1.2 |--------| 1.4 |          | 1.6 | -------
```

Internal help for this function in Asterisk 1.4:

```
     -= Info about function 'GROUP' =-

[Syntax]
GROUP([category])

[Synopsis]
Gets or sets the channel group.

[Description]
Gets or sets the channel group.
```

Diff of the internal help from Asterisk 1.2 to 1.4:

None

Diff of the internal help from Asterisk 1.4 to 1.6:

None

See also group_count, group_list, group_match_count.

C.24 GROUP_COUNT()

```
GROUP_COUNT([group[@category]])
```

Returns the number of channels in the specified group. If no group is specified, the group of the current channel is assumed.

```
; Query outgoing for number of channels:
exten => 123,1,Set(foo=${GROUP_COUNT(outgoing)})
```

Asterisk versions:

```
--------| 1.2 |--------| 1.4 |--------| 1.6 |--------
```

Internal help for this function in Asterisk 1.4:

```
    -= Info about function 'GROUP_COUNT' =-

[Syntax]
GROUP_COUNT([groupname][@category])

[Synopsis]
Counts the number of channels in the specified group

[Description]
Calculates the group count for the specified group, or uses the
channel's current group if not specifed (and non-empty).
```

Diff of the internal help from Asterisk 1.2 to 1.4:

None

Diff of the internal help from Asterisk 1.4 to 1.6:

None

See also group, group_list, group_match_count.

C.25 GROUP_LIST()

```
GROUP_LIST()
```

Returns a space-separated list of all the groups set for the current channel.

```
exten => 123,1,Set(foo=${GROUP_LIST()})
```

Asterisk versions:

```
--------| 1.2 |--------| 1.4 |--------| 1.6 |--------
```

Internal help for this function in Asterisk 1.4:

```
    -= Info about function 'GROUP_LIST' =-

[Syntax]
GROUP_LIST()

[Synopsis]
Gets a list of the groups set on a channel.

[Description]
Gets a list of the groups set on a channel.
```

Diff of the internal help from Asterisk 1.2 to 1.4:

None

Diff of the internal help from Asterisk 1.4 to 1.6:

None

See also group, group_count, group_match_count.

C.26 GROUP_MATCH_COUNT()

```
GROUP_MATCH_COUNT(pattern [@ category])
```

Returns the number of channels in groups matching the specified pattern.

```
; Query for the number of channels in groups group[1-4]:
exten => 123,1,Set(foo=${GROUP_MATCH_COUNT(group[1-4])})
```

Asterisk versions:

```
--------| 1.2 |--------| 1.4 |--------| 1.6 |--------
```

Internal help for this function in Asterisk 1.4:

```
    -= Info about function 'GROUP_MATCH_COUNT' =-

[Syntax]
GROUP_MATCH_COUNT(groupmatch[@category])

[Synopsis]
Counts the number of channels in the groups matching the specified pattern

[Description]
Calculates the group count for all groups that match the specified
pattern.
Uses standard regular expression matching (see regex(7)).
```

Diff of the internal help from Asterisk 1.2 to 1.4:

None

Diff of the internal help from Asterisk 1.4 to 1.6:

None

See also `group, group_count, group_list`.

C.27 IAXPEER()

```
IAXPEER(peername[:field])
```

Returns information about an IAX peer. The *peername* can be replaced with CURRENTCHANNEL to specify the current channel. The *field* is one of the following:

- **ip**
  (default) The IP address of the peer
- **status**
  Peer status (when `qualify=yes`)
- **mailbox**
  The configured mailbox
- **context**
  The configured context
- **expire**
  The expiry time (in UNIX time) for the connection

- **dynamic**

  Whether the connection is dynamic (yes|no).

- **callerid_name**

  The configured CID name

- **callerid_num**

  The configured CID number

- **codecs**

  The accessible codecs

- **codec[x]**

  Preferred codec number x (beginning with 0)

```
; Query the IP address of peer1:
exten => 123,1,Set(foo=${IAXPEER(peer1:ip)})
```

Asterisk versions:

```
--------| 1.2 |--------| 1.4 |--------| 1.6 |--------
```

Internal help for this function in Asterisk 1.4:

```
      -= Info about function 'IAXPEER' =-

[Syntax]
IAXPEER(<peername|CURRENTCHANNEL > [|item])

[Synopsis]
Gets IAX peer information

[Description]
If peername specified, valid items are:
-  ip (default)      The IP address.
-  status            The peer's status (if qualify=yes)
-  mailbox           The configured mailbox.
-  context           The configured context.
▪  expire            The epoch time of the next expire
-  dynamic           Is it dynamic? (yes/no).
-  callerid_name     The configured Caller ID name.
-  callerid_num      The configured Caller ID number.
-  codecs            The configured codecs.
-  codec[x]          Preferred codec index number 'x' (beginning with
                     zero).
If CURRENTCHANNEL specified, returns IP address of current channel
```

Diff of the internal help from Asterisk 1.2 to 1.4:

```
--- in Asterisk 1.2
+++ in Asterisk 1.4
@@ -1,7 +1,7 @@
     -= Info about function 'IAXPEER' =-
```

```
    [Syntax]
-   IAXPEER(<peername|CURRENTCHANNEL>[:item])
+   IAXPEER(<peername|CURRENTCHANNEL>[|item])

    [Synopsis]
    Gets IAX peer information
```

Diff of the internal help from Asterisk 1.4 to 1.6:

```
--- in Asterisk 1.4
+++ in Asterisk 1.6
@@ -1,7 +1,7 @@
        -= Info about function 'IAXPEER' =-

    [Syntax]
-   IAXPEER(<peername|CURRENTCHANNEL>[|item])
+   IAXPEER(<peername|CURRENTCHANNEL>[,item])

    [Synopsis]
    Gets IAX peer information
```

See also **sippeer**.

C.28 IF()

IF(*expression?trueVal:falseVal*)

Returns a value depending on a condition. If the condition is true, the value follow-ing ? is returned, otherwise the value following : is returned.

```
; If ${Var}=123, return 5, otherwise return 9:
exten => 123,1,Set(foo=${IF($[${Var} = 123]?5:9)})
```

Asterisk versions:

```
--------| 1.2 |--------| 1.4 |--------| 1.6 |--------
```

Internal help for this function in Asterisk 1.4:

```
        -= Info about function 'IF' =-

[Syntax]
IF(<expr > ?[<true >][:<false >])

[Synopsis]
Conditional: Returns the data following '?' if true else the data
following ':'
[Description]
Not available
```

Diff of the internal help from Asterisk 1.2 to 1.4:

None

Diff of the internal help from Asterisk 1.4 to 1.6:

```
--- in Asterisk 1.4
+++ in Asterisk 1.6
@@ -4,7 +4,7 @@
   IF(<expr > ?[<true >][:<false >])

   [Synopsis]
Conditional: Returns the data following '?' if true else the data
following ':'
+ Conditional: Returns the data following '?' if true, else the data
following ':'

   [Description]
   Not available
```

See also `iftime`, and in Appendix B, see `execif`, `gotoif`, `gotoiftime`.

C.29 IFTIME()

```
IFTIME(time-condition?trueVal:falseVal)
```

Returns a value depending on the time condition.

The *time-condition* follows the format *time/dayofweek/date/month*; each parameter may also be a range separated by –, or contain the wildcard *. Time is given in 24-hour format (e.g., 08:00-18:00), weekdays and month names are three-letter English language abbreviations (mon, tue, wed, thu, fri, sat, sun and jan, feb, mar, apr, may, jun, jul, aug, sep, oct, nov, dec)

```
; Valid from 8 a.m. to 6 p.m., Mondays, 1st to the 15th of December:
exten => 123,1,Set(foo 0{IFTIME(08:00 18:00|mon|1-15|dec?5;0)})

; Valid every Saturday and Sunday:
exten => 123,1,Set(foo=${IFTIME(*|sat-sun|*|*?5:0)})
```

Asterisk versions:

```
--------| 1.2 |--------| 1.4 |--------| 1.6 |--------
```

Internal help for this function in Asterisk 1.4:

```
   -= Info about function 'IFTIME' =-

[Syntax]
IFTIME(<timespec > ?[<true >][:<false >])
```

```
[Synopsis]
Temporal Conditional: Returns the data following '?' if true else the
data following ':'

[Description]
Not available
```

Diff of the internal help from Asterisk 1.2 to 1.4:

None

Diff of the internal help from Asterisk 1.4 to 1.6:

```
--- in Asterisk 1.4
+++ in Asterisk 1.6
@@ -4,7 +4,7 @@
   IFTIME(<timespec > ?[<true >][:<false >])

   [Synopsis]
-  Temporal Conditional: Returns the data following '?' if true else the
data following ':'
+  Temporal Conditional: Returns the data following '?' if true, else the
data following ':'

   [Description]
   Not available
```

See also **if**, and in Appendix B, see **execif**, **gotoif**, **gotoiftime**.

C.30 ISNULL()

```
ISNULL(value)
```

Returns 1 if value is null, otherwise returns 0.

```
exten => 123,1,Set(foo=${ISNULL(${Var1})})
```

Asterisk versions:

```
--------| 1.2 |--------| 1.4 |--------| 1.6 |--------
```

Internal help for this function in Asterisk 1.4:

```
    -= Info about function 'ISNULL' =-

[Syntax]
ISNULL(<data >)

[Synopsis]
NULL Test: Returns 1 if NULL or 0 otherwise
```

```
[Description]
Not available
```

Diff of the internal help from Asterisk 1.2 to 1.4:

None

Diff of the internal help from Asterisk 1.4 to 1.6:

None

C.31 KEYPADHASH()

```
KEYPADHASH(string)
```

Transforms an alphabetic string to digits according to the standard telephone key-pad letter assignments. This enables quick conversion of vanity numbers to actual numbers; e.g., 1-800-BADHAIR (1-800-2234247).

1	2 ABC	3 DEF
4 GHI	5 JKL	6 MNO
7 PQRS	8 TUV	9 WXYZ
*	0	#

```
exten => 123,1,Set(foo=${KEYPADHASH(BADHAIR)}) ; returns 2234247
```

Asterisk versions:

```
    |       |          |  1.4  |--------|  1.6  |--------
```

Internal help for this function in Asterisk 1.4:

```
        -= Info about function 'KEYPADHASH' =-

[Syntax]
KEYPADHASH(<string >)

[Synopsis]
Hash the letters in the string into the equivalent keypad numbers.

[Description]
Example: ${KEYPADHASH(Les)} returns "537"
```

Diff of the internal help from Asterisk 1.2 to 1.4:

Not available in Asterisk 1.2

Diff of the internal help from Asterisk 1.4 to 1.6:

None

C.32 LANGUAGE()

```
LANGUAGE()
```

Reads/sets the language of the current channel. This setting determines, among other things, which audio files are played. If the language is set to de and **Playback(tt-weasels)** is run in the dialplan, Asterisk will play de/tt-weasels, if it exists, and similarly for **SayDigits()** and other applications which rely on prerecorded audio files.

```
; Query:
exten => 123,1,Set(foo=${LANGUAGE()})

; Set Spanish:
exten => 123,1,Set(LANGUAGE()=es)
```

Important: **LANGUAGE()** is removed as of Asterisk 1.6 and was replaced with the **CHANNEL()** function:

```
Set(CHANNEL(language)=de)
```

Asterisk versions:

```
--------| 1.2 |--------| 1.4 |           |        |
```

Internal help for this function in Asterisk 1.4:

```
      -= Info about function 'LANGUAGE' =-

[Syntax]
LANGUAGE()

[Synopsis]
Gets or sets the channel's language.

[Description]
Deprecated. Use CHANNEL(language) instead.
```

Diff of the internal help from Asterisk 1.2 to 1.4:

```
--- in Asterisk 1.2
+++ in Asterisk 1.4
```

```
@@ -7,11 +7,4 @@
   Gets or sets the channel's language.

   [Description]
-  Gets or sets the channel language. This information is used for the
-  syntax in generation of numbers, and to choose a natural language file
-  when available. For example, if language is set to 'fr' and the file
-  'demo-congrats' is requested to be played, if the file
-  'fr/demo-congrats' exists, then it will play that file, and if not
-  will play the normal 'demo-congrats'. For some language codes,
-  changing the language also changes the syntax of some Asterisk
-  functions, like SayNumber.
+  Deprecated. Use CHANNEL(language) instead.
```

Diff of the internal help from Asterisk 1.4 to 1.6:

Not available in Asterisk 1.6

C.33 LEN()

```
LEN(string)
```

Returns the length of *string*.

```
; If ${test} is "Hello World"
exten => 123,1,Set(foo=${LEN(${test})})
; returns 11
```

Asterisk versions:

```
--------| 1.2 |--------| 1.4 |--------| 1.6 |--------
```

Internal help for this function in Asterisk 1.4:

```
   -= Info about function 'LEN' =-

[Syntax]
LEN(<string >)

[Synopsis]
Returns the length of the argument given

[Description]
Not available
```

Diff of the internal help from Asterisk 1.2 to 1.4:

None

Diff of the internal help from Asterisk 1.4 to 1.6:

None

C.34 MATH()

```
MATH(number1 operator number2 [, typeofresult])
```

Calculates simple mathematical expressions. Allowed operators are +, -, /, *, <, >, <=, > =, ==, % (modulo). The *typeofresult* may be f, float (default), i, int (integer), h, hex (hexadecimal), c, char (byte output).

```
; Calculate 3*8 as an integer:
exten => 123,1,Set(i=${MATH(3*8,int)})
```

Asterisk versions:

```
--------| 1.2 |--------| 1.4 |--------| 1.6 |--------
```

Internal help for this function in Asterisk 1.4:

```
     -= Info about function 'MATH' =-

[Syntax]
MATH(<number1 > <op > <number 2 > [,<type_of_result >])

[Synopsis]
Performs Mathematical Functions

[Description]
Perform calculation on number 1 to number 2. Valid ops are:
    +,-,/,*,%%%%,<, >, > =,<=,==
and behave as their C equivalents.
<type_of_result > - wanted type of result:
        f, float - float(default)
        i, int - integer,
        h, hex - hex,
        c, char - char
Example: Set(i=${MATH(123%%%%16,int)}) - sets var i=11
```

Diff of the internal help from Asterisk 1.2 to 1.4:

None

Diff of the internal help from Asterisk 1.4 to 1.6:

```
--- in Asterisk 1.4
+++ in Asterisk 1.6
@@ -1,14 +1,14 @@
     -= Info about function 'MATH' =-
```

```
      [Syntax]
  -   MATH(<number1 > <op > <number 2 > [,<type_of_result >])
  +   MATH(<number1 > <op > <number2 > [,<type_of_result >])

      [Synopsis]
      Performs Mathematical Functions

      [Description]
  -   Perform calculation on number 1 to number 2. Valid ops are:
  -       +,-,/,*,%%%%,<, >, > =,<=,==
  +   Perform calculation on number1 to number2. Valid ops are:
  +       +,-,/,*,%%%%,<<, > >,^,AND,OR,XOR,<, >, > =,<=,==
      and behave as their C equivalents.
      <type_of_result > - wanted type of result:
              f, float - float(default)
```

C.35 MD5()

```
MD5(string)
```

Calculates the MD5 hash (checksum) of a string (returns in hexadecimal format).

```
      exten => 123,1,Set(foo=${MD5(${string})})
```

Asterisk versions:

```
--------| 1.2 |--------| 1.4 |--------| 1.6 |--------
```

Internal help for this function in Asterisk 1.4:

```
      -= Info about function 'MD5' =-

  [Syntax]
  MD5(<data>)

  [Synopsis]
  Computes an MD5 digest

  [Description]
  Not available
```

Diff of the internal help from Asterisk 1.2 to 1.4:

None

Diff of the internal help from Asterisk 1.4 to 1.6:

None

C.36 MUSICCLASS()

```
MUSICCLASS(class)
```

Reads/sets the music-on-hold class.

```
; Query:
exten => 123,1,Set(foo=${MUSICCLASS()})

; Set to "HeavyMetal":
exten => 123,1,Set(MUSICCLASS()=HeavyMetal)
```

Important: Deprecated as of Asterisk 1.4. Use **Set(CHANNEL(musicclass)= default)** instead. See channel.

Asterisk versions:

```
--------| 1.2 |--------| 1.4 |         |        |
```

Internal help for this function in Asterisk 1.4:

```
     -= Info about function 'MUSICCLASS' =-

[Syntax]
MUSICCLASS()

[Synopsis]
Read or Set the MusicOnHold class

[Description]
Deprecated. Use CHANNEL(musicclass) instead.
```

Diff of the internal help from Asterisk 1.2 to 1.4:

```
--- in Asterisk 1.2
+++ in Asterisk 1.4
@@ -7,4 +7,4 @@
   Read or Set the MusicOnHold class

   [Description]
-  This function will read or set the music on hold class for a channel.
+  Deprecated. Use CHANNEL(musicclass) instead.
```

Diff of the internal help from Asterisk 1.4 to 1.6:

Not available in Asterisk 1.6

C.37 ODBC_SQL()

ODBC_SQL(*SQL-query*)

Executes the specified SQL query and returns the result, if any.

```
; Query:
exten => 123,1,Set(Name=${ODBC_SQL(SELECT name FROM list WHERE
number='123')})

; Set:
exten => 123,1,Set(ODBC_SQL(UPDATE list SET name='Robert' WHERE
number='123'))
```

Asterisk versions:

```
 |        |           | 1.4 |--------| 1.6 |--------
```

Internal help for this function in Asterisk 1.4:

```
      -= Info about function 'ODBC_SQL' =-

[Syntax]
ODBC_SQL(<arg1 > [...[,<argN >]])

[Synopsis]
Runs the referenced query with the specified arguments

[Description]
Runs the following query, as defined in func_odbc.conf, performing
substitution of the arguments into the query as specified by ${ARG1},
${ARG2}, ... ${ARGn}. This function may only be read, not set.

SQL:
${ARG1}
```

Diff of the internal help from Asterisk 1.2 to 1.4:

Not available in Asterisk 1.2

Diff of the internal help from Asterisk 1.4 to 1.6:

None

C.38 ODBC_USER_DATABASE()

ODBC_USER_DATABASE(*var1*[,*var2*[,...]])

Runs the SQL query defined in `func_odbc.conf` and returns the result, if any. The values defined in `func_odbc.conf`, such as $\{VAL1\}$, $\{VAL2\}$, ..., $\{ARG1\}$, $\{ARG2\}$, ... are replaced by the corresponding values provided when the function is called.

```
func_odbc.conf:

[USER_DATABASE]
dsn=my_database
read=SELECT name FROM list WHERE number='${ARG1}'
write=UPDATE list SET name=${ARG1} WHERE number='${VAL1}'

extensions.conf:

; Query (read):
exten => 123,1,Set(Name=${ODBC_USER_DATABASE(${EXTEN})})

; Update (write):
exten => 123,1,Set(ODBC_USER_DATABASE(${CALLERID(name)})=1000)
```

C.39 QUEUEAGENTCOUNT()

In Asterisk 1.2. Asterisk 1.4 users see **QUEUE_MEMBER_COUNT()**.

```
QUEUEAGENTCOUNT(queue)
```

Returns the number of agents (as opposed to number of callers) in the specified *queue*.

```
; Number of agents in "supportqueue":
exten => 123,1,Set(foo=${QUEUEAGENTCOUNT(supportqueue)})
```

Asterisk versions:

```
--------| 1.2 |--------| 1.4 |              |        |
```

Internal help for this function in Asterisk 1.4:

```
      -= Info about function 'QUEUEAGENTCOUNT' =-

[Syntax]
QUEUEAGENTCOUNT(<queuename >)

[Synopsis]
Count number of agents answering a queue

[Description]
Returns the number of members currently associated with the specified
queue.
This function is deprecated. You should use QUEUE_MEMBER_COUNT()
instead.
```

Diff of the internal help from Asterisk 1.2 to 1.4:

```
--- in Asterisk 1.2
+++ in Asterisk 1.4
@@ -7,4 +7,5 @@
   Count number of agents answering a queue

   [Description]
-  Not available
+  Returns the number of members currently associated with the
specified queue.
+  This function is deprecated. You should use QUEUE_MEMBER_COUNT()
instead.
```

Diff of the internal help from Asterisk 1.4 to 1.6:

Not available in Asterisk 1.6

C.40 QUEUE_MEMBER_COUNT()

In Asterisk 1.4. Asterisk 1.2 users see **QUEUEAGENTCOUNT()**.

```
QUEUE_MEMBER_COUNT(queue)
```

Returns the number of agents (and/or members, which may be devices rather than logged-in users) in the specified *queue*.

```
; Number of members in "supportqueue":
exten => 123,1,Set(foo=${QUEUE_MEMBER_COUNT(supportqueue)})
```

Asterisk versions:

```
  |      |          |  1.4  |--------|  1.6  |--------
```

Internal help for this function in Asterisk 1.4:

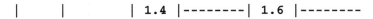

```
    -= Info about function  QUEUE_MEMBER_COUNT '

[Syntax]
QUEUE_MEMBER_COUNT(<queuename >)

[Synopsis]
Count number of members answering a queue

[Description]
Returns the number of members currently associated with the specified
queue.
```

Diff of the internal help from Asterisk 1.2 to 1.4:

Not available in Asterisk 1.2

Diff of the internal help from Asterisk 1.4 to 1.6:

```
--- in Asterisk 1.4
+++ in Asterisk 1.6
@@ -8,3 +8,5 @@

  [Description]
  Returns the number of members currently associated with the specified
queue.
+
+  This function has been deprecated in favor of the QUEUE_MEMBER
function.
```

C.41 QUEUE_MEMBER_LIST()

QUEUE_MEMBER_LIST(*queue*)

Returns a comma-delimited list of the members in the specified *queue*.

```
; Agents in "supportqueue":
exten => 123,1,Set(foo=${QUEUE_MEMBER_LIST(supportqueue)})
; Returns, for example, 5,8,33
```

Asterisk versions:

```
    |     |         | 1.4 |--------| 1.6 |--------
```

Internal help for this function in Asterisk 1.4:

```
    -= Info about function 'QUEUE_MEMBER_LIST' =-

[Syntax]
QUEUE_MEMBER_LIST(<queuename >)

[Synopsis]
Returns a list of interfaces on a queue

[Description]
Returns a comma-separated list of members associated with the specified
queue.
```

Diff of the internal help from Asterisk 1.2 to 1.4:

Not available in Asterisk 1.2

Diff of the internal help from Asterisk 1.4 to 1.6:

None

C.42 QUOTE()

QUOTE (*string*)

Quotes a string exactly, escaping embedded quotation marks if necessary.

```
; If ${var} is >>The "Asterisk"-PBX<<
exten => 123,1,Set(foo=${QUOTE(${var})})
; returns >>The \"Asterisk\"-PBX<<
```

Asterisk versions:

```
  |       |        |  1.4  |--------|  1.6  |--------
```

Internal help for this function in Asterisk 1.4:

```
        -= Info about function 'QUOTE' =-

[Syntax]
QUOTE(<string>)

[Synopsis]
Quotes a given string, escaping embedded quotes as necessary

[Description]
Not available
```

Diff of the internal help from Asterisk 1.2 to 1.4:

Not available in Asterisk 1.2

Diff of the internal help from Asterisk 1.4 to 1.6:

None

C.43 RAND()

RAND(*min, max*)

Returns a randomly generated number between min and max inclusive. The default for min is 0, for max the default is the largest integer supported by the system (usually 2147483647).

```
; Choose a random number between 1 and 10 (inclusive):
exten => 123,1,Set(coincidence=${RAND(1,10)})

; Game of chance:
exten => 123,1,GotoIf($[${RAND(0,100)}< 25]?won:lost)
exten => won,1,Playback(won)
```

```
exten => won,n,Goto(123,1)
exten => lost,1,Playback(lost)
exten => lost,n,Goto(123,1)
```

Note: If you are using versions prior to Asterisk 1.4, use the application **Random()**.

Asterisk versions:

```
  |      |              |  1.4  |--------|  1.6  |--------
```

Internal help for this function in Asterisk 1.4:

```
      -= Info about function 'RAND' =-

[Syntax]
RAND([min][|max])

[Synopsis]
Choose a random number in a range

[Description]
Choose a random number between min and max. Min defaults to 0, if not
specified, while max defaults to RAND_MAX (2147483647 on many systems).
  Example: Set(junky=${RAND(1|8)});
  Sets junky to a random number between 1 and 8, inclusive.
```

Diff of the internal help from Asterisk 1.2 to 1.4:

Not available in Asterisk 1.2

Diff of the internal help from Asterisk 1.4 to 1.6:

```
--- in Asterisk 1.4
+++ in Asterisk 1.6
@@ -1,7 +1,7 @@
     -= Info about function 'RAND' =-

   [Syntax]
-  RAND([min][|max])
+  RAND([min][,max])

   [Synopsis]
   Choose a random number in a range
@@ -9,5 +9,5 @@
   [Description]
   Choose a random number between min and max. Min defaults to 0, if not
   specified, while max defaults to RAND_MAX (2147483647 on many systems).
-    Example: Set(junky=${RAND(1|8)});
+    Example: Set(junky=${RAND(1,8)});
     Sets junky to a random number between 1 and 8, inclusive.
```

See also **random** in Appendix B.

C.44 REGEX()

```
REGEX("expression" string)
```

Returns 1, if *string* matches the regular *expression* expression, otherwise returns 0. The regular expression may include ^ (matches the beginning) and $ (matches the end). Variables are evaluated first.

The parser in Asterisk 1.2 does not behave consistently and can be confused by expressions containing special characters such as $ or angle brackets. An ugly workaround is to define a variable (for example ${dollar}) and have it contain the special character (for example, $).

```
; Test to see if the string "b3" matches the regular expression
"[abc][0-9]":
exten => 123,1,Set(foo=${REGEX("[abc][0-9]" b3)}) ; returns 1

; Test to see if ${str} ends in 0, for Asterisk 1.4:
exten => 123,1,Set(foo=${REGEX("0$" ${str})})

; in Asterisk 1.2, using the workaround described above:
exten => 123,1,Set(foo=${REGEX("0${dollar}" ${str})})
```

Asterisk versions:

```
--------| 1.2 |--------| 1.4 |--------| 1.6 |--------
```

Internal help for this function in Asterisk 1.4:

```
    -= Info about function 'REGEX' =-

[Syntax]
REGEX("<regular expression >" <data >)

[Synopsis]
Regular Expression

[Description]
Returns 1 if data matches regular expression, or 0 otherwise.
Please note that the space following the double quotes separating the
regex from the data is optional and if present, is skipped. If a space
is desired at the beginning of the data, then put two spaces there;
the second will not be skipped.
```

Diff of the internal help from Asterisk 1.2 to 1.4:

```
--- in Asterisk 1.2
+++ in Asterisk 1.4
@@ -4,7 +4,10 @@
   REGEX("<regular expression >" <data >)
```

```
     [Synopsis]
 -   Regular Expression: Returns 1 if data matches regular expression.
 +   Regular Expression

     [Description]
 -   Not available
 +   Returns 1 if data matches regular expression, or 0 otherwise.
 +   Please note that the space following the double quotes separating
 the regex from the data is optional and if present, is skipped. If a
 space is desired at the beginning of the data, then put two spaces
 there; the second will not be skipped.
```

Diff of the internal help from Asterisk 1.4 to 1.6:

None

C.45 SET()

```
SET(variablename = expression)
```

Can be used inside nested expressions to set variables to the desired value. (Not to be confused with the application **Set()** ! This is the cause of much grief!)

```
; Set ${a}, ${b}, ${c}, and ${d} to 8:
exten => 123,1,Set(a=${SET(b=${SET(c=${SET(d=8)})})})

; In the interest of readability and comprehension, it is
; usually better to write one or two more lines
```

Asterisk versions:

```
--------| 1.2 |--------| 1.4 |--------| 1.6 |--------
```

Internal help for this function in Asterisk 1.4:

```
     -= Info about function 'SET' =-

[Syntax]
SET(<varname > =[<value >])

[Synopsis]
SET assigns a value to a channel variable

[Description]
Not available
```

Diff of the internal help from Asterisk 1.2 to 1.4:

None

Diff of the internal help from Asterisk 1.4 to 1.6:

None

See also set in Appendix B.

C.46 SHA1()

SHA1(*string*)

Calculates the SHA1 hash (checksum) of a string (returns hexadecimal).

```
; Calculate the SHA1 hash of "Hello World":
exten => 123,1,Set(sha1hash=${SHA1(Hello World)})
```

Asterisk versions:

```
    |      |          | 1.4 |--------| 1.6 |--------
```

Internal help for this function in Asterisk 1.4:

```
      -= Info about function 'SHA1' =-

[Syntax]
SHA1(<data >)

[Synopsis]
Computes a SHA1 digest

[Description]
Generate a SHA1 digest via the SHA1 algorythm.
  Example: Set(sha1hash=${SHA1(junky)})
  Sets the asterisk variable sha1hash to the string
'60fa5675b9303eb62f99a9cd47f9f5837d18f9a0'
  which is known as his hash
```

Diff of the internal help from Asterisk 1.2 to 1.4:

Not available in Asterisk 1.2

Diff of the internal help from Asterisk 1.4 to 1.6:

None

C.47 SIPCHANINFO()

SIPCHANINFO (*field*)

Returns information about the current SIP channel. The field may be one of the following:

- **peerip**
  The IP address of the peer

- **recvip**
  The source IP address of the peer

- **from**
  The URI from the From: header

- **uri**
  The URI from the Contact: header

- **useragent**
  The user agent

- **peername**
  The name of the peer

  ```
  ; Query the name of a SIP peer:
  exten => 123,1,Set(foo=${SIPCHANINFO(peername)})
  ```

Asterisk versions:

```
--------| 1.2 |--------| 1.4 |--------| 1.6 |--------
```

Internal help for this function in Asterisk 1.4:

```
        -= Info about function 'SIPCHANINFO' =-

[Syntax]
SIPCHANINFO(item)

[Synopsis]
Gets the specified SIP parameter from the current channel

[Description]
Valid items are:
- peerip           The IP address of the peer.
- recvip           The source IP address of the peer.
- from             The URI from the From: header.
```

```
- uri                    The URI from the Contact: header.
- useragent              The useragent.
- peername               The name of the peer.
- t38passthrough         1 if T38 is offered or enabled in this channel,
                         otherwise 0
```

Diff of the internal help from Asterisk 1.2 to 1.4:

```
--- in Asterisk 1.2
+++ in Asterisk 1.4
@@ -14,3 +14,4 @@
    - uri                    The URI from the Contact: header.
    - useragent              The useragent.
    - peername               The name of the peer.
+   - t38passthrough         1 if T38 is offered or enabled in this channel,
                             otherwise 0
```

Diff of the internal help from Asterisk 1.4 to 1.6:

None

C.48 SIPPEER()

```
SIPPEER(peername[,field])
```

Returns information about a SIP peer. The `field` may be one of the following:

- **ip**

  The IP address of the peer (default)

- **mailbox**

  The configured mailbox

- **context**

  The configured context

- **expire**

  The expiry time (in UNIX time) for the connection

- **dynamic**

  Whether `dynamic` is set (yes | no).

- **callerid_name**

  The configured CID name

- **callerid_num**

  The configured CID number

- **codecs**

  Available codecs

- **status**

  The status (when `qualify=yes` is set)

- **regexten**

  The registration extension

- **limit**

  Maximum number of calls

- **curcalls**

  Number of current calls (only if a limit is set)

- **language**

  The default language for this peer

- **useragent**

  The useragent of the peer

- **codec[x]**

  Preferred codec number x (beginning with 0)

- **accountcode**

  The billing account code in the CDR for conversations with this peer

```
; The IP address of peer 2001:
exten => 123,1,Set(sip_ip=${SIPPEER(2001,ip)})
; the preferred codec of the peer:
exten => 123,n,Set(sip_ip=${SIPPEER(2001,codec[0])})
```

Asterisk versions:

```
--------| 1.2 |--------| 1.4 |--------| 1.6 |--------
```

Internal help for this function in Asterisk 1.4:

```
    -= Info about function 'SIPPEER' =-

[Syntax]
SIPPEER(<peername>[|item])

[Synopsis]
Gets SIP peer information

[Description]
Valid items are:
- ip (default)      The IP address.
- mailbox           The configured mailbox.
```

```
- context          The configured context.
- expire           The epoch time of the next expire.
- dynamic          Is it dynamic? (yes/no).
- callerid_name    The configured Caller ID name.
- callerid_num     The configured Caller ID number.
- codecs           The configured codecs.
- status           Status (if qualify=yes).
- regexten         Registration extension
- limit            Call limit (call-limit)
- curcalls         Current amount of calls
                   Only available if call-limit is set
- language         Default language for peer
- accountcode      Account code for this peer
- useragent        Current user agent id for peer
- codec[x]         Preferred codec index number 'x' (beginning
                   with zero).
```

Diff of the internal help from Asterisk 1.2 to 1.4:

```
--- in Asterisk 1.2
+++ in Asterisk 1.4
@@ -1,7 +1,7 @@
    -= Info about function 'SIPPEER' =-

  [Syntax]
- SIPPEER(<peername>[:item])
+ SIPPEER(<peername>[|item])

  [Synopsis]
  Gets SIP peer information
```

Diff of the internal help from Asterisk 1.4 to 1.6:

```
--- in Asterisk 1.4
+++ in Asterisk 1.6
@@ -1,7 +1,7 @@
    -= Info about function 'SIPPEER' =-

  [Syntax]
- SIPPEER(<peername>[|item])
+ SIPPEER(<peername>[,item])

  [Synopsis]
  Gets SIP peer information
@@ -9,19 +9,24 @@
  [Description]
  Valid items are:
  -ip (default)      The IP address.
+ - port            The port number
  - mailbox         The configured mailbox.
  - context         The configured context.
  - expire          The epoch time of the next expire.
```

```
        - dynamic              Is it dynamic? (yes/no).
        - callerid_name        The configured Caller ID name.
        - callerid_num         The configured Caller ID number.
    +   - callgroup            The configured Callgroup.
    +   - pickupgroup          The configured Pickupgroup.
        - codecs               The configured codecs.
        - status               Status (if qualify=yes).
        - regexten             Registration extension
        - limit                Call limit (call-limit)
    +   - busylevel            Configured call level for signaling busy
        - curcalls             Current amount of calls
                               Only available if call-limit is set
        - language             Default language for peer
        - accountcode          Account code for this peer
        - useragent            Current user agent id for peer
    +   - chanvar[name]        A channel variable configured with setvar for
                               this peer.
        - codec[x]             Preferred codec index number 'x' (beginning
                               with zero).
```

See also **iaxpeer**.

C.49 SIP_HEADER()

```
SIP_HEADER(headername[,number])
```

Retrieves the specified SIP header. You are not likely to need this unless you have a
thorough understanding of the SIP protocol. Because some headers can appear
more than once in a SIP packet (such as `Via`, for example) you can specify which
instance of the header you want to see with *number*.

```
; Query the TO header:
exten => 123,1,Set(DN=${SIP_HEADER(TO):5})
exten => 123,2,Set(DN=${CUT(DN,@,1)})
```

Asterisk versions:

```
--------| 1.2 |--------| 1.4 |--------| 1.6 |--------
```

Internal help for this function in Asterisk 1.4:

```
    -= Info about function 'SIP_HEADER' =-

[Syntax]
SIP_HEADER(<name>[,<number>])

[Synopsis]
Gets the specified SIP header
```

```
[Description]
Since there are several headers (such as Via) which can occur multiple
times, SIP_HEADER takes an optional second argument to specify which
header with that name to retrieve. Headers start at offset 1.
```

Diff of the internal help from Asterisk 1.2 to 1.4:

```
--- in Asterisk 1.2
+++ in Asterisk 1.4
@@ -1,10 +1,12 @@
     -= Info about function 'SIP_HEADER' =-

  [Syntax]
-  SIP_HEADER(<name>)
+  SIP_HEADER(<name>[,<number>])

  [Synopsis]
  Gets the specified SIP header

  [Description]
-  Not available
+  Since there are several headers (such as Via) which can occur multiple
+  times, SIP_HEADER takes an optional second argument to specify which
   header with that name to retrieve. Headers start at offset 1.
```

Diff of the internal help from Asterisk 1.4 to 1.6:

None

See also sipaddheader in Appendix B.

C.50 SORT()

```
SORT(key1:value1[,key2:value2 [,...]])
```

Processes a list of keys and values and returns a comma-separated list of the keys
sorted based on their floating-point values.

```
; Sort a list:
exten =>
123,1,Set(foo=${SORT(four:4|half:.5|hundred:100|pi:3.14|e:2.71828|minusone:-1)})
; foo is now "minusone,half,e,pi,four,hundred"
```

Asterisk versions:

```
--------| 1.2 |--------| 1.4 |--------| 1.6 |--------
```

Internal help for this function in Asterisk 1.4:

```
    -= Info about function 'SORT' =-

[Syntax]
SORT(key1:val1[...][,keyN:valN])

[Synopsis]
Sorts a list of key/vals into a list of keys, based upon the vals

[Description]
Takes a comma-separated list of keys and values, each separated by a
colon, and returns a comma-separated list of the keys, sorted by their
values. Values will be evaluated as floating-point numbers.
```

Diff of the internal help from Asterisk 1.2 to 1.4:

None

Diff of the internal help from Asterisk 1.4 to 1.6:

None

C.51 STAT()

STAT(*flag, filename*)

Returns status information about a file (compare the shell commands **test** and **stat**). The *filename* refers to an inode, so it can be a directory or a specific file. The *flag* can be one of the following:

- **d**

  Tests to see whether *filename* is a directory.

- **e**

  Tests if the file exists.

- **f**

  Tests if *filename* is a regular file (as opposed to a special file, such as a block special file, character special file, symbolic link, named pipe, or socket).

- **m**

  Returns the mode of *filename* (octal); i.e. the permissions (e.g., 0754).

- **s**

  Returns the file size in bytes.

- **A**

  Returns the last access time (in UNIX time).

- **C**

  Returns the last inode change time (in UNIX time).

- **M**

  Returns the last modified time (in UNIX time).

```
; See when /etc/crontab was last changed:
exten => 123,1,Set(foo=${STAT(M,/etc/crontab)})
```

Asterisk versions:

```
    |        |              | 1.4 |--------| 1.6 |--------
```

Internal help for this function in Asterisk 1.4:

```
      -= Info about function 'STAT' =-

[Syntax]
STAT(<flag >,<filename >)

[Synopsis]
Does a check on the specified file

[Description]
flag may be one of the following:
  d - Checks if the file is a directory
  e - Checks if the file exists
  f - Checks if the file is a regular file
  m - Returns the file mode (in octal)
  s - Returns the size (in bytes) of the file
  A - Returns the epoch at which the file was last accessed
  C - Returns the epoch at which the inode was last changed
  M - Returns the epoch at which the file was last modified
```

Diff of the internal help from Asterisk 1.2 to 1.4:

Not available in Asterisk 1.2

Diff of the internal help from Asterisk 1.4 to 1.6:

None

C.52 STRFTIME()

```
STRFTIME([unixtime][,[timezone][, format]])
```

Returns a date and time in the specified format. If *unixtime* is not provided, the
current time is used. The default *timezone* is the system default time zone. Possible
time zones may be found in /usr/share/zoneinfo. The format placeholders are
the same as those for the C function **strftime()** (see **man strftime**), the default
is %c; i.e., the locale-dependent date-time format.

```
; Date/time in format YYYY-MM-DD HH:MM:SS
exten => 123,1,Set(time=${STRFTIME(${EPOCH},America/Los_Angeles,
↳"%Y-%m-%d %H:%M:%S")})
```

Asterisk versions:

```
--------| 1.2 |--------| 1.4 |--------| 1.6 |--------
```

Internal help for this function in Asterisk 1.4:

```
    -= Info about function 'STRFTIME' =-

[Syntax]
STRFTIME([<epoch>][|[timezone][|format]])

[Synopsis]
Returns the current date/time in a specified format.

[Description]
Not available
```

Diff of the internal help from Asterisk 1.2 to 1.4:

```
--- in Asterisk 1.2
+++ in Asterisk 1.4
@@ -1,7 +1,7 @@
    -= Info about function 'STRFTIME' =-

  [Syntax]
- STRFTIME([<epoch>][,[timezone][,format]])
+ STRFTIME([<epoch>][|[timezone][|format]])

  [Synopsis]
  Returns the current date/time in a specified format.
```

Diff of the internal help from Asterisk 1.4 to 1.6:

```
--- in Asterisk 1.4
+++ in Asterisk 1.6
@@ -1,10 +1,16 @@
     -= Info about function 'STRFTIME' =-

  [Syntax]
- STRFTIME([<epoch>][|[timezone][|format]])
+ STRFTIME([<epoch>][,[timezone][,format]])

  [Synopsis]
  Returns the current date/time in a specified format.

  [Description]
- Not available
+ STRFTIME sports all of the same formats as the underlying C function
+ strftime(3) - see the man page for details. It also supports the
+ following format:
+   %%%%[n]q - fractions of a second, with leading zeroes. For example,
  %%%%3q will give milliseconds and %%%%1q will give tenths of a second.
  The default is to output milliseconds (n=3). The common case is to use
  it in combination with %%%%S, as in "%%%%S.%%%%3q".
```

C.53 STRPTIME()

```
STRPTIME(datetime|timezone|format)
```

Converts a formatted date and time string into a UNIX timestamp.

```
; Save the date/time in the format YYYY-MM-DD HH:MM:SS in the variable
↳${time}:
exten => 123,1,Set(time=${STRFTIME(${EPOCH},America/Los_Angeles,
↳"%Y-%m-%d %H:%M:%S")})
; Convert ${time} into Unix time:
exten => 123,n,Set(timestamp=${STRPTIME(${time}|America/Los_Angeles
↳|%Y-%m-%d %H:%M:%S)}
```

Asterisk versions:

```
    |       |       | 1.4 |--------| 1.6 |--------
```

Internal help for this function in Asterisk 1.4:

```
     -= Info about function 'STRPTIME' =-

[Syntax]
STRPTIME(<datetime>|<timezone>|<format >)
```

```
[Synopsis]
Returns the epoch of the arbitrary date/time string structured as
described in the format.

[Description]
This is useful for converting a date into an EPOCH time, possibly to
pass to an application like SayUnixTime or to calculate the difference
between two date strings.

Example:
  ${STRPTIME(2006-03-01 07:30:35|America/Chicago|%%%%Y-%%%%m-%%%%d
↳%%%%H:%%%%M:%%%%S)}returns 114
```

Diff of the internal help from Asterisk 1.2 to 1.4:

Not available in Asterisk 1.2

Diff of the internal help from Asterisk 1.4 to 1.6:

```
--- in Asterisk 1.4
+++ in Asterisk 1.6
@@ -1,7 +1,7 @@
    -= Info about function 'STRPTIME' =-

  [Syntax]
-  STRPTIME(<datetime>|<timezone>|<format>)
+  STRPTIME(<datetime>,<timezone>,<format>)

  [Synopsis]
  Returns the epoch of the arbitrary date/time string structured
as described in the format.
@@ -12,4 +12,4 @@
  date strings.

  Example:
-    ${STRPTIME(2006-03-01 07:30:35|America/Chicago|%%%%Y-%%%%m-%%%%d
↳%%%%H:%%%%M:%%%%S)} returns
+    ${STRPTIME(2006-03-01 07:30:35,America/Chicago,%%%%Y-%%%%m-%%%%d
↳%%%%H:%%%%M:%%%%S)} returns
```

See also strftime.

C.54 TIMEOUT()

TIMEOUT(*type*)

Reads/sets a timeout on the channel. The following types are permitted:

- **absolute**

  The absolute, maximum duration of a call. Once reached, the call is passed to the extension T, if it exists, or hung up. A value of 0 is the same as no timeout. When this function is called, the existing setting is reset and overwritten. The timeout counter starts when this function is called, not when the call begins.

- **digit**

  The maximum time allowed between entry of digits. If exceeded, user input is deemed to have finished. If the resulting extension does not exist, the call is passed to the extension i (invalid), if it exists, or hung up. The default is 5 seconds.

- **response**

  The maximum time to wait for input from a user. If the user does not enter an extension, the call is passed to extension t (timeout), if it exists, or the call is hung up. Default: 10 seconds.

```
; Check the absolute timeout:
exten => 123,1,Set(foo=${TIMEOUT(absolute)})

; Limit call duration to a maximum of 60 seconds:
exten => 123,1,Set(TIMEOUT(absolute)=60)
exten => 123,n,Dial(SIP/${EXTEN})
exten => T,1,Playback(sorry-dude)
exten => T,n,Playback(buh-bye)
exten => T,n,Hangup()
```

Asterisk versions:

```
--------| 1.2 |--------| 1.4 |--------| 1.6 |--------
```

Internal help for this function in Asterisk 1.4:

```
    -= Info about function 'TIMEOUT' =-

[Syntax]
TIMEOUT(timeouttype)

[Synopsis]
Gets or sets timeouts on the channel.
```

```
[Description]
Gets or sets various channel timeouts. The timeouts that can be
manipulated are:

absolute: The absolute maximum amount of time permitted for a call.
          A setting of 0 disables the timeout.

digit:    The maximum amount of time permitted between digits when
          the user is typing in an extension. When this timeout expires,
          after the user has started to type in an extension, the
          extension will be considered complete, and will be
          interpreted. Note that if an extension typed in is valid,
          it will not have to timeout to be tested, so typically at
          the expiry of this timeout, the extension will be considered
          invalid (and thus control would be passed to the 'i'
          extension, or if it doesn't exist the call would be
          terminated). The default timeout is 5 seconds.

response: The maximum amount of time permitted after falling through a
          series of priorities for a channel in which the user may
          begin typing an extension. If the user does not type an
          extension in this amount of time, control will pass to the
          't'extension if it exists, and if not the call would be
          terminated. The default timeout is 10 seconds.
```

Diff of the internal help from Asterisk 1.2 to 1.4:

None

Diff of the internal help from Asterisk 1.4 to 1.6:

```
--- in Asterisk 1.4
+++ in Asterisk 1.6
@@ -4,14 +4,14 @@
    TIMEOUT(timeouttype)

    [Synopsis]
-  Gets or sets timeouts on the channel.
+  Gets or sets timeouts on the channel. Timeout values are in seconds.

    [Description]
    Gets or sets various channel timeouts. The timeouts that can be
    manipulated are:

    absolute: The absolute maximum amount of time permitted for a call. A
-             setting of 0 disables the timeout.
+             setting of 0 disables the timeout.

    digit:    The maximum amount of time permitted between digits when the
              user is typing in an extension. When this timeout expires,
@@ -25,8 +25,8 @@
              terminated). The default timeout is 5 seconds.
```

```
response: The maximum amount of time permitted after falling through a
-              series of priorities for a channel in which the user may
-              begin typing an extension. If the user does not type an
-              extension in this amount of time, control will pass to the
-              't' extension if it exists, and if not the call would be
-              terminated. The default timeout is 10 seconds.
+              series of priorities for a channel in which the user may
+              begin typing an extension. If the user does not type an
+              extension in this amount of time, control will pass to the
+              't' extension if it exists, and if not the call would be
+              terminated. The default timeout is 10 seconds.
```

C.55 TXTCIDNAME()

```
TXTCIDNAME(number[,zone-suffix])
```

Looks up the CID name of the caller in DNS (via a TXT-Record).

```
exten => 123,1,Set(callername=${TXTCIDNAME(9755557346)})
```

The zone-suffix option was added in Asterisk 1.6. If not provided, the default is e164.arpa.

Asterisk versions:

```
--------| 1.2 |--------| 1.4 |--------| 1.6 |--------
```

Internal help for this function in Asterisk 1.4:

```
        -= Info about function 'TXTCIDNAME' =-

[Syntax]
TXTCIDNAME(<number>)

[Synopsis]
TXTCIDNAME looks up a caller name via DNS

[Description]
This function looks up the given phone number in DNS to retrieve
the caller id name. The result will either be blank or be the value
found in the TXT record in DNS.
```

Diff of the internal help from Asterisk 1.2 to 1.4:

None

Diff of the internal help from Asterisk 1.4 to 1.6:

```
--- in Asterisk 1.4
+++ in Asterisk 1.6
```

```
@@ -1,7 +1,7 @@
    -= Info about function 'TXTCIDNAME' =-

  [Syntax]
-  TXTCIDNAME(<number>)
+  TXTCIDNAME(<number>[,zone-suffix])

  [Synopsis]
  TXTCIDNAME looks up a caller name via DNS
@@ -9,4 +9,4 @@
  [Description]
  This function looks up the given phone number in DNS to retrieve
  the caller id name. The result will either be blank or be the value
-  found in the TXT record in DNS.
+  found in the TXT record in DNS. The default zone-suffix is e164.arpa.
```

C.56 URIDECODE()

URIDECODE(*string*)

Decodes a URI encoded string. See **URIENCODE()**.

```
; Decode "www.example.com/?page=Hello%20World":
exten => 123,1,Set(foo=${URIDECODE("Hello%20World")})
; Returns "Hello World"
```

Asterisk versions:

```
--------| 1.2 |--------| 1.4 |--------| 1.6 |--------
```

Internal help for this function in Asterisk 1.4:

```
    -= Info about function 'URIDECODE' =-

[Syntax]
URIDECODE(<data>)

[Synopsis]
Decodes a URI-encoded string according to RFC 2396.

[Description]
Not available
```

Diff of the internal help from Asterisk 1.2 to 1.4:

```
--- in Asterisk 1.2
+++ in Asterisk 1.4
@@ -4,7 +4,7 @@
   URIDECODE(<data>)
```

```
    [Synopsis]
-   Decodes an URI-encoded string.
+   Decodes a URI-encoded string according to RFC 2396.

    [Description]
    Not available
```

Diff of the internal help from Asterisk 1.4 to 1.6:

None

C.57 URIENCODE()

URIENCODE(*string*)

URI-encodes a string, so that characters not normally allowed in a URL are replaced with escape sequences following the format %XX, where XX is the hexadecimal bytecode of the character.

```
; Encode "Hello World":
exten => 123,1,Set(foo=${URIENCODE("Hello World")})
; Returns "Hello%20World"
```

Asterisk versions:

```
--------| 1.2 |--------| 1.4 |--------| 1.6 |--------
```

Internal help for this function in Asterisk 1.4:

```
        -= Info about function 'URIENCODE' =-

[Syntax]
URIENCODE(<data>)

[Synopsis]
Encodes a string to URI-safe encoding according to RFC 2396.

[Description]
Not available
```

Diff of the internal help from Asterisk 1.2 to 1.4:

```
--- in Asterisk 1.2
+++ in Asterisk 1.4
@@ -4,7 +4,7 @@
   URIENCODE(<data>)

   [Synopsis]
-  Encodes a string to URI-safe encoding.
+  Encodes a string to URI-safe encoding according to RFC 2396.
```

```
[Description]
Not available
```

Diff of the internal help from Asterisk 1.4 to 1.6:

None

C.58 VMCOUNT()

```
VMCOUNT(VM-box [@ context][| folder])
```

Returns the number of voicemail messages in the specified mailbox. The default context is default, the default folder is INBOX.

```
; Query for the number of messages in mailbox 456:
exten => 123,1,Answer()
exten => 123,n,Set(count=${VMCOUNT(456)})
exten => 123,n,Playback(vm-youhave)            ; "You have"
exten => 123,n,GotoIf($[${count} = 0]?none:new)

exten => 123,10(none),Playback(vm-no)          ; "no"
exten => 123,n,Goto(continue)

exten => 123,20(new),SayNumber($COUNT)         ; count
exten => 123,n,Goto(continue)

exten => 123,30(continue),Playback(vm-INBOX) ; "new"
exten => 123,n,Playback(vm-messages)          ; "messages"
exten => 123,n,Playback(vm-goodbye)           ; "Goodbye!"
exten => 123,n,Hangup()
```

Asterisk versions:

```
--------| 1.2 |--------| 1.4 |--------| 1.6 |--------
```

Internal help for this function in Asterisk 1.4:

```
    -= Info about function 'VMCOUNT' =-

[Syntax]
VMCOUNT(vmbox[@context][|folder])

[Synopsis]
Counts the voicemail in a specified mailbox

[Description]
   context - defaults to "default"
   folder  - defaults to "INBOX"
```

Diff of the internal help from Asterisk 1.2 to 1.4:

None

Diff of the internal help from Asterisk 1.4 to 1.6:

```
--- in Asterisk 1.4
+++ in Asterisk 1.6
@@ -1,7 +1,7 @@
    -= Info about function 'VMCOUNT' =-

  [Syntax]
- VMCOUNT(vmbox[@context][|folder])
+ VMCOUNT(vmbox[@context][,folder])

  [Synopsis]
  Counts the voicemail in a specified mailbox
```

See also mailboxexists in Appendix B.

AGI Command Summary

The following commands, which enable an Asterisk Gateway Interface (AGI) script (see Chapter 16, "Asterisk Gateway Interface") to interact with Asterisk, are taken from the Asterisk internal help and provided here for your convenience. Additional information is provided for commands that merit it. Most of the command names derive from their application (see Appendix B, "Dialplan Applications") or function (see Appendix C, "Dialplan Functions") equivalents. Remember that you must enclose empty parameters in double quotation marks (" "). The help detail Runs Dead: Yes|No indicates whether the command can be used on a dead (i.e., unanswered or hung-up) channel. For more information, see Appendix B.

D.1 ANSWER

Asterisk versions:

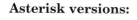

```
--------| 1.2 |--------| 1.4 |--------| 1.6 |--------
```

Internal help for this AGI command in Asterisk 1.4:

```
Usage: ANSWER
       Answers channel if not already in answer state. Returns -1 on
channel failure, or 0 if successful.
```

Diff of the internal help from Asterisk 1.2 to 1.4:

None

Diff of the internal help from Asterisk 1.4 to 1.6:

```
--- in Asterisk 1.4
+++ in Asterisk 1.6
@@ -1,3 +1,4 @@
   Usage: ANSWER
          Answers channel if not already in answer state. Returns -1 on
   channel failure, or 0 if successful.
+  Runs Dead : No
```

D.2 ASYNCAGI BREAK

Asterisk versions:

```
    |        |           |           |           | 1.6 |--------
```

Internal help for this AGI command in Asterisk 1.6:

```
Usage: ASYNCAGI BREAK
       Break the Async AGI loop.
Runs Dead : No
```

Diff of the internal help from Asterisk 1.4 to 1.6:

Not available in Asterisk 1.4

D.3 CHANNEL STATUS

Asterisk versions:

```
--------| 1.2 |--------| 1.4 |--------| 1.6 |--------
```

Internal help for this AGI command in Asterisk 1.4:

```
Usage: CHANNEL STATUS [<channelname>]
       Returns the status of the specified channel.
If no channel name is given the returns the status of the
current channel. Return values:
 0 Channel is down and available
 1 Channel is down, but reserved
 2 Channel is off hook
 3 Digits (or equivalent) have been dialed
 4 Line is ringing
```

```
5 Remote end is ringing
6 Line is up
7 Line is busy
```

Diff of the internal help from Asterisk 1.2 to 1.4:

None

Diff of the internal help from Asterisk 1.4 to 1.6:

```
--- in Asterisk 1.4
+++ in Asterisk 1.6
@@ -10,3 +10,4 @@
      5 Remote end is ringing
      6 Line is up
      7 Line is busy
+    Runs Dead : No
```

D.4 CONTROL STREAM FILE

Asterisk versions:

```
--------| 1.2 |--------| 1.4 |--------| 1.6 |--------
```

Internal help for this AGI command in Asterisk 1.4:

```
Usage: CONTROL STREAM FILE <filename> <escape digits> [skipms] [ffchar]
[rewchr] [pausechr]
        Send the given file, allowing playback to be controled by the
given digits, if any. Use double quotes for the digits if you wish none
to be permitted. Returns 0 if playback completes without a digit
being pressed, or the ASCII numerical value of the digit if one was
pressed, or -1 on error or if the channel was disconnected. Remember,
the file extension must not be included in the filename.

Note: ffchar and rewchar default to * and # respectively.
```

Diff of the internal help from Asterisk 1.2 to 1.4:

None

Diff of the internal help from Asterisk 1.4 to 1.6:

```
--- in Asterisk 1.4
+++ in Asterisk 1.6
@@ -7,3 +7,4 @@
    extension must not be included in the filename.

    Note: ffchar and rewchar default to * and # respectively.
+    Runs Dead : No
```

D.5 DATABASE DEL

Asterisk versions:

```
--------| 1.2 |--------| 1.4 |--------| 1.6 |--------
```

Internal help for this AGI command in Asterisk 1.4:

```
Usage: DATABASE DEL <family> <key>
       Deletes an entry in the Asterisk database for a
given family and key.
Returns 1 if successful, 0 otherwise.
```

Diff of the internal help from Asterisk 1.2 to 1.4:

None

Diff of the internal help from Asterisk 1.4 to 1.6:

```
--- in Asterisk 1.4
+++ in Asterisk 1.6
@@ -2,3 +2,4 @@
         Deletes an entry in the Asterisk database for a
    given family and key.
    Returns 1 if successful, 0 otherwise.
+    Runs Dead : Yes
```

D.6 DATABASE DELTREE

Asterisk versions:

```
--------| 1.2 |--------| 1.4 |--------| 1.6 |--------
```

Internal help for this AGI command in Asterisk 1.4:

```
Usage: DATABASE DELTREE <family> |keytree|
       Deletes a family or specific keytree within a family
in the Asterisk database.
Returns 1 if successful, 0 otherwise.
```

Diff of the internal help from Asterisk 1.2 to 1.4:

None

Diff of the internal help from Asterisk 1.4 to 1.6:

```
--- in Asterisk 1.4
+++ in Asterisk 1.6
```

```
@@ -2,3 +2,4 @@
        Deletes a family or specific keytree within a family
    in the Asterisk database.
    Returns 1 if successful, 0 otherwise.
+   Runs Dead : Yes
```

D.7 DATABASE GET

Asterisk versions:

```
--------| 1.2 |--------| 1.4 |--------| 1.6 |--------
```

Internal help for this AGI command in Asterisk 1.4:

```
Usage: DATABASE GET <family> <key>
        Retrieves an entry in the Asterisk database for a
given family and key.
Returns 0 if <key> is not set. Returns 1 if <key>
is set and returns the variable in parentheses.
Example return code: 200 result=1 (testvariable)
```

Diff of the internal help from Asterisk 1.2 to 1.4:

None

Diff of the internal help from Asterisk 1.4 to 1.6:

```
--- in Asterisk 1.4
+++ in Asterisk 1.6
@@ -4,3 +4,4 @@
    Returns 0 if <key> is not set. Returns 1 if <key>
    is set and returns the variable in parentheses.
    Example return code: 200 result=1 (testvariable)
+   Runs Dead : Yes
```

D.8 DATABASE PUT

Asterisk versions:

```
--------| 1.2 |--------| 1.4 |--------| 1.6 |--------
```

Internal help for this AGI command in Asterisk 1.4:

```
Usage: DATABASE PUT <family> <key> <value>
        Adds or updates an entry in the Asterisk database for a
given family, key, and value.
Returns 1 if successful, 0 otherwise.
```

Diff of the internal help from Asterisk 1.2 to 1.4:

None

Diff of the internal help from Asterisk 1.4 to 1.6:

```
--- in Asterisk 1.4
+++ in Asterisk 1.6
@@ -2,3 +2,4 @@
          Adds or updates an entry in the Asterisk database for a
     given family, key, and value.
     Returns 1 if successful, 0 otherwise.
+    Runs Dead : Yes
```

D.9 EXEC

Executes a dialplan application. The arguments (called options below) are passed to the application. Multiple arguments are separated with a pipe (|) character.
   Here is a simple example using **Dial()**:

```
EXEC Dial Zap/g1/4035551212
```

Asterisk versions:

```
--------| 1.2 |--------| 1.4 |--------| 1.6 |--------
```

Internal help for this AGI command in Asterisk 1.4:

```
Usage: EXEC <application> <options>
       Executes <application> with given <options>.
Returns whatever the application returns, or -2 on failure to find
application
```

Diff of the internal help from Asterisk 1.2 to 1.4:

None

Diff of the internal help from Asterisk 1.4 to 1.6:

```
--- in Asterisk 1.4
+++ in Asterisk 1.6
@@ -1,3 +1,4 @@
   Usage: EXEC <application> <options>
          Executes <application> with given <options>.
   Returns whatever the application returns, or -2 on failure to find
   application
+  Runs Dead : Yes
```

D.10 GET DATA

This is comparable to the **Background()** application in extensions.conf. Plays the audio file with the name *filename* and waits for DTMF input. This may be limited with *max_digits*; the maximum wait time is set with *timeout*.

Asterisk versions:

```
--------| 1.2 |--------| 1.4 |--------| 1.6 |--------
```

Internal help for this AGI command in Asterisk 1.4:

```
Usage: GET DATA <file to be streamed> [timeout] [max digits]
       Stream the given file, and recieve DTMF data. Returns the digits
       received from the channel at the other end.
```

Diff of the internal help from Asterisk 1.2 to 1.4:

None

Diff of the internal help from Asterisk 1.4 to 1.6:

```
--- in Asterisk 1.4
+++ in Asterisk 1.6
@@ -1,3 +1,4 @@
    Usage: GET DATA <file to be streamed> [timeout] [max digits]
           Stream the given file, and recieve DTMF data. Returns the
           digits received from the channel at the other end.
+   Runs Dead : No
```

D.11 GET FULL VARIABLE

Asterisk versions:

```
--------| 1.2 |--------| 1.4 |--------| 1.6 |--------
```

Internal help for this AGI command in Asterisk 1.4:

```
Usage: GET FULL VARIABLE <variablename> [<channel name>]
       Returns 0 if <variablename> is not set or channel does not exist.
Returns 1
if <variablename>  is set and returns the variable in parenthesis.
Understands complex variable names and builtin variables, unlike GET
VARIABLE.
Example return code: 200 result=1 (testvariable)
```

Diff of the internal help from Asterisk 1.2 to 1.4:

None

Diff of the internal help from Asterisk 1.4 to 1.6:

```
--- in Asterisk 1.4
+++ in Asterisk 1.6
@@ -3,3 +3,4 @@
     if <variablename> is set and returns the variable in parenthesis.
Understands complex variable names and builtin variables, unlike GET
VARIABLE.
     Example return code: 200 result=1 (testvariable)
+    Runs Dead : Yes
```

D.12 GET OPTION

Asterisk versions:

```
--------| 1.2 |--------| 1.4 |--------| 1.6 |--------
```

Internal help for this AGI command in Asterisk 1.4:

```
Usage: GET OPTION <filename> <escape_digits> [timeout]
       Behaves similar to STREAM FILE but used with a timeout option.
```

Diff of the internal help from Asterisk 1.2 to 1.4:

None

Diff of the internal help from Asterisk 1.4 to 1.6:

```
--- in Asterisk 1.4
+++ in Asterisk 1.6
@@ -1,2 +1,3 @@
     Usage: GET OPTION <filename> <escape_digits> [timeout]
            Behaves similar to STREAM FILE but used with a timeout option.
+    Runs Dead : No
```

D.13 GET VARIABLE

Similar to GET FULL VARIABLE, but does not understand complex or system variables.

Tip: If in doubt, use GET FULL VARIABLE.

Asterisk versions:

```
--------| 1.2 |--------| 1.4 |--------| 1.6 |--------
```

Internal help for this AGI command in Asterisk 1.4:

```
Usage: GET VARIABLE <variablename>
        Returns 0 if <variablename> is not set. Returns 1 if <variablename>
is set and returns the variable in parentheses.
example return code: 200 result=1 (testvariable)
```

Diff of the internal help from Asterisk 1.2 to 1.4:

None

Diff of the internal help from Asterisk 1.4 to 1.6:

```
--- in Asterisk 1.4
+++ in Asterisk 1.6
@@ -2,3 +2,4 @@
          Returns 0 if <variablename> is not set. Returns 1 if
          <variablename>
    is set and returns the variable in parentheses.
    example return code: 200 result=1 (testvariable)
+   Runs Dead : Yes
```

D.14 GOSUB

Asterisk versions:

```
    |     |        |     |        | 1.6 |--------
```

Internal help for this AGI command in Asterisk 1.6:

```
Usage: GOSUB <context> <extension> <priority> [<optional-argument>]
   Cause the channel to execute the specified dialplan subroutine,
returning to the dialplan with execution of a Return()
Runs Dead : No
```

Diff of the internal help from Asterisk 1.4 to 1.6:

Not available in Asterisk 1.4

D.15 HANGUP

Asterisk versions:

```
--------| 1.2 |--------| 1.4 |--------| 1.6 |--------
```

Internal help for this AGI command in Asterisk 1.4:

```
Usage: HANGUP [<channelname>]
       Hangs up the specified channel.
If no channel name is given, hangs up the current channel
```

Diff of the internal help from Asterisk 1.2 to 1.4:

None

Diff of the internal help from Asterisk 1.4 to 1.6:

```
--- in Asterisk 1.4
+++ in Asterisk 1.6
@@ -1,3 +1,4 @@
   Usage: HANGUP [<channelname>]
          Hangs up the specified channel.
   If no channel name is given, hangs up the current channel
+   Runs Dead : No
```

D.16 NOOP

Does nothing, but appended text (NoOP [*text*]) will be printed out in the CLI. Always returns 0.

Asterisk versions:

```
--------| 1.2 |--------| 1.4 |--------| 1.6 |--------
```

Internal help for this AGI command in Asterisk 1.4:

```
Usage: NoOp
       Does nothing.
```

Diff of the internal help from Asterisk 1.2 to 1.4:

None

Diff of the internal help from Asterisk 1.4 to 1.6:

```
--- in Asterisk 1.4
+++ in Asterisk 1.6
```

```
@@ -1,2 +1,3 @@
    Usage: NoOp
            Does nothing.
+ Runs Dead : Yes
```

D.17 RECEIVE CHAR

Asterisk versions:

--------| 1.2 |--------| 1.4 |--------| 1.6 |--------

Internal help for this AGI command in Asterisk 1.4:

```
Usage: RECEIVE CHAR <timeout>
        Receives a character of text on a channel. Specify timeout to be
the maximum time to wait for input in milliseconds, or 0 for infinite.
Most channels do not support the reception of text. Returns the decimal
value of the character if one is received, or 0 if the channel does not
support text reception. Returns -1 only on error/hangup.
```

Diff of the internal help from Asterisk 1.2 to 1.4:

None

Diff of the internal help from Asterisk 1.4 to 1.6:

```
--- in Asterisk 1.4
+++ in Asterisk 1.6
@@ -4,3 +4,4 @@
    do not support the reception of text. Returns the decimal value of the
character if one is received, or 0 if the channel does not support text
reception. Returns -1 only on error/hangup.
+    Runs Dead : No
```

D.18 RECEIVE TEXT

Asterisk versions:

--------| 1.2 |--------| 1.4 |--------| 1.6 |--------

Internal help for this AGI command in Asterisk 1.4:

```
Usage: RECEIVE TEXT <timeout>
        Receives a string of text on a channel. Specify timeout to be the
maximum time to wait for input in milliseconds, or 0 for infinite. Most
channels do not support the reception of text. Returns -1 for failure or
1 for success, and the string in parentheses.
```

Diff of the internal help from Asterisk 1.2 to 1.4:

None

Diff of the internal help from Asterisk 1.4 to 1.6:

```
--- in Asterisk 1.4
+++ in Asterisk 1.6
@@ -2,3 +2,4 @@              Receives a string of text on a channel. Specify timeout to be
the maximum time to wait for input in milliseconds, or 0 for infinite.
Most channels do not support the reception of text. Returns -1 for
failure or 1 for success, and the string in parentheses.
+    Runs Dead : No
```

D.19 RECORD FILE

Asterisk versions:

```
--------| 1.2 |--------| 1.4 |--------| 1.6 |--------
```

Internal help for this AGI command in Asterisk 1.4:

```
Usage: RECORD FILE <filename> <format> <escape digits> <timeout> \
                               [offset samples] [BEEP] [s=silence]
        Record to a file until a given dtmf digit in the sequence is
received. Returns -1 on hangup or error. The format will specify what
kind of file will be recorded. The timeout is the maximum record time in
milliseconds, or -1 for no timeout. "Offset samples" is optional, and, if
provided, will seek to the offset without exceeding the end of the file.
"silence" is the number of seconds of silence allowed before the function
returns despite the lack of dtmf digits or reaching timeout. Silence
value must be preceeded by "s=" and is also optional.
```

Diff of the internal help from Asterisk 1.2 to 1.4:

None

Diff of the internal help from Asterisk 1.4 to 1.6:

```
--- in Asterisk 1.4
+++ in Asterisk 1.6
@@ -8,3 +8,4 @@
    of seconds of silence allowed before the function returns despite the
    lack of dtmf digits or reaching timeout. Silence value must be
    preceeded by "s=" and is also optional.
+    Runs Dead : No
```

D.20 SAY ALPHA

Asterisk versions:

```
--------| 1.2 |--------| 1.4 |--------| 1.6 |--------
```

Internal help for this AGI command in Asterisk 1.4:

```
Usage: SAY ALPHA <number> <escape digits>
        Say a given character string, returning early if any of the given
DTMF digits are received on the channel. Returns 0 if playback completes
without a digit being pressed, or the ASCII numerical value of the digit
if one was pressed or -1 on error/hangup.
```

Diff of the internal help from Asterisk 1.2 to 1.4:

None

Diff of the internal help from Asterisk 1.4 to 1.6:

```
--- in Asterisk 1.4
+++ in Asterisk 1.6
@@ -3,3 +3,4 @@
     are received on the channel. Returns 0 if playback completes without
     a digit being pressed, or the ASCII numerical value of the digit if one
     was pressed or -1 on error/hangup.
+    Runs Dead : No
```

D.21 SAY DATE

Asterisk versions:

```
--------| 1.2 |--------| 1.4 |--------| 1.6 |--------
```

Internal help for this AGI command in Asterisk 1.4:

```
Usage: SAY DATE <date> <escape digits>
        Say a given date, returning early if any of the given DTMF
digits are received on the channel. <date> is number of seconds elapsed
since 00:00:00 on January 1, 1970, Coordinated Universal Time (UTC).
Returns 0 if playback completes without a digit being pressed, or the
ASCII numerical value of the digit if one was pressed or -1 on
error/hangup.
```

Diff of the internal help from Asterisk 1.2 to 1.4:

None

Diff of the internal help from Asterisk 1.4 to 1.6:

```
--- in Asterisk 1.4
+++ in Asterisk 1.6
@@ -4,3 +4,4 @@
    on January 1, 1970, Coordinated Universal Time (UTC). Returns 0 if
    playback completes without a digit being pressed, or the ASCII
    numerical value of the digit if one was pressed or -1 on
    error/hangup.
+   Runs Dead : No
```

D.22 SAY DATETIME

This command follows this format:

```
SAY DATETIME timestamp [escape_digits] [format] [timezone]
```

It speaks the date specified. The timestamp is the number of seconds since January 1, 1970, at 00:00:00. The format can include the following values:

filename	A sound file
${VAR}	A variable
A or a	Day of the week (e.g. Monday or Tuesday)
B, b or h	Name of the month (e.g., January)
d or e	Ordinal number of the day of the month (e.g., first, second, third)
Y	Year
I or i	Hour (12-hour format)
H	Hour (24-hour format); 07 is called out as zero-seven
k	Hour (24-hour format); 07 is called out as seven
M	Minutes
P or p	a.m. or p.m.
Q	Today, yesterday, or the contents of ABdY
q	(Says nothing for today), yesterday, the day of the week, or the contents of ABdY
R	Time, including minutes, in 24-hour format

The *timezone* is in the same format as that used in /etc/asterisk/voicemail.conf.

Output can be interrupted with escape_digits. If you omit these parameters, you must use placeholders in the form of double quotes (" ").

Asterisk versions:

```
--------| 1.2 |--------| 1.4 |--------| 1.6 |--------
```

Internal help for this AGI command in Asterisk 1.4:

```
Usage: SAY DATETIME <time> <escape digits> [format] [timezone]
        Say a given time, returning early if any of the given DTMF
digits are received on the channel. <time> is number of seconds
elapsed since 00:00:00 on January 1, 1970, Coordinated Universal
Time (UTC). [format] is the format the time should be said in.
See voicemail.conf (defaults to "ABdY 'digits/at' IMp"). Acceptable
values for [timezone] can be found in /usr/share/zoneinfo. Defaults
to machine default. Returns 0 if playback completes without a digit
being pressed, or the ASCII numerical value of the digit if one was
pressed or -1 on error/hangup.
```

Diff of the internal help from Asterisk 1.2 to 1.4:

None

Diff of the internal help from Asterisk 1.4 to 1.6:

```
--- in Asterisk 1.4
+++ in Asterisk 1.6
@@ -7,3 +7,4 @@
   /usr/share/zoneinfo. Defaults to machine default. Returns 0 if
   playback completes without a digit being pressed, or the ASCII
   numerical value of the digit if one was pressed or -1 on
   error/hangup.
+   Runs Dead : No
```

D.23 SAY DIGITS

Asterisk versions:

```
--------| 1.2 |--------| 1.4 |--------| 1.6 |--------
```

Internal help for this AGI command in Asterisk 1.4:

```
Usage: SAY DIGITS <number> <escape digits>
        Say a given digit string, returning early if any of the given
        DTMF digits are received on the channel. Returns 0 if playback
        completes without a digit being pressed, or the ASCII numerical
        value of the digit if one was pressed or -1 on error/hangup.
```

Diff of the internal help from Asterisk 1.2 to 1.4:

None

Diff of the internal help from Asterisk 1.4 to 1.6:

```
--- in Asterisk 1.4
+++ in Asterisk 1.6
@@ -3,3 +3,4 @@
```

```
    are received on the channel. Returns 0 if playback completes
    without a digit being pressed, or the ASCII numerical value of the
    digit if one was pressed or -1 on error/hangup.
+   Runs Dead : No
```

D.24 SAY NUMBER

Says a number (e.g., 123 is spoken as one hundred and twenty-three).

Asterisk versions:

```
--------| 1.2 |--------| 1.4 |--------| 1.6 |--------
```

Internal help for this AGI command in Asterisk 1.4:

```
Usage: SAY NUMBER <number> <escape digits>
       Say a given number, returning early if any of the given
       DTMF digits are received on the channel. Returns 0 if playback
       completes without a digit being pressed, or the ASCII numerical
       value of the digit if one was pressed or -1 on error/hangup.
```

Diff of the internal help from Asterisk 1.2 to 1.4:

None

Diff of the internal help from Asterisk 1.4 to 1.6:

```
--- in Asterisk 1.4
+++ in Asterisk 1.6
@@ -1,5 +1,6 @@
-   Usage:   SAY NUMBER <number> <escape digits>
+   Usage:   SAY NUMBER <number> <escape digits> [gender]
            Say a given number, returning early if any of the given
            DTMF digits are received on the channel. Returns 0 if
            playback completes without a digit being pressed, or the
            ASCII numerical value of the digit if one was pressed or
            -1 on error/hangup.
+   Runs Dead : No
```

D.25 SAY PHONETIC

Says a given string using the NATO phonetic alphabet (e.g., a is said alpha, b as bravo).

Asterisk versions:

```
--------| 1.2 |--------| 1.4 |--------| 1.6 |--------
```

Internal help for this AGI command in Asterisk 1.4:

```
Usage: SAY PHONETIC <string> <escape digits>
       Say a given character string with phonetics, returning early
       if any of the given DTMF digits are received on the channel.
       Returns 0 if playback completes without a digit pressed, the
       ASCII numerical value of the digit if one was pressed, or -1
       on error/hangup.
```

Diff of the internal help from Asterisk 1.2 to 1.4:

None

Diff of the internal help from Asterisk 1.4 to 1.6:

```
--- in Asterisk 1.4
+++ in Asterisk 1.6
@@ -3,3 +3,4 @@
    given DTMF digits are received on the channel. Returns 0 if
    playback completes without a digit pressed, the ASCII numerical
    value of the digit if one was pressed, or -1 on error/hangup.
+   Runs Dead : No
```

D.26 SAY TIME

Asterisk versions:

```
--------| 1.2 |--------| 1.4 |--------| 1.6 |--------
```

Internal help for this AGI command in Asterisk 1.4:

```
Usage: SAY TIME <time> <escape digits>
       Say a given time, returning early if any of the given DTMF
       digits are received on the channel. <time> is number of seconds
       elapsed since 00:00:00 on January 1, 1970, Coordinated Universal
       Time (UTC). Returns 0 if playback completes without a digit being
       pressed, or the ASCII numerical value of the digit if one was
       pressed or -1 on error/hangup.
```

Diff of the internal help from Asterisk 1.2 to 1.4:

None

Diff of the internal help from Asterisk 1.4 to 1.6:

```
--- in Asterisk 1.4
+++ in Asterisk 1.6
@@ -4,3 +4,4 @@
```

on January 1, 1970, Coordinated Universal Time (UTC). Returns 0
if playback completes without a digit being pressed, or the ASCII
numerical value of the digit if one was pressed or -1 on error/hangup.
+   Runs Dead : No

D.27 SEND IMAGE

Asterisk versions:

--------| 1.2 |--------| 1.4 |--------| 1.6 |--------

Internal help for this AGI command in Asterisk 1.4:

```
Usage: SEND IMAGE <image>
       Sends the given image on a channel. Most channels do not
       support the transmission of images. Returns 0 if image is sent,
       or if the channel does not support image transmission. Returns
       -1 only on error/hangup. Image names should not include
       extensions.
```

Diff of the internal help from Asterisk 1.2 to 1.4:

None

Diff of the internal help from Asterisk 1.4 to 1.6:

```
--- in Asterisk 1.4
+++ in Asterisk 1.6
@@ -3,3 +3,4 @@
     transmission of images. Returns 0 if image is sent, or if the
     channel does not support image transmission. Returns -1 only on
     error/hangup. Image names should not include extensions.
+    Runs Dead : No
```

D.28 SEND TEXT

Asterisk versions:

--------| 1.2 |--------| 1.4 |--------| 1.6 |--------

Internal help for this AGI command in Asterisk 1.4:

```
Usage: SEND TEXT "<text to send>"
       Sends the given text on a channel. Most channels do not support
       the transmission of text. Returns 0 if text is sent, or if the
       channel does not support text transmission. Returns -1 only on
       error/hangup. Text consisting of greater than one word should be
       placed in quotes since the command only accepts a single argument.
```

Diff of the internal help from Asterisk 1.2 to 1.4:

None

Diff of the internal help from Asterisk 1.4 to 1.6:

```
--- in Asterisk 1.4
+++ in Asterisk 1.6
@@ -4,3 +4,4 @@
    support text transmission. Returns -1 only on error/hangup. Text
    consisting of greater than one word should be placed in quotes since
    the command only accepts a single argument.
+   Runs Dead : No
```

D.29 SET AUTOHANGUP

Asterisk versions:

```
--------| 1.2 |--------| 1.4 |--------| 1.6 |--------
```

Internal help for this AGI command in Asterisk 1.4:

```
Usage: SET AUTOHANGUP <time>
       Cause the channel to automatically hangup at <time> seconds
       in the future. Of course it can be hungup before then as well.
       Setting to 0 will cause the autohangup feature to be disabled on
       this channel.
```

Diff of the internal help from Asterisk 1.2 to 1.4:

None

Diff of the internal help from Asterisk 1.4 to 1.6:

```
--- in Asterisk 1.4
+++ in Asterisk 1.6
@@ -2,3 +2,4 @@
       Cause the channel to automatically hangup at <time> seconds
       in the future. Of course it can be hungup before then as well.
       Setting to 0 will cause the autohangup feature to be disabled
       on this channel.
+   Runs Dead : No
```

D.30 SET CALLERID

Asterisk versions:

```
--------| 1.2 |--------| 1.4 |--------| 1.6 |--------
```

Internal help for this AGI command in Asterisk 1.4:

```
Usage: SET CALLERID <number>
        Changes the callerid of the current channel.
```

Diff of the internal help from Asterisk 1.2 to 1.4:

None

Diff of the internal help from Asterisk 1.4 to 1.6:

```
--- in Asterisk 1.4
+++ in Asterisk 1.6
@@ -1,2 +1,3 @@
    Usage: SET CALLERID <number>
            Changes the callerid of the current channel.
+    Runs Dead : No
```

D.31 SET CONTEXT

Sets the context to go to after the AGI script has finished.

Warning: This does not check to see whether the context exists. If you provide an invalid context, the call will be terminated after the AGI script finishes.

Asterisk versions:

```
--------| 1.2 |--------| 1.4 |--------| 1.6 |--------
```

Internal help for this AGI command in Asterisk 1.4:

```
Usage: SET CONTEXT <desired context>
        Sets the context for continuation upon exiting the application.
```

Diff of the internal help from Asterisk 1.2 to 1.4:

None

Diff of the internal help from Asterisk 1.4 to 1.6:

```
--- in Asterisk 1.4
+++ in Asterisk 1.6
@@ -1,2 +1,3 @@
    Usage: SET CONTEXT <desired context>
            Sets the context for continuation upon exiting the application.
+    Runs Dead : No
```

D.32 SET EXTENSION

Sets the extension to go to after the AGI script has finished.

Warning: This does not check to see whether the extension exists. If you provide an invalid extension, the call will be terminated after the AGI script finishes.

Asterisk versions:

```
--------| 1.2 |--------| 1.4 |--------| 1.6 |--------
```

Internal help for this AGI command in Asterisk 1.4:

```
Usage: SET EXTENSION <new extension>
       Changes the extension for continuation upon exiting
       the application.
```

Diff of the internal help from Asterisk 1.2 to 1.4:

None

Diff of the internal help from Asterisk 1.4 to 1.6:

```
--- in Asterisk 1.4
+++ in Asterisk 1.6
@@ -1,2 +1,3 @@
   Usage: SET EXTENSION <new extension>
          Changes the extension for continuation upon exiting the
          application.
+   Runs Dead : No
```

D.33 SET MUSIC

Asterisk versions:

```
--------| 1.2 |--------| 1.4 |--------| 1.6 |--------
```

Internal help for this AGI command in Asterisk 1.4:

```
Usage: SET MUSIC ON <on|off> <class>
       Enables/Disables the music on hold generator. If <class> is
       not specified, then the default music on hold class will be used.
       Always returns 0.
```

Diff of the internal help from Asterisk 1.2 to 1.4:

None

Diff of the internal help from Asterisk 1.4 to 1.6:

```
--- in Asterisk 1.4
+++ in Asterisk 1.6
@@ -2,3 +2,4 @@
            Enables/Disables the music on hold generator. If <class> is
    not specified, then the default music on hold class will be used.
    Always returns 0.
+    Runs Dead : No
```

D.34 SET PRIORITY

Sets the priority to go to after the AGI script has finished.

Warning: This does not check to see whether the priority exists. If you provide
an invalid priority, the call will be terminated after the AGI script finishes.

Asterisk versions:

```
--------| 1.2 |--------| 1.4 |--------| 1.6 |--------
```

Internal help for this AGI command in Asterisk 1.4:

```
Usage: SET PRIORITY <priority>
        Changes the priority for continuation upon exiting the application.
        The priority must be a valid priority or label.
```

Diff of the internal help from Asterisk 1.2 to 1.4:

None

Diff of the internal help from Asterisk 1.4 to 1.6:

```
--- in Asterisk 1.4
+++ in Asterisk 1.6
@@ -1,3 +1,4 @@
   Usage: SET PRIORITY <priority>
        Changes the priority for continuation upon exiting
        the application.
        The priority must be a valid priority or label.
+    Runs Dead : No
```

D.35 SET VARIABLE

Asterisk versions:

```
--------| 1.2 |--------| 1.4 |--------| 1.6 |--------
```

Internal help for this AGI command in Asterisk 1.4:

```
    Usage: SET VARIABLE <variablename> <value>
```

Diff of the internal help from Asterisk 1.2 to 1.4:

None

Diff of the internal help from Asterisk 1.4 to 1.6:

```
    --- in Asterisk 1.4
    +++ in Asterisk 1.6
    @@ -1 +1,2 @@
        Usage: SET VARIABLE <variablename> <value>
    +    Runs Dead : Yes
```

D.36 STREAM FILE

Important: STREAM FILE does not always work properly with languages other than English. In this case, you may use EXEC Playback *filename*.

Asterisk versions:

```
--------| 1.2 |--------| 1.4 |--------| 1.6 |--------
```

Internal help for this AGI command in Asterisk 1.4:

```
    Usage: STREAM FILE <filename> <escape digits> [sample offset]
            Send the given file, allowing playback to be interrupted by the
            given digits, if any. Use double quotes for the digits if you wish
            none to be permitted. If sample offset is provided then the audio
            will seek to sample offset before play starts. Returns 0 if
            playback completes without a digit being pressed, or the ASCII
            numerical value of the digit if one was pressed, or -1 on error or
            if the channel was disconnected. Remember, the file extension must
            not be included in the filename.
```

Diff of the internal help from Asterisk 1.2 to 1.4:

None

Diff of the internal help from Asterisk 1.4 to 1.6:

```
--- in Asterisk 1.4
+++ in Asterisk 1.6
@@ -6,3 +6,4 @@
    being pressed, or the ASCII numerical value of the digit if one was
    pressed, or -1 on error or if the channel was disconnected.
    Remember, the file extension must not be included in the filename.
+   Runs Dead : No
```

D.37 TDD MODE

Activates or deactivates the transmission and reception of Telephone Device for the Deaf (TDD) signals on the active channel. So far, only Zap channels support TDD.

Asterisk versions:

```
--------| 1.2 |--------| 1.4 |--------| 1.6 |--------
```

Internal help for this AGI command in Asterisk 1.4:

```
Usage: TDD MODE <on|off>
       Enable/Disable TDD transmission/reception on a channel.
       Returns 1 if successful, or 0 if channel is not TDD-capable.
```

Diff of the internal help from Asterisk 1.2 to 1.4:

None

Diff of the internal help from Asterisk 1.4 to 1.6:

```
--- in Asterisk 1.4
+++ in Asterisk 1.6
@@ -1,3 +1,4 @@
    Usage: TDD MODE <on|off>
       Enable/Disable TDD transmission/reception on a channel
       Returns 1 if successful, or 0 if channel is not TDD-capable.
+   Runs Dead : No
```

D.38 VERBOSE

Asterisk versions:

```
--------| 1.2 |--------| 1.4 |--------| 1.6 |--------
```

Internal help for this AGI command in Asterisk 1.4:

```
Usage: VERBOSE <message> <level>
       Sends <message> to the console via verbose message system.
<level> is the the verbose level (1-4)
Always returns 1.
```

Diff of the internal help from Asterisk 1.2 to 1.4:

None

Diff of the internal help from Asterisk 1.4 to 1.6:

```
--- in Asterisk 1.4
+++ in Asterisk 1.6
@@ -2,3 +2,4 @@
          Sends <message> to the console via verbose message system.
    <level> is the the verbose level (1-4)
    Always returns 1.
+   Runs Dead : Yes
```

D.39 WAIT FOR DIGIT

Asterisk versions:

```
--------| 1.2 |--------| 1.4 |--------| 1.6 |--------
```

Internal help for this AGI command in Asterisk 1.4:

```
Usage: WAIT FOR DIGIT <timeout>
       Waits up to 'timeout' milliseconds for channel to receive a DTMF
       digit.
Returns -1 on channel failure, 0 if no digit is received in the timeout,
or the numerical value of the ascii of the digit if one is received. Use
-1 for the timeout value if you desire the call to block indefinitely.
```

Diff of the internal help from Asterisk 1.2 to 1.4:

None

Diff of the internal help from Asterisk 1.4 to 1.6:

```
--- in Asterisk 1.4
+++ in Asterisk 1.6
@@ -3,3 +3,4 @@
    Returns -1 on channel failure, 0 if no digit is received in the
    timeout, or the numerical value of the ascii of the digit if one
    is received. Use -1 for the timeout value if you desire the call
    to block indefinitely.
+   Runs Dead : No
```

AMI Command Summary

These command summaries, taken from the Asterisk internal help, are provided here for your convenience. Commands go in the `Action` field when transmitted to the AMI interface.

E.1 AbsoluteTimeout

Asterisk versions:

```
--------| 1.2 |--------| 1.4 |--------| 1.6 |--------
```

Internal help for this AMI command in Asterisk 1.4:

```
Action: AbsoluteTimeout
Synopsis: Set Absolute Timeout
Privilege: call,all
Description: Hangup a channel after a certain time.
Variables: (Names marked with * are required)
        *Channel: Channel name to hangup
        *Timeout: Maximum duration of the call (sec)
Acknowledges set time with 'Timeout Set' message
```

Diff of the internal help from Asterisk 1.2 to 1.4:

None

Diff of the internal help from Asterisk 1.4 to 1.6:

```
--- in Asterisk 1.4
+++ in Asterisk 1.6
@@ -1,6 +1,6 @@
    Action: AbsoluteTimeout
    Synopsis: Set Absolute Timeout
-   Privilege: call,all
+   Privilege: system,call,all
    Description: Hangup a channel after a certain time.
    Variables: (Names marked with * are required)
            *Channel: Channel name to hangup
```

E.2 AgentCallbackLogin

Asterisk versions:

```
--------| 1.2 |--------| 1.4 |              |      |
```

Internal help for this AMI command in Asterisk 1.4:

```
Action: AgentCallbackLogin
Synopsis: Sets an agent as logged in by callback
Privilege: agent,all
Description: Sets an agent as logged in with callback.
Variables: (Names marked with * are required)
        *Agent: Agent ID of the agent to login
        *Exten: Extension to use for callback
        Context: Context to use for callback
        AckCall: Set to 'true' to require an acknowledgement by '#'
        when agent is called back
        WrapupTime: the minimum amount of time after disconnecting
        before the caller can receive a new call
```

Diff of the internal help from Asterisk 1.2 to 1.4:

None

Diff of the internal help from Asterisk 1.4 to 1.6:

Not available in Asterisk 1.6

E.3 AgentLogoff

Asterisk versions:

```
--------| 1.2 |--------| 1.4 |--------| 1.6 |--------
```

Internal help for this AMI command in Asterisk 1.4:

```
Action: AgentLogoff
Synopsis: Sets an agent as no longer logged in
Privilege: agent,all
Description: Sets an agent as no longer logged in.
Variables: (Names marked with * are required)
        *Agent: Agent ID of the agent to log off
        Soft: Set to 'true' to not hangup existing calls
```

Diff of the internal help from Asterisk 1.2 to 1.4:

None

Diff of the internal help from Asterisk 1.4 to 1.6:

None

E.4 Agents

Asterisk versions:

```
--------| 1.2 |--------| 1.4 |--------| 1.6 |--------
```

Internal help for this AMI command in Asterisk 1.4:

```
Action: Agents
Synopsis: Lists agents and their status
Privilege: agent,all
Description: Will list info about all possible agents.
Variables: NONE
```

Diff of the internal help from Asterisk 1.2 to 1.4:

None

Diff of the internal help from Asterisk 1.4 to 1.6:

None

E.5 AGI

Asterisk versions:

```
     |      |         |      |          | 1.6 |--------
```

Internal help for this AMI command in Asterisk 1.6:

```
Action: AGI
Synopsis: Add an AGI command to execute by Async AGI
Privilege: call,all
Description: Add an AGI command to the execute queue of the channel
             in Async AGI
Variables:
  *Channel: Channel that is currently in Async AGI
  *Command: Application to execute
   CommandID: comand id. This will be sent back in CommandID header
of AsyncAGI exec event notification
```

Diff of the internal help from Asterisk 1.4 to 1.6:

Not available in Asterisk 1.4

E.6 Atxfer

Asterisk versions:

```
    |        |          |          |        | 1.6 |--------
```

Internal help for this AMI command in Asterisk 1.6:

```
Action: Atxfer
Synopsis: Attended transfer
Privilege: call,all
Description: Attended transfer.
Variables: (Names marked with * are required)
        *Channel: Transferer's channel
        *Exten: Extension to transfer to
        *Context: Context to transfer to
        *Priority: Priority to transfer to
        ActionID: Optional Action id for message matching.
```

Diff of the internal help from Asterisk 1.4 to 1.6:

Not available in Asterisk 1.4

E.7 Bridge

Asterisk versions:

```
    |        |          |          |        | 1.6 |--------
```

Internal help for this AMI command in Asterisk 1.6:

```
Action: Bridge
Synopsis: Bridge two channels already in the PBX
Privilege: call,all
Description: Bridge together two channels already in the PBX
Variables: ( Headers marked with * are required )
   *Channel1: Channel to Bridge to Channel2
   *Channel2: Channel to Bridge to Channel1
        Tone: (Yes|No) Play courtesy tone to Channel 2
```

Diff of the internal help from Asterisk 1.4 to 1.6:

Not available in Asterisk 1.4

E.8 Challenge

Asterisk versions:

```
   |       |         |       |       | 1.6 |--------
```

Internal help for this AMI command in Asterisk 1.6:

```
Action: Challenge
Synopsis: Generate Challenge for MD5 Auth
Privilege: <none>
```

Diff of the internal help from Asterisk 1.4 to 1.6:

Not available in Asterisk 1.4

E.9 ChangeMonitor

Asterisk versions:

```
--------| 1.2 |--------| 1.4 |--------| 1.6 |--------
```

Internal help for this AMI command in Asterisk 1.4:

```
Action: ChangeMonitor
Synopsis: Change monitoring filename of a channel
Privilege: call,all
Description: The 'ChangeMonitor' action may be used to change the file
   started by a previous 'Monitor' action. The following parameters may
   be used to control this:
   Channel    - Required. Used to specify the channel to record.
   File       - Required. Is the new name of the file created in the
                monitor spool directory.
```

Diff of the internal help from Asterisk 1.2 to 1.4:

None

Diff of the internal help from Asterisk 1.4 to 1.6:

None

E.10 Command

Asterisk versions:

```
--------| 1.2 |--------| 1.4 |--------| 1.6 |--------
```

Internal help for this AMI command in Asterisk 1.4:

```
Action: Command
Synopsis: Execute Asterisk CLI Command
Privilege: command,all
Description: Run a CLI command.
Variables: (Names marked with * are required)
      *Command: Asterisk CLI command to run
       ActionID: Optional Action id for message matching.
```

Diff of the internal help from Asterisk 1.2 to 1.4:

None

Diff of the internal help from Asterisk 1.4 to 1.6:

None

E.11 CoreSettings

Asterisk versions:

```
      |     |       |     |       | 1.6 |--------
```

Internal help for this AMI command in Asterisk 1.6:

```
Action: CoreSettings
Synopsis: Show PBX core settings (version etc)
Privilege: system,reporting,all
Description: Query for Core PBX settings.
Variables: (Names marked with * are optional)
      *ActionID: ActionID of this transaction
```

Diff of the internal help from Asterisk 1.4 to 1.6:

Not available in Asterisk 1.4

E.12 CoreShowChannels

Asterisk versions:

```
    |       |           |       |           | 1.6 |--------
```

Internal help for this AMI command in Asterisk 1.6:

```
Action: CoreShowChannels
Synopsis: List currently active channels
Privilege: system,reporting,all
Description: List currently defined channels and some information
           about them.
Variables:
        ActionID: Optional Action id for message matching.
```

Diff of the internal help from Asterisk 1.4 to 1.6:

Not available in Asterisk 1.4

E.13 CoreStatus

Asterisk versions:

```
    |       |           |       |           | 1.6 |--------
```

Internal help for this AMI command in Asterisk 1.6:

```
Action: CoreStatus
Synopsis: Show PBX core status variables
Privilege: system,reporting,all
Description: Query for Core PBX status.
Variables: (Names marked with * are optional)
        *ActionID: ActionID of this transaction
```

Diff of the internal help from Asterisk 1.4 to 1.6:

Not available in Asterisk 1.4

E.14 CreateConfig

Asterisk versions:

```
     |        |          |        |          |  1.6  |--------
```

Internal help for this AMI command in Asterisk 1.6:

```
Action: CreateConfig
Synopsis: Creates an empty file in the configuration directory
Privilege: config,all
Description: A 'CreateConfig' action will create an empty file in the
configuration directory. This action is intended to be used before an
UpdateConfig action.
Variables
   Filename:    The configuration filename to create (e.g. foo.conf)
```

Diff of the internal help from Asterisk 1.4 to 1.6:

Not available in Asterisk 1.4

E.15 DAHDIDialOffhook

Asterisk versions:

```
--------|  1.2*|--------|  1.4*|--------|  1.6  |--------
```

(* different name)

Internal help for this AMI command in Asterisk 1.6:

```
Action: DAHDIDialOffhook
Synopsis: Dial over DAHDI channel while offhook
Privilege: <none>
```

Diff of the internal help from Asterisk 1.2 (zapdialoffhook) to 1.4 (zapdialoffhook):

None

Diff of the internal help from Asterisk 1.4 (zapdialoffhook) to 1.6:

```
--- in Asterisk 1.4
+++ in Asterisk 1.6
@@ -1,3 +1,3 @@
-  Action: ZapDialOffhook
-  Synopsis: Dial over Zap channel while offhook
+  Action: DAHDIDialOffhook
+  Synopsis: Dial over DAHDI channel while offhook
   Privilege: <none>
```

E.16 DAHDIDNDoff

Asterisk versions:

```
--------| 1.2*|--------| 1.4*|--------| 1.6 |--------
```

(* different name)

Internal help for this AMI command in Asterisk 1.6:

```
Action: DAHDIDNDoff
Synopsis: Toggle DAHDI channel Do Not Disturb status OFF
Privilege: <none>
```

Diff of the internal help from Asterisk 1.2 (zapdndoff) to 1.4 (zapdndoff):

None

Diff of the internal help from Asterisk 1.4 (zapdndoff) to 1.6:

```
--- in Asterisk 1.4
+++ in Asterisk 1.6
@@ -1,3 +1,3 @@
- Action: ZapDNDoff
- Synopsis: Toggle Zap channel Do Not Disturb status OFF
+ Action: DAHDIDNDoff
+ Synopsis: Toggle DAHDI channel Do Not Disturb status OFF
  Privilege: <none>
```

E.17 DAHDIDNDon

Asterisk versions:

```
--------| 1.2*|--------| 1.4*|--------| 1.6 |--------
```

(* different name)

Internal help for this AMI command in Asterisk 1.6:

```
Action: DAHDIDNDon
Synopsis: Toggle DAHDI channel Do Not Disturb status ON
Privilege: <none>
```

Diff of the internal help from Asterisk 1.2 (zapdndon) to 1.4 (zapdndon):

None

Diff of the internal help from Asterisk 1.4 (zapdndon) to 1.6:

```
--- in Asterisk 1.4
+++ in Asterisk 1.6
@@ -1,3 +1,3 @@
-  Action: ZapDNDon
-  Synopsis: Toggle Zap channel Do Not Disturb status ON
+  Action: DAHDIDNDon
+  Synopsis: Toggle DAHDI channel Do Not Disturb status ON
   Privilege: <none>
```

E.18 DAHDIHangup

Asterisk versions:

```
--------| 1.2*|--------| 1.4*|--------| 1.6 |--------
```

(* different name)

Internal help for this AMI command in Asterisk 1.6:

```
Action: DAHDIHangup
Synopsis: Hangup DAHDI Channel
Privilege: <none>
```

Diff of the internal help from Asterisk 1.2 (zaphangup) to 1.4 (zaphangup):

None

Diff of the internal help from Asterisk 1.4 (zaphangup) to 1.6:

```
--- in Asterisk 1.4
+++ in Asterisk 1.6
@@ -1,3 +1,3 @@
-  Action: ZapHangup
-  Synopsis: Hangup Zap Channel
+  Action: DAHDIHangup
+  Synopsis: Hangup DAHDI Channel
   Privilege: <none>
```

E.19 DAHDIRestart

Asterisk versions:

```
    |      |        | 1.4*|--------| 1.6 |--------
```

(* different name)

Internal help for this AMI command in Asterisk 1.6:

```
Action: DAHDIRestart
Synopsis: Fully Restart DAHDI channels (terminates calls)
Privilege: <none>
```

Diff of the internal help from Asterisk 1.2 to 1.4:

Not available in Asterisk 1.2

Diff of the internal help from Asterisk 1.4 (zaprestart) to 1.6:

```
--- in Asterisk 1.4
+++ in Asterisk 1.6
@@ -1,3 +1,3 @@
-  Action: ZapRestart
-  Synopsis: Fully Restart zaptel channels (terminates calls)
+  Action: DAHDIRestart
+  Synopsis: Fully Restart DAHDI channels (terminates calls)
   Privilege: <none>
```

E.20 DAHDIShowChannels

Asterisk versions:

```
--------| 1.2*|--------| 1.4*|--------| 1.6 |--------
```

(* different name)

Internal help for this AMI command in Asterisk 1.6:

```
Action: DAHDIShowChannels
Synopsis: Show status DAHDI channels
Privilege: <none>
```

Diff of the internal help from Asterisk 1.2 (zapshowchannels) to 1.4 (zapshowchannels):

None

Diff of the internal help from Asterisk 1.4 (zapshowchannels) to 1.6:

```
--- in Asterisk 1.4
+++ in Asterisk 1.6
@@ -1,3 +1,3 @@
-  Action: ZapShowChannels
-  Synopsis: Show status zapata channels
+  Action: DAHDIShowChannels
+  Synopsis: Show status DAHDI channels
   Privilege: <none>
```

E.21 DAHDITransfer

Asterisk versions:

```
--------| 1.2*|--------| 1.4*|--------| 1.6 |--------
```

(* different name)

Internal help for this AMI command in Asterisk 1.6:

```
Action: DAHDITransfer
Synopsis: Transfer DAHDI Channel
Privilege: <none>
```

Diff of the internal help from Asterisk 1.2 (zaptransfer) to 1.4 (zaptransfer):

None

Diff of the internal help from Asterisk 1.4 (zaptransfer) to 1.6:

```
--- in Asterisk 1.4
+++ in Asterisk 1.6
@@ -1,3 +1,3 @@
-  Action: ZapTransfer
-  Synopsis: Transfer Zap Channel
+  Action: DAHDITransfer
+  Synopsis: Transfer DAHDI Channel
   Privilege: <none>
```

E.22 DBDel

Asterisk versions:

```
  |      |        |       |       | 1.6 |--------
```

Internal help for this AMI command in Asterisk 1.6:

```
Action: DBDel
Synopsis: Delete DB Entry
Privilege: system,all
```

Diff of the internal help from Asterisk 1.4 to 1.6:

Not available in Asterisk 1.4

E.23 DBDelTree

Asterisk versions:

```
     |        |            |          |         | 1.6 |--------
```

Internal help for this AMI command in Asterisk 1.6:

```
Action: DBDelTree
Synopsis: Delete DB Tree
Privilege: system,all
```

Diff of the internal help from Asterisk 1.4 to 1.6:

Not available in Asterisk 1.4

E.24 DBGet

Asterisk versions:

```
--------| 1.2 |--------| 1.4 |--------| 1.6 |--------
```

Internal help for this AMI command in Asterisk 1.4:

```
Action: DBGet
Synopsis: Get DB Entry
Privilege: system,all
```

Diff of the internal help from Asterisk 1.2 to 1.4:

None

Diff of the internal help from Asterisk 1.4 to 1.6:

```
--- in Asterisk 1.4
+++ in Asterisk 1.6
@@ -1,3 +1,3 @@
   Action: DBGet
   Synopsis: Get DB Entry
-  Privilege: system,all
+  Privilege: system,reporting,all
```

E.25 DBPut

Asterisk versions:

`--------| 1.2 |--------| 1.4 |--------| 1.6 |--------`

Internal help for this AMI command in Asterisk 1.4:

```
Action: DBPut
Synopsis: Put DB Entry
Privilege: system,all
```

Diff of the internal help from Asterisk 1.2 to 1.4:

None

Diff of the internal help from Asterisk 1.4 to 1.6:

None

E.26 Events

Asterisk versions:

`--------| 1.2 |--------| 1.4 |--------| 1.6 |--------`

Internal help for this AMI command in Asterisk 1.4:

```
Action: Events
Synopsis: Control Event Flow
Privilege: <none>
Description: Enable/Disable sending of events to this manager
   client.
Variables:
        EventMask: 'on' if all events should be sent,
                   'off' if no events should be sent,
                   'system,call,log' to select which flags events should
                   have to be sent.
```

Diff of the internal help from Asterisk 1.2 to 1.4:

None

Diff of the internal help from Asterisk 1.4 to 1.6:

None

E.27 ExtensionState

Asterisk versions:

```
--------| 1.2 |--------| 1.4 |--------| 1.6 |--------
```

Internal help for this AMI command in Asterisk 1.4:

```
Action: ExtensionState
Synopsis: Check Extension Status
Privilege: call,all
Description: Report the extension state for given extension.
    If the extension has a hint, will use devicestate to check
    the status of the device connected to the extension.
Variables: (Names marked with * are required)
        *Exten: Extension to check state on
        *Context: Context for extension
        ActionId: Optional ID for this transaction
Will return an "Extension Status" message.
The response will include the hint for the extension and the status.
```

Diff of the internal help from Asterisk 1.2 to 1.4:

None

Diff of the internal help from Asterisk 1.4 to 1.6:

```
--- in Asterisk 1.4
+++ in Asterisk 1.6
@@ -1,6 +1,6 @@
   Action: ExtensionState
   Synopsis: Check Extension Status
-  Privilege: call,all
+  Privilege: call,reporting,all
   Description: Report the extension state for given extension.
     If the extension has a hint, will use devicestate to check
     the status of the device connected to the extension.
```

E.28 GetConfig

Asterisk versions:

```
     |      |       | 1.4 |--------| 1.6 |--------
```

Internal help for this AMI command in Asterisk 1.4:

```
Action: GetConfig
Synopsis: Retrieve configuration
```

```
Privilege: config,all
Description: A 'GetConfig' action will dump the contents of a
configuration file by category and contents.
Variables:
        Filename: Configuration filename (e.g. foo.conf)
```

Diff of the internal help from Asterisk 1.2 to 1.4:

Not available in Asterisk 1.2

Diff of the internal help from Asterisk 1.4 to 1.6:

```
--- in Asterisk 1.4
+++ in Asterisk 1.6
@@ -1,7 +1,8 @@
   Action: GetConfig
   Synopsis: Retrieve configuration
-  Privilege: config,all
+  Privilege: system,config,all
   Description: A 'GetConfig' action will dump the contents of a configuration
-  file by category and contents.
-  Variables:
-    Filename: Configuration filename (e.g. foo.conf)
+  file by category and contents or optionally by specified category
   only.
+  Variables: (Names marked with * are required)
+    *Filename: Configuration filename (e.g. foo.conf)
+    Category: Category in configuration file
```

E.29 GetConfigJSON

Asterisk versions:

```
    |        |           |        |        | 1.6 |--------
```

Internal help for this AMI command in Asterisk 1.6:

```
Action: GetConfigJSON
Synopsis: Retrieve configuration (JSON format)
Privilege: system,config,all
Description: A 'GetConfigJSON' action will dump the contents of a
configuration file by category and contents in JSON format. This only
makes sense to be used using rawman over the HTTP interface.
Variables:
   Filename: Configuration filename (e.g. foo.conf)
```

Diff of the internal help from Asterisk 1.4 to 1.6:

Not available in Asterisk 1.4

E.30 Getvar

Asterisk versions:

```
--------| 1.2 |--------| 1.4 |--------| 1.6 |--------
```

Internal help for this AMI command in Asterisk 1.4:

```
Action: Getvar
Synopsis: Gets a Channel Variable
Privilege: call,all
Description: Get the value of a global or local channel variable.
Variables: (Names marked with * are required)
        Channel: Channel to read variable from
        *Variable: Variable name
        ActionID: Optional Action id for message matching.
```

Diff of the internal help from Asterisk 1.2 to 1.4:

None

Diff of the internal help from Asterisk 1.4 to 1.6:

```
--- in Asterisk 1.4
+++ in Asterisk 1.6
@@ -1,6 +1,6 @@
   Action: Getvar
   Synopsis: Gets a Channel Variable
-  Privilege: call,all
+  Privilege: call,reporting,all
   Description: Get the value of a global or local channel variable.
   Variables: (Names marked with * are required)
         Channel: Channel to read variable from
```

E.31 Hangup

Asterisk versions:

```
--------| 1.2 |--------| 1.4 |--------| 1.6 |--------
```

Internal help for this AMI command in Asterisk 1.4:

```
Action: Hangup
Synopsis: Hangup Channel
Privilege: call,all
Description: Hangup a channel
Variables:
        Channel: The channel name to be hungup
```

Diff of the internal help from Asterisk 1.2 to 1.4:

None

Diff of the internal help from Asterisk 1.4 to 1.6:

```
--- in Asterisk 1.4
+++ in Asterisk 1.6
@@ -1,6 +1,6 @@
   Action: Hangup
   Synopsis: Hangup Channel
-  Privilege: call,all
+  Privilege: system,call,all
   Description: Hangup a channel
   Variables:
           Channel: The channel name to be hungup
```

E.32 IAXnetstats

Asterisk versions:

```
--------| 1.2 |--------| 1.4 |--------| 1.6 |--------
```

Internal help for this AMI command in Asterisk 1.4:

```
Action: IAXnetstats
Synopsis: Show IAX Netstats
Privilege: <none>
```

Diff of the internal help from Asterisk 1.2 to 1.4:

None

Diff of the internal help from Asterisk 1.4 to 1.6:

```
--- in Asterisk 1.4
+++ in Asterisk 1.6
@@ -1,3 +1,3 @@
   Action: IAXnetstats
   Synopsis: Show IAX Netstats
-  Privilege: <none>
+  Privilege: system,reporting,all
```

E.33 IAXpeerlist

Asterisk versions:

```
   |    |        |    |        | 1.6 |--------
```

Internal help for this AMI command in Asterisk 1.6:

```
Action: IAXpeerlist
Synopsis: List IAX Peers
Privilege: system,reporting,all
```

Diff of the internal help from Asterisk 1.4 to 1.6:

Not available in Asterisk 1.4

E.34 IAXpeers

Asterisk versions:

```
--------| 1.2 |--------| 1.4 |--------| 1.6 |--------
```

Internal help for this AMI command in Asterisk 1.4:

```
Action: IAXpeers
Synopsis: List IAX Peers
Privilege: <none>
```

Diff of the internal help from Asterisk 1.2 to 1.4:

None

Diff of the internal help from Asterisk 1.4 to 1.6:

```
--- in Asterisk 1.4
+++ in Asterisk 1.6
@@ -1,3 +1,3 @@
   Action: IAXpeers
   Synopsis: List IAX Peers
-  Privilege: <none>
+  Privilege: system,reporting,all
```

E.35 JabberSend

Asterisk versions:

```
    |     |        | 1.4 |--------| 1.6 |--------
```

Internal help for this AMI command in Asterisk 1.4:

```
Action: JabberSend
Synopsis: Sends a message to a Jabber Client
Privilege: system,all
Description: Sends a message to a Jabber Client.
```

```
Variables:
  Jabber:        Client or transport Asterisk uses to connect to JABBER.
  ScreenName:    User Name to message.
  Message:       Message to be sent to the buddy
```

Diff of the internal help from Asterisk 1.2 to 1.4:

Not available in Asterisk 1.2

Diff of the internal help from Asterisk 1.4 to 1.6:

None

E.36 ListCategories

Asterisk versions:

```
    |        |          |         |          | 1.6 |--------
```

Internal help for this AMI command in Asterisk 1.6:

```
Action: ListCategories
Synopsis: List categories in configuration file
Privilege: config,all
Description: A 'ListCategories' action will dump the categories in
a given file.
Variables
    Filename: Configuration filename (e.g., foo.conf)
```

Diff of the internal help from Asterisk 1.4 to 1.6:

Not available in Asterisk 1.4

E.37 ListCommands

Asterisk versions:

```
--------| 1.2 |--------| 1.4 |--------| 1.6 |--------
```

Internal help for this AMI command in Asterisk 1.4:

```
Action: ListCommands
Synopsis: List available manager commands
Privilege: <none>
Description: Returns the action name and synopsis for every
    action that is available to the user
Variables: NONE
```

Diff of the internal help from Asterisk 1.2 to 1.4:

None

Diff of the internal help from Asterisk 1.4 to 1.6:

None

E.38 Login

Asterisk versions:

```
    |        |            |         |        | 1.6 |--------
```

Internal help for this AMI command in Asterisk 1.6:

```
Action: Login
Synopsis: Login Manager
Privilege: <none>
```

Diff of the internal help from Asterisk 1.4 to 1.6:

Not available in Asterisk 1.4

E.39 Logoff

Asterisk versions:

```
--------| 1.2 |--------| 1.4 |--------| 1.6 |--------
```

Internal help for this AMI command in Asterisk 1.4:

```
Action: Logoff
Synopsis: Logoff Manager
Privilege: <none>
Description: Logoff this manager session
Variables: NONE
```

Diff of the internal help from Asterisk 1.2 to 1.4:

None

Diff of the internal help from Asterisk 1.4 to 1.6:

None

E.40 MailboxCount

Asterisk versions:

```
--------| 1.2 |--------| 1.4 |--------| 1.6 |--------
```

Internal help for this AMI command in Asterisk 1.4:

```
Action: MailboxCount
Synopsis: Check Mailbox Message Count
Privilege: call,all
Description: Checks a voicemail account for new messages.
Variables: (Names marked with * are required)
        *Mailbox: Full mailbox ID <mailbox>@<vm-context>
         ActionID: Optional ActionID for message matching.
Returns number of new and old messages.
        Message: Mailbox Message Count
        Mailbox: <mailboxid>
        NewMessages: <count>
        OldMessages: <count>
```

Diff of the internal help from Asterisk 1.2 to 1.4:

None

Diff of the internal help from Asterisk 1.4 to 1.6:

```
--- in Asterisk 1.4
+++ in Asterisk 1.6
@@ -1,12 +1,13 @@
   Action: MailboxCount
   Synopsis: Check Mailbox Message Count
-  Privilege: call,all
+  Privilege: call,reporting,all
   Description: Checks a voicemail account for new messages.
   Variables: (Names marked with * are required)
           *Mailbox: Full mailbox ID <mailbox>@<vm-context>
            ActionID: Optional ActionID for message matching.
-  Returns number of new and old messages.
+  Returns number of urgent, new and old messages.
           Message: Mailbox Message Count
           Mailbox: <mailboxid>
+          UrgentMessages: <count>
           NewMessages: <count>
           OldMessages: <count>
```

E.41 MailboxStatus

Asterisk versions:

```
--------| 1.2 |--------| 1.4 |--------| 1.6 |--------
```

Internal help for this AMI command in Asterisk 1.4:

```
Action: MailboxStatus
Synopsis: Check Mailbox
Privilege: call,all
Description: Checks a voicemail account for status.
Variables: (Names marked with * are required)
        *Mailbox: Full mailbox ID <mailbox>@<vm-context>
        ActionID: Optional ActionID for message matching.
Returns number of messages.
        Message: Mailbox Status
        Mailbox: <mailboxid>
        Waiting: <count>
```

Diff of the internal help from Asterisk 1.2 to 1.4:

None

Diff of the internal help from Asterisk 1.4 to 1.6:

```
--- in Asterisk 1.4
+++ in Asterisk 1.6
@@ -1,6 +1,6 @@
   Action: MailboxStatus
   Synopsis: Check Mailbox
-  Privilege: call,all
+  Privilege: call,reporting,all
   Description: Checks a voicemail account for status.
   Variables: (Names marked with * are required)
        *Mailbox: Full mailbox ID <mailbox>@<vm-context>
```

E.42 MeetmeList

Asterisk versions:

```
    |     |      |     |       | 1.6 |--------
```

Internal help for this AMI command in Asterisk 1.6:

```
Action: MeetmeList
Synopsis: List participants in a conference
Privilege: reporting,all
```

```
Description: Lists all users in a particular MeetMe conference.
MeetmeList will follow as separate events, followed by a final event called
MeetmeListComplete.
Variables:
    *ActionId: <id>
    *Conference: <confno>
```

Diff of the internal help from Asterisk 1.4 to 1.6:

Not available in Asterisk 1.4

E.43 MeetmeMute

Asterisk versions:

```
|     |          | 1.4 |--------| 1.6 |--------
```

Internal help for this AMI command in Asterisk 1.4:

```
Action: MeetmeMute
Synopsis: Mute a Meetme user
Privilege: call,all
```

Diff of the internal help from Asterisk 1.2 to 1.4:

Not available in Asterisk 1.2

Diff of the internal help from Asterisk 1.4 to 1.6:

None

E.44 MeetmeUnmute

Asterisk versions:

```
|     |          | 1.4 |--------| 1.6 |--------
```

Internal help for this AMI command in Asterisk 1.4:

```
Action: MeetmeUnmute
Synopsis: Unmute a Meetme user
Privilege: call,all
```

Diff of the internal help from Asterisk 1.2 to 1.4:

Not available in Asterisk 1.2

Diff of the internal help from Asterisk 1.4 to 1.6:

None

E.45 ModuleCheck

Asterisk versions:

```
  |       |         |       |         |  1.6  |--------
```

Internal help for this AMI command in Asterisk 1.6:

```
Action: ModuleCheck
Synopsis: Check if module is loaded
Privilege: system,all
Description: Checks if Asterisk module is loaded
Variables:
   ActionID: <id>        Action ID for this transaction. Will be returned.
   Module: <name>        Asterisk module name (not including extension)

Will return Success/Failure
For success returns, the module revision number is included.
```

Diff of the internal help from Asterisk 1.4 to 1.6:

Not available in Asterisk 1.4

E.46 ModuleLoad

Asterisk versions:

```
  |       |         |       |         |  1.6  |--------
```

Internal help for this AMI command in Asterisk 1.6:

```
Action: ModuleLoad
Synopsis: Module management
Privilege: system,all
Description: Loads, unloads or reloads an Asterisk module in a running
system.
Variables:
   ActionID: <id>        Action ID for this transaction. Will be returned.
   Module: <name>        Asterisk module name (including .so extension)
                         or subsystem identifier:
                             cdr, enum, dnsmgr, extconfig, manager,
                             rtp, http
```

```
LoadType: load | unload | reload
                    The operation to be done on module
  If no module is specified for a reload loadtype, all modules are reloaded
```

Diff of the internal help from Asterisk 1.4 to 1.6:

Not available in Asterisk 1.4

E.47 Monitor

Asterisk versions:

```
--------| 1.2 |--------| 1.4 |--------| 1.6 |--------
```

Internal help for this AMI command in Asterisk 1.4:

```
Action: Monitor
Synopsis: Monitor a channel
Privilege: call,all
Description: The 'Monitor' action may be used to record the audio on a
   specified channel. The following parameters may be used to control
   this:
   Channel    - Required. Used to specify the channel to record.
   File       - Optional. Is the name of the file created in the
                monitor spool directory. Defaults to the same name
                as the channel (with slashes replaced with dashes).
   Format     - Optional. Is the audio recording format. Defaults
                to "wav".
   Mix        - Optional. Boolean parameter as to whether to mix
                the input and output channels together after the
                recording is finished.
```

Diff of the internal help from Asterisk 1.2 to 1.4:

None

Diff of the internal help from Asterisk 1.4 to 1.6:

None

E.48 Originate

Asterisk versions:

```
--------| 1.2 |--------| 1.4 |--------| 1.6 |--------
```

Internal help for this AMI command in Asterisk 1.4:

```
Action: Originate
Synopsis: Originate Call
Privilege: call,all
Description: Generates an outgoing call to a Extension/Context/Priority or
    Application/Data
Variables: (Names marked with * are required)
        *Channel: Channel name to call
        Exten: Extension to use (requires 'Context' and 'Priority')
        Context: Context to use (requires 'Exten' and 'Priority')
        Priority: Priority to use (requires 'Exten' and 'Context')
        Application: Application to use
        Data: Data to use (requires 'Application')
        Timeout: How long to wait for call to be answered (in ms)
        CallerID: Caller ID to be set on the outgoing channel
        Variable: Channel variable to set, multiple Variable: headers
                are allowed
        Account: Account code
        Async: Set to 'true' for fast origination
```

Diff of the internal help from Asterisk 1.2 to 1.4:

None

Diff of the internal help from Asterisk 1.4 to 1.6:

```
--- in Asterisk 1.4
+++ in Asterisk 1.6
@@ -1,6 +1,6 @@
  Action: Originate
  Synopsis: Originate Call
- Privilege: call,all
+ Privilege: originate,all
  Description: Generates an outgoing call to a Extension/Context/Priority
    or Application/Data
  Variables: (Names marked with * are required)
```

E.49 Park

Asterisk versions:

```
    |       |           |  1.4  |--------|  1.6  |--------
```

Internal help for this AMI command in Asterisk 1.4:

```
Action: Park
Synopsis: Park a channel
Privilege: call,all
```

```
Description: Park a channel.
Variables: (Names marked with * are required)
        *Channel: Channel name to park
        *Channel2: Channel to announce park info to (and return
                to if timeout)
        Timeout: Number of milliseconds to wait before callback.
```

Diff of the internal help from Asterisk 1.2 to 1.4:

Not available in Asterisk 1.2

Diff of the internal help from Asterisk 1.4 to 1.6:

None

E.50 ParkedCalls

Asterisk versions:

```
--------| 1.2 |--------| 1.4 |--------| 1.6 |--------
```

Internal help for this AMI command in Asterisk 1.4:

```
Action: ParkedCalls
Synopsis: List parked calls
Privilege: <none>
```

Diff of the internal help from Asterisk 1.2 to 1.4:

None

Diff of the internal help from Asterisk 1.4 to 1.6:

None

E.51 PauseMonitor

Asterisk versions:

```
    |     |        | 1.4 |--------| 1.6 |--------
```

Internal help for this AMI command in Asterisk 1.4:

```
Action: PauseMonitor
Synopsis: Pause monitoring of a channel
```

```
Privilege: call,all
Description: The 'PauseMonitor' action may be used to temporarily
             stop the recording of a channel. The following parameters may
             be used to control this:
    Channel       - Required. Used to specify the channel to record.
```

Diff of the internal help from Asterisk 1.2 to 1.4:

Not available in Asterisk 1.2

Diff of the internal help from Asterisk 1.4 to 1.6:

None

E.52 Ping

Asterisk versions:

```
--------| 1.2 |--------| 1.4 |--------| 1.6 |--------
```

Internal help for this AMI command in Asterisk 1.4:

```
Action: Ping
Synopsis: Keepalive command
Privilege: <none>
Description: A 'Ping' action will ellicit a 'Pong' response. Used
             to keep the manager connection open.
Variables: NONE
```

Diff of the internal help from Asterisk 1.2 to 1.4:

```
--- in Asterisk 1.2
+++ in Asterisk 1.4
@@ -1,5 +1,6 @@
   Action: Ping
   Synopsis: Keepalive command
   Privilege: <none>
-  Description: A 'Ping' action will elicit a 'Pong' response.
                Used to keep the manager connection open.
+  Description: A 'Ping' action will elicit a 'Pong' response.
                Used to keep the manager connection open.
   Variables: NONE
```

Diff of the internal help from Asterisk 1.4 to 1.6:

None

E.53 PlayDTMF

Asterisk versions:

```
    |      |           |  1.4  |--------|  1.6  |--------
```

Internal help for this AMI command in Asterisk 1.4:

```
Action: PlayDTMF
Synopsis: Play DTMF signal on a specific channel.
Privilege: call,all
Description: Plays a dtmf digit on the specified channel.
Variables: (all are required)
        Channel: Channel name to send digit to
        Digit: The dtmf digit to play
```

Diff of the internal help from Asterisk 1.2 to 1.4:

Not available in Asterisk 1.2

Diff of the internal help from Asterisk 1.4 to 1.6:

None

E.54 QueueAdd

Asterisk versions:

```
--------|  1.2  |--------|  1.4  |--------|  1.6  |--------
```

Internal help for this AMI command in Asterisk 1.4:

```
Action: QueueAdd
Synopsis: Add interface to queue.
Privilege: agent,all
```

Diff of the internal help from Asterisk 1.2 to 1.4:

None

Diff of the internal help from Asterisk 1.4 to 1.6:

None

E.55 QueueLog

Asterisk versions:

Internal help for this AMI command in Asterisk 1.6:

```
Action: QueueLog
Synopsis: Adds custom entry in queue_log
Privilege: agent,all
```

Diff of the internal help from Asterisk 1.4 to 1.6:

Not available in Asterisk 1.4

E.56 QueuePause

Asterisk versions:

Internal help for this AMI command in Asterisk 1.4:

```
Action: QueuePause
Synopsis: Makes a queue member temporarily unavailable
Privilege: agent,all
```

Diff of the internal help from Asterisk 1.2 to 1.4:

None

Diff of the internal help from Asterisk 1.4 to 1.6:

None

E.57 QueuePenalty

Asterisk versions:

Internal help for this AMI command in Asterisk 1.6:

```
Action: QueuePenalty
Synopsis: Set the penalty for a queue member
Privilege: agent,all
```

Diff of the internal help from Asterisk 1.4 to 1.6:

Not available in Asterisk 1.4

E.58 QueueRemove

Asterisk versions:

```
--------| 1.2 |--------| 1.4 |--------| 1.6 |--------
```

Internal help for this AMI command in Asterisk 1.4:

```
Action: QueueRemove
Synopsis: Remove interface from queue.
Privilege: agent,all
```

Diff of the internal help from Asterisk 1.2 to 1.4:

None

Diff of the internal help from Asterisk 1.4 to 1.6:

None

E.59 QueueRule

Asterisk versions:

```
   |     |       |     |       | 1.6 |--------
```

Internal help for this AMI command in Asterisk 1.6:

```
Action: QueueRule
Synopsis: Queue Rules
Privilege: <none>
```

Diff of the internal help from Asterisk 1.4 to 1.6:

Not available in Asterisk 1.4

E.60 Queues

Asterisk versions:

```
--------| 1.2 |--------| 1.4 |--------| 1.6 |--------
```

Internal help for this AMI command in Asterisk 1.4:

```
Action: Queues
Synopsis: Queues
Privilege: <none>
```

Diff of the internal help from Asterisk 1.2 to 1.4:

None

Diff of the internal help from Asterisk 1.4 to 1.6:

None

E.61 QueueStatus

Asterisk versions:

```
--------| 1.2 |--------| 1.4 |--------| 1.6 |--------
```

Internal help for this AMI command in Asterisk 1.4:

```
Action: QueueStatus
Synopsis: Queue Status
Privilege: <none>
```

Diff of the internal help from Asterisk 1.2 to 1.4:

None

Diff of the internal help from Asterisk 1.4 to 1.6:

None

E.62 QueueSummary

Asterisk versions:

```
   |      |         |      |         | 1.6 |--------
```

Internal help for this AMI command in Asterisk 1.6:

```
Action: QueueSummary
Synopsis: Queue Summary
Privilege: <none>
```

Diff of the internal help from Asterisk 1.4 to 1.6:

Not available in Asterisk 1.4

E.63 Redirect

Asterisk versions:

```
--------| 1.2 |--------| 1.4 |--------| 1.6 |--------
```

Internal help for this AMI command in Asterisk 1.4:

```
Action: Redirect
Synopsis: Redirect (transfer) a call
Privilege: call,all
Description: Redirect (transfer) a call.
Variables: (Names marked with * are required)
        *Channel: Channel to redirect
        ExtraChannel: Second call leg to transfer (optional)
        *Exten: Extension to transfer to
        *Context: Context to transfer to
        *Priority: Priority to transfer to
        ActionID: Optional Action id for message matching.
```

Diff of the internal help from Asterisk 1.2 to 1.4:

None

Diff of the internal help from Asterisk 1.4 to 1.6:

None

E.64 Reload

Asterisk versions:

```
    |     |        |     |        | 1.6 |--------
```

Internal help for this AMI command in Asterisk 1.6:

```
Action: Reload
Synopsis: Send a reload event
Privilege: system,config,all
Description: Send a reload event.
Variables: (Names marked with * are optional)
        *ActionID: ActionID of this transaction
        *Module: Name of the module to reload
```

Diff of the internal help from Asterisk 1.4 to 1.6:

Not available in Asterisk 1.4

E.65 SendText

Asterisk versions:

```
    |      |         |         |         | 1.6 |--------
```

Internal help for this AMI command in Asterisk 1.6:

```
Action: SendText
Synopsis: Send text message to channel
Privilege: call,all
Description: Sends A Text Message while in a call.
Variables: (Names marked with * are required)
        *Channel: Channel to send message to
        *Message: Message to send
        ActionID: Optional Action id for message matching.
```

Diff of the internal help from Asterisk 1.4 to 1.6:

Not available in Asterisk 1.4

E.66 SetCDRUserField

Asterisk versions:

```
--------| 1.2 |--------| 1.4 |         |       |
```

Internal help for this AMI command in Asterisk 1.4:

```
Action: SetCDRUserField
Synopsis: Set the CDR UserField
Privilege: call,all
```

Diff of the internal help from Asterisk 1.2 to 1.4:

None

Diff of the internal help from Asterisk 1.4 to 1.6:

Not available in Asterisk 1.6

E.67 Setvar

Asterisk versions:

```
--------| 1.2 |--------| 1.4 |--------| 1.6 |--------
```

Internal help for this AMI command in Asterisk 1.4:

```
Action: Setvar
Synopsis: Set Channel Variable
Privilege: call,all
Description: Set a global or local channel variable.
Variables: (Names marked with * are required)
        Channel: Channel to set variable for
        *Variable: Variable name
        *Value: Value
```

Diff of the internal help from Asterisk 1.2 to 1.4:

None

Diff of the internal help from Asterisk 1.4 to 1.6:

None

E.68 ShowDialPlan

Asterisk versions:

```
    |     |        |          |      | 1.6 |--------
```

Internal help for this AMI command in Asterisk 1.6:

```
Action: ShowDialPlan
Synopsis: List dialplan
Privilege: config,reporting,all
Description: Show dialplan contexts and extensions.
Be aware that showing the full dialplan may take a lot of capacity
Variables:
    ActionID: <id>              Action ID for this AMI transaction (optional)
    Extension: <extension>      Extension (Optional)
    Context: <context>          Context (Optional)
```

Diff of the internal help from Asterisk 1.4 to 1.6:

Not available in Asterisk 1.4

E.69 SIPnotify

Asterisk versions:

| | | | | | 1.6 |--------

Internal help for this AMI command in Asterisk 1.6:

```
Action: SIPnotify
Synopsis: Send a SIP notify
Privilege: system,all
Description: Sends a SIP Notify event
All parameters for this event must be specified in the body of this
  request via multiple Variable: name=value sequences.
Variables:
 *Channel: <peername>       Peer to receive the notify. Required.
 *Variable: <name>=<value>  At least one variable pair must be specified.
  ActionID: <id>            Action ID for this transaction.
                            Will be returned.
```

Diff of the internal help from Asterisk 1.4 to 1.6:

Not available in Asterisk 1.4

E.70 SIPpeers

Asterisk versions:

--------| 1.2 |--------| 1.4 |--------| 1.6 |--------

Internal help for this AMI command in Asterisk 1.4:

```
Action: SIPpeers
Synopsis: List SIP peers (text format)
Privilege: system,all
Description: Lists SIP peers in text format with details on current
             status.
Variables:
   ActionID: <id> Action ID for this transaction. Will be returned.
```

Diff of the internal help from Asterisk 1.2 to 1.4:

None

Diff of the internal help from Asterisk 1.4 to 1.6:

```
--- in Asterisk 1.4
+++ in Asterisk 1.6
```

```
@@ -1,6 +1,8 @@
   Action: SIPpeers
   Synopsis: List SIP peers (text format)
-  Privilege: system,all
+  Privilege: system,reporting,all
   Description: Lists SIP peers in text format with details on current
                status.
+  Peerlist will follow as separate events, followed by a final event
+    called
+  PeerlistComplete.
   Variables:
      ActionID: <id> Action ID for this transaction. Will be returned.
```

E.71 SIPqualifypeer

Asterisk versions:

Internal help for this AMI command in Asterisk 1.6:

```
Action: SIPqualifypeer
Synopsis: Show SIP peer (text format)
Privilege: system,reporting,all
Description: Show one SIP peer with details on current status.
Variables:
   Peer: <name>          The peer name you want to check.
   ActionID: <id>         Optional action ID for this AMI transaction.
```

Diff of the internal help from Asterisk 1.4 to 1.6:

Not available in Asterisk 1.4

E.72 SIPshowpeer

Asterisk versions:

-------- | 1.2 |-------- | 1.4 |-------- | 1.6 |--------

Internal help for this AMI command in Asterisk 1.4:

```
Action: SIPshowpeer
Synopsis: Show SIP peer (text format)
Privilege: system,all
Description: Show one SIP peer with details on current status.
Variables:
   Peer: <name>          The peer name you want to check.
   ActionID: <id>         Optional action ID for this AMI transaction.
```

Diff of the internal help from Asterisk 1.2 to 1.4:

```
--- in Asterisk 1.2
+++ in Asterisk 1.4
@@ -2,7 +2,6 @@
   Synopsis: Show SIP peer (text format)
   Privilege: system,all
   Description: Show one SIP peer with details on current status.
-    The XML format is under development, feedback welcome! /oej
   Variables:
      Peer: <name>           The peer name you want to check.
      ActionID: <id>         Optional action ID for this AMI transaction.
```

Diff of the internal help from Asterisk 1.4 to 1.6:

```
--- in Asterisk 1.4
+++ in Asterisk 1.6
@@ -1,6 +1,6 @@
   Action: SIPshowpeer
   Synopsis: Show SIP peer (text format)
-  Privilege: system,all
+  Privilege: system,reporting,all
   Description: Show one SIP peer with details on current status.
   Variables:
      Peer: <name> The peer name you want to check.
```

E.73 SIPshowregistry

Asterisk versions:

```
    |      |        |        |         |  1.6  |--------
```

Internal help for this AMI command in Asterisk 1.6:

```
Action: SIPshowregistry
Synopsis: Show SIP registrations (text format)
Privilege: system,reporting,all
Description: Lists all registration requests and status
Registrations will follow as separate events, followed by a final
   event called.
RegistrationsComplete.
Variables:
   ActionID: <id>          Action ID for this transaction.
                           Will be returned.
```

Diff of the internal help from Asterisk 1.4 to 1.6:

Not available in Asterisk 1.4

E.74 Status

Asterisk versions:

`--------| 1.2 |--------| 1.4 |--------| 1.6 |--------`

Internal help for this AMI command in Asterisk 1.4:

```
Action: Status
Synopsis: Lists channel status
Privilege: call,all
```

Diff of the internal help from Asterisk 1.2 to 1.4:

None

Diff of the internal help from Asterisk 1.4 to 1.6:

```
--- in Asterisk 1.4
+++ in Asterisk 1.6
@@ -1,3 +1,10 @@
   Action: Status
   Synopsis: Lists channel status
-  Privilege: call,all
+  Privilege: system,call,reporting,all
+  Description: Lists channel status along with requested channel vars.
+  Variables: (Names marked with * are required)
+          *Channel: Name of the channel to query for status
+          Variables: Comma ',' separated list of variables to include
+          ActionID: Optional ID for this transaction
+  Will return the status information of each channel along with the
+  value for the specified channel variables.
```

E.75 StopMonitor

Asterisk versions:

`--------| 1.2 |--------| 1.4 |--------| 1.6 |--------`

Internal help for this AMI command in Asterisk 1.4:

```
Action: StopMonitor
Synopsis: Stop monitoring a channel
Privilege: call,all
Description: The 'StopMonitor' action may be used to end a previously
            started 'Monitor' action. The only parameter is 'Channel',
            the name of the channel monitored.
```

Diff of the internal help from Asterisk 1.2 to 1.4:

None

Diff of the internal help from Asterisk 1.4 to 1.6:

None

E.76 UnpauseMonitor

Asterisk versions:

```
    |      |            | 1.4 |--------| 1.6 |--------
```

Internal help for this AMI command in Asterisk 1.4:

```
Action: UnpauseMonitor
Synopsis: Unpause monitoring of a channel
Privilege: call,all
Description: The 'UnpauseMonitor' action may be used to re-enable
             recording of a channel after calling PauseMonitor. The
             following parameters may be used to control this:
   Channel    - Required. Used to specify the channel to record.
```

Diff of the internal help from Asterisk 1.2 to 1.4:

Not available in Asterisk 1.2

Diff of the internal help from Asterisk 1.4 to 1.6:

None

E.77 UpdateConfig

Asterisk versions:

```
    |      |            | 1.4 |--------| 1.6 |--------
```

Internal help for this AMI command in Asterisk 1.4:

```
Action: UpdateConfig
Synopsis: Update basic configuration
Privilege: config,all
Description: A 'UpdateConfig' action will modify, create, or delete
configuration elements in Asterisk configuration files.
```

```
Variables (X's represent 6 digit number beginning with 000000):
   SrcFilename:   Configuration filename to read(e.g. foo.conf)
   DstFilename:   Configuration filename to write(e.g. foo.conf)
   Reload:        Whether or not a reload should take place (or name of
                  specific module)
   Action-XXXXXX: Action to Take
(NewCat,RenameCat,DelCat,Update,Delete,Append)
   Cat-XXXXXX:    Category to operate on
   Var-XXXXXX:    Variable to work on
   Value-XXXXXX:  Value to work on
   Match-XXXXXX:  Extra match required to match line
```

Diff of the internal help from Asterisk 1.2 to 1.4:

Not available in Asterisk 1.2

Diff of the internal help from Asterisk 1.4 to 1.6:

```
--- in Asterisk 1.4
+++ in Asterisk 1.6
@@ -7,8 +7,9 @@
      SrcFilename:   Configuration filename to read(e.g. foo.conf)
      DstFilename:   Configuration filename to write(e.g. foo.conf)
      Reload:        Whether or not a reload should take place (or name
                     of specific module)
-     Action-XXXXXX: Action to Take
(NewCat,RenameCat,DelCat,Update,Delete,Append)
+     Action-XXXXXX: Action to Take
(NewCat,RenameCat,DelCat,EmptyCat,Update,Delete,Append,Insert)
      Cat-XXXXXX:    Category to operate on
      Var-XXXXXX:    Variable to work on
      Value-XXXXXX:  Value to work on
      Match-XXXXXX:  Extra match required to match line
+     Line-XXXXXX:   Line in category to operate on (used with delete
      and insert actions)
```

E.78 UserEvent

Asterisk versions:

```
    |       |           | 1.4 |--------| 1.6 |--------
```

Internal help for this AMI command in Asterisk 1.4:

```
Action: UserEvent
Synopsis: Send an arbitrary event
Privilege: user,all
Description: Send an event to manager sessions.
```

```
Variables: (Names marked with * are required)
       *UserEvent: EventStringToSend
       Header1: Content1
       HeaderN: ContentN
```

Diff of the internal help from Asterisk 1.2 to 1.4:

Not available in Asterisk 1.2

Diff of the internal help from Asterisk 1.4 to 1.6:

None

E.79 VoicemailUsersList

Asterisk versions:

```
  |       |            |       |       | 1.6 |--------
```

Internal help for this AMI command in Asterisk 1.6:

```
Action: VoicemailUsersList
Synopsis: List All Voicemail User Information
Privilege: call,reporting,all
```

Diff of the internal help from Asterisk 1.4 to 1.6:

Not available in Asterisk 1.4

E.80 WaitEvent

Asterisk versions:

```
  |       |          | 1.4 |--------| 1.6 |--------
```

Internal help for this AMI command in Asterisk 1.4:

```
Action: WaitEvent
Synopsis: Wait for an event to occur
Privilege: <none>
Description: A 'WaitEvent' action will elicit a 'Success' response.
Whenever a manager event is queued. Once WaitEvent has been called on
   an HTTP manager session, events will be generated and queued.
Variables:
   Timeout: Maximum time to wait for events
```

Diff of the internal help from Asterisk 1.2 to 1.4:

Not available in Asterisk 1.2

Diff of the internal help from Asterisk 1.4 to 1.6:

```
--- in Asterisk 1.4
+++ in Asterisk 1.6
@@ -5,4 +5,4 @@
    a manager event is queued. Once WaitEvent has been called on an HTTP
    manager session, events will be generated and queued.
    Variables:
-       Timeout: Maximum time to wait for events
+       Timeout: Maximum time (in seconds) to wait for events, -1 means
                 forever.
```

E.81 ZapDialOffhook

Asterisk versions:

```
--------| 1.2 |--------| 1.4 |--------| 1.6*|--------
```

(* different name)

Internal help for this AMI command in Asterisk 1.4:

```
Action: ZapDialOffhook
Synopsis: Dial over Zap channel while offhook
Privilege: <none>
```

Diff of the internal help from Asterisk 1.2 to 1.4:

None

Diff of the internal help from Asterisk 1.4 to 1.6 (dahdidialoffhook):

```
--- in Asterisk 1.4
+++ in Asterisk 1.6
@@ -1,3 +1,3 @@
-  Action: ZapDialOffhook
-  Synopsis: Dial over Zap channel while offhook
+  Action: DAHDIDialOffhook
+  Synopsis: Dial over DAHDI channel while offhook
   Privilege: <none>
```

E.82 ZapDNDoff

Asterisk versions:

```
--------| 1.2 |--------| 1.4 |--------| 1.6*|--------
```

(* different name)

Internal help for this AMI command in Asterisk 1.4:

```
Action: ZapDNDoff
Synopsis: Toggle Zap channel Do Not Disturb status OFF
Privilege: <none>
```

Diff of the internal help from Asterisk 1.2 to 1.4:

None

Diff of the internal help from Asterisk 1.4 to 1.6 (dahdidndoff):

```
--- in Asterisk 1.4
+++ in Asterisk 1.6
@@ -1,3 +1,3 @@
- Action: ZapDNDoff
- Synopsis: Toggle Zap channel Do Not Disturb status OFF
+ Action: DAHDIDNDoff
+ Synopsis: Toggle DAHDI channel Do Not Disturb status OFF
  Privilege: <none>
```

E.83 ZapDNDon

Asterisk versions:

```
--------| 1.2 |--------| 1.4 |--------| 1.6*|--------
```

(* different name)

Internal help for this AMI command in Asterisk 1.4:

```
Action: ZapDNDon
Synopsis: Toggle Zap channel Do Not Disturb status ON
Privilege: <none>
```

Diff of the internal help from Asterisk 1.2 to 1.4:

None

Diff of the internal help from Asterisk 1.4 to 1.6 (dahdidndon):

```
--- in Asterisk 1.4
+++ in Asterisk 1.6
@@ -1,3 +1,3 @@
-  Action: ZapDNDon
-  Synopsis: Toggle Zap channel Do Not Disturb status ON
+  Action: DAHDIDNDon
+  Synopsis: Toggle DAHDI channel Do Not Disturb status ON
   Privilege: <none>
```

E.84 ZapHangup

Asterisk versions:

```
--------| 1.2 |--------| 1.4 |--------| 1.6*|--------
```

(* different name)

Internal help for this AMI command in Asterisk 1.4:

```
Action: ZapHangup
Synopsis: Hangup Zap Channel
Privilege: <none>
```

Diff of the internal help from Asterisk 1.2 to 1.4:

None

Diff of the internal help from Asterisk 1.4 to 1.6 (dahdihangup):

```
--- in Asterisk 1.4
+++ in Asterisk 1.6
@@ -1,3 +1,3 @@
-  Action: ZapHangup
-  Synopsis: Hangup Zap Channel
+  Action: DAHDIHangup
+  Synopsis: Hangup DAHDI Channel
   Privilege: <none>
```

E.85 ZapRestart

Asterisk versions:

```
   |     |        | 1.4 |--------| 1.6*|--------
```

(* different name)

Internal help for this AMI command in Asterisk 1.4:

```
Action: ZapRestart
Synopsis: Fully Restart zaptel channels (terminates calls)
Privilege: <none>
```

Diff of the internal help from Asterisk 1.2 to 1.4:

Not available in Asterisk 1.2

Diff of the internal help from Asterisk 1.4 to 1.6 (dahdirestart):

```
--- in Asterisk 1.4
+++ in Asterisk 1.6
@@ -1,3 +1,3 @@
-  Action: ZapRestart
-  Synopsis: Fully Restart zaptel channels (terminates calls)
+  Action: DAHDIRestart
+  Synopsis: Fully Restart DAHDI channels (terminates calls)
   Privilege: <none>
```

E.86 ZapShowChannels

Asterisk versions:

```
--------| 1.2 |--------| 1.4 |--------| 1.6*|--------
```

(* different name)

Internal help for this AMI command in Asterisk 1.4:

```
Action: ZapShowChannels
Synopsis: Show status zapata channels
Privilege: <none>
```

Diff of the internal help from Asterisk 1.2 to 1.4:

None

Diff of the internal help from Asterisk 1.4 to 1.6 (dahdishowchannels):

```
--- in Asterisk 1.4
+++ in Asterisk 1.6
@@ -1,3 +1,3 @@
-  Action: ZapShowChannels
-  Synopsis: Show status zapata channels
+  Action: DAHDIShowChannels
+  Synopsis: Show status DAHDI channels
   Privilege: <none>
```

E.87 ZapTransfer

Asterisk versions:

```
--------| 1.2 |--------| 1.4 |--------| 1.6*|--------
```

(* different name)

Internal help for this AMI command in Asterisk 1.4:

```
Action: ZapTransfer
Synopsis: Transfer Zap Channel
Privilege: <none>
```

Diff of the internal help from Asterisk 1.2 to 1.4:

None

Diff of the internal help from Asterisk 1.4 to 1.6 (dahditransfer):

```
--- in Asterisk 1.4
+++ in Asterisk 1.6
@@ -1,3 +1,3 @@
-  Action: ZapTransfer
-  Synopsis: Transfer Zap Channel
+  Action: DAHDITransfer
+  Synopsis: Transfer DAHDI Channel
   Privilege: <none>
```

Configuration Templates

Configuration files such as `sip.conf`, `iax.conf`, and so on can have hundreds of entries; such files are difficult to maintain.

Take a typical `sip.conf`, for example:

```
[201]
username=201
secret=1111
context=default
type=friend
qualify=yes
host=dynamic
canreinvite=no

[202]
username=202
secret=2222
context=default
type=friend
qualify=yes
host=dynamic
canreinvite=no

[203]
username=203
secret=3333
context=default
type=friend
qualify=yes
host=dynamic
canreinvite=no
```

There is another way. Asterisk offers the little-known support for templates! Using a template, our `sip.conf` would look like this instead:

```
[my-phones](!)            ; This entry is the template
context=default
type=friend
qualify=yes
host=dynamic
canreinvite=no

[201](my-phones)          ; Station 201
username=201
secret=1111

[202](my-phones)          ; Station 202
username=202
secret=2222

[203](my-phones)          ; Station 203
username=203
secret=3333
```

This is particularly useful when you have groups or classes of stations with very similar characteristics; that is, in cases where it isn't possible to put all the common parameters in the [general] section. Even in this small and simple example, we've managed to save a few lines and centralize future changes to the "class" my-phones.

F.1 Creating Templates

In principle, any section can serve as a template for other entries; using the tag (!) tells the parser that this section is to be used *exclusively* as a template. Note that no spaces are permitted between the square brackets or the parentheses. Templates may also be based on other templates.

```
[my-example-template](!)
context=default
type=friend
qualify=yes
host=dynamic
canreinvite=no
```

F.2 Using Templates

Use templates by entering the template name (*without* spaces) immediately after the section title. The contents of the template are interpreted first, then the other lines in the section. Sections can inherit multiple templates, as follows:

```
[201](my-phones,sales)    ; inherits "my-phones" and "sales"
username=201
secret=1111
```

Parameters in the template may be overwritten with parameters in the section, if necessary.

F.2.1 Example

```
[stations](!)            ; Template "stations"
type=friend
qualify=yes
dtmfmode=rfc2833
disallow=all
allow=alaw

[snom](!,stations)       ; Template "snom", inherits "stations"
dtmfmode=inband

[linksys](!,stations)    ; Template "linksys", inherits "stations"
qualify=no

[snom1](snom)            ; Station "snom1", inherits "snom"
username=101
secret=123

[snom2](snom)            ; Station "snom2", inherits "snom"
username=102
secret=123

[linksys1](linksys)      ; Station "linksys1", inherits "linksys"
username=103
secret=123
```

You can find additional examples at www.voip-info.org/wiki/index.php?page=Asterisk+config+template.

Upgrading from 1.4 to 1.6

$\mathbf{T}$his appendix explains, using the contents of UPGRADE-1.6.txt along with some additional examples and commentary, what you need to consider when upgrading from Asterisk 1.4 to 1.6.

G.1 AEL

- Macros are now implemented underneath with the **Gosub()** application. (Heaven help you if you wrote code depending on any aspect of this!) Prior to 1.6, macros were implemented with the **Macro()** app, which provided a nice feature of autoreturning. The compiler will do its best to insert a **Return()** app call at the end of your macro if you did not include it, but really, you should make sure that all execution paths within your macros end in return; (as in the following example):

```
macro my-macro () {
    // do something
    return;
}
```

- The conf2ael program is introduced in this release; it is in a rather crude state, but deemed useful for making a first pass at converting extensions.conf code into Asterisk Extensions Language (AEL). The application will become more intelligent as it is developed.

G.2 Core

- The `languageprefix` option in `asterisk.conf` is now deprecated, and the default sound file layout for non-English sounds is the "new style" layout introduced in Asterisk 1.4 (and used by the automatic sound file installer in the Makefile).

- The `ast_expr2` stuff has been modified to handle floating-point numbers. Numbers of the format D.D are now acceptable input for the expr parser, where D is a string of base-10 digits. All math is now done in "long double," if it is available on your compiler/architecture. This was halfway between a bug fix (because the MATH func returns fp by default) and an enhancement. Also, for those counting on, or needing, integer operations, a series of functions were also added to the expr language, to allow several styles of rounding/truncation, along with a set of common floating-point operations, like sin, cos, tan, log, pow, and so on. The ability to call external functions such as `CDR()` and so forth was also added, without having to use the `${...}` notation.

- The delimiter passed to applications has been changed to the comma (,), because that is what people are used to using within `extensions.conf`. If you are using real-time extensions, you will need to translate your existing dialplan to use this separator. To use a literal comma, you need merely to escape it with a backslash (\). Another possible side effect is that you may need to remove the obscene level of backslashing that was necessary for the dialplan to work correctly in 1.4 and earlier versions. This should make writing dialplans less painful in the future, albeit with the pain of a one-time conversion. If you would like to avoid this conversion immediately, set `pbx_realtime=1.4` in the `[compat]` section of `asterisk.conf`. After transitioning, set `pbx_realtime=1.6` in the same section.

- For the same reason as above, you may set `res_agi=1.4` in the `[compat]` section of `asterisk.conf` to continue to use the `|` delimiter in the passed arguments of Asterisk Gateway Interface (AGI) applications. After converting to use the `,` delimiter, change this option to `res_agi=1.6`.

- As a side-effect of the application delimiter change, many places that used to need quotes in order to get the proper meaning are no longer required. You now only need to quote strings in configuration files if you literally want quotation marks within a string.

- The `logger.conf` option `rotatetimestamp` has been deprecated in favor of `rotatestrategy`. This new option supports a rotate strategy that more closely mimics the system logger in terms of file rotation.

- The concise versions of various CLI commands are now deprecated. We recommend using the manager interface (AMI) for application integration with Asterisk instead of using **asterisk -rx "*command*"**.

G.3 Voicemail

- The voicemail configuration values maxmessage and minmessage have been changed to maxsecs and minsecs to clarify their purpose and to make them more distinguishable from maxmsgs, which sets folder size. The old variables will continue to work in this version, albeit with a deprecation warning.
- If you use any interface for modifying voicemail aside from the built-in dialplan applications, the option pollmailboxes must be set in the general section of voicemail.conf for message waiting indication (MWI) to work properly; the polling frequency is set with pollfreq. This is because voicemail notification is now event based rather than polling based. The channel drivers are no longer responsible for constantly manually checking mailboxes for changes so that they can send MWI information to users. Examples of situations that would require this option are web interfaces to voicemail or an e-mail client in the case of using Internet Message Access Protocol (IMAP) storage.

G.4 Dialplan Applications and Functions

- **ChanIsAvail()** now has a t option, which allows the specified device to be queried for state without consulting the channel drivers. This performs mostly a ChanExists sort of function.
- **ChannelRedirect()** will not terminate the channel that fails to do a channelredirect as it has done previously. Instead, CHANNELREDIRECT_STATUS will reflect if the attempt was successful.
- **SetCallerPres()** has been replaced with the **CALLERPRES()** dialplan function and is now deprecated.
- **DISA()**'s fifth argument is now an options argument. If you have previously used NOANSWER in this argument, you'll need to convert that to the new option n.

- **Macro()** is now deprecated. If you need subroutines, you should use the **Gosub()/Return()** applications. To replace **MacroExclusive()**, we have introduced dialplan functions **LOCK()**, **TRYLOCK()**, and **UNLOCK()**. You may use these functions in any location where you desire to ensure that only one channel is executing that path at any one time. The **Macro()** applications are deprecated for performance reasons. However, because **Macro()** has been around for a long time and so many dialplans depend heavily on it, for the sake of backward compatibility it will not be removed. It is also worth noting that using both **Macro()** and **GoSub()** at the same time is *heavily* discouraged.

- **Read()** now sets a READSTATUS variable on exit. It does *not* automatically return -1 (and hang up) anymore on error. If you want to hang up on error, you need to do so explicitly in your dialplan.

- **Privacy()** no longer uses privacy.conf, so any options must be specified directly in the application arguments.

- **MusicOnHold()** application now has duration parameter which allows specifying timeout in seconds.

- **WaitMusicOnHold()** application is now deprecated in favor of extended MusicOnHold.

- **SetMusicOnHold()** is now deprecated. You should use Set(CHANNEL (musicclass)=...) instead.

- The arguments in **ExecIf()** changed a bit, to be more like other applications. The old syntax is

  ```
  ExecIf(expression,application,arguments)
  ```

  The new syntax is

  ```
  ExecIf(<expression>?application_if_true(args):application_if
  _false(args))
  ```

- The behavior of the **Set()** application now depends on a compatibility option, set in asterisk.conf. To use the old 1.4 behavior, which allowed **Set()** to take multiple key/value pairs, set app_set=1.4 in [compat] in asterisk .conf. To use the new behavior, which permits variables to be set with embedded commas, set app_set=1.6 in [compat] in asterisk.conf. Note that you can have both behaviors at the same time, if you switch to using MSet() if you want the old behavior.

- The function **QUEUE_MEMBER_COUNT()** has been deprecated in favor of the **QUEUE_MEMBER()** function. For more information, issue a **show function QUEUE_MEMBER** from the CLI.

G.5 CDR

- The cdr_sqlite module has been marked as deprecated in favor of cdr_sqlite3_custom. It will potentially be removed from the tree after Asterisk 1.6 is released.

- The cdr_odbc module now uses res_odbc to manage its connections. The username and password parameters in cdr_odbc.conf, therefore, are no longer used. The dsn parameter now points to an entry in res_odbc.conf.

- The uniqueid field in the core Asterisk structure has been changed from a maximum 31-character field to a 149-character field, to account for all possible values the systemname prefix could be. In the past, if the systemname was too long, the uniqueid would have been truncated.

- The cdr_tds module now supports all versions of FreeTDS that contain the db-lib front end. It will also now log the userfield variable if the target database table contains a column for it.

G.6 Audio Formats

- format_wav: The GAIN preprocessor definition and source code that used it is removed. This change was made in response to user complaints of choppiness or the clipping of loud signal peaks. To increase the volume of voicemail messages, use the volgain option in voicemail.conf.

- Previously, the Asterisk source code distribution included the iLBC encoder/decoder source code, from Global IP Solutions (http://www.gipscorp.com). This code is not licensed for distribution, and therefore has been removed from the Asterisk source code distribution. If you want to use codec_ilbc to support iLBC channels in Asterisk, you can run the contrib/scripts/get_ilbc_source.sh script to download the source and put it in the proper place in the Asterisk build tree. Once that is done, you can follow your normal steps of building Asterisk. You will need to run menuselect and enable the iLBC codec in the Codec Translators category, then make and make install.

G.7 Channel Drivers

- **SIP:** A small upgrade to support the Record button on the SNOM360, which sends a sip INFO message with a Record: on or Record: off header. If Asterisk is set up (through the `automon` parameter in `features.conf`) to accept One Touch Monitor requests (by default, via *1), the user-configured dialpad sequence is generated, and recording can be started and stopped via this button. The filenames and formats are all controlled via the normal mechanisms. If the user has not configured the automon feature, the normal `415 Unsupported media type` is returned, and nothing is done.

- The `call-limit` option in `sip.conf` is marked as deprecated. It still works in this version of Asterisk, but will be removed in the following version. Please use the **GROUP()**, **GROUP_COUNT()**, and **GROUP_MATCH_COUNT()** functions in the dialplan to enforce call limits. The `limitonpeer` configuration option is now renamed to `counteronpeer`.

- The `username` option is now renamed to `defaultuser` to match `defaultip`. These are used only before registration to call a peer with the URI `sip:defaultuser@defaultip`. The `username` setting still works, but is deprecated and will not work in the next version of Asterisk.

- **chan_local.c:** The comma delimiter inside the channel name has been changed to a semicolon, to make the Local channel driver compatible with the comma-delimiter change in applications.

- **H323:** The `tos` setting has changed name to `tos_audio` and `cos` to `cos_audio` to be compatible with settings in `sip.conf`. The `tos` and `cos` configuration is deprecated and will stop working in the next release of Asterisk.

- **Console:** A new console channel driver, `chan_console`, has been added to Asterisk. This new module can not be loaded at the same time as `chan_alsa` or `chan_oss`. The default `modules.conf` loads only one of them (`chan_oss` by default). So, unless you have modified your `modules.conf` to not use the autoload option, you will need to modify `modules.conf` to add another `noload` line to ensure that only one of these three modules gets loaded.

- **DAHDI:** The `chan_zap` module that supported public switched telephone network (PSTN) interfaces using Zaptel has been renamed to `chan_dahdi`, and only supports the DAHDI telephony driver package for PSTN interfaces. See the `Zaptel-to-DAHDI.txt` file for more details on this transition.

- The `msdstrip` option has been deprecated because it provides no value over the method of stripping digits in the dialplan using variable substring syntax.

G.8 Configuration

- **dundi.conf:** The `tos` parameter changed to use new values. Old values like `lowdelay`, `lowcost`, and `other` are not acceptable now. Look into `doc/tex/qos.tex` for more information about this parameter.

- **queues.conf:** The `queue-lessthan` sound file option is no longer available, and the `queue-round-seconds` option no longer takes `1` as a valid parameter.

G.9 Asterisk Manager Interface

- Manager has been upgraded to version 1.1 with a lot of changes. Check `doc/manager_1_1.txt` for information.

- The **IAXpeers** command output has been changed to more closely resemble the output of the **SIPpeers** command.

- The `cdr_manager` module now reports at the `cdr` level, not at `call` as it was before. You may need to change your `manager.conf` to add the level to existing AMI users, if they want to see the CDR events generated.

- The **Originate** command now requires `write` permission. For **Originate** with the `Application` parameter, you need the additional `system` privilege if you want to do anything that calls out to a subshell.

H

From Zaptel to DAHDI

Digium has used the name Zaptel for its pubic switched telephone network (PSTN) drivers for some time. In 2006, the trademark owner notified Digium that the name was, in fact, trademarked. In response, Digium has renamed the driver package to DAHDI as of 2008. Kevin Fleming, Digium's director of software development, explained this in the following blog posted in May 2008.

```
Zaptel Project Being Renamed to DAHDI
—kpfleming, May 19, 2008

Approximately two years ago, the owner of the trademark ZapTel (for
telephony purposes) contacted Digium and notified us that the name
was in fact trademarked. His company owns the ZapTel trademark and
sells telephone calling cards, and would prefer that Internet
searches for "Zaptel cards" not return products unrelated to their
trademark.

In an effort to rectify this situation, since that time we have had
many discussions internally at Digium about what we should do,
including ensuring that any new name we chose would be something
that was available to be trademarked (as Asterisk is), so that we
wouldn't have a recurrence of this situation. The owner of the
trademark has been very accommodating as we worked through this
process, but it is now time for us to take action.

Beginning immediately, we are going to work on renaming Zaptel to
DAHDI, which stands for Digium Asterisk Hardware Device Interface.
When DAHDI 2.0.0 (the first version) is released, it will contain
nearly all the functionality of Zaptel 1.4, except for Linux
```

kernel 2.4 support, devfs support, and drivers for some very
outdated (and no longer available) Digium hardware (torisa and
wcusb). Zaptel 1.2 will continue to be released in new versions as
bugs are found and fixed, but when DAHDI 2.0.0 is released there
will be no more releases of Zaptel 1.4.

Asterisk 1.6.0 will use DAHDI exclusively (it will not have
support for Zaptel), although this will be done in a highly
backward compatible way so that disruption to existing dialplans
and configurations will be minimized.

Asterisk 1.4 will continue to have support for Zaptel, although it
will be enhanced to also transparently support DAHDI instead, and
the documentation (and default configuration files) will encourage
new users to use DAHDI instead of Zaptel.

Asterisk 1.2 will be unaffected by these changes.

It is unfortunate that all of us have to bear the inconvenience of
this change, but we will try to ensure that it is as easy for
users to switch from Zaptel to DAHDI as we can make it.

For continuing updates on this process, including additional
upgrade notes and other information, please monitor the page at:

www.asterisk.org/zaptel-to-dahdi

The switch has made writing a current and consistent book on
Asterisk even more difficult than it already was, and we can't
avoid using both terms interchangeably, at least for this edition.
The most current material may be found at http://www.the-asterisk-
book.com.

Digium also provides additional information on the change at
www.asterisk.org/zaptel-to-dahdi.

H.1 DAHDI

H.1.1 Dial()

Anybody using DAHDI, be it in Asterisk 1.6 or Asterisk 1.4.22 and above, should
note that any references to **Dial(Zap/...)** in the dialplan must be changed to
Dial(DAHDI/...).

This behavior can be configured for the sake of backward compatibility of dialplans. To use the old terms, set the following in the [options] section of /etc/
asterisk/asterisk.conf:

```
dahdichanname = no
```

H.1.2 Kernel Modules (.ko)

All the former Zaptel kernel modules are also getting new names, as shown in Table H.1.

The names for the kernel modules for individual interface cards have not changed.

H.1.3 Files

The configuration files and directories have changed accordingly, as shown in Table H.2.

Warning: Setting `dahdichanname` = `no` as described previously will make Asterisk read the configuration from `zapata.conf` rather than `chan_dahdi.conf`.

Table H.1 Kernel Module Renaming

Zaptel	DAHDI
zaptel	dahdi
ztd-eth	dahdi_dynamic_eth
ztd-loc	dahdi_dynamic_loc
ztdummy	dahdi_dummy
ztdynamic	dahdi_dynamic
zttranscode	dahdi_transcode

Table H.2 Configuration File Renaming

Zaptel	DAHDI
/etc/zaptel.conf	/etc/dahdi/system.conf
/etc/asterisk/zapata.conf	/etc/asterisk/chan_dahdi.conf

H.1.4 Shell Programs

The new shell programs and scripts have these corresponding names, as shown in Table H.3.

H.1.5 Asterisk Modules

Similarly, the corresponding Asterisk modules, which provide the **Zap...()** applications, have new names, as shown in Table H.4.

For dialplan backward compatibility, the commands **Zap...()** and the new **DAHDI...()** are interchangeable. You can find more information about the applications in Appendix B, "Dialplan Applications."

Table H.3 Shell Programs

Zaptel	DAHDI
ztcfg	dahdi_cfg
ztmonitor	dahdi_monitor
ztscan	dahdi_scan
ztspeed	dahdi_speed
zttest	dahdi_test
zttool	dahdi_tool
zapconf	dahdi_genconf

Table H.4 Asterisk Modules

Zaptel	DAHDI
chan_zap.so	chan_dahdi.so
app_zapbarge.so	app_dahdibarge.so
app_zapras.so	app_dahdiras.so
app_zapscan.so	app_dahdiscan.so

H.2 Interview with Kevin P. Fleming

In January 2009, Stefan Wintermeyer interviewed Kevin P. Fleming, Digium's director of software development, on the transition from Zaptel to DAHDI and the evolution of ISDN support in Asterisk:

Q: *Does DAHDI still need mISDN for the B410P card or does it come with support for Euro-ISDN BRI by itself now?*

A: You are confusing two different things; DAHDI is the same layer as Zaptel was. It provides hardware drivers and a limited amount of signaling abstraction (along with some other features), but is not at all involved in protocol-level work. `chan_dahdi` is the channel driver (which used to be called `chan_zap`), which contains protocol level support, including analog signaling, CAS, PRI, and SS7 (in Asterisk 1.6).

Zaptel and mISDN were not at all related, and the same is true of DAHDI and mISDN. The only reference to mISDN in Zaptel was a Makefile target called `b410p`, which allowed people to download, build, and install the mISDN and mISDNuser packages in an automated fashion, but this was just a convenience for users, it was not related to Zaptel. In DAHDI, this is no longer present, and mISDN users must follow normal processes to obtain and install mISDN.

Now, when it gets to the channel driver layer, things get a little more complicated. Asterisk 1.4 had `chan_zap`, which talks to Zaptel, and `chan_misdn`, which talks to mISDN. They don't support any hardware in common, so for Asterisk 1.4 users, support of EuroISDN BRI is still handled by `chan_misdn`. This is still true even in recent Asterisk 1.4 releases where `chan_zap` has been renamed to `chan_dahdi`, and now talks to DAHDI instead of Zaptel. Even though DAHDI now contains a driver for Digium's B410P BRI card, the Asterisk 1.4 `chan_dahdi` channel driver does not support EuroISDN BRI.

In Asterisk 1.6 the picture is a little different: The `chan_misdn` channel driver is still present, and users who prefer mISDN or have hardware only supported by mISDN are still welcome to use it. The Asterisk 1.6 `chan_dahdi` channel driver works only with DAHDI, *and* it contains protocol-level support for EuroISDN BRI. This means that Asterisk 1.6 users can use `chan_dahdi` for their EuroISDN BRI connections, over any BRI card with a DAHDI driver (see below for more details).

Q: *Would you call it stable and ready for production?*

A: There are two answers to that question: The EuroISDN BRI protocol support in `chan_dahdi` (which actually comes from `libpri`) is both stable and ready for production. It has been through extensive telecom certification testing and will be supported for BRI interoperability just as it has always been for PRI interoperability.

   The DAHDI driver for the Digium B410P is very new, and is working quite well for a large number of users. However, some open issues are being worked on (visible in our open source issue tracker on bugs.digium.com), primarily related to ISDN L1 layer behavior with various telco operators in Europe. It is certainly stable and fully supported by Digium's development team, but there are probably environments where it is not yet ready for production until these issues are resolved.

Q: *Will Euro-ISDN BRI support be available in 1.4 and Zaptel, too, or only in 1.6 and DAHDI?*

A: The B410P driver will never be part of Zaptel (as there will be no more Zaptel releases anyway), and the EuroISDN BRI support in `chan_dahdi` will not be added to Asterisk 1.4, as we don't add new features to that release branch. However, the `chan_dahdi` changes required to provide EuroISDN BRI support are relatively minor, so I expect that one or more community members will publish "backport" patches so that users can use the DAHDI B410P driver and the libpri EuroISDN BRI support in Asterisk 1.4.

Q: *Will DAHDI and Zaptel support generic Cologne HFC-S BRI cards or is it B410P specific?*

A: The DAHDI driver for the B410P card was written specifically for that card, but because the card is based on an HFC-4S, of course the driver could be modified to support cards from other manufacturers. We would be happy to accept patches through our normal patch submission process (on bugs.digium.com) that add support for generic or manufacturer-specific versions of the HFC-S cards, if community members wish to provide them.

Q: *Why do I need the libpri PRI stack for a BRI card (as it has been announced on asterisk-users)?*

A: Because ISDN PRI (both ANSI and Euro) and EuroISDN BRI are very nearly the same protocol; in essence, `libpri` has the wrong name, and it should instead be something like `libisdn`, signifying that it is really an ISDN protocol stack (Q.921 and Q.931, among others), not specific to a particular interface type.

Q: *Does the BRI support in DAHDI include both PtP (point to point) / PtmP (point to multipoint) and TE/NT mode? If the limitation is not in DAHDI, does libpri support all of these modes?*

A: The B410P driver in DAHDI supports TE and NT mode on the HFC4-S chip, but at that level there is no difference (the hardware layer) between PtP and PtMP. `libpri` currently supports being a TE over either PtP or PtMP links, or being an NT over PtP links. It does not currently support acting as an NT over PtMP links, but of course patches are welcome.

Q: *Will Asterisk continue to support* `chan_misdn?`

A: Yes, we will continue to support it (as will the community members who help support it now), but our primary ISDN-related development efforts will continue to be on `libpri/chan_dahdi`.

IAX vs. SIP

O ne question you need to answer, especially when it comes to the question of connecting two Asterisk servers through a trunk, is whether to use IAX or SIP. Opinion is divided over which is the better option, but IAX was written to overcome some of the limitations of SIP. For your convenience, we include an e-mail to the Asterisk users list from Mark Spencer, in which he describes the advantages of IAX:

```
Date Mon, 5 Jul 2004 18:59:52-0500 (CDT)
Author Mark Spencer markster@digium.com
Citetitle [Asterisk-Users] Re: iax or sip
Biblioid http://lists.digium.com/pipermail/asterisk-users/
2004-July/046272.html

[...] let me summarize some differences between SIP and IAX,
and it might help you make a decision about what is best for you.

1) IAX is more efficient on the wire than RTP for *any* number of
calls, *any* codec. The benefit is anywhere from 2.4k for a single
call to approximately tripling the number of calls per megabit for
G.729 when measured to the MAC level when running trunk mode.

2) IAX is information-element encoded rather than ASCII encoded.
This makes implementations substantially simpler and more robust
to buffer overrun attacks since absolutely no text parsing or
interpretation is required. The IAXy runs its entire IP stack, IAX
stack, TDM interface, echo canceller, and callerid generation in 4k
of heap and stack and 64k of flash. Clearly this demonstrates the
implementation efficiency of its design. The size of IAX signaling
```

packets is phenomenally smaller than those of SIP, but that is
generally not a concern except with large numbers of clients
frequently registering. Generally speaking, IAX2 is more efficient
in its encoding, decoding and verifying information, and it would be
extremely difficult for an author of an IAX implementation to
somehow be incompatible with another implementation since so little
is left to interpretation.

3) IAX has a very clear Layer 2 and Layer 3 separation, meaning that
both signaling and audio have defined states, are robustly
transmitted in a consistent fashion, and that when one end of the
call abruptly disappears, the call WILL terminate in a timely
fashion, even if no more signaling and/or audio is received. SIP
does not have such a mechanism, and its reliability from a signaling
perspective is obviously very poor and clumsy requiring additional
standards beyond the core RF3261 [here he means RFC 3261, http://
tools.ietf.org/html/rfc3261].

4) IAX's unified signaling and audio paths permit it to
transparently navigate NATs and provide a firewall administrator
only a *single* port to have to open to permit its use. It requires
an IAX client to know absolutely nothing about the network that it
is on to operate. More clearly stated, there is *never* a situation
that can be created with a firewall in which IAX can complete a call
and not be able to pass audio (except of course if there was
insufficient bandwidth).

5) IAX's authenticated transfer system allows you to transfer audio
and call control off a server-in-the-middle in a robust fashion such
that if the two endpoints cannot see one another for any reason, the
call continues through the central server.

6) IAX clearly separates Caller*ID from the authentication mechanism
of the user. SIP does not have a clear method to do this unless
Remote-Party-ID is used.

7) SIP is an IETF standard. While there is some fledgling
documentation courtesy Frank Miller, IAX is not a published standard
at this time.[1]

8) IAX allows an endpoint to check the validity of a phone number to
know whether the number is complete, may be complete, or is complete
but could be longer. There is no way to completely support this in
SIP.[2]

---

1. IAX2 was submitted as RFC 5456 (www.rfc-editor.org/authors/rfc5456.txt), although
as an informational RFC it remains unofficial.

2. This is no longer strictly true.

9) IAX always sends DTMF out of band so there is never any confusion
about what method is used.

10) IAX support transmission of language and context, which are
useful in an Asterisk environment. That's pretty much all that comes
to mind at the moment.

Mark

In defense of SIP, Mark later wrote:

I guess there must be some advantages to SIP (or we should call the
writers of it stupid).

So here a few questions to elaborate how IAX handles:

1) Bandwidth indications

2) New codecs

3) Extensibility

4) Call hold and other complex scenarios

5) Video telephony

I have gotten the impression that this has all been better arranged in SIP.

GNU Free Documentation License

Version 1.3, 3 November 2008

0 PREAMBLE

The purpose of this License is to make a manual, textbook, or other functional and
useful document "free" in the sense of freedom: to assure everyone the effective
freedom to copy and redistribute it, with or without modifying it, either commercially or noncommercially. Secondarily, this License preserves for the author and
publisher a way to get credit for their work, while not being considered responsible
for modifications made by others.

This License is a kind of "copyleft", which means that derivative works of the document must themselves be free in the same sense. It complements the GNU General
Public License, which is a copyleft license designed for free software.

We have designed this License in order to use it for manuals for free software, because free software needs free documentation: a free program should come with manuals providing the same freedoms that the software does. But this License is not limited to software manuals; it can be used for any textual work, regardless of subject matter or whether it is published as a printed book. We recommend this License principally for works whose purpose is instruction or reference.

1 APPLICABILITY AND DEFINITIONS

This License applies to any manual or other work, in any medium, that contains a notice placed by the copyright holder saying it can be distributed under the terms of this License. Such a notice grants a world-wide, royalty-free license, unlimited in duration, to use that work under the conditions stated herein. The "Document", below, refers to any such manual or work. Any member of the public is a licensee, and is addressed as "you". You accept the license if you copy, modify or distribute the work in a way requiring permission under copyright law.

A "Modified Version" of the Document means any work containing the Document or a portion of it, either copied verbatim, or with modifications and/or translated into another language.

A "Secondary Section" is a named appendix or a front-matter section of the Document that deals exclusively with the relationship of the publishers or authors of the Document to the Document's overall subject (or to related matters) and contains nothing that could fall directly within that overall subject. (Thus, if the Document is in part a textbook of mathematics, a Secondary Section may not explain any mathematics.) The relationship could be a matter of historical connection with the subject or with related matters or of legal, commercial, philosophical, ethical or political position regarding them.

The "Invariant Sections" are certain Secondary Sections whose titles are designated, as being those of Invariant Sections, in the notice that says that the Document is released under this License. If a section does not fit the above definition of Secondary then it is not allowed to be designated as Invariant. The Document may contain zero Invariant Sections. If the Document does not identify any Invariant Sections then there are none.

The "Cover Texts" are certain short passages of text that are listed, as Front-Cover Texts or Back-Cover Texts, in the notice that says that the Document is released under this License. A Front-Cover Text may be at most 5 words, and a Back-Cover Text may be at most 25 words.

A "Transparent" copy of the Document means a machine-readable copy, represented in a format whose specification is available to the general public, that is suitable for revising the document straightforwardly with generic text editors or (for images composed of pixels) generic paint programs or (for drawings) some widely available drawing editor, and that is suitable for input to text formatters or for automatic translation to a variety of formats suitable for input to text formatters. A copy made in an otherwise Transparent file format whose markup, or absence of markup, has been arranged to thwart or discourage subsequent modification by readers is not Transparent. An image format is not Transparent if used for any substantial amount of text. A copy that is not "Transparent" is called "Opaque".

Examples of suitable formats for Transparent copies include plain ASCII without markup, Texinfo input format, LaTeX input format, SGML or XML using a publicly available DTD, and standard-conforming simple HTML, PostScript or PDF designed for human modification. Examples of transparent image formats include PNG, XCF and JPG. Opaque formats include proprietary formats that can be read and edited only by proprietary word processors, SGML or XML for which the DTD and/or processing tools are not generally available, and the machine-generated HTML, PostScript or PDF produced by some word processors for output purposes only.

The "Title Page" means, for a printed book, the title page itself, plus such following pages as are needed to hold, legibly, the material this License requires to appear in the title page. For works in formats which do not have any title page as such, "Title Page" means the text near the most prominent appearance of the work's title, preceding the beginning of the body of the text.

The "publisher" means any person or entity that distributes copies of the Document to the public.

A section "Entitled XYZ" means a named subunit of the Document whose title either is precisely XYZ or contains XYZ in parentheses following text that translates XYZ in another language. (Here XYZ stands for a specific section name mentioned below, such as "Acknowledgements", "Dedications", "Endorsements", or "History".) To "Preserve the Title" of such a section when you modify the Document means that it remains a section "Entitled XYZ" according to this definition.

The Document may include Warranty Disclaimers next to the notice which states that this License applies to the Document. These Warranty Disclaimers are considered to be included by reference in this License, but only as regards disclaiming warranties: any other implication that these Warranty Disclaimers may have is void and has no effect on the meaning of this License.

2 VERBATIM COPYING

You may copy and distribute the Document in any medium, either commercially or noncommercially, provided that this License, the copyright notices, and the license notice saying this License applies to the Document are reproduced in all copies, and that you add no other conditions whatsoever to those of this License. You may not use technical measures to obstruct or control the reading or further copying of the copies you make or distribute. However, you may accept compensation in exchange for copies. If you distribute a large enough number of copies you must also follow the conditions in section 3.

You may also lend copies, under the same conditions stated above, and you may publicly display copies.

3 COPYING IN QUANTITY

If you publish printed copies (or copies in media that commonly have printed covers) of the Document, numbering more than 100, and the Document's license notice requires Cover Texts, you must enclose the copies in covers that carry, clearly and legibly, all these Cover Texts: Front-Cover Texts on the front cover, and Back-Cover Texts on the back cover. Both covers must also clearly and legibly identify you as the publisher of these copies. The front cover must present the full title with all words of the title equally prominent and visible. You may add other material on the covers in addition. Copying with changes limited to the covers, as long as they preserve the title of the Document and satisfy these conditions, can be treated as verbatim copying in other respects.

If the required texts for either cover are too voluminous to fit legibly, you should put the first ones listed (as many as fit reasonably) on the actual cover, and continue the rest onto adjacent pages.

If you publish or distribute Opaque copies of the Document numbering more than 100, you must either include a machine-readable Transparent copy along with each Opaque copy, or state in or with each Opaque copy a computer-network location from which the general network-using public has access to download using public-standard network protocols a complete Transparent copy of the Document, free of added material. If you use the latter option, you must take reasonably prudent steps, when you begin distribution of Opaque copies in quantity, to ensure that this Transparent copy will remain thus accessible at the stated location until at least one year after the last time you distribute an Opaque copy (directly or through your agents or retailers) of that edition to the public.

It is requested, but not required, that you contact the authors of the Document well before redistributing any large number of copies, to give them a chance to provide you with an updated version of the Document.

4 MODIFICATIONS

You may copy and distribute a Modified Version of the Document under the conditions of sections 2 and 3 above, provided that you release the Modified Version under precisely this License, with the Modified Version filling the role of the Document, thus licensing distribution and modification of the Modified Version to whoever possesses a copy of it. In addition, you must do these things in the Modified Version:

A. Use in the Title Page (and on the covers, if any) a title distinct from that of the Document, and from those of previous versions (which should, if there were any, be listed in the History section of the Document). You may use the same title as a previous version if the original publisher of that version gives permission.

B. List on the Title Page, as authors, one or more persons or entities responsible for authorship of the modifications in the Modified Version, together with at least five of the principal authors of the Document (all of its principal authors, if it has fewer than five), unless they release you from this requirement.

C. State on the Title page the name of the publisher of the Modified Version, as the publisher.

D. Preserve all the copyright notices of the Document.

E. Add an appropriate copyright notice for your modifications adjacent to the other copyright notices.

F. Include, immediately after the copyright notices, a license notice giving the public permission to use the Modified Version under the terms of this License, in the form shown in the Addendum below.

G. Preserve in that license notice the full lists of Invariant Sections and required Cover Texts given in the Document's license notice.

H. Include an unaltered copy of this License.

I. Preserve the section Entitled "History", Preserve its Title, and add to it an item stating at least the title, year, new authors, and publisher of the Modified Version as given on the Title Page. If there is no section Entitled "History" in the Document, create one stating the title, year, authors, and publisher of the Document as given on its Title Page, then add an item describing the Modified Version as stated in the previous sentence.

J. Preserve the network location, if any, given in the Document for public access to a Transparent copy of the Document, and likewise the network locations given in the Document for previous versions it was based on. These may be placed in the "History" section. You may omit a network location for a work that was published at least four years before the Document itself, or if the original publisher of the version it refers to gives permission.

K. For any section Entitled "Acknowledgements" or "Dedications", Preserve the Title of the section, and preserve in the section all the substance and tone of each of the contributor acknowledgements and/or dedications given therein.

L. Preserve all the Invariant Sections of the Document, unaltered in their text and in their titles. Section numbers or the equivalent are not considered part of the section titles.

M. Delete any section Entitled "Endorsements". Such a section may not be included in the Modified Version.

N. Do not retitle any existing section to be Entitled "Endorsements" or to conflict in title with any Invariant Section.

O. Preserve any Warranty Disclaimers.

If the Modified Version includes new front-matter sections or appendices that qualify as Secondary Sections and contain no material copied from the Document, you may at your option designate some or all of these sections as invariant. To do this, add their titles to the list of Invariant Sections in the Modified Version's license notice. These titles must be distinct from any other section titles.

You may add a section Entitled "Endorsements", provided it contains nothing but endorsements of your Modified Version by various parties—for example, statements of peer review or that the text has been approved by an organization as the authoritative definition of a standard.

You may add a passage of up to five words as a Front-Cover Text, and a passage of up to 25 words as a Back-Cover Text, to the end of the list of Cover Texts in the Modified Version. Only one passage of Front-Cover Text and one of Back-Cover Text may be added by (or through arrangements made by) any one entity. If the Document already includes a cover text for the same cover, previously added by you or by arrangement made by the same entity you are acting on behalf of, you may not add another; but you may replace the old one, on explicit permission from the previous publisher that added the old one.

The author(s) and publisher(s) of the Document do not by this License give permission to use their names for publicity for or to assert or imply endorsement of any Modified Version.

5 COMBINING DOCUMENTS

You may combine the Document with other documents released under this License, under the terms defined in section 4 above for modified versions, provided that you include in the combination all of the Invariant Sections of all of the original documents, unmodified, and list them all as Invariant Sections of your combined work in its license notice, and that you preserve all their Warranty Disclaimers.

The combined work need only contain one copy of this License, and multiple identical Invariant Sections may be replaced with a single copy. If there are multiple Invariant Sections with the same name but different contents, make the title of each such section unique by adding at the end of it, in parentheses, the name of the original author or publisher of that section if known, or else a unique number. Make the same adjustment to the section titles in the list of Invariant Sections in the license notice of the combined work.

In the combination, you must combine any sections Entitled "History" in the various original documents, forming one section Entitled "History"; likewise combine any sections Entitled "Acknowledgements", and any sections Entitled "Dedications". You must delete all sections Entitled "Endorsements".

6 COLLECTIONS OF DOCUMENTS

You may make a collection consisting of the Document and other documents released under this License, and replace the individual copies of this License in the various documents with a single copy that is included in the collection, provided that you follow the rules of this License for verbatim copying of each of the documents in all other respects.

You may extract a single document from such a collection, and distribute it individually under this License, provided you insert a copy of this License into the extracted document, and follow this License in all other respects regarding verbatim copying of that document.

7 AGGREGATION WITH INDEPENDENT WORKS

A compilation of the Document or its derivatives with other separate and independent documents or works, in or on a volume of a storage or distribution medium, is called an "aggregate" if the copyright resulting from the compilation is not used to limit the legal rights of the compilation's users beyond what the individual works permit. When the Document is included in an aggregate, this License does not apply to the other works in the aggregate which are not themselves derivative works of the Document.

If the Cover Text requirement of section 3 is applicable to these copies of the Document, then if the Document is less than one half of the entire aggregate, the Document's Cover Texts may be placed on covers that bracket the Document within the aggregate, or the electronic equivalent of covers if the Document is in electronic form. Otherwise they must appear on printed covers that bracket the whole aggregate.

8 TRANSLATION

Translation is considered a kind of modification, so you may distribute translations of the Document under the terms of section 4. Replacing Invariant Sections with translations requires special permission from their copyright holders, but you may include translations of some or all Invariant Sections in addition to the original versions of these Invariant Sections. You may include a translation of this License, and all the license notices in the Document, and any Warranty Disclaimers, provided that you also include the original English version of this License and the original versions of those notices and disclaimers. In case of a disagreement between the translation and the original version of this License or a notice or disclaimer, the original version will prevail.

If a section in the Document is Entitled "Acknowledgements", "Dedications", or "History", the requirement (section 4) to Preserve its Title (section 1) will typically require changing the actual title.

9 TERMINATION

You may not copy, modify, sublicense, or distribute the Document except as expressly provided under this License. Any attempt otherwise to copy, modify, sublicense, or distribute it is void, and will automatically terminate your rights under this License.

However, if you cease all violation of this License, then your license from a particular copyright holder is reinstated (a) provisionally, unless and until the copyright holder explicitly and finally terminates your license, and (b) permanently, if the copyright holder fails to notify you of the violation by some reasonable means prior to 60 days after the cessation.

Moreover, your license from a particular copyright holder is reinstated permanently if the copyright holder notifies you of the violation by some reasonable means, this is the first time you have received notice of violation of this License (for any work) from that copyright holder, and you cure the violation prior to 30 days after your receipt of the notice.

Termination of your rights under this section does not terminate the licenses of parties who have received copies or rights from you under this License. If your rights have been terminated and not permanently reinstated, receipt of a copy of some or all of the same material does not give you any rights to use it.

10 FUTURE REVISIONS OF THIS LICENSE

The Free Software Foundation may publish new, revised versions of the GNU Free Documentation License from time to time. Such new versions will be similar in spirit to the present version, but may differ in detail to address new problems or concerns. See http://www.gnu.org/copyleft/.

Each version of the License is given a distinguishing version number. If the Document specifies that a particular numbered version of this License "or any later version" applies to it, you have the option of following the terms and conditions either of that specified version or of any later version that has been published (not as a draft) by the Free Software Foundation. If the Document does not specify a version number of this License, you may choose any version ever published (not as a draft) by the Free Software Foundation. If the Document specifies that a proxy can decide which future versions of this License can be used, that proxy's public statement of acceptance of a version permanently authorizes you to choose that version for the Document.

11 RELICENSING

"Massive Multiauthor Collaboration Site" (or "MMC Site") means any World Wide Web server that publishes copyrightable works and also provides prominent facilities for anybody to edit those works. A public wiki that anybody can edit is an example of such a server. A "Massive Multiauthor Collaboration" (or "MMC") contained in the site means any set of copyrightable works thus published on the MMC site.

"CC-BY-SA" means the Creative Commons Attribution-Share Alike 3.0 license published by Creative Commons Corporation, a not-for-profit corporation with a principal place of business in San Francisco, California, as well as future copyleft versions of that license published by that same organization.

"Incorporate" means to publish or republish a Document, in whole or in part, as part of another Document.

An MMC is "eligible for relicensing" if it is licensed under this License, and if all works that were first published under this License somewhere other than this MMC, and subsequently incorporated in whole or in part into the MMC, (1) had no cover texts or invariant sections, and (2) were thus incorporated prior to November 1, 2008.

The operator of an MMC Site may republish an MMC contained in the site under CC-BY-SA on the same site at any time before August 1, 2009, provided the MMC is eligible for relicensing.

Index

FREE Online Edition

Your purchase of *Practical Asterisk 1.4 and 1.6* includes access to a free online edition for 45 days through the Safari Books Online subscription service. Nearly every Addison-Wesley Professional book is available online through Safari Books Online, along with more than 5,000 other technical books and videos from publishers such as Cisco Press, Exam Cram, IBM Press, O'Reilly, Prentice Hall, Que, and Sams.

SAFARI BOOKS ONLINE allows you to search for a specific answer, cut and paste code, download chapters, and stay current with emerging technologies.

Activate your FREE Online Edition at
www.informit.com/safarifree

> **STEP 1:** Enter the coupon code: PMLWNCB.

> **STEP 2:** New Safari users, complete the brief registration form.
> Safari subscribers, just log in.

If you have difficulty registering on Safari or accessing the online edition,
please e-mail customer-service@safaribooksonline.com